AF897422

WILEY

Shipment Summary

Thanks for ordering from John Wiley & Sons.
An invoice will be mailed separately.
Contact Customer Care with questions: 1--800–225–5945

Shipped To:
INGRAM BOOK GROUP LLC – LAVERGNE
1 INGRAM BLVD
LA VERGNE, TN 37086
United States
X

Publisher Order # 330982841

Purchase Order # NZ180XLN

Shipment date:	Ship Method:
9/3/2026	Consolidation Wiley Ingram

Qty.	SKU	Description
1	139435536X	Resorbable Composites for Bioimplants and Internal Fixation Devices

Bin: D6

Resorbable Composites for Bioimplants and Fixation Devices

Scrivener Publishing
100 Cummings Center, Suite 541J
Beverly, MA 01915-6106

Publishers at Scrivener
Martin Scrivener (martin@scrivenerpublishing.com)
Phillip Carmical (pcarmical@scrivenerpublishing.com)

Resorbable Composites for Bioimplants and Fixation Devices

Edited by

Arbind Prasad
Mechanical Engineering Department, Katihar Engineering College, (Under Department of Science, Technology and Technical Education, Government of Bihar), Katihar, Bihar, India

Scrivener Publishing

WILEY

This edition first published 2026 by John Wiley & Sons, Inc., 111 River Street, Hoboken, NJ 07030, USA and Scrivener Publishing LLC, 100 Cummings Center, Suite 541J, Beverly, MA 01915, USA
© 2026 Scrivener Publishing LLC
For more information about Scrivener publications please visit www.scrivenerpublishing.com.

All rights reserved. No part of this publication may be reproduced, stored in a retrieval system, or transmitted, in any form or by any means, electronic, mechanical, photocopying, recording, or otherwise, except as permitted by law. Advice on how to obtain permission to reuse material from this title is available at http://www.wiley.com/go/permissions.

Wiley Global Headquarters
111 River Street, Hoboken, NJ 07030, USA

For details of our global editorial offices, customer services, and more information about Wiley products visit us at www.wiley.com.

The manufacturer's authorized representative according to the EU General Product Safety Regulation is Wiley-VCH GmbH, Boschstr. 12, 69469 Weinheim, Germany, e-mail: Product_Safety@wiley.com.

Limit of Liability/Disclaimer of Warranty
While the publisher and authors have used their best efforts in preparing this work, they make no representations or warranties with respect to the accuracy or completeness of the contents of this work and specifically disclaim all warranties, including without limitation any implied warranties of merchantability or fitness for a particular purpose. No warranty may be created or extended by sales representatives, written sales materials, or promotional statements for this work. The fact that an organization, website, or product is referred to in this work as a citation and/or potential source of further information does not mean that the publisher and authors endorse the information or services the organization, website, or product may provide or recommendations it may make. This work is sold with the understanding that the publisher is not engaged in rendering professional services. The advice and strategies contained herein may not be suitable for your situation. You should consult with a specialist where appropriate. Neither the publisher nor authors shall be liable for any loss of profit or any other commercial damages, including but not limited to special, incidental, consequential, or other damages. Further, readers should be aware that websites listed in this work may have changed or disappeared between when this work was written and when it is read.

Library of Congress Cataloging-in-Publication Data

ISBN 978-1-394-35536-5

Front cover images courtesy of Wikimedia Commons
Cover design by Russell Richardson

Set in size of 11pt and Minion Pro by Manila Typesetting Company, Makati, Philippines

Printed in the USA

10 9 8 7 6 5 4 3 2 1

Contents

Preface

The development of implants based on bio-based materials requires a deep understanding of synthesis, properties, and processing to enable the fabrication of customized biomedical devices that are both efficient and cost-effective. This book bridges the knowledge gap by presenting extensive information on manufacturing techniques, fabrication challenges, and solutions encountered throughout production and processing.

Modern biomedical applications—including tissue engineering, orthopedic implants, drug delivery, and biofluids—are increasingly utilizing bio-based resources for their superior properties and biocompatibility. Materials such as cellulose, silk, chitosan, and natural gums are advancing the field by providing sustainable alternatives to conventional manufacturing processes.

This book delivers comprehensive insights into synthesis, processing, characterization, optimization, modeling, and design considerations. It explores additive manufacturing techniques for creating implants, scaffolds, and biomedical devices from naturally sourced biomaterials, emphasizing performance, precision, and clinical relevance.

Key features include:

- The latest advances in resorbable composites for hydrogels, bone grafting, and biofluid interaction;
- Detailed studies of emerging manufacturing processes and their impact on composite performance;
- Thorough analysis of the tribological, mechanical, degradation, and biological properties of resorbable materials;
- Dedicated chapters on hydrogels, drug-eluting systems, modeling, surface coating, and clinical translation;
- Exploration of cost-effectiveness and advanced scaffolding approaches for bone regeneration.

The book is organized into fifteen chapters. Chapter 1 introduces the fundamentals of resorbable composites for implants and fixation devices. Chapter 2 focuses on synthesis, processing, and characterization. Chapter 3 examines optimization, modeling, and design parameters for 3D-printed implants. Chapters 4 through 6 address the role of biofluids, implant design, and mechanical performance in orthopedic applications. Chapters 7 to 9 cover hydrogel and scaffold fabrication, including the potential of 3D printing in fixation devices. Chapter 10 investigates advanced scaffolding for bone regeneration, while Chapter 11 presents degradation studies. Chapters 12 and 13 analyze biological and tribological properties. Chapter 14 explores surface coating and clinical translation, and Chapter 15 discusses the integration of artificial intelligence in implant development.

This comprehensive resource provides researchers, practitioners, and students with the knowledge to design and manufacture bioresorbable implants using modern additive manufacturing techniques. It offers practical insights into design optimization, modeling, biopolymer development, and translational applications in both research and industry. By presenting the latest developments in resorbable composite technology, this book lays the foundation for future innovations in biomedical science and engineering.

Editor
Dr. Arbind Prasad

Acknowledgments

We express our gratitude to Martin Scrivener and the team at Scrivener Publishing for their suggestions and support during completion of this book. We are grateful to all contributors and reviewers for their illuminating views on each chapter that follows.

"This book is dedicated to all engineers, doctors, researchers, health professionals and academicians..."

Editor
Dr. Arbind Prasad

Part I

FUNDAMENTALS, PROCESSING AND DESIGN

1

Introduction to Resorbable Composites for Implants and Fixation Devices

Atanu Kumar Paul[1]*, Gourhari Chakraborty[2], Arbind Prasad[3] and Bidyanand Mahto[4]

[1]*Department of Chemical Engineering, National Institute of Technology Andhra Pradesh, Tadepalligudem, Andhra Pradesh, India*
[2]*School of Chemical and Biotechnology, SASTRA Deemed University, Thirumalaisamudram, Tamil Nadu, India*
[3]*Mechanical Engineering Department, Katihar Engineering College (Under Department of Science, Technology and Technical Education, Government of Bihar), Katihar, Bihar, India*
[4]*Government Engineering College, Vaishali (Under Department of Science, Technology and Technical Education, Government of Bihar), Vaishali, Bihar, India*

Abstract

This introductory chapter provides a comprehensive foundation for understanding resorbable composites and their expanding role in biomedical implants and fixation devices. It begins by contextualizing the emergence of resorbable biomaterials as viable alternatives to permanent metallic and polymeric implants, emphasizing the clinical motivation to eliminate secondary surgeries, reduce long-term complications, and support natural tissue regeneration. Fundamental principles of composite design are reviewed, including the interplay between matrix materials, reinforcement phases, interfacial bonding, and structure–property relationships that collectively govern mechanical performance and degradation behavior.

A central component of the chapter is the classification and description of materials commonly employed in resorbable composite systems, such as biodegradable polymers, bioactive ceramics, calcium-based fillers, and functional reinforcements. Their contribution to mechanical integrity, bioactivity, and controlled resorption is evaluated from materials science and biomedical engineering perspectives. The chapter also examines degradation mechanisms, biocompatibility

**Corresponding author*: atanukpaul@outlook.com

Arbind Prasad (ed.) Resorbable Composites for Bioimplants and Fixation Devices, (3–48) © 2026 Scrivener Publishing LLC

factors, and host responses, highlighting the need for predictable resorption kinetics and non-toxic by-products that integrate safely into metabolic pathways. Processing and fabrication approaches, including melt processing, solvent-based techniques, and additive manufacturing, are surveyed to demonstrate how microstructural control and surface modification influence clinical performance. Regulatory frameworks, sterilization constraints, and quality standards are outlined to underscore the translational considerations necessary for widespread adoption.

Finally, the chapter addresses clinical indications and current applications in orthopedics, maxillofacial surgery, and tissue fixation, while identifying emerging research trends such as nanocomposite systems, bioactive modifications, and smart degradation control. Through this holistic overview, the chapter establishes the scientific and technical basis required to explore subsequent chapters in greater detail, positioning resorbable composites as an evolving class of biomaterials that align with modern demands for patient-centric, regenerative, and minimally invasive therapeutic solutions.

Keywords: Resorbable composites, biodegradable polymers, bioactive ceramics, degradation mechanisms, biocompatibility, orthopedic fixation, additive manufacturing, regenerative implants

1.1 Overview of Resorbable Biomaterials

1.1.1 Defining Resorbability and Degradation Behavior

Resorbable biomaterials are designed to degrade and be eliminated from the body, eliminating the need for secondary surgical removal procedures [1]. These materials undergo degradation through multiple chemical and biological mechanisms involving cellular processes that reduce and remove the material from the body [1]. The degradation behavior varies across different material categories, including synthetic polymers, natural polymers, and bioresorbable ceramics, each with distinct degradation mechanisms [2]. Resorbable biomaterials encompass metals, ceramics, and polymers, with evaluation methods including *in vitro* physicochemical and biological tests, as well as *in vivo* assessments [2, 3]. For tissue engineering applications, these materials are categorized into fossil-based materials (PCL, PVA, PU), natural materials (HA, PLA, chitosan, collagen), and hybrid biomaterials (PCL/PLA, PCL/collagen combinations) [4, 5]. Essential properties include biocompatibility, bioactivity, biodegradability, and non-toxicity to ensure appropriate host response and effective tissue regeneration [4].

1.1.2 Historical Evolution of Resorbable Implants

The historical evolution of resorbable implants spans several decades, with significant developments in both materials and applications. Early resorbable polymers containing glycolide, lactide, ε-caprolactone, and trimethylene carbonate were developed primarily between the 1960s-1990s, requiring extensive scientific evaluation to establish their safety for medical device applications [6]. Clinical implementation has demonstrated success across multiple specialties, with resorbable distraction devices being used in craniofacial surgery since 2002, showing predictable results in mandibular lengthening procedures [7]. The field has expanded to include biodegradable orthopedic implants utilizing various biomaterials including polymers, metals, and bioceramics, manufactured through advanced techniques like additive manufacturing [8]. Recent innovations have extended to bioresorbable electronic implants incorporating active components such as biosignal sensors, therapeutic actuators, and wireless power systems, representing a significant advancement from passive structural support to active diagnostic and therapeutic capabilities [9].

1.1.3 Comparison with Permanent and Biostable Implant Materials

Research comparing permanent and bioresorbable implant materials demonstrates comparable clinical efficacy with distinct advantages for each approach. In orbital floor fracture reconstruction, bioresorbable and permanent implants showed equivalent outcomes for diplopia, enophthalmos, and ocular motility restriction across all fracture sizes, with bioresorbable implants offering the advantage of gradual force transfer during healing [10, 11]. Biodegradable implants provide temporary support matching tissue formation rates and eliminate secondary removal surgeries, though they initially trigger inflammatory responses that typically cease upon material degradation [12]. Permanent titanium implants achieve direct bone bonding without fibrous tissue formation, providing superior long-term stability compared to materials that develop intervening soft tissue layers leading to loosening [13]. Material selection depends on biocompatibility, with stable ceramics and metals showing excellent tissue adhesion, while polymers similar to biological structures face dissolution challenges [14]. Figure 1.1 discusses the primary motivations for adopting resorbable implants, highlighting the limitations of permanent implants, the avoidance of secondary removal surgeries, and their ability to support natural healing and controlled biodegradation. The illustration also emphasizes the clinical advantages of minimally invasive care and improved patient recovery associated with resorbable composite systems.

1.1.4 Objectives of this Chapter

The primary objective of this chapter is to provide a comprehensive introductory framework for understanding resorbable composites and their application in bioimplants and fixation devices. Specifically, this chapter aims to:

- Establish fundamental concepts related to resorbable biomaterials, including definitions of resorbability, degradation behavior, and their evolution in comparison with permanent and biostable implant materials.
- Present the principles of composite design for biomedical applications, emphasizing structure–property relationships, mechanical performance, and toughness mechanisms relevant to load-bearing and fixation devices.
- Classify and critically discuss the major material systems used in resorbable composites, including biodegradable polymers, bioactive ceramics and glasses, and natural or

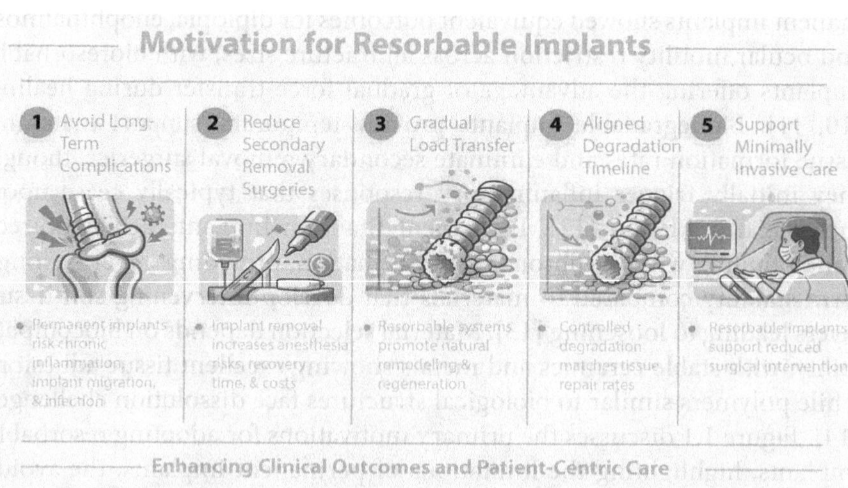

synthetic reinforcements, with attention to their individual and synergistic roles.
- Examine degradation mechanisms, biocompatibility, and host response, highlighting hydrolytic and enzymatic pathways, tissue integration, and toxicological and biosafety considerations essential for clinical success.
- Review key processing and fabrication techniques, such as melt processing, solution-based methods, additive manufacturing, and surface functionalization, and their influence on microstructure, performance, and translational feasibility.
- Define the performance requirements for implant and fixation applications, including mechanical integrity, fatigue resistance, degradation kinetics, load transfer, sterilization, and packaging constraints.
- Discuss clinical and surgical considerations, encompassing implantation sites, resorption profiles in orthopedic and maxillofacial applications, and the regulatory frameworks governing resorbable medical devices.
- Highlight emerging trends and research opportunities, including bioactive and smart composites, nanocomposite systems with controlled degradation, and future directions toward personalized and regenerative implant technologies.

Through these objectives, the chapter is intended to establish a coherent scientific and clinical foundation that supports deeper exploration of resorbable composite materials, processing strategies, and applications in subsequent chapters.

1.2 Fundamentals of Composite Design for Biomedical Use

1.2.1 Key Components: Matrices and Reinforcements

Biomedical composites are engineered materials that combine two or more constituent materials with different physical, chemical, and mechanical properties to achieve superior performance for medical applications [15]. These materials offer significant advantages over traditional metals, polymers, or ceramics alone by providing tailored properties that individual materials cannot achieve [16, 17]. Early work demonstrated the potential of biomedical composites using poly(2-hydroxyethyl-methacrylate) as a

Table 1.1 Essential scaffold properties and functional requirements for tissue engineering applications.

Scaffold attribute	Functional requirement
Biocompatibility	The scaffold must be non-toxic and biologically safe, enabling normal cellular functions while supporting osteoconduction, osteoinduction, osteogenesis, and vascularization.
Biodegradability/ Bioresorbability	The degradation behavior should be synchronized with new tissue formation, allowing adjustable resorption rates and facilitating controlled drug release or incorporation of bioactive molecules.
Pore Size and Geometry	Adequate porosity is required to permit nutrient diffusion and cell proliferation, with pore dimensions exceeding 100 μm and preferably within 200–350 μm for bone tissue ingrowth; however, increased multiscale porosity may compromise mechanical stability.
Mechanical Performance	The scaffold should possess sufficient strength to endure physiological loads and stresses while closely replicating the mechanical behavior of native tissue.

matrix with polyester resin reinforcing fibers, successfully matching human tendon behavior in uniaxially oriented composites and creating laminates for plastic surgery applications [18]. Modern biomedical composites are designed by carefully selecting biocompatible matrix and reinforcement materials, with their properties determined by constituent properties, volume fractions, and arrangements [15]. Applications span orthopedics, dentistry, drug delivery, and emerging fields like cancer theranostics [16, 17]. Table 1.1 discusses the critical scaffold properties and the corresponding functional requirements necessary to support biocompatibility, controlled degradation, structural integrity, and effective tissue regeneration in tissue engineering applications.

1.2.2 Structure–Property Relationships

Biomedical composites represent engineered materials combining two or more constituent components with distinct physical, chemical, and mechanical properties to overcome limitations of individual metallic, polymeric, and ceramic biomaterials [15, 16]. These materials can be designed to satisfy specific geometrical, structural, mechanical, and chemical requirements for targeted biomedical applications [15, 19]. The characteristic properties of composites result from individual constituent properties, their volume fractions, and arrangements within the material system [15, 19]. Applications span orthopedics, dentistry, drug delivery, and advanced technologies like cancer theranostics [16]. Polymer-silicate nanocomposites, including clay-based silicate and bioactive glass nanoparticles, show particular promise for diagnostic devices, tissue regeneration, and drug delivery systems [20]. Understanding polymer-nanoparticle interactions is crucial for controlling structure-property relationships within biomedical constraints [20, 21]. Despite extensive research, commercial production remains limited, requiring careful consideration of intended goals and limitations [19]. Table 1.2 discusses representative biodegradable polymers by summarizing their origins, key material properties, dominant degradation mechanisms, typical application areas, and inherent limitations.

1.2.3 Mechanisms of Mechanical Strength and Toughness

Resorbable composites for load-bearing applications demonstrate promising mechanical properties through strategic material selection and fiber architecture optimization. Bioresorbable glass fiber-reinforced poly(l-lactide-co-dl-lactide) composites exhibit strength retention that can be tailored by modifying glass fiber oxide composition, with degradation occurring through combined surface and bulk erosion mechanisms [35]. Poly(glycolic acid) scaffolds coated with poly(glycolide-co-caprolactone) elastomer achieve enhanced mechanical strength comparable to metallic stents while maintaining complete resorbability within 18 months [36]. Iron-doped phosphate glass fiber-reinforced PLA composites reach maximum values of 10 GPa modulus and 120 MPa strength, though continuous fiber arrangements show accelerated degradation due to wicking effects compared to discontinuous random mat configurations [37]. Optimized silk fibroin-hydroxyapatite-polylactic acid composites achieve exceptional properties of 13.7 GPa flexural modulus and 437 MPa strength, representing among the highest values reported for bioresorbable composites [38]. Figure 1.2 discusses the progressive evolution of resorbable implant technologies, beginning with early homopolymer systems and advancing to the incorporation of bioactive fillers such as calcium phosphates and bioglass. The figure further highlights the transition toward composite designs that integrate polymers with bioactive reinforcements to achieve improved mechanical performance and enhanced bioactivity.

Table 1.2 Summary of representative biodegradable polymers, their sources, degradation mechanisms, key properties, applications, and limitations.

Polymer	Origin	Salient properties	Primary degradation route	Typical uses	Key constraints	Ref.
Cellulose	Plant-derived	High mechanical robustness, renewable and abundant	Enzyme-mediated breakdown	Flexible substrates, sensing platforms, paper-based electronics	Cost-intensive processing, prone to moisture uptake	[22–24]
Chitosan	Animal-derived	Biologically compatible with inherent antimicrobial activity	Enzymatic and hydrolytic degradation	Wound dressings, tissue scaffolds, biosensing materials	Limited mechanical strength	[25, 26]
Polylactic Acid (PLA)	Synthetic	Adjustable properties with optical clarity	Hydrolytic cleavage	Packaging materials, additive manufacturing, biomedical implants	Low thermal resistance and brittle behavior	[27, 28]

(*Continued*)

Table 1.2 Summary of representative biodegradable polymers, their sources, degradation mechanisms, key properties, applications, and limitations. (*Continued*)

Polymer	Origin	Salient properties	Primary degradation route	Typical uses	Key constraints	Ref.
Poly(lactic-co-glycolic acid) (PLGA)	Synthetic	Biocompatible and approved for controlled release systems	Hydrolysis	Drug delivery carriers, transient electronic devices	Acidic by-products during degradation	[29, 30]
Polycaprolactone (PCL)	Synthetic	Highly flexible with prolonged degradation timeline	Hydrolytic degradation	Tissue scaffolds, biomedical adhesives, surgical sutures	Slow resorption rate	[31, 32]
Polyhydroxyalkanoates (PHAs)	Microbially synthesized	Elastomeric nature, hydrophobicity, strong mechanical performance	Enzymatic degradation	Implantable devices, actuators, controlled drug delivery systems	Reduced flexibility	[33, 34]

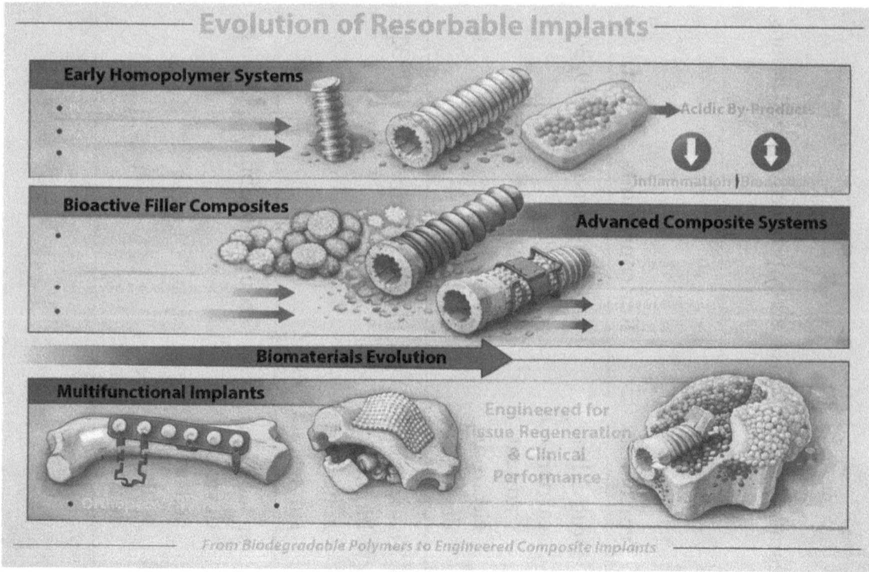

1.3 Material Classes Utilized in Resorbable Composites

1.3.1 Polymer Constituents and Biodegradable Polymers

Biodegradable polymers, particularly polylactic acid (PLA), polyglycolic acid (PGA), and polycaprolactone (PCL), serve as essential matrix materials in biomedical applications due to their favorable biocompatibility and controlled degradation properties [39, 40]. These polymers degrade primarily through hydrolysis and enzymatic activity, breaking down into biocompatible by-products that are safely metabolized or excreted by the body [39, 40]. PLA degrades into lactic acid, carbon dioxide, and water, which are metabolized intracellularly or excreted through urine and breath, with a median half-life of 30 weeks [40]. The degradation characteristics and mechanical properties of these polymers can be engineered by controlling molecular structure, crystallinity, and copolymer ratios to suit specific applications [39]. These materials are extensively used in tissue engineering scaffolds, drug delivery systems, and orthopedic devices, providing temporary structural support while gradually resorbing as new tissue forms [41].

1.3.2 Bioactive Ceramics and Glass Fillers

Bioactive ceramics and glass fillers significantly enhance polymer matrix composites for bone tissue engineering applications. These inorganic fillers, including hydroxyapatite, calcium phosphates, and bioactive glasses, provide both mechanical reinforcement and bioactivity to polymer matrices [42]. The incorporation of bioactive glass particles into bioabsorbable polymers modifies degradation kinetics and material morphology, with optimal filler content being application-dependent [43]. Comparative studies demonstrate that different bioactive fillers affect composite properties differently: while hydroxyapatite and $CaCO_3$ maintain matrix integrity during processing, bioactive glasses can cause premature degradation but enhance water uptake and bioactivity [44, 45]. In polycaprolactone composites, calcium phosphate-based ceramics and bioactive glasses address the polymer's limitations by improving mechanical properties, controlling degradation rates, and enhancing bioactivity, making these hybrid biomaterials suitable for bone regeneration applications [46].

1.3.3 Natural and Synthetic Fiber Reinforcements

Natural and synthetic fibers serve as crucial reinforcement phases in biomedical composite systems for tissue engineering applications. Natural fibers offer biocompatibility, non-toxicity, and good mechanical performance while being environmentally beneficial [47]. Widely used natural polymers, including chitosan and collagen, along with synthetic polymers like polycaprolactone and poly(lactic-co-glycolic) acid, are incorporated with nanofillers to create nanocomposite biomaterials that support bone tissue regeneration [48]. These composite materials combine bioresorbable polymers with bioactive inorganic phases to achieve tailored physical, biological, and mechanical properties with predictable degradation behavior [49]. Smart composite biomaterials can be designed to mimic native tissue properties and provide controllable delivery of therapeutic molecules, with responsiveness to various stimuli enabling multifunctional scaffolds for tissue regenerative applications [50]. The composite approach effectively addresses individual material limitations while optimizing performance for clinical tissue engineering.

1.4 Degradation Mechanisms and Biocompatibility

1.4.1 Hydrolytic and Enzymatic Degradation Pathways

Hydrolytic and enzymatic degradation pathways are fundamental mechanisms controlling the breakdown of composite biomaterials. Hydrolytic degradation occurs through water-mediated scission of polymer bonds, particularly affecting condensation-type bonds like esters, ethers, and amides in monomer resins [51]. For synthetic polymers, chemical hydrolysis represents the primary degradation mode, with rates controllable through molecular weight and crystallinity modifications [52]. Enzymatic degradation provides an accelerated pathway through specific biological catalysts. Embedding enzymes like Candida antarctica Lipase B within polymer matrices can dramatically regulate bioresorption rates by creating multiple enzyme-surface interfaces throughout the material [53]. Different polymer compositions respond distinctly to enzymatic versus oxidative degradation - conventional polycaprolactone shows greater susceptibility to hydrolytic enzymatic degradation, while supramolecular variants are more prone to oxidative breakdown [54]. These complementary degradation mechanisms ultimately dictate material longevity and resorption profiles, enabling tailored biomaterial design for specific biomedical applications. Figure 1.3 discusses the key degradation mechanisms of resorbable

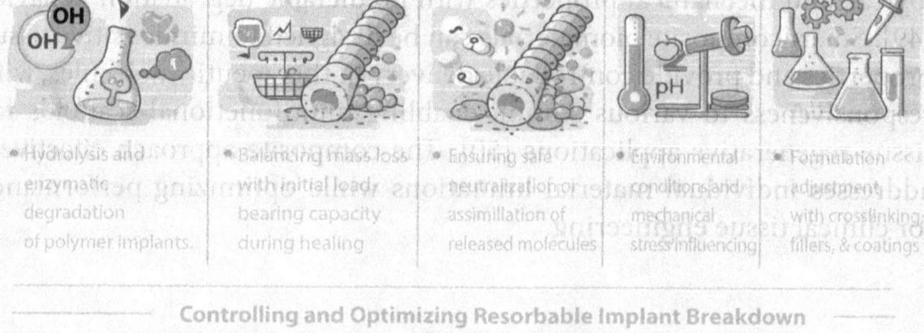

implants, illustrating how hydrolysis and enzymatic activity contribute to material breakdown and influence the balance between mass loss and mechanical strength retention during healing. The figure also highlights the metabolic assimilation of degradation by-products, demonstrating how compounds released during resorption are buffered and incorporated into physiological pathways to ensure biocompatibility.

1.4.2 Tissue Response and Host Integration

The tissue response to implanted biomaterials involves a complex cascade of inflammatory, proliferative, and remodeling phases that determine biocompatibility success. Titanium dental implants demonstrate exceptional biocompatibility through controlled osseointegration processes, where implant surfaces recruit and differentiate osteogenic cells while maintaining transient, controlled immune responses mediated by M2-polarized macrophages [55]. Optimal biocompatibility is characterized by minimal inflammation and either thin fibrous encapsulation or mesenchymal tissue ingrowth into porous structures, requiring materials that are non-toxic, non-antigenic, and appropriately sized with adequate pore dimensions to admit immune cells [56]. The integration process depends critically on adhesion molecules and cytokines, particularly osteopontin, which functions both as an adhesion molecule and growth factor in wound healing and tissue integration. Understanding these histologic events in both normal and compromised osseous conditions is fundamental for successful implant case selection and outcomes [57].

1.4.3 Toxicological and Biosafety Considerations

Toxicological and biosafety evaluation of resorbable composites requires a comprehensive assessment of cytotoxicity, leachable by-products, and systemic effects to ensure clinical safety. Medical implants and their degradation products must meet regulatory biocompatibility standards, with potential for dose-dependent local and systemic exposures creating risks for hypersensitivity reactions [58]. Dental composites release various chemical components that can cause toxic, allergic, and mutagenic effects, though clinical relevance remains questionable due to small amounts and unique oral conditions. Conventional dental resin composites exhibit cytotoxic potential and undergo biodegradation, with residual monomer release creating synergistic "degradation-toxicity" effects that compromise restoration longevity [59]. However, properly designed composites can demonstrate excellent safety profiles, as shown by PLGA/bioactive glass composites that exhibited no genetic toxicity, systemic toxicity, hemolytic reactions, or pyrogen responses in comprehensive *in-vivo* testing [60].

1.5 Processing and Fabrication Techniques

1.5.1 Melt Processing and Solution-Based Methods

Research demonstrates that melt processing and solution-based methods provide versatile approaches for fabricating resorbable polymer composites with controlled properties. Gel casting techniques enable the production of thick-section materials and blending of polymers with particulate fillers, allowing adjustment of degradation rates through polymer type and morphology variations [61]. Processing technologies include extrusion, injection molding, compression, and solvent casting, each offering specific advantages for different product requirements [62]. Melt mixing, extrusion, and injection molding are established techniques for manufacturing biodegradable polymers, with compositing and blending approaches providing promising property improvements [63]. Melt-based additive manufacturing techniques, including extrusion-based printing and selective laser sintering, enable customization of biodegradable scaffolds with designer microstructures for tailored mechanical and biological properties [64]. These scalable fabrication routes offer tunable processing parameters to achieve controlled microstructure and performance in resorbable composite components.

1.5.2 Additive Manufacturing Approaches

Additive manufacturing (AM) has emerged as a transformative approach for fabricating biomaterials with complex geometries and patient-specific designs. AM enables layer-by-layer construction directly from CAD files, offering complexity-for-free manufacturing and batch-size-indifference for fully customized medical devices [65, 66]. Key AM technologies, including stereolithography, powder bed fusion, and extrusion-based bioprinting, allow precise fabrication of metamaterials with tailored porous structures such as truss, triply periodic minimal surfaces, and functionally graded lattices [67, 68]. These approaches enable control over porosity, architecture, and mechanical properties essential for biomedical applications. Stereolithography has successfully produced patient-specific scaffolds containing up to 60wt% β-tricalcium phosphate in poly(trimethylene carbonate) composites, demonstrating enhanced mechanical reinforcement and surface bioactivity [69]. The combination of AM with diverse material

Table 1.3 Comparative overview of additive manufacturing techniques, key advantages, and limitations for tissue engineering applications.

Technique	Advantages	Limitations	Ref.
Fused Deposition Modeling (FDM)	Enables cost-effective fabrication of complex architectures with adequate mechanical integrity and supports printing with multiple materials.	Exhibits directional dependence of properties, inherent pore formation, and susceptibility to nozzle clogging.	[70]
Direct Pellet Multi-Extrusion Printing	Eliminates the need for filament preparation, allows multi-material processing, and supports intricate structural designs with satisfactory strength.	Characterized by anisotropic behavior, porous structures, and frequent extrusion blockage issues.	[71]
Robocasting	Operates without thermal degradation and is well suited for highly loaded or paste-based material systems.	Often results in limited mechanical performance, residual solvent presence, porous constructs, and prolonged fabrication durations.	[72]
Electrospinning and Melt Electrospinning Writing (MEW)	Produces highly porous networks with aligned fibrous architectures, large surface area, controlled fiber placement, and favorable mechanical response.	Involves solvent residue concerns, reliance on high-voltage equipment, and constraints on achievable geometric complexity.	[73, 74]

(Continued)

Table 1.3 Comparative overview of additive manufacturing techniques, key advantages, and limitations for tissue engineering applications. (*Continued*)

Technique	Advantages	Limitations	Ref.
Selective Laser Sintering (SLS)	Offers high design freedom and resolution with minimal material wastage, while eliminating the requirement for additional support structures.	Associated with high operational costs, extended processing times, residual thermal stresses, post-processing requirements, and rough surface texture.	[75]
Stereolithography (SLA)	Produces parts with excellent surface finish and reliable mechanical properties.	Requires specialized photosensitive materials, involves higher costs, and necessitates post-curing steps to remove uncured resin remnants from scaffolds.	[73]

palettes, from biodegradable polymers to bioactive ceramics, advances the fabrication of resorbable composites with unprecedented property combinations for tissue engineering and regenerative medicine applications. Table 1.3 discusses the major additive manufacturing techniques used in tissue engineering, highlighting their key advantages, inherent limitations, and representative literature references for comparative evaluation.

1.5.3 Surface Modification and Functionalization Strategies

Surface modification strategies for dental and orthopedic implants have emerged as critical approaches to enhance osseointegration, biocompatibility, and antibacterial performance. Various techniques, including sand blasting, acid etching, plasma etching, plasma spray deposition, sputtering, and cathodic arc deposition, have been extensively studied to improve implant performance [76, 77]. These modifications address key clinical challenges such as failed osseointegration, marginal bone resorption, and peri-implantitis that contribute to implant failure [78]. Surface engineering approaches encompass both passive modifications that alter surface texture and chemistry to prevent bacterial adhesion, and active strategies incorporating antimicrobial agents for sustained infection control [79]. Nanotechnology has enabled the development of nanoscale coatings like TiO_2 and ZnO that promote osseointegration while inhibiting biofilm formation through techniques such as plasma spraying, ion implantation, and plasma electrolytic oxidation [79]. Research indicates that combining multiple coating types often provides superior results, though optimal surface characteristics for bone integration remain to be fully elucidated [80]. Figure 1.4 discusses the principal fabrication and processing methods used for resorbable implants, illustrating how melt processing, solvent casting, and additive manufacturing enable the formation of bulk structures and patient-specific designs. The figure also highlights various surface modification strategies - such as plasma treatment, bioactive coatings, and chemical functionalization - that enhance implant bioactivity and integration without altering bulk properties. Table 1.4 discusses a comparative overview of metal-filled PLA biocomposites fabricated *via* additive manufacturing, highlighting the influence of metal type, loading, particle size, and processing conditions on mechanical, thermal, biological, and functional performance improvements.

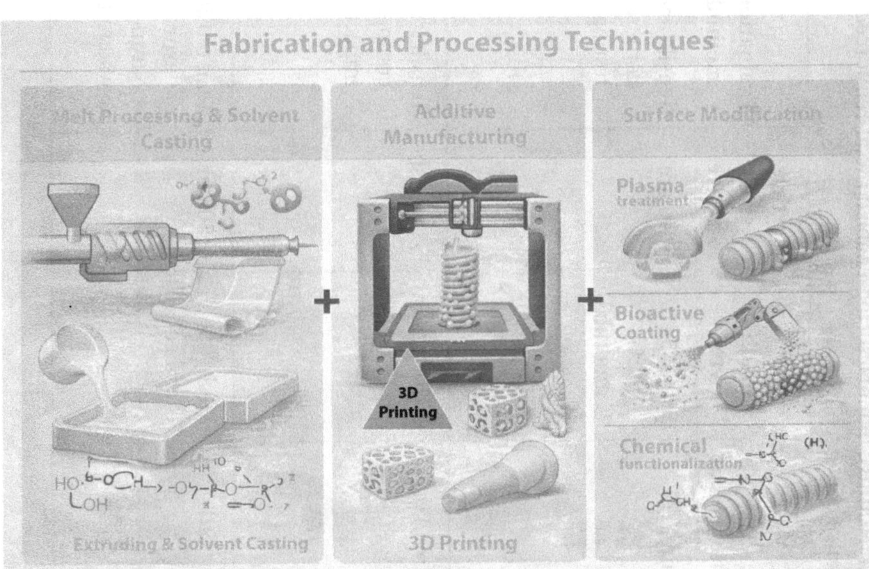

Table 1.4 Additive manufacturing-enabled metal-reinforced PLA biocomposites: composition, processing parameters, and performance enhancements.

Metal reinforcement	Metal loading (wt%)	Particle size range	Evaluation methods	Comparison framework	Observed enhancements	Ref.
Titanium (Ti)	5 (Ti), 5–10 (PEG-modified)	≈ 40 μm	Tensile and compression testing	Neat PLA vs. Ti/PLA/PEG systems	Substantial ductility improvement without compromising strength; improved surface wettability, print quality, roughness control, and enhanced cell viability	[81]
Titanium (Ti)	0–20	325 mesh	Tensile, compression, impact	PLA compared with PLA–Ti composites	Mechanical performance maximized around 10 vol% Ti with superior impact resistance and improved *in vitro* biocompatibility	[82]
Copper (Cu)	Mesh embedded	≈ 30 μm	Tear, tensile, flexural, hardness, water uptake	PLA vs. Cu-reinforced PLA *via* HFDM	Strategic Cu mesh integration increased strength and interfacial integrity, supported by microstructural evidence	[83]

(Continued)

Table 1.4 Additive manufacturing-enabled metal-reinforced PLA biocomposites: composition, processing parameters, and performance enhancements. (*Continued*)

Metal reinforcement	Metal loading (wt%)	Particle size range	Evaluation methods	Comparison framework	Observed enhancements	Ref.
Copper (Cu)	14	20–30 μm	Compression and flexural tests	Effect of FDM processing conditions	Elevated nozzle and bed temperatures improved the compressive and flexural strength of Cu-filled PLA	[84]
316L Stainless Steel	5–15	20–50 μm	Compression testing	PLA vs. PLA–316L composites	Higher compressive strength and modulus at 10–15 vol%; reduced thermal expansion and controlled surface roughness with comparable degradation behavior	[85]
316L Stainless Steel	—	50–150 μm	Tensile joint strength	PLA–316L joints vs. neat PLA	Mechanical interlocking dominated joint performance; optimized joint design achieved ~96% strength of pure PLA	[85]

(*Continued*)

Table 1.4 Additive manufacturing-enabled metal-reinforced PLA biocomposites: composition, processing parameters, and performance enhancements. (*Continued*)

Metal reinforcement	Metal loading (wt%)	Particle size range	Evaluation methods	Comparison framework	Observed enhancements	Ref.
Magnesium (Mg)	1–7 (optimum ~3)	≈ 50 μm	Tensile and compression tests	PLA vs. PLA–Mg systems	At 3 wt% Mg, tensile strength and modulus improved markedly; excessive Mg caused agglomeration and weak interfaces	[86]
Magnesium (Mg)	5–50	< 50 μm	Nanoindentation (hardness, modulus)	Neat PLA vs. Mg-filled PLA	Elastic modulus and hardness increased significantly; good dispersion enhanced printability and thermal stability, especially with interfacial modifiers	[87]
Silver (Ag)	4	Flakes (5 × 90 μm)	Tensile and compression tests	PLA vs. Ag/PLA nanocomposites	Balanced mechanical performance combined with strong antibacterial functionality	[88]

1.6 Performance Requirements for Implant and Fixation Applications

1.6.1 Mechanical Performance and Fatigue Behavior

Resorbable composites for bone fixation applications must provide adequate mechanical support during healing while gradually transferring load to regenerating tissue. Heimbach *et al.* developed high-performance composites using degummed silk fibroin, hydroxyapatite, and polylactic acid, achieving flexural modulus and strength of 13.7 GPa and 437 MPa, respectively, among the highest reported for bioresorbable materials [38]. These properties far exceed fixation device requirements while maintaining resorbability. Daniels *et al.* reviewed biodegradable polymers and composites, finding that while unreinforced polymers achieved only 5% of stainless steel stiffness, carbon fiber reinforcement increased this to 62% [89]. However, biodegradable composites with inorganic fibers showed rapid strength loss, highlighting fiber-matrix coupling challenges. Researcher emphasized that biodegradable polymers undergo hydrolysis-controlled degradation with progressive mechanical property reduction, while fatigue and creep damage contribute to performance decline [90]. Heimbach *et al.* demonstrated that silk fibroin-based composites exhibit excellent initial biocompatibility alongside superior mechanical properties for load-bearing applications [38].

1.6.2 Degradation Kinetics and Load Transfer

Biodegradable implant materials must achieve synchronized degradation and load transfer to prevent premature failure or stress shielding. Previous studies have demonstrated that elastomer-toughened composite scaffolds can exhibit mechanically stable behavior through three characteristic degradation phases – lag, log, and plateau – which can be aligned with natural bone healing rates to ensure sufficient initial support [91]. Other investigations have shown that magnesium-based biomaterials possess mechanical properties comparable to bone, yet require precise control of degradation to maintain an effective balance between material loss and retained strength during tissue regeneration [92]. Further research indicates that combining bioceramics with biodegradable polymers can enhance mechanical properties and degradation resistance while reducing foreign body reactions [93]. Additional findings confirm that bioresorbable glass fiber-reinforced composites can be engineered for controllable strength retention through tailored fiber compositions, allowing optimization of surface and bulk erosion mechanisms for specific clinical applications [35].

1.6.3 Sterilization and Packaging Requirements

Sterilization and packaging of polymer-based implantable medical devices are critical for maintaining material integrity and ensuring patient safety. Polymeric materials are increasingly used in implantable devices due to their biodegradable and biocompatible properties, making sterilization methods crucial for device performance [94, 95]. Traditional gamma sterilization in air environments can cause oxidation in polyethylene, leading to embrittlement and compromised mechanical integrity, prompting manufacturers to adopt alternative methods, including non-radiation sterilization and gamma irradiation in inert environments [96]. The stability and integrity of polymeric materials over extended periods are essential, as hydrolysis and oxidation reactions can lead to polymer degradation, affecting composite material behavior [97]. Biocompatible packaging polymers must provide adequate gas and water permeability barriers while maintaining sufficient tensile strength to protect electronic circuits from moisture and ions throughout the implant's lifetime [98, 99]. Figure 1.5 discusses the major clinical applications of resorbable implants, illustrating their use in orthopedic fixation systems such as screws and plates, maxillofacial reconstruction devices for craniofacial stabilization, and soft tissue anchors for ligament and tendon repair. The figure emphasizes how these resorbable

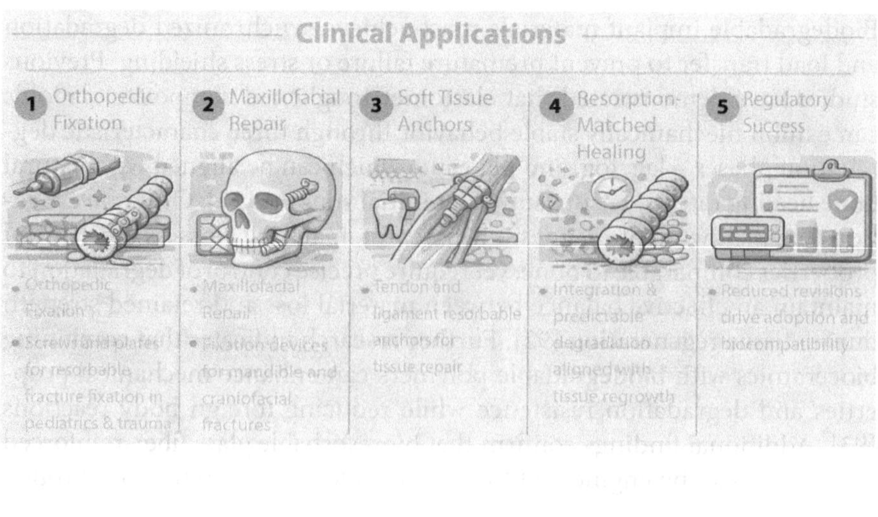

Table 1.5 Comprehensive comparison of biopolymer-based composite systems for electronic, energy storage, sensing, and electrochemical applications.

Biopolymer system	Application area	Processing/ fabrication method	Key performance outcomes	Functional/ material characteristics	Key remarks	Ref.
Starch/metal halide	Gel electrolyte for electronics	Extrusion processing	Electrical conductivity enhanced to ~10^{-6} S/cm	Moisture-sensitive conductivity influenced by water uptake	Plasticized starch provides stable ion transport	[100]
Starch/inorganic salt	Flexible electronic sensors	Thermogelation	Self-healing efficiency ~98% for 2–3 h	Hydrogel network with excellent mechanical recovery	Hydrogen bonding supports flexible device structures	[101]
Starch/ montmorillonite/ ionic liquid	Ion-conductive nanocomposite	Solution casting	Ionic conductivity ~10^{-3} S/cm	Hydrophilic IL enhances interfacial bonding	Improved charge transport efficiency	[102]
Starch/ionic liquid acetate	Electroactive polymer films	Compression molding	Ionic conductivity ~0.012 S/cm	Reduced thermal stability (~30 K)	Higher humidity and moderate temperature improve conductivity	[103]

(Continued)

Table 1.5 Comprehensive comparison of biopolymer-based composite systems for electronic, energy storage, sensing, and electrochemical applications. (*Continued*)

Biopolymer system	Application area	Processing/ fabrication method	Key performance outcomes	Functional/ material characteristics	Key remarks	Ref.
Chitosan/PVA/ silica	Proton exchange membrane	Solution casting	Proton conductivity 5.08×10^{-4} S/cm at 100 °C	Silica loading controls water uptake and proton transport	Enhanced membrane hydration (~55.7%)	[104]
Chitosan/PVA/SSA	Proton-conducting membranes	Solution casting	Peak conductivity 3.9×10^{-3} S/cm	Improved oxidative and hydrolytic resistance	Nanoparticles enhance electrochemical durability	[105]
Chitosan/ silicotungstic acid	Fuel cell membranes	Ionotropic gelation	High proton conductivity with power density up to 268 mW/cm^2	Heteropolyacid retained within the membrane matrix	Efficient O_2 reduction confirmed by EIS analysis	[106]

(*Continued*)

Table 1.5 Comprehensive comparison of biopolymer-based composite systems for electronic, energy storage, sensing, and electrochemical applications. (*Continued*)

Biopolymer system	Application area	Processing/ fabrication method	Key performance outcomes	Functional/ material characteristics	Key remarks	Ref.
Cellulose nanocrystal/ protic ionic liquid	Fuel cell electrolyte membrane	Solution casting	Proton conductivity 10^{-4}–10^{-3} S/cm at 120–160 °C	Nanoscale morphology governs defect density and aggregation	Suitable for non-humidified, high-temperature fuel cells	[107]
Carboxylated chitosan/acrylic acid network	Supercapacitor electrode	Graft copolymerization	Ionic conductivity 7.82×10^{-2} S/cm; energy density 4.39 Wh/kg	Stable electrochemical behavior across pH conditions	Applicable in acidic, neutral, and alkaline media	[108]
Starch/$CaCl_2$/ glycerol	Self-powered strain and pressure sensors	Solution blending	Sensitivity 1.537 kPa^{-1} with stability over 200 cycles	Highly stretchable hydrogel (elongation ~208%)	Maintains sensing accuracy under repeated deformation	[109]

(*Continued*)

Table 1.5 Comprehensive comparison of biopolymer-based composite systems for electronic, energy storage, sensing, and electrochemical applications. (*Continued*)

Biopolymer system	Application area	Processing/ fabrication method	Key performance outcomes	Functional/ material characteristics	Key remarks	Ref.
PLA/PEG	Separator for supercapacitors	Phase inversion + RF plasma	Conductivity increased to 6.6×10^{-5} S/cm; resistance ~1.9 Ω	Highly porous, hydrophilic membrane structure	Plasma treatment improves electrolyte wettability	[110]
Chitosan/ hydroxyapatite	Thermal insulation materials	Crosslinking and freeze-drying	Thermal conductivity reduced to 37.43 mW/mK	High mechanical strength and specific modulus	Suitable for lightweight insulation systems	[111]

devices offer structural support during healing while eliminating the need for secondary removal procedures and promoting improved patient outcomes across diverse surgical fields. Table 1.5 discusses a comprehensive comparison of biopolymer-based composite systems by outlining their applications in electronics, energy storage, sensing, and electrochemical devices, along with associated processing methods, key performance metrics, material characteristics, and representative references.

1.7 Clinical and Surgical Considerations

1.7.1 Typical Implantation Sites and Use Cases

Resorbable composites have demonstrated substantial clinical utility across multiple surgical specialties, particularly in pediatric applications where the elimination of secondary implant removal procedures represents a significant clinical advantage. In craniofacial surgery, these materials have shown favorable safety and efficacy profiles, with clinical studies reporting successful outcomes in large pediatric cohorts treated with polylactic and polyglycolic acid–based copolymer fixation systems, achieving complete material resorption within approximately one year and exhibiting minimal postoperative complications [112]. Subsequent large-scale clinical investigations in craniosynostosis management have further corroborated these findings, demonstrating high success rates and low incidences of adverse events [113]. Beyond pediatric use, resorbable fixation technologies have been effectively applied in orthognathic surgery and maxillofacial trauma management, where they provide adequate temporary mechanical stability while mitigating risks associated with permanent metallic hardware, such as stress shielding and implant migration [114, 115]. Collectively, clinical evidence has established key paradigms for resorbable fixation systems, highlighting their progression from early orthopedic applications to widespread adoption in craniomaxillofacial surgery and underscoring their potential to reduce surgical morbidity by obviating the need for secondary removal interventions [116].

1.7.2 Resorption Profiles in Orthopedic and Maxillofacial Use

Biodegradable implants for orthopedic and maxillofacial applications offer significant advantages by eliminating the need for secondary removal surgeries and potentially stimulating healing through gradual load transfer to regenerating tissue [117, 118]. However, these materials face critical challenges in balancing mechanical strength requirements with appropriate degradation profiles. Traditional biodegradable devices have been successful primarily in low load-bearing applications, with polymers like polylactide, polyglycolide, and polydioxanone showing good biocompatibility but limited mechanical properties [117, 119]. In maxillofacial surgery, bioresorbable plate systems address limitations of titanium implants, including palpability and imaging interference, though issues remain regarding mechanical strength and degradation rates [120, 121]. The fundamental challenge lies in achieving optimal implant design that maintains sufficient strength during early healing phases while degrading harmoniously with bone remodeling, requiring careful consideration of anatomical location, load-bearing demands, and tissue-specific healing rates [118, 119].

1.7.3 Regulatory Requirements and Standards

The regulatory landscape for resorbable implants and medical devices is governed by comprehensive international standards and guidelines that ensure patient safety and device efficacy. Key regulatory frameworks include International Organization for Standardization (ISO) guidelines, particularly ISO-10993 for biocompatibility assessment, U.S. Pharmacopeial Convention standards, ASTM International specifications, and European Conformity marking requirements [122, 123]. These standards mandate rigorous evaluation of biocompatibility, mechanical strength, corrosion resistance, and bioactivity based on patient exposure type and duration [123, 124]. Manufacturing processes significantly impact material properties and regulatory compliance, with quality assurance requirements extending to material powder properties, printing parameters, and post-processing considerations, including cleaning, finishing, and sterilization [124, 125]. Advanced manufacturing techniques like 3D printing introduce unique regulatory challenges due to their customizability and novel build processes, requiring specialized design control models for FDA approval [125]. Regulatory evaluation culminates in integrated risk assessments based on comprehensive testing data [123]. Figure 1.6 discusses the primary advantages of resorbable composite implants, highlighting their ability to eliminate secondary removal surgeries, reduce long-term complications associated with permanent hardware, and promote enhanced tissue regeneration. The illustration visually contrasts traditional complications with the biological integration achieved by resorbable systems, emphasizing how controlled degradation supports healing while minimizing clinical risks.

1.8 Emerging Trends and Research Opportunities

1.8.1 Bioactive and Smart Composite Platforms

Smart bioactive composite platforms represent a rapidly advancing field focused on developing biomaterials that actively interact with physiological environments. These systems combine biopolymers and bioceramics to create instructive matrices that stimulate cellular responses crucial for tissue regeneration [50]. Smart Bioactive Glass Composites (SBGC) exemplify this approach by responding to endogenous disease microenvironments or exogenous physical stimuli, enabling active treatment of tissue defects in complex conditions, including tumors, infections, and diabetes [126]. Key design strategies involve modulating bulk and surface properties to mimic native tissues while incorporating therapeutic molecules for sustainable, controllable delivery systems responsive to stimuli such as pH, temperature, ionic strength, and magnetism [50]. The integration of artificial intelligence and machine learning is accelerating material discovery and property optimization, advancing these intelligent therapeutic platforms toward clinical translation despite ongoing challenges in scalability, reproducibility, and regulatory approval [127]. Figure 1.7 discusses the major limitations and challenges associated with resorbable composite implants, illustrating mechanical strength trade-offs, variability in degradation behavior, and the stringent regulatory requirements governing clinical approval.

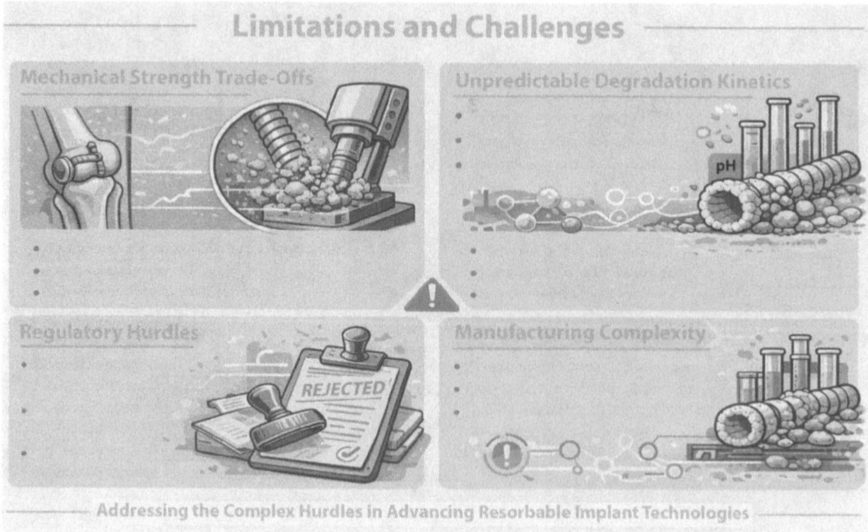

The figure emphasizes how these challenges collectively impact implant reliability, predictability, and pathways to clinical translation.

1.8.2 Nanocomposites and Controlled Degradation Systems

Nanocomposites represent a promising class of materials for tissue engineering applications by integrating nanoscale fillers within polymer matrices to achieve enhanced mechanical performance and precisely controlled degradation behavior. Experimental investigations have shown that polyampholyte-based nanocomposites incorporating laponite clay nanoparticles can be engineered to exhibit tunable degradation profiles spanning from days to several months, while simultaneously improving mechanical toughness and maintaining high levels of cytocompatibility [128, 129]. Studies on aliphatic polyester–based nanocomposites have further demonstrated that synergistic interactions between biodegradable polymer matrices and nanofillers lead to property enhancements that are strongly dependent on nanoparticle type, loading concentration, and morphological characteristics [130]. Such nanocomposite systems allow fine regulation of thermal, electrical, and biodegradation properties while supporting critical cellular functions, including adhesion, viability, and differentiation. Owing to their nanoscale interfacial features that closely mimic native biological structures, nanocomposites are increasingly recognized as effective bioactive platforms capable of integrating multifunctional attributes such as magnetic responsiveness and controlled drug delivery for both bone and soft tissue regeneration [131].

1.8.3 Future Directions in Personalized and Regenerative Implants

Additive manufacturing (AM) has emerged as a transformative technology for personalized implants and regenerative medicine, enabling patient-specific design and biomimetic structures. AM technologies, particularly selective laser melting and electron beam melting, can fabricate complex lattice structures that replicate trabecular bone architecture, enhancing load transfer and promoting osseointegration while reducing stress shielding [132]. These technologies surpass conventional manufacturing in precision, customization, and structural complexity, with end-to-end workflows coupling computer-aided design with patient-specific anatomical data [133]. AM enables free-form manufacturing and mass customization, serving as a critical technology for precision biomaterials tailored to individual patient needs, including tissue substitutes, drug delivery systems, and medical devices [134]. Current personalized tissue engineering approaches focus on cell-based therapies and scaffold customization through 3D printing, though challenges remain in timing for cell

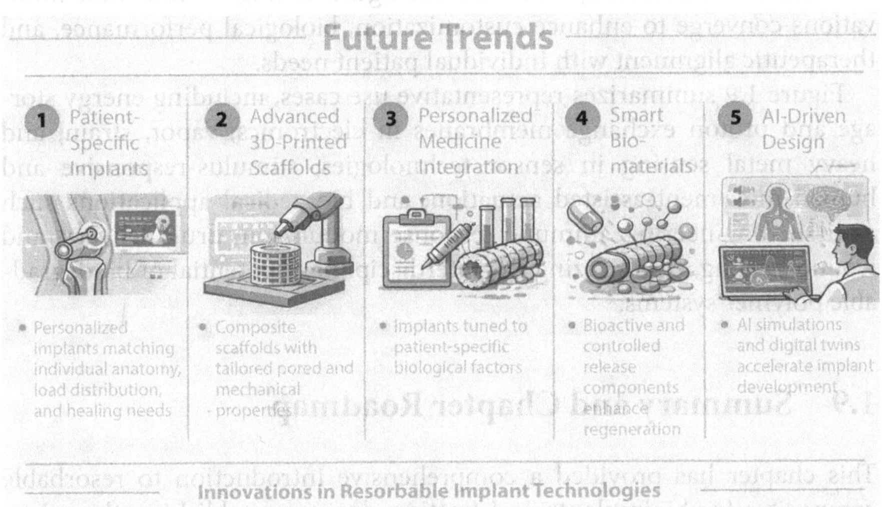

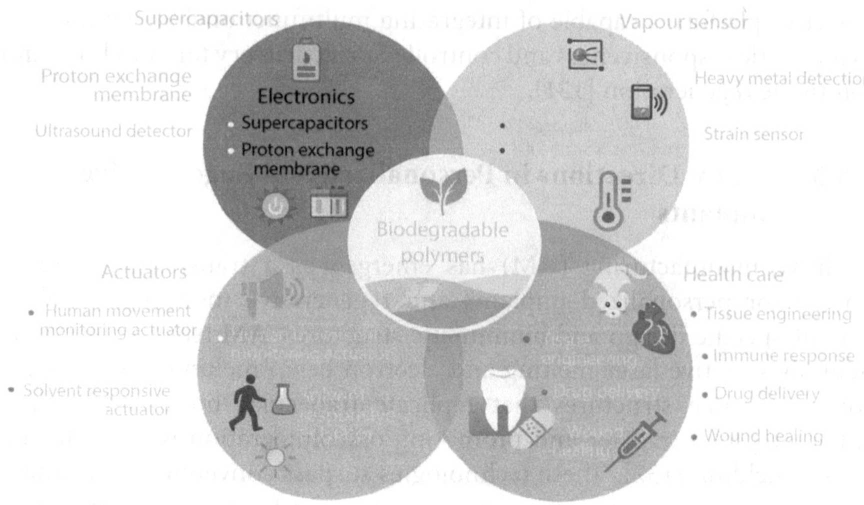

expansion and creating off-the-shelf solutions for immediate clinical availability [135, 136]. Figure 1.8 discusses emerging future trends in resorbable composite implants, illustrating advancements in patient-specific implant design, the use of 3D-printed composite scaffolds, and the integration of personalized medicine approaches. The figure highlights how these innovations converge to enhance customization, biological performance, and therapeutic alignment with individual patient needs.

Figure 1.9 summarizes representative use cases, including energy storage and proton exchange membranes in electronics; vapor, strain, and heavy metal sensing in sensor technologies; stimulus-responsive and human-movement-assisted actuation; and biomedical applications such as tissue engineering, immune response modulation, drug delivery, and wound healing, emphasizing the interdisciplinary potential of biodegradable polymer systems.

1.9 Summary and Chapter Roadmap

This chapter has provided a comprehensive introduction to resorbable composites for bioimplants and fixation devices, establishing the scientific, technological, and clinical context necessary for understanding their growing importance in modern biomedical engineering. The discussion began with an overview of resorbable biomaterials, defining key concepts related to resorbability, degradation behavior, and biocompatibility, while tracing the historical evolution of resorbable implants from early polymeric systems to advanced multifunctional platforms. A critical comparison with permanent and biostable implant materials highlighted the clinical motivations for resorbable solutions, particularly the elimination of secondary removal surgeries, reduced long-term complications, and improved alignment with natural tissue regeneration processes.

Fundamental principles of composite design were then examined, emphasizing the roles of matrices, reinforcements, and interfacial interactions in governing structure–property relationships. The chapter detailed how mechanical strength, toughness, and fatigue behavior can be tailored through material selection and architectural control, enabling resorbable composites to meet demanding load-bearing requirements. Subsequently, major material classes utilized in resorbable composites were systematically reviewed, including biodegradable polymers, bioactive ceramics and glass fillers, and natural or synthetic fiber reinforcements, with attention to their synergistic contributions to mechanical integrity, bioactivity, and controlled degradation.

Degradation mechanisms and biocompatibility considerations were discussed in depth, focusing on hydrolytic and enzymatic pathways, host tissue response, and toxicological safety. These sections underscored the importance of predictable resorption kinetics and non-toxic degradation by-products to ensure safe and effective clinical performance. The chapter further reviewed key processing and fabrication techniques, such as melt processing, solution-based methods, additive manufacturing, and surface modification strategies, demonstrating how manufacturing choices influence microstructure, functional performance, and translational viability.

Performance requirements for implant and fixation applications were addressed by integrating mechanical performance, degradation kinetics, load transfer, sterilization, and packaging considerations. Clinical and surgical perspectives were then explored, highlighting typical implantation sites, resorption profiles in orthopedic and maxillofacial applications, and the regulatory frameworks governing resorbable medical devices. Finally, emerging trends and research opportunities were introduced, including smart and bioactive composite platforms, nanocomposites with controlled degradation, and future directions toward personalized and regenerative implants.

Chapter Roadmap

Building on this foundational overview, subsequent chapters will delve deeper into specific material systems, advanced processing strategies, and detailed structure–property–performance relationships of resorbable composites. Later chapters will focus on application-specific design considerations, *in vitro* and *in vivo* evaluation methods, regulatory pathways, and case studies demonstrating clinical translation. Collectively, this progression aims to guide the reader from fundamental principles to advanced research and practical implementation of resorbable composites in next-generation biomedical implants and fixation devices.

AI disclosure statement: The author confirms that no artificial intelligence (AI)-generated content or AI-assisted writing tools were used in the preparation of this chapter (including summaries and the abstract). All text, analysis, and conclusions are the result of the author's original work.

References

1. Burg, K.J.L. and Orr, D.E., An overview of bioresorbable materials, in: *Degrad. Rate Bioresorbable Mater. Predict. Eval.*, pp. 3–18, Elsevier, Cambridge, United Kingdom, 2008, https://doi.org/10.1533/9781845695033.1.3.
2. Buchanan, F., *Degradation rate of bioresorbable materials: Prediction and evaluation*, Woodhead Publishing Limited, Cambridge, United Kingdom, 2008, https://doi.org/10.1533/9781845695033.
3. Das, D., Zhang, Z., Winkler, T., Mour, M., Gunter, C., Morlock, M., Machens, H.G., Schilling, A.F., Bioresorption and degradation of biomaterials, in: *Adv. Biochem. Eng. Biotechnol.*, pp. 317–333, Springer, Berlin and Heidelberg, Germany, 2012, https://doi.org/10.1007/10_2011_119.
4. Vach Agocsova, S., Culenova, M., Birova, I., Omanikova, L., Moncmanova, B., Danisovic, L., Ziaran, S., Bakos, D., Alexy, P., Resorbable Biomaterials Used for 3D Scaffolds in Tissue Engineering: A Review. *Mater. (Basel)*, 16, 4267, 2023, https://doi.org/10.3390/ma16124267.
5. Girish, B., Rakshith, G.R., Paul, A.K., Raja, V.K., Chakraborty, G., Hydrolytic and soil degradation of cellulosic material (paper): optimization of parameters using ANN and RSM. *Polym. Bull.*, 81, 12893–12920, 2024, https://doi.org/10.1007/s00289-024-05319-5.
6. Pappalardo, D., Mathisen, T., Finne-Wistrand, A., Biocompatibility of Resorbable Polymers: A Historical Perspective and Framework for the Future. *Biomacromolecules*, 20, 1465–1477, 2019, https://doi.org/10.1021/acs.biomac.9b00159.

7. Burstein, F.D., Resorbable distraction of the mandible: Technical evolution and clinical experience. *J. Craniofac. Surg.*, 19, 637–643, 2008, https://doi.org/10.1097/SCS.0b013e31816b6c8f.
8. Kotteda, A.K., Saikumar, T., Shreegan, A.A., Advancements in Biodegradable Orthopaedic Implants: A Literature Review. *Indian J. Orthop.*, 59, 1090–1100, 2025, https://doi.org/10.1007/s43465-025-01445-y.
9. Cha, G.D., Kang, D., Lee, J., Kim, D.H., Bioresorbable Electronic Implants: History, Materials, Fabrication, Devices, and Clinical Applications. *Adv. Healthc. Mater.*, 8, 2019, https://doi.org/10.1002/adhm.201801660.
10. Paul, A.K., Prasad, A., Mahto, B., Kumar, A., Singh, P., Introduction to bioenergy and waste to energy, in: *Biodegrad. Waste Process. Sustain. Dev.*, pp. 1–33, CRC Press, Boca Raton, 2024, https://doi.org/10.1201/9781003502012-1.
11. Seen, S., Young, S.M., Teo, S.J., Lang, S.S., Amrith, S., Lim, T.C., Sundar, G., Permanent Versus Bioresorbable Implants in Orbital Floor Blowout Fractures. *Ophthal. Plast. Reconstr. Surg.*, 34, 536–543, 2018, https://doi.org/10.1097/IOP.0000000000001077.
12. Amini, A.R., Wallace, J.S., Nukavarapu, S.P., Short-term and long-term effects of orthopedic biodegradable implants. *J. Long-Term Eff. Med. Implants*, 21, 93–122, 2011, https://doi.org/10.1615/JLongTermEffMedImplants.v21.i2.10.
13. Holt, G.R., Parel, S.M., Branemark, P.I., Osseointegrated Titanium Implants. *Facial Plast. Surg.*, 3, 113–124, 1986, https://doi.org/10.1055/s-2008-1064831.
14. Moghadasi, K., Isa, M.S.M., Ariffin, M.A., Jamil, M.Z.M., Raja, S., Wu, B., Yamani, M., Muhamad, M.R.B., Yusof, F., Jamaludin, M.F., Karim, M.S. bin Ab, Razak, B.A., bin Yusoff, N., A review on biomedical implant materials and the effect of friction stir based techniques on their mechanical and tribological properties. *J. Mater. Res. Technol.*, 17, 1054–1121, 2022, https://doi.org/10.1016/j.jmrt.2022.01.050.
15. Egbo, M.K., A fundamental review on composite materials and some of their applications in biomedical engineering. *J. King Saud Univ. Eng. Sci.*, 33, 557–568, 2021, https://doi.org/10.1016/j.jksues.2020.07.007.
16. Wang, M. and Zhao, Q., Biomedical composites, in: *Encycl. Biomed. Eng.*, pp. 34–52, Elsevier, Amsterdam, The Netherlands, 2019, https://doi.org/10.1016/B978-0-12-801238-3.99868-4.
17. Zhang, G., Lu, H., Mamidwar, S., Wang, M., Composites, in: *Biomater. Sci. An Introd. to Mater. Med.*, pp. 415–429, Elsevier, Amsterdam, The Netherlands, 2020, https://doi.org/10.1016/B978-0-12-816137-1.00029-5.
18. Migliaresi, C. and Nicolais, L., Composite materials for biomedical applications. *Int. J. Artif. Organs*, 3, 114–118, 1980, https://doi.org/10.1177/039139888000300213.
19. Nagabushanam, M., Devade, K., Reddy, G.A., Goud, B.N., Sayed, R.M., Sood, S., Sonia, P., Advance biomedical engineering – A fundamental review of composite materials and its applications. *Mater. Today Proc.*, 2023, https://doi.org/10.1016/j.matpr.2023.08.216.

20. Wu, C.J., Gaharwar, A.K., Schexnailder, P.J., Schmidt, G., Development of biomedical polymer-silicate nanocomposites: A materials science perspective. *Mater. (Basel).*, 3, 2986–3005, 2010, https://doi.org/10.3390/ma3052986.
21. Bhattacharjee, M., Sai, R.M., Chakraborty, G., Paul, A.K., Degradation and Toxicological Studies of Polymeric Nanocomposites, in: *Resorbable Polym. Bioimplants Fixat. Devices*, pp. 283–314, Springer Nature Singapore, Singapore, 2025, https://doi.org/10.1007/978-981-96-9306-1_12.
22. Singh, N., Dewangan, K., Srivastava, A., Azad, U.P., Plant-based cellulose paper matrices for electro-optical sensing: Advances in fabrication and applications. *Ind. Crops Prod.*, 235, 121679, 2025, https://doi.org/10.1016/j.indcrop.2025.121679.
23. Zahra, D., Zarei, M., Hwang, J., Lee, E., Lee, S.G., Oxidation-resistant and highly sensitive cellulose paper pressure sensor for wearable electronics. *Carbohydr. Polym. Technol. Appl.*, 9, 100672, 2025, https://doi.org/10.1016/j.carpta.2025.100672.
24. Jaiswal, A.K., Kumar, V., Jansson, E., Huttunen, O.H., Yamamoto, A., Vikman, M., Khakalo, A., Hiltunen, J., Behfar, M.H., Biodegradable Cellulose Nanocomposite Substrate for Recyclable Flexible Printed Electronics. *Adv. Electron. Mater.*, 9, 2023, https://doi.org/10.1002/aelm.202201094.
25. Nowroozi, N., Faraji, S., Nouralishahi, A., Shahrousvand, M., Biological and structural properties of graphene oxide/curcumin nanocomposite incorporated chitosan as a scaffold for wound healing application. *Life Sci.*, 264, 118640, 2021, https://doi.org/10.1016/j.lfs.2020.118640.
26. Choudhary, P., Ramalingam, B., Das, S.K., Fabrication of Chitosan-Reinforced Multifunctional Graphene Nanocomposite as Antibacterial Scaffolds for Hemorrhage Control and Wound-Healing Application. *ACS Biomater. Sci. Eng.*, 6, 5911–5929, 2020, https://doi.org/10.1021/acsbiomaterials.0c00923.
27. Kalva, S.N., Uysal, I., Khan, S.A., Z., Evis, Tezcaner, A., Keskin, D., Hegazy, A., Ibrahim, T., Kariyal, A.P., Velasquez, C.A., Koç, M., Performance evaluation of 3D-printed PLA composites doped with WE43 magnesium alloy for bone tissue engineering applications. *Biomater. Adv.*, 177, 214414, 2025, https://doi.org/10.1016/j.bioadv.2025.214414.
28. Alonso-Fernández, I., Haugen, H.J., López-Peña, M., González-Cantalapiedra, A., Muñoz, F., Use of 3D-printed polylactic acid/bioceramic composite scaffolds for bone tissue engineering in preclinical *in vivo* studies: A systematic review. *Acta Biomater.*, 168, 1–21, 2023, https://doi.org/10.1016/j.actbio.2023.07.013.
29. Seo, I.H., Hassan, R.U., Ryu, B., Koh, W.G., Ryu, W.H., Electrohydrodynamic Printing of Biodegradable PLGA Micro-Patterns on 3D Polymer Structures. *Adv. Mater. Technol.*, 9, 2024, https://doi.org/10.1002/admt.202400230.
30. Weng, W.H., Wang, C.Y., Yan, Z.Y., Lee, H.T., Kao, C.Y., Chang, C.W., Isolation and characterizations of multidrug-resistant human cancer cells by a biodegradable nano-sensor. *Biosens. Bioelectron.*, 249, 115985, 2024, https://doi.org/10.1016/j.bios.2023.115985.

31. Kirbay, F.O. and Odaci, D., Electrospun Poly-ϵ-caprolactone/Poly-l-lysine (PCL/PLL) Nanofibers as an Emergent Material for the Preparation of Electrochemical Immunosensor to Detect Serum Amyloid A. *ACS Appl. Polym. Mater.*, 6, 3778–3786, 2024, https://doi.org/10.1021/acsapm.3c03002.
32. Tang, C., Shen, Y., Xing, Y., Wu, Y., Zhang, M., Zhang, H., Zhao, S., Zhou, Z., Sun, Y., Mo, X., Wang, W., 3D-Printed Stents Loaded with Panax notoginseng Saponin for Promoting Re-endothelialization and Reducing Local Inflammation in the Carotid Artery of Rabbits. *ACS Biomater. Sci. Eng.*, 10, 6483–6497, 2024, https://doi.org/10.1021/acsbiomaterials.4c00925.
33. Sim, E.J., Sakthiabirami, K., Jung, T.G., Jo, S.K., Lee, S.J., Xu, P., Sung, H.K., Lee, E.H., Yoon, K.C., Kang, K.S., Jeong, Y.H., Thordarson, D.B., Lee, C.S., Kim, S.M., Kang, C., Song, J.H., Park, W.H., Bioabsorbable PLA/PHA composite pins for orthopedic applications: *In vitro* and *in vivo* study. *J. Ind. Eng. Chem.*, 146, 861–870, 2025, https://doi.org/10.1016/j.jiec.2025.03.023.
34. Petousis, M., David, C., Sagris, D., Nasikas, N.K., Papadakis, V., Argyros, A., Efstratiadis, V.S., Gaganatsiou, A., Michailidis, N., Vidakis, N., Reinforced PHA/CNC Biocomposites in Extrusion-Based Additive Manufacturing. *ACS Omega*, 10, 36613–36630, 2025, https://doi.org/10.1021/acsomega.5c05743.
35. Lehtonen, T.J., Tuominen, J.U., Hiekkanen, E., Resorbable composites with bioresorbable glass fibers for load-bearing applications. *in vitro* degradation and degradation mechanism. *Acta Biomater.*, 9, 4868–4877, 2013, https://doi.org/10.1016/j.actbio.2012.08.052.
36. Sharma, U., Concagh, D., Core, L., Kuang, Y., You, C., Pham, Q., Zugates, G., Busold, R., Webber, S., Merlo, J., Langer, R., Whitesides, G.M., Palasis, M., The development of bioresorbable composite polymeric implants with high mechanical strength. *Nat. Mater.*, 17, 96–102, 2018, https://doi.org/10.1038/NMAT5016.
37. Ahmed, I., Jones, I.A., Parsons, A.J., Bernard, J., Farmer, J., Scotchford, C.A., Walker, G.S., Rudd, C.D., Composites for bone repair: Phosphate glass fibre reinforced PLA with varying fibre architecture. *J. Mater. Sci. Mater. Med.*, 22, 1825–1834, 2011, https://doi.org/10.1007/s10856-011-4361-0.
38. Heimbach, B., Tonyali, B., Zhang, D., Wei, M., High performance resorbable composites for load-bearing bone fixation devices. *J. Mech. Behav. Biomed. Mater.*, 81, 1–9, 2018, https://doi.org/10.1016/j.jmbbm.2018.01.031.
39. Athanasiou, K.A., Agrawal, C.M., Barber, F.A., Burkhart, S.S., Orthopaedic applications for PLA-PGA biodegradable polymers. *Arthroscopy*, 14, 726–737, 1998, https://doi.org/10.1016/S0749-8063(98)70099-4.
40. da Silva, D., Kaduri, M., Poley, M., Adir, O., Krinsky, N., Shainsky-Roitman, J., Schroeder, A., Biocompatibility, biodegradation and excretion of polylactic acid (PLA) in medical implants and theranostic systems. *Chem. Eng. J.*, 340, 9–14, 2018, https://doi.org/10.1016/j.cej.2018.01.010.
41. Hutmacher, D.W., Goh, J.C.H., Teoh, S.H., An introduction to biodegradable materials for tissue engineering applications. *Ann. Acad. Med. Singapore*, 30, 183–191, 2001, https://doi.org/10.47102/annals-acadmedsg.v30n2p183.

42. Kaur, G., Kumar, V., Baino, F., Mauro, J.C., Pickrell, G., Evans, I., Bretcanu, O., Mechanical properties of bioactive glasses, ceramics, glass-ceramics and composites: State-of-the-art review and future challenges. *Mater. Sci. Eng., C*, 104, 2019, https://doi.org/10.1016/j.msec.2019.109895.
43. Niemelä, T., Niiranen, H., Kellomäki, M., Self-reinforced composites of bioabsorbable polymer and bioactive glass with different bioactive glass contents. Part II: *In vitro* degradation. *Acta Biomater.*, 4, 156–164, 2008, https://doi.org/10.1016/j.actbio.2007.06.007.
44. Paul, A.K., Borkotoky, S.S., Prasad, A., Nanohydroxyapatite-Based Composite Materials and Processing, in: *Adv. Mater. Manuf. Tech. Biomed. Appl.*, pp. 157–183, Wiley, Beverly, United States, 2023, https://doi.org/10.1002/9781394166985.ch7.
45. Simpson, R.L., Nazhat, S.N., Blaker, J.J., Bismarck, A., Hill, R., Boccaccini, A.R., Hansen, U.N., Amis, A.A., A comparative study of the effects of different bioactive fillers in PLGA matrix composites and their suitability as bone substitute materials: A thermo-mechanical and *in vitro* investigation. *J. Mech. Behav. Biomed. Mater.*, 50, 277–289, 2015, https://doi.org/10.1016/j.jmbbm.2015.06.008.
46. Hajiali, F., Tajbakhsh, S., Shojaei, A., Fabrication and Properties of Polycaprolactone Composites Containing Calcium Phosphate-Based Ceramics and Bioactive Glasses in Bone Tissue Engineering: A Review. *Polym. Rev.*, 58, 164–207, 2018, https://doi.org/10.1080/15583724.2017.1332640.
47. Purnomo, Setyarini, P.H., Cahyandari, D., Potential Natural Fiber-Reinforced Composite for Biomedical Application. *IOP Conf. Ser. Mater. Sci. Eng.*, 494, 012018, 2019, https://doi.org/10.1088/1757-899X/494/1/012018.
48. Bharadwaz, A. and Jayasuriya, A.C., Recent trends in the application of widely used natural and synthetic polymer nanocomposites in bone tissue regeneration. *Mater. Sci. Eng., C*, 110, 110698, 2020, https://doi.org/10.1016/j.msec.2020.110698.
49. Boccaccini, A.R. and Blaker, J.J., Bioactive composite materials for tissue engineering scaffolds. *Expert Rev. Med. Devices*, 2, 303–317, 2005, https://doi.org/10.1586/17434440.2.3.303.
50. Pérez, R.A., Won, J.E., Knowles, J.C., Kim, H.W., Naturally and synthetic smart composite biomaterials for tissue regeneration. *Adv. Drug Deliv. Rev.*, 65, 471–496, 2013, https://doi.org/10.1016/j.addr.2012.03.009.
51. Santerre, J.P., Shajii, L., Leung, B.W., Relation of dental composite formulations to their degradation and the release of hydrolyzed polymeric-resin-derived products. *Crit. Rev. Oral Biol. Med.*, 12, 136–151, 2001, https://doi.org/10.1177/10454411010120020401.
52. Hajebi, S., Mohammadi Nasr, S.A., Rabiee, N., Bagherzadeh, M., Ahmadi, S., Rabiee, M., Tahriri, M., Tayebi, L., Hamblin, M.R., Bioresorbable composite polymeric materials for tissue engineering applications. *Int. J. Polym. Mater.*

Polym. Biomater., 70, 926–940, 2021, https://doi.org/10.1080/00914037.2020.1765365.
53. Ganesh, M. and Gross, R., Embedding enzymes to control biomaterial lifetime, in: *ACS Symp. Ser.*, pp. 375–384, 2010, https://doi.org/10.1021/bk-2010-1043.ch025.
54. Brugmans, M.C.P., Söntjens, S.H.M., Cox, M.A.J., Nandakumar, A., Bosman, A.W., Mes, T., Janssen, H.M., Bouten, C.V.C., Baaijens, F.P.T., Driessen-Mol, A., Hydrolytic and oxidative degradation of electrospun supramolecular biomaterials: *In vitro* degradation pathways. *Acta Biomater.*, 27, 21–31, 2015, https://doi.org/10.1016/j.actbio.2015.08.034.
55. Sarbu, I., Bida, C., Forna, D.A., Focsaneanu, S., Dimofte, A.-R., Stan, G., Earar, K., Biocompatibility of Titanium Dental Implants: Influences on Osseointegration and Local Immunity. *Rom. J. Oral Rehabil.*, 17, 330–341, 2025, https://doi.org/10.62610/rjor.2025.2.17.30.
56. Morehead, J.M. and Holt, G.R., Soft-tissue response to synthetic biomaterials. *Otolaryngol. Clin. North Am.*, 27, 195–201, 1994, https://doi.org/10.1016/s0030-6665(20)30723-4.
57. Babu, R.S.A. and Ogle, O., Tissue Response: Biomaterials, Dental Implants, and Compromised Osseous tissue. *Dent. Clin. North Am.*, 59, 305–315, 2015, https://doi.org/10.1016/j.cden.2014.10.010.
58. Barchowsky, A., Systemic and Immune Toxicity of Implanted Materials, in: *Biomater. Sci. An Introd. to Mater. Med.*, pp. 791–799, Elsevier, Cambridge, United Kingdom, 2020, https://doi.org/10.1016/B978-0-12-816137-1.00051-9.
59. Gong, B., Zhang, H., Li, S., Lin, Y., Xu, S., Jia, B., Lu, Y., Liu, Z., Strategies for enhancing biocompatibility, bioactivity, and resistance to biodegradation in dental resin composites: A narrative review. *Biomater. Adv.*, 179, 214496, 2026, https://doi.org/10.1016/j.bioadv.2025.214496.
60. Zhou, Z., Liu, L., Liu, Q., Huang, T., Zeng, W., Yi, Q., Yang, Z., Cao, D., Evaluation of biological properties *in-vivo* of poly(l-Lactide-Coglycolide) composites containing bioactive glass. *Polym. Polym. Compos.*, 21, 79–84, 2013, https://doi.org/10.1177/096739111302100204.
61. Coombes, A.G.A. and Heckman, J.D., Gel casting of resorbable polymers. 1. Processing and applications. *Biomaterials*, 13, 217–224, 1992, https://doi.org/10.1016/0142-9612(92)90187-S.
62. Cardon, L.K., Ragaert, K.J., Koster, R.P., Design and fabrication of biocomposites, in: *Biomed. Compos.*, pp. 25–43, Elsevier, Cambridge, United Kingdom, 2009, https://doi.org/10.1533/9781845697372.1.25.
63. Maurya, A.K., de Souza, F.M., Dawsey, T., Gupta, R.K., Biodegradable polymers and composites: Recent development and challenges. *Polym. Compos.*, 45, 2896–2918, 2024, https://doi.org/10.1002/pc.28023.
64. Meng, Z., He, J., Li, J., Su, Y., Li, D., Melt-based, solvent-free additive manufacturing of biodegradable polymeric scaffolds with designer microstructures for tailored mechanical/biological properties and clinical applications.

Virtual Phys. Prototyp., 15, 417–444, 2020, https://doi.org/10.1080/17452759.2020.1808937.

65. Bose, S., Ke, D., Sahasrabudhe, H., Bandyopadhyay, A., Additive manufacturing of biomaterials. *Prog. Mater Sci.*, 93, 45–111, 2018, https://doi.org/10.1016/j.pmatsci.2017.08.003.
66. Zadpoor, A.A., Design for additive bio-manufacturing: From patient-specific medical devices to rationally designed meta-biomaterials. *Int. J. Mol. Sci.*, 18, 1607, 2017, https://doi.org/10.3390/ijms18081607.
67. Kumar, M.H., Paul, A.K., Chakraborty, G., Utilization of gelatin, collagen, and chondroitin sulfate in 3D bioprinting, in: *Handb. 3D Print. Biomed. Appl.*, pp. 165–187, CRC Press, Boca Raton, 2025, https://doi.org/10.1201/9781003521792-10.
68. Chen, J., Chen, J., Wang, H., He, L., Huang, B., Dadbakhsh, S., Bartolo, P., Fabrication and development of mechanical metamaterials *via* additive manufacturing for biomedical applications: a review. *Int. J. Extrem. Manuf.*, 7, 012001, 2025, https://doi.org/10.1088/2631-7990/ad88e3.
69. Dienel, K.E.G., Van Bochove, B., Seppälä, J.V., Additive Manufacturing of Bioactive Poly(trimethylene carbonate)/β-Tricalcium Phosphate Composites for Bone Regeneration. *Biomacromolecules*, 21, 366–375, 2020, https://doi.org/10.1021/acs.biomac.9b01272.
70. Liu, Z., Wang, Y., Wu, B., Cui, C., Guo, Y., Yan, C., A critical review of fused deposition modeling 3D printing technology in manufacturing polylactic acid parts. *Int. J. Adv. Manuf. Technol.*, 102, 2877–2889, 2019, https://doi.org/10.1007/s00170-019-03332-x.
71. Netto, JM.J., Idogava, H.T., Santos, LE.F., de C. Silveira, Z., Romio, P., Alves, J.L., Screw-assisted 3D printing with granulated materials: a systematic review. *Int. J. Adv. Manuf. Technol.*, 115, 2711–2727, 2021, https://doi.org/10.1007/s00170-021-07365-z.
72. Balani, S.B., Ghaffar, S.H., Chougan, M., Pei, E., Şahin, E., Processes and materials used for direct writing technologies: A review. *Results Eng.*, 11, 100257, 2021, https://doi.org/10.1016/j.rineng.2021.100257.
73. Rahmati, M., Mills, D.K., Urbanska, A.M., Saeb, M.R., Venugopal, J.R., Ramakrishna, S., Mozafari, M., Electrospinning for tissue engineering applications. *Prog. Mater Sci.*, 117, 100721, 2021, https://doi.org/10.1016/j.pmatsci.2020.100721.
74. Kade, J.C. and Dalton, P.D., Polymers for Melt Electrowriting. *Adv. Healthc. Mater.*, 10, 2021, https://doi.org/10.1002/adhm.202001232.
75. Lupone, F., Padovano, E., Casamento, F., Badini, C., Process phenomena and material properties in selective laser sintering of polymers: A review. *Mater. (Basel).*, 15, 183, 2022, https://doi.org/10.3390/ma15010183.
76. Paul, A.K., Chakraborty, G., Prasad, A., Toxicology Assessment of Polymeric Material for Implants, in: *Resorbable Polym. Bioimplants Fixat. Devices*, pp. 349–369, Springer Nature Singapore, Singapore, 2025, https://doi.org/10.1007/978-981-96-9306-1_14.

77. Mandracci, P., Mussano, F., Rivolo, P., Carossa, S., Surface treatments and functional coatings for biocompatibility improvement and bacterial adhesion reduction in dental implantology. *Coatings*, 6, 7, 2016, https://doi.org/10.3390/coatings6010007.
78. Dong, H., Liu, H., Zhou, N., Li, Q., Yang, G., Chen, L., Mou, Y., Surface modified techniques and emerging functional coating of dental implants. *Coatings*, 10, 1–25, 2020, https://doi.org/10.3390/coatings10111012.
79. Mishchenko, O., Volchykhina, K., Maksymov, D., Manukhina, O., Pogorielov, M., Pavlenko, M., Iatsunskyi, I., Advanced Strategies for Enhancing the Biocompatibility and Antibacterial Properties of Implantable Structures. *Mater. (Basel).*, 18, 822, 2025, https://doi.org/10.3390/ma18040822.
80. Nobles, K.P., Janorkar, A.V., Williamson, R.S., Surface modifications to enhance osseointegration–Resulting material properties and biological responses. *J. Biomed. Mater. Res. Part B Appl. Biomater.*, 109, 1909–1923, 2021, https://doi.org/10.1002/jbm.b.34835.
81. Asadollahi, M., Gerashi, E., Zohrevand, M., Zarei, M., Sayedain, S.S., Alizadeh, R., Labbaf, S., Atari, M., Improving mechanical properties and biocompatibility of 3D printed PLA by the addition of PEG and titanium particles, using a novel incorporation method. *Bioprinting*, 27, e00228, 2022, https://doi.org/10.1016/j.bprint.2022.e00228.
82. Lee, J., Lee, H., Cheon, K.H., Park, C., Jang, T.S., Kim, H.E., Do Jung, H., Fabrication of poly(lactic acid)/Ti composite scaffolds with enhanced mechanical properties and biocompatibility *via* fused filament fabrication (FFF)–based 3D printing. *Addit. Manuf.*, 30, 100883, 2019, https://doi.org/10.1016/j.addma.2019.100883.
83. Butt, J., Oxford, P., Sadeghi-Esfahlani, S., Ghorabian, M., Shirvani, H., Hybrid Manufacturing and Mechanical Characterization of Cu/PLA Composites. *Arab. J. Sci. Eng.*, 45, 9339–9356, 2020, https://doi.org/10.1007/s13369-020-04778-y.
84. Balamurugan, K., Venkata Pavan, M., Ahamad Ali, S.K., Kalusuraman, G., Compression and flexural study on PLA-Cu composite filament using FDM. *Mater. Today Proc.*, 44, 1687–1691, 2021, https://doi.org/10.1016/j.matpr.2020.11.858.
85. Jiang, D. and Ning, F., Fused filament fabrication of biodegradable PLA/316L composite scaffolds: Effects of metal particle content. *Procedia Manuf.*, 48, 755–762, 2020, https://doi.org/10.1016/j.promfg.2020.05.110.
86. Ali, F., Kalva, S.N., Koç, M., Additive Manufacturing of Polymer/Mg-Based Composites for Porous Tissue Scaffolds. *Polym. (Basel).*, 14, 5460, 2022, https://doi.org/10.3390/polym14245460.
87. Ferrández-Montero, A., Lieblich, M., Benavente, R., González-Carrasco, J.L., Ferrari, B., Study of the matrix-filler interface in PLA/Mg composites manufactured by Material Extrusion using a colloidal feedstock. *Addit. Manuf.*, 33, 101142, 2020, https://doi.org/10.1016/j.addma.2020.101142.

88. Alam, F., Shukla, V.R., Varadarajan, K.M., Kumar, S., Microarchitected 3D printed polylactic acid (PLA) nanocomposite scaffolds for biomedical applications. *J. Mech. Behav. Biomed. Mater.*, 103, 103576, 2020, https://doi.org/10.1016/j.jmbbm.2019.103576.
89. Daniels, A.U., Chang, M.K., Andriano, K.P., Mechanical properties of biodegradable polymers and composites proposed for internal fixation of bone. *J. Appl. Biomater.*, 1, 57–78, 1990, https://doi.org/10.1002/jab.770010109.
90. Paul, A.K. and Das, S., Biolubricants and its application in engineering, in: *Sustain. Bio-Based Compos.*, pp. 271–296, De Gruyter, Berlin, Germany, 2024, https://doi.org/10.1515/9783111321530-013.
91. Chen, Q.Z., Quinn, J.M.W., Thouas, G.A., Zhou, X., Komesaroff, P.A., Bone-like elastomer-toughened scaffolds with degradability kinetics matching healing rates of injured bone. *Adv. Eng. Mater.*, 12, 2010, https://doi.org/10.1002/adem.201080002.
92. Jhamb, S.K., Goyal, A., Pandey, A., Marwaha, J., Matai, J., Degradation behaviour of magnesium alloy and its composite used as a biomaterial. *E3S Web Conf*, vol. 309, p. 01085, 2021, https://doi.org/10.1051/e3sconf/202130901085.
93. Paramsothy, M. and Ramakrishna, S., Biodegradable Materials for Clinical Applications: A Review. *Rev. Adv. Sci. Eng.*, 4, 221–238, 2016, https://doi.org/10.1166/rase.2015.1097.
94. Prasad, A. and Paul, A.K., *Biodegradable waste processing for sustainable developments*, CRC Press, Boca Raton, 2024, https://doi.org/10.1201/9781003502012.
95. Tipnis, N.P. and Burgess, D.J., Sterilization of implantable polymer-based medical devices: A review. *Int. J. Pharm.*, 544, 455–460, 2018, https://doi.org/10.1016/j.ijpharm.2017.12.003.
96. Bargmann, L.S., Bargmann, B.C., Collier, J.P., Currier, B.H., Mayor, M.B., Current sterilization and packaging methods for polyethylene. *Clin. Orthop. Relat. Res.*, 369, 49–58, 1999, https://doi.org/10.1097/00003086-199912000-00006.
97. Lyu, S.P. and Untereker, D., Degradability of polymers for implantable biomedical devices. *Int. J. Mol. Sci.*, 10, 4033–4065, 2009, https://doi.org/10.3390/ijms10094033.
98. Paul, A.K., Chakraborty, G., Borkotoky, S.S., Prasad, A., Degradation Studies of Bioabsorbable Polymeric Composites, in: *Bioabsorbable Polym. Drug Carriers Fixat. Devices*, pp. 105–128, Apple Academic Press, New York, 2025, https://doi.org/10.1201/9781003569817-5.
99. Teo, A.J.T., Mishra, A., Park, I., Kim, Y.J., Park, W.T., Yoon, Y.J., Polymeric Biomaterials for Medical Implants and Devices. *ACS Biomater. Sci. Eng.*, 2, 454–472, 2016, https://doi.org/10.1021/acsbiomaterials.5b00429.
100. Finkenstadt, V.L. and Willett, J.L., Electroactive materials composed of starch. *J. Polym. Environ.*, 12, 43–46, 2004, https://doi.org/10.1023/B:JOOE.0000010049.33284.08.

101. Hou, J., Liu, M., Zhang, H., Song, Y., Jiang, X., Yu, A., Jiang, L., Su, B., Healable green hydrogen bonded networks for circuit repair, wearable sensor and flexible electronic devices. *J. Mater. Chem. A*, 5, 13138–13144, 2017, https://doi.org/10.1039/c7ta03100a.
102. Ning, W., Xingxiang, Z., Xuechen, W., Haihui, L., Communications: Ionic liquids modified montmorillonite/ thermoplastic starch nanocomposites as ionic conducting biopolymer. *Macromol. Res.*, 17, 285–288, 2009, https://doi.org/10.1007/BF03218863.
103. Zhang, B., Xie, F., Shamshina, J.L., Rogers, R.D., McNally, T., Wang, D.K., Halley, P.J., Truss, R.W., Zhao, S., Chen, L., Facile Preparation of Starch-Based Electroconductive Films with Ionic Liquid. *ACS Sustain. Chem. Eng.*, 5, 5457–5467, 2017, https://doi.org/10.1021/acssuschemeng.7b00788.
104. Rosli, N.A.H., Loh, K.S., Wong, W.Y., Lee, T.K., Ahmad, A., Hybrid composite membrane of phosphorylated chitosan/ poly (Vinyl alcohol)/silica as a proton exchange membrane. *Membr. (Basel).*, 11, 675, 2021, https://doi.org/10.3390/membranes11090675.
105. Kamjornsupamitr, T., Sangthumchai, T., Saejueng, P., Sumranjit, J., Hunt, A.J., Budsombat, S., Composite proton conducting membranes from chitosan, poly(vinyl alcohol) and sulfonic acid-functionalized silica nanoparticles. *Int. J. Hydrogen Energy*, 46, 2479–2490, 2021, https://doi.org/10.1016/j.ijhydene.2020.10.062.
106. Di Franco, F., Zaffora, A., Burgio, G., Santamaria, M., Performance of H2-fed fuel cell with chitosan/silicotungstic acid membrane as proton conductor. *J. Appl. Electrochem.*, 50, 333–341, 2020, https://doi.org/10.1007/s10800-019-01394-z.
107. Danyliv, O., Strach, M., Nechyporchuk, O., Nypelö, T., Martinelli, A., Self-Standing, Robust Membranes Made of Cellulose Nanocrystals (CNCs) and a Protic Ionic Liquid: Toward Sustainable Electrolytes for Fuel Cells. *ACS Appl. Energy Mater.*, 4, 6474–6485, 2021, https://doi.org/10.1021/acsaem.1c00452.
108. Zhang, Q., Zhao, L., Yang, H., Kong, L., Ran, F., Alkali-tolerant polymeric gel electrolyte membrane based on cross-linked carboxylated chitosan for supercapacitors. *J. Memb. Sci.*, 629, 119083, 2021, https://doi.org/10.1016/j.memsci.2021.119083.
109. Ma, C., Xie, F., Wei, L., Zheng, C., Liu, X., Wang, L., Liu, P., All-Starch-Based Hydrogel for Flexible Electronics: Strain-Sensitive Batteries and Self-Powered Sensors. *ACS Sustain. Chem. Eng.*, 10, 6724–6735, 2022, https://doi.org/10.1021/acssuschemeng.2c00872.
110. Vargun, E., Ozaltin, K., Fei, H., Harea, E., Vilčáková, J., Kazantseva, N., Saha, P., Biodegradable porous polylactic acid film as a separator for supercapacitors. *J. Appl. Polym. Sci.*, 137, 2020, https://doi.org/10.1002/app.49270.
111. Zhu, J., Xiong, R., Zhao, F., Peng, T., Hu, J., Xie, L., Xie, H., Wang, K., Jiang, C., Lightweight, High-Strength, and Anisotropic Structure Composite Aerogel Based on Hydroxyapatite Nanocrystal and Chitosan with Thermal Insulation

and Flame Retardant Properties. *ACS Sustain. Chem. Eng.*, 8, 71–83, 2020, https://doi.org/10.1021/acssuschemeng.9b03953.

112. Eppley, B.L., Sadove, A.M., Havlik, R.J., Resorbable plate fixation in pediatric craniofacial surgery. *Plast. Reconstr. Surg.*, 100, 1–13, 1997, https://doi.org/10.1097/00006534-199707000-00001.
113. Ahmad, N., Lyles, J., Panchal, J., Outcomes and complications based on experience with resorbable plates in pediatric craniosynostosis patients. *J. Craniofac. Surg.*, 19, 855–860, 2008, https://doi.org/10.1097/SCS.0b013e31816ae358.
114. Paul, A.K., Migration Studies of Biodegradable Composites, in: *Biodegrad. Compos. Packag. Appl.*, pp. 193–212, CRC Press, Boca Raton, 2022, https://doi.org/10.1201/9781003227908-11.
115. Mohamed-Hashem, I.K. and Mitchell, D.A., Resorbable implants (plates and screws) in orthognathic surgery. *J. Orthod.*, 27, 198–199, 2000, https://doi.org/10.1093/ortho/27.2.198.
116. Pietrzak, W.S. and Eppley, B.L., Resorbable polymer fixation for craniomaxillofacial surgery: Development and engineering paradigms. *J. Craniofac. Surg.*, 11, 575–585, 2000, https://doi.org/10.1097/00001665-200011060-00011.
117. Claes, L. and Ignatius, A., Entwicklung neuer biodegradabler Implantate. *Chirurg*, 73, 990–996, 2002, https://doi.org/10.1007/s00104-002-0543-0.
118. Chandra, G. and Pandey, A., Biodegradable bone implants in orthopedic applications: a review. *Biocybern. Biomed. Eng.*, 40, 596–610, 2020, https://doi.org/10.1016/j.bbe.2020.02.003.
119. Sheikh, Z., Najeeb, S., Khurshid, Z., Verma, V., Rashid, H., Glogauer, M., Biodegradable materials for bone repair and tissue engineering applications. *Mater. (Basel).*, 8, 5744–5794, 2015, https://doi.org/10.3390/ma8095273.
120. Patel, S., Sanjay, M.S.S.S., Chakraborty, G., Paul, A.K., Nanocomposites/Nanomaterials and the Risk Management of Biofuels Production, in: *Nanomater. as a Catal. Biofuels Prod.*, pp. 151–184, Springer Nature Singapore, Singapore, 2025, https://doi.org/10.1007/978-981-96-1706-7_7.
121. Kanno, T., Sukegawa, S., Furuki, Y., Nariai, Y., Sekine, J., Overview of innovative advances in bioresorbable plate systems for oral and maxillofacial surgery. *Jpn. Dent. Sci. Rev.*, 54, 127–138, 2018, https://doi.org/10.1016/j.jdsr.2018.03.003.
122. Schuh, J.A.C.L. and Funk, K.A., Compilation of International Standards and Regulatory Guidance Documents for Evaluation of Biomaterials, Medical Devices, and 3-D Printed and Regenerative Medicine Products. *Toxicol. Pathol.*, 47, 344–357, 2019, https://doi.org/10.1177/0192623318804121.
123. Gad, S.C., Standards and methods for assessing the safety and biocompatibility of biomaterials, in: *Charact. Biomater.*, pp. 285–306, Elsevier, Berlin, Germany, 2012, https://doi.org/10.1533/9780857093684.285.
124. Singh, A.B., Khandelwal, C., Dangayach, G.S., Revolutionizing healthcare materials: Innovations in processing, advancements, and challenges for enhanced medical device integration and performance. *J. Micromanuf.*, 2025, https://doi.org/10.1177/25165984241256234.

125. Morrison, R.J., Kashlan, K.N., Flanangan, C.L., Wright, J.K., Green, G.E., Hollister, S.J., Weatherwax, K.J., Regulatory Considerations in the Design and Manufacturing of Implantable 3D-Printed Medical Devices. *Clin. Transl. Sci.*, 8, 594–600, 2015, https://doi.org/10.1111/cts.12315.
126. Yang, Y., Qiu, Y., Lin, C., Chen, X., Zhao, F., Stimulus-responsive smart bioactive glass composites for repair of complex tissue defects. *Theranostics*, 15, 1760–1786, 2025, https://doi.org/10.7150/thno.104944.
127. Chakrapani, G., Vijay Kumar, V., Zare, M., Next-Generation Biomaterials: Design Strategies, Clinical Translation, and the Rise of Intelligent Therapeutic Platforms. *ACS Omega*, 10, 46272–46293, 2025, https://doi.org/10.1021/acsomega.5c06034.
128. Kumar, M.H., Paul, A.K., Prasad, A., Chakraborty, G., Rheological Studies of Resorbable Polymeric Nanocomposites, in: *Resorbable Polym. Bioimplants Fixat. Devices*, pp. 163–183, Springer Nature Singapore, Singapore, 2025, https://doi.org/10.1007/978-981-96-9306-1_7.
129. Jain, M. and Matsumura, K., Polyampholyte- and nanosilicate-based soft bionanocomposites with tailorable mechanical and cell adhesion properties. *J. Biomed. Mater. Res. Part A*, 104, 1379–1386, 2016, https://doi.org/10.1002/jbm.a.35672.
130. Armentano, I., Gigli, M., Morena, F., Argentati, C., Torre, L., Martino, S., Recent advances in nanocomposites based on aliphatic polyesters: Design, synthesis, and applications in regenerative medicine. *Appl. Sci.*, 8, 1452, 2018, https://doi.org/10.3390/app8091452.
131. Bramhill, J., Ross, S., Ross, G., Bioactive nanocomposites for tissue repair and regeneration: A review. *Int. J. Environ. Res. Public Health*, 14, 66, 2017, https://doi.org/10.3390/ijerph14010066.
132. Jaouher, L., Barkaoui, A., Aouadi, K., Sallem, H., Biomimetic and personalized optimization of additive-manufactured metallic bone implants: Design, simulation, and clinical outcomes. *Int. J. Bioprint.*, 025270261, 2025, https://doi.org/10.36922/ijb025270261.
133. Wu, P., Liu, X., Guo, Z., Lau, L., Personalization and Precision: Innovative Applications and Future Challenges of Additive Manufacturing in Orthopedic Implants. *J. Orthop. Res.*, 2025, https://doi.org/10.1002/jor.70082.
134. Guzzi, E.A. and Tibbitt, M.W., Additive Manufacturing of Precision Biomaterials. *Adv. Mater.*, 32, 2020, https://doi.org/10.1002/adma.201901994.
135. Chakraborty, G. and Paul, A.K., 3D Printing in Biomedical Applications: Techniques and Emerging Trends, in: *Adv. Mater. Manuf. Tech. Biomed. Appl.*, pp. 275–301, Wiley, Beverly, United States, 2023, https://doi.org/10.1002/9781394166985.ch11.
136. Neves, L.S., Rodrigues, M.T., Reis, R.L., Gomes, M.E., Current approaches and future perspectives on strategies for the development of personalized tissue engineering therapies. *Expert Rev. Precis. Med. Drug Dev.*, 1, 93–108, 2016, https://doi.org/10.1080/23808993.2016.1140004.

2

Synthesis, Processing, and Characterization of the Resorbable Composites

Kandala Chandra Sekhar[1], Atanu Kumar Paul[1] and Gourhari Chakraborty[2]*

[1]*Department of Chemical Engineering, National Institute of Technology, Andhra Pradesh, India*

[2]*School of Chemical and Biotechnology, SASTRA Deemed University, Tamil Nadu, India*

Abstract

The chapter titled Synthesis, Processing, and Characterization of Resorbable Composites provides a comprehensive overview of the development and evaluation of resorbable materials for biomedical use. It covers various synthesis methodologies for polymers such as poly(octamethylene citrate) (POC) and poly(L-lactide-co-D,L-lactide) (PLDLA), noting POC's enhanced degradation resistance due to improved crosslinking and lower hydrophilicity. The chapter explores multiple fabrication techniques, including chemical precipitation, electrospinning, and sol-gel methods, with emphasis on controlling factors like the Ca/P ratio and crystallinity to closely mimic natural apatite structures. It further examines advanced processing approaches such as microemulsion, microwave irradiation, and 3D printing, highlighting their roles in achieving phase purity, particle uniformity, and customized scaffold design. Characterization techniques—including mechanical testing, X-ray diffraction (XRD), scanning electron microscopy (SEM), and degradation studies in phosphate-buffered saline—are discussed in the context of predicting *in vivo* behavior. The chapter concludes by linking these materials and methods to practical applications like bone grafts, drug delivery systems, and bioactive implants, underlining the importance of integrating design, processing, and performance for successful clinical translation.

***Keywords*:** Resorbable composites, biomedical polymers, synthesis techniques, scaffold fabrication, material characterization

**Corresponding author*: gourhari115@gmail.com

Arbind Prasad (ed.) Resorbable Composites for Bioimplants and Fixation Devices, (49–96) © 2026 Scrivener Publishing LLC

2.1 Introduction

2.1.1 Overview of Resorbable Composites

Resorbable composites have gained significant attention in biomedical applications due to their unique properties and degradation mechanisms. These materials, often combining bioresorbable polymers with reinforcing elements like glass fibers, offer tailorable mechanical properties and degradation rates [1, 2]. Key applications include orthopedic implants, cardiac stents, and tissue engineering scaffolds [2, 3]. The degradation of these composites typically involves a combination of surface and bulk erosion, with mechanisms influenced by factors such as material composition and environmental conditions [1, 4]. Importantly, the incorporation of reinforcing elements like phosphate glasses can suppress polymer embrittlement mechanisms, allowing for more controlled degradation and gradual transfer of load to healing tissue [2]. Ongoing research focuses on optimizing these materials for specific biomedical applications, aiming to enhance their efficacy and minimize patient discomfort [5–8]. In 1988, a National Science Foundation workshop described the term tissue engineering. It defined it as "The application of principles and methods of engineering and life sciences towards a fundamental understanding of structural and functional relationships in both normal and mammalian tissues and the development of biological substitutes to restore and maintain tissue functions". A wide range of biomaterials such as bio-polymers, bio-ceramics, and other bio-inorganics, bioactive molecules, are utilized in this emerging field of bioengineering [6, 8].

Resorbable composite materials have gained significant attention in orthopedic and tissue engineering applications due to their ability to combine bioactive and bioresorbable properties [9]. These materials, which include polymers, ceramics, and their composites, offer tailored mechanical properties and degradation behavior suitable for various clinical applications [10]. Bioactive ceramics like hydroxyapatite and tricalcium phosphate, when combined with bioresorbable polymers, can enhance mechanical strength and bioactivity [9, 11]. Magnesium alloys show promise for load-bearing applications during the initial stages of recovery [10]. The synergistic combination of bioceramics and biodegradable polymers improves mechanical properties, degradation resistance, and biocompatibility [10]. These composite biomaterials can mimic natural tissues and restore the mechanical functions of damaged tissue [12]. The development of these materials has led to new surgical applications not possible with conventional metals and plastics [13].

2.1.2 Importance in Biomedical and Environmental Applications

Resorbable composites play crucial roles in advancing sustainable biomedical treatments and environmental waste reduction [14]. These materials, often based on bioresorbable polymers like bacterial cellulose, collagen, and silk, offer unique structural and biological properties for various biomedical applications, including wound healing, tissue engineering, and drug delivery [3]. Bioresorbable biocomposites can be reinforced with natural fibers, nanoparticles, or bioactive ceramics to enhance their efficacy [15]. Recent advancements in nanocomposites have demonstrated promising clinical and environmental applications, showcasing advantages such as versatility, biocompatibility, and biodegradability [16]. Furthermore, the development of biocomposites from renewable and sustainable resources, including bio-sourced and recycled materials, has led to innovations in lightweight automotive parts and sustainable packaging. These advancements contribute to reduced dependence on virgin petro-based materials and support global sustainability goals [17].

Resorbable biomaterials play a crucial role in tissue engineering, offering biocompatibility, biodegradability, and functional performance [18]. These materials, including natural polymers like collagen and synthetic ones like polylactides, provide scaffolds for cell growth and tissue regeneration. They undergo controlled degradation in the body, eliminating the need for removal surgeries [4]. Resorbable polymers have diverse applications in wound healing, artificial organs, and drug delivery systems [3]. Their properties can be modified to suit specific clinical needs, such as matching the mechanical properties of target tissues [18]. The development of these materials has revolutionized biomedical applications, from simple sutures to complex tissue-engineered constructs [19]. Ongoing research focuses on improving their performance and expanding their applications in regenerative medicine [5, 8, 20].

Polymeric scaffolds made from resorbable materials have gained significant attention in biomedical applications, particularly for tissue engineering and drug delivery. These scaffolds offer biocompatibility, biodegradability, and favorable mechanical properties [21]. They can be engineered to support cell adhesion, proliferation, and differentiation, making them ideal for tissue regeneration [22]. Biodegradable polymers, both natural and synthetic, are being extensively studied for various applications, including temporary implants, wound healing, and controlled drug release [23]. In cardiovascular applications, bioresorbable polymeric scaffolds have shown promise in promoting positive vessel remodeling and

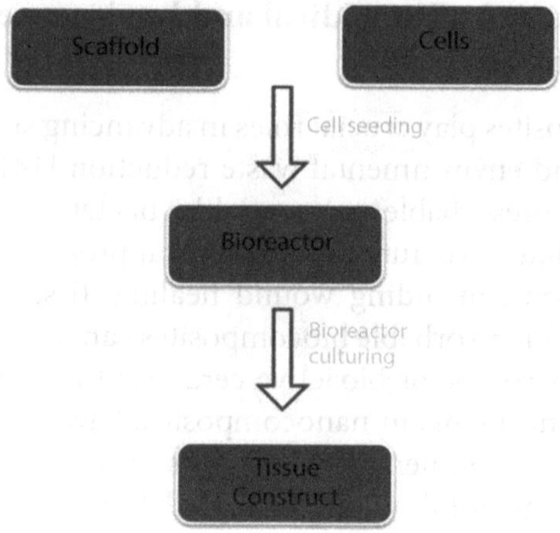

eliminating long-term complications associated with permanent implants [24]. The ability to manipulate degradation rates and functionalize these materials enhances their versatility in meeting specific medical requirements. However, challenges remain in translating research into clinically viable and marketable products, necessitating further advancements in polymer technology and fabrication methods [22, 25, 26]. Figure 2.1 discusses the sequential process involved in tissue engineering, emphasizing the integration of scaffolds and cells, followed by cell seeding into a bioreactor and subsequent bioreactor-based culturing, ultimately leading to the formation of a functional tissue construct [27].

2.1.3 Scope and Objectives of the Chapter

This chapter aims to provide a comprehensive understanding of the materials, methods, and considerations involved in the design and evaluation of resorbable composites for biomedical applications, particularly in bioimplants and fixation devices. The scope encompasses the selection of suitable resorbable polymers and bioactive fillers, the implementation of diverse synthesis and processing techniques, and the application of advanced characterization methods to evaluate structural, mechanical, and biological properties.

The key objectives of this chapter are:

- **To explore the fundamental concepts and classifications of resorbable composites**, highlighting their relevance in tissue engineering, regenerative medicine, and environmentally sustainable biomedical applications.
- **To present an in-depth review of the raw materials used in resorbable composites**, including polymers, bioactive ceramics, and functional additives, with an emphasis on their biocompatibility, degradability, and mechanical integrity.
- **To examine various synthesis strategies and fabrication techniques**, such as solvent casting, *in situ* polymerization, electrospinning, and 3D printing, focusing on how these methods influence the final properties of the composite materials.
- **To describe advanced processing techniques and the effect of processing parameters** on morphology, porosity, mechanical strength, and degradation behavior of the composites.
- **To introduce key characterization tools and techniques** for evaluating the structural, thermal, mechanical, and biological performance of resorbable composites, including *in vitro* and *in vivo* assessments.
- **To establish structure–property relationships** that govern the behavior of resorbable composites during service, with particular attention to degradation kinetics, interface characteristics, and functional performance.
- **To discuss current biomedical and environmental applications**, such as scaffolds for tissue engineering, drug delivery platforms, and temporary implants, and highlight real-world implementations.
- **To identify existing challenges and future research directions**, including scalability, regulatory requirements, and innovations in composite design and fabrication technologies.

By addressing these objectives, the chapter seeks to bridge the gap between material development and clinical translation, offering valuable insights for researchers, engineers, and clinicians working in the field of bioresorbable materials and biomedical device engineering.

2.2 Synthesis Protocols of Resorbable Polymers

Bioresorbable polymers have gained prominence in medical device development due to their biocompatibility and ability to avoid post-healing device removal [28]. The synthesis and processing of these polymers significantly influence their physicochemical properties and biodegradation behavior. Crystallinity plays a crucial role in preserving chain distribution and physical properties, while sterilization methods can impact these characteristics [28]. Various factors affect bioresorption, including processing, sterilization, storage, and porous structure. Analytical methods are essential for characterizing polymer matrices and monitoring degradation, involving molecular weight determination, chemical structure analysis, and surface analysis [8]. *In-vitro* and *in-vivo* test methods, as well as degradation process modeling, are employed to evaluate bioresorbability. Understanding these aspects is crucial for selecting appropriate biodegradable polymers and optimizing their performance in medical applications [8, 20, 29–31].

Recent studies have explored various synthesis protocols for resorbable polymer-based nanocomposites and their effects on nanoparticle dispersion, interfacial bonding, and resultant properties. Improved dispersion techniques, such as high-power sonication combined with dual asymmetric centrifugal mixing, have been shown to enhance nanoparticle distribution in polymer matrices, affecting optical properties and cell responses [32]. Different compounding methods, including solvent evaporation and twin screw extrusion, can yield similar mechanical properties in dry conditions, but these properties significantly decrease when tested under physiological conditions [33]. *In situ* polymerization of biodegradable copolyesters with hydroxyapatite has resulted in nanocomposites with improved mechanical properties, bioactivity, and controlled biodegradation [21, 34, 35]. The choice of crosslinking agents, such as methyl methacrylate or N-vinyl pyrrolidone, can also influence mechanical strength and degradation rates. Overall, these studies highlight the importance of synthesis protocols in determining the final properties of resorbable polymer-based nanocomposites [26, 36–42].

This summary explores various synthesis strategies for resorbable polymer-ceramic composites in biomedical applications. *In-situ* polymerization (ISP) of polycaprolactone/phosphate glass fiber composites demonstrated superior mechanical properties and degradation resistance compared to conventional laminate stacking [43]. Solvent casting techniques were used to create bioplastic composites with hydroxyapatite and dried grains, showing potential for bone tissue engineering [44]. Polylactic acid (PLA) composites with calcium phosphate glass particles exhibited enhanced mechanical properties and improved biological behavior [45]. These polymer-ceramic composites offer advantages over metal implants, including eliminating stress shielding and reducing the need for revision surgery [46]. The microstructure, degradation behavior, and mechanical performance of these composites are significantly influenced by the synthesis method, with ISP and solvent casting showing promising results for creating scaffolds with controlled porosity and surface properties.

2.2.1 Chitosan

Chitosan, a versatile biopolymer derived from chitin, has gained significant attention in biomedical applications due to its biocompatibility, biodegradability, and antimicrobial properties [47]. Various synthesis protocols, including gelation, electrospinning, and 3D printing, influence the structural and mechanical properties of chitosan-based scaffolds [48]. These scaffolds can be tailored for diverse applications such as tissue engineering, drug delivery, and wound healing [49]. The incorporation of bioactive compounds and nanoparticles can enhance therapeutic effects and control drug release [48]. Chitosan's polycationic nature and reactive functional groups allow for the formation of various structures like nanoparticles, hydrogels, nanofibers, and films [50]. However, challenges remain in improving scaffold design, mechanical properties, and understanding interactions with host tissues [48]. Overall, chitosan-based biomaterials show great potential for advancing biomedical science and clinical applications.

Chitosan's physicochemical properties and biomedical applications are significantly influenced by its degree of deacetylation (DD), molecular weight (MW), and preparation methods. Higher DD and MW increase water absorption capacity, water vapor transmission rate, and porosity in chitosan films [51]. For osteochondral tissue regeneration, chitosan with lower MW, lower DD, and intact mineral content shows better bone and cartilage regeneration [52]. Other factors affecting chitosan's biomedical activity include polydispersity, crystallinity, acetylation pattern, and solubility [53]. Microwave irradiation can decrease chitosan's MW without affecting DD, leading to reduced crystallinity and swelling ratio. Lower MW chitosan scaffolds demonstrate decreased cell growth compared to higher MW scaffolds [54]. These findings highlight the importance of carefully controlling chitosan's physicochemical properties to optimize its performance in various biomedical applications such as wound healing and tissue engineering.

Chitin and its derivative chitosan are versatile biopolymers with significant potential in biomedical applications due to their biocompatibility, biodegradability, and antibacterial properties [48, 55]. Various synthesis protocols, including gelation, electrospinning, and 3D printing, can be employed to create scaffolds with tailored properties for tissue engineering, drug delivery, and wound healing [48, 56]. The choice of chitin type, solvent system, and fabrication technique significantly influences scaffold microarchitecture and performance [56]. Composite scaffolds incorporating other natural polymers like cellulose, alginate, and agarose have shown promise in tissue engineering applications [57]. These scaffolds support cell adhesion, proliferation, and differentiation in various tissue types, including bone, cartilage, skin, and nerves [48]. Future research should focus on improving scaffold design, mechanical properties, and understanding interactions with host tissues to fully harness the potential of chitin-based materials in biomedical applications [8, 48, 58].

2.2.2 Alginates

Alginates, natural polymers derived from brown algae, have gained significant attention in biomaterial design for tissue engineering, drug delivery, and scaffold fabrication [59]. Their biocompatibility, biodegradability, and ability to form hydrogels through various crosslinking methods make them versatile for biomedical applications [60]. Alginate-based scaffolds exhibit excellent cell proliferation and osteoblast differentiation, making them suitable for bone tissue regeneration [61]. The cross-linking process plays a crucial role in scaffold design, allowing for the creation of biomimetic hydrogels with improved mechanical properties through chemical modification or ionic combination with other polymers [62]. Alginates can be processed into various forms, including injectable gels, porous scaffolds, and micro/nanoparticles, expanding their potential uses in wound healing, cell transplantation, and drug delivery [60]. Their unique properties and versatility position alginates as valuable assets in advancing therapeutic and diagnostic interventions [59].

Alginate, a natural polysaccharide derived from brown algae, has gained significant attention in biomedical applications due to its biocompatibility and biodegradability [63]. Its versatility stems from its ability to form hydrogels, microspheres, and other scaffolding materials through various crosslinking methods [62]. The cross-linking process plays a crucial role in determining the mechanical properties and degradation of alginate-based biomaterials [64]. Chemical modifications and combinations with other polymers can further enhance alginate's functionality for tissue engineering, drug delivery, and regenerative medicine applications [63]. Recent advancements have focused on tailoring alginate's properties, such as molecular weight and monomer composition, to optimize its performance in specific biomedical contexts [59]. The ability to form pH-responsive gels and chelate with divalent cations makes alginate a promising candidate for developing advanced drug delivery systems and tissue engineering scaffolds [59, 65].

2.2.3 Starches

Starch-based biomaterials have gained significant attention in tissue engineering, drug delivery, and scaffold fabrication due to their biocompatibility and biodegradability [66, 67]. The incorporation of starch in scaffold design improves porosity and mechanical strength, enhancing their effectiveness for tissue engineering applications [66]. Starch-based hydrogels serve as effective drug delivery systems, providing controlled release of pharmaceutical compounds over extended periods [68]. These hydrogels also support cell growth and tissue regeneration by mimicking the extracellular matrix [68]. The versatility of starch allows for tailored modifications through grafting and other chemistries, enabling the development of high-performance biomaterials for various biomedical applications [69]. Starch-based materials have been successfully used in wound dressings, bone replacement implants, and diagnostic imaging, demonstrating their broad potential in the biomedical field [67, 69].

Starch-based polymers have emerged as versatile materials for biomedical applications, including tissue engineering, drug delivery, and scaffold fabrication. Their biocompatibility, biodegradability, and sustainability make them attractive for various synthesis protocols [69]. Starch can be blended with other materials like hydroxyapatite to improve scaffold properties or grafted with specific substituents to tailor its performance for particular applications [66, 69]. Starch-based microparticles have shown promise in bone tissue engineering, demonstrating bioactivity, good biological performance, and controlled release capabilities for bioactive molecules [70]. Furthermore, starch can be combined with synthetic polymers to create hybrid materials with enhanced functionality [71]. These diverse synthesis approaches allow for the modification of physicochemical properties, enabling the development of starch-based materials with tailored characteristics for specific biomedical applications [69–71].

2.2.4 Cellulose

Cellulose and its derivatives have emerged as versatile biomaterials for various biomedical applications due to their biocompatibility, biodegradability, and modifiable properties [72]. These materials show promise in tissue engineering, particularly for bone regeneration, where cellulose-based composite scaffolds can deliver growth factors and antibiotics to promote tissue repair [73]. Cellulose derivatives can be tailored through modification techniques like oxidation and esterification, enhancing their performance in drug delivery systems, wound dressings, and biosensors [72, 74]. The diverse morphological forms of cellulose, including fibers, nanofibrils, and nanocrystals, allow for the creation of materials with varied microstructures and properties suitable for specific biomedical applications [72]. Despite its potential, cellulose's use in biomedicine is still evolving, with ongoing research focusing on optimizing its structure and expanding its applications in personalized medicine and intelligent healthcare devices [72, 74].

2.2.5 Gelatin

Gelatin, a biodegradable and biocompatible protein derived from collagen, plays a crucial role in tissue engineering, drug delivery, and scaffold fabrication [75, 76]. Its excellent biocompatibility, biodegradability, and retention of informational signals like RGD sequences promote cell adhesion and proliferation [75]. Gelatin can be easily modified through chemical reactions and physical methods to enhance its mechanical strength and bioactivity [75, 77]. Various gelatin derivatives, such as methacryloyl gelatin and thiolated gelatin, exhibit improved physicochemical and biological properties for biomedical applications [76]. Gelatin-based nanofibers have shown potential in drug delivery systems and tissue engineering, providing controlled release of therapeutic agents and an ideal matrix for cell proliferation [78]. Recent advancements in manufacturing techniques, like 3D printing, have further expanded the possibilities for creating functional gelatin-based materials for medical purposes [77].

Gelatin-based biomaterials have gained significant attention in tissue engineering and drug delivery due to their biocompatibility, biodegradability, and similarity to the extracellular matrix [79, 80]. Gelatin methacryloyl (GelMA) hydrogels, in particular, offer tunable mechanical properties and can be microfabricated into controlled architectures for various biomedical applications [79]. To overcome gelatin's limitations, such as poor mechanical properties and enzymatic degradation sensitivity, researchers have developed composite scaffolds by combining gelatin with polysaccharides, synthetic polymers, and nanoparticles [81, 82]. These composites exhibit improved mechanical strength, stability, and functionality, making them suitable for diverse tissue engineering applications, including bone, cartilage, cardiac, and nerve tissue regeneration [82]. Gelatin-based nanoparticles have also shown promise in drug and gene delivery systems, offering versatile preparation techniques and the ability to modify their surface for targeted delivery [80].

2.3 Processing of Resorbable Composites

Processing of bio-nanocomposites involves incorporating nanoparticles within the biopolymeric matrix. This produces one composite material which is at the nano level. Generally, when homogeneous Nano level suspension of nanoparticles is formed, the properties of the biopolymers are altered. When there is a proper interaction between the nanoparticle and that of the polymer used, the properties of the biopolymer are improved. Hence, ensuring appropriate dispersion of the nanoparticles in the biopolymer matrices is very much necessary in any of the processing techniques employed. Optimizing the process along with maintaining the structural integrity of the nanoparticles plays a key role [35, 58, 83–86].

There are many techniques which are available for processing nanocomposites namely:

2.3.1 Solution-Casting Method

This simple method is used for the preparation of polymeric nanocomposite materials and works based on a solvent system [35]. Three steps are involved in this method. The filler is initially dispersed in an appropriate solvent by stirring or sonication. The same solvent is then used to dissolve the polymer that is needed. Finally, both the filler and polymer solution are mixed at appropriate temperatures. Composites can be obtained by either casting the solution mixture or by precipitation. A research group prepared chitosan gold nanocomposite film by pouring chitosan-based gold nanoparticles dispersion into petri plates and drying it in a hot air oven at 60 °C until the solvent was fully evaporated [87].

2.3.2 *In-Situ* Technique

This method is effectively used to obtain composites having a uniform distribution of filler material. Fillers are first dispersed in monomer or monomer solution at experimental temperature before adding curing agents for polymerization. This can be performed either with or without solvent [87]. Polymerization is then initiated using heat, radiation, or any other suitable initiator [35]. Composites prepared through this method have enhanced properties. Synthesis of nanomaterials may take place within the reaction mixture at the same reaction conditions. A research group synthesized chitosan-grafted polyacrylonitrile silver nanocomposite material *via in-situ* chemical reduction of silver ions in graft co-polymerization of acrylonitrile onto chitosan [87].

2.3.3 Electrospinning Technique

This is a relatively straightforward technique. It mainly involves three components a high-voltage power supply, a spinneret, and a collector drum. The spinneret is connected to a syringe containing a solution of polymer and nanomaterial. When a voltage is between 1 - 30 kV, a highly electrified droplet of solution forms at the tip of the nozzle. There is an even distribution of the induced charges across the droplet's surface, and electrostatic repulsion stretches the droplet and counteracts the surface tension [21]. This happens when sufficiently high voltage is applied, the electrified jet undergoes a stretching and whipping process leading to the formation of long, thin threads [87]. As the jet travels, evaporation of solvent takes place and the nanofibers are then deposited onto the collector drum beneath the spinneret. The characteristics of nanofibers are influenced by both the properties and process conditions [40]. This method is effective in obtaining polymer composites in the form of fibers with diameters ranging from tens of nanometers to a few micrometers. One group of researchers developed electrospun chitosan/polyethylene oxide fiber mats reinforced with chitin nanocrystals using this method [7]. These fibers exhibit a high surface area-to-volume ratio and excellent mechanical properties. Notably, their ability to mimic the extracellular matrix (ECM) of various tissues creates a highly supportive environment for cells. As a result, recent research has focused on using this technique to develop biodegradable nanocomposites [87, 88].

2.3.4 Freeze-Drying

This method known as lyophilization, is a dehydration technique used to obtain polymeric composite scaffolds with varying porosity and pore size. It involves an emulsion/freeze-drying process [7]. Initially, a polymer is dissolved into a solvent with added water leading to the formation of a uniform mixture. This mixture is then frozen at different temperatures and at different rates which results in the synthesis of ice crystals. Molecules undergo aggregation in interstitial spaces. Under low pressure and temperature, frozen material is dried, removing the dispersed water and solvent. This results in a porous polymer structure. However, the resulting scaffolds typically have small pores with irregular porosity and are unsuitable for handling mechanical stress [40]. One group of researchers developed chitosan–halloysite nanotube nanocomposite scaffolds by combining solution mixing with freeze-drying techniques [7].

2.3.5 Foam Replica Method

In this method, a porous polymeric foam is impregnated with an aqueous suspension until all pores are filled. Excess suspension is then removed by passing the foam through a roller or centrifuge, leaving a thin coating layer over the original structure. The coated foam is then left to dry. Subsequently, the foams are heated at temperatures ranging from 300 to 800 °C, with heating rates kept below 1 °C per minute. This slow heating ensures gradual decomposition and diffusion of the polymer template. As a result, a porous structure is formed. However, the mechanical strength of the structure may be reduced due to the formation of cracked struts during foam decomposition [40].

2.3.6 Phase Separation

The principle of this technique is based on phase separation to decrease the system with free energy [40]. Porous polymeric membranes have been developed using this technique, which enables the fabrication of three-dimensional nanofibrous scaffolds. The macroporous structures can be tailored to suit various applications. Biodegradable aliphatic polyesters can be used to prepare scaffolds using a technique whose diameter varies from 50 to 500 nm. Polymeric solution is thermodynamically separated into polymer components and polymer-poor components. On dissolving a polymer in solution, induction of phase separation takes place. This phase separation can either be started thermally or by the addition of a non-solvent to the polymer solution. This results in gel formation. The solvent is then extracted from the gel using water, and the gel is cooled to a temperature below the polymer's glass transition temperature (Tg). Finally, nanofibrous scaffolds are obtained by freeze-drying the gel under vacuum. The desired architecture can be achieved by adding various porogens to the polymer solution, allowing researchers to tailor pore size and interconnectivity by adjusting the concentration, size, and geometry of the porogens. This method does not require specialized equipment and ensures easy batch-to-batch consistency. Additionally, the mechanical properties of the scaffolds can be tuned by altering the polymer or porogen concentration. However, a drawback of this technique is the formation of pores with non-uniform diameters [40].

2.3.7 Solvent Casting and Particulate Leaching

A 3D porous scaffold can be synthesized by combining solvent-casting and particulate-leaching methods. In this process, the polymer is first dissolved in an organic solvent, after which porogen granules such as sugar, inorganic salts, or paraffin spheres are dispersed into the solution. Solidification of the polymer occurs as the solvent is removed. Subsequently, water is used to leach out the porogen particles, resulting in the formation of a scaffold with pores. The pore size of the scaffold can be controlled by adjusting the size of the salt crystals used, while porosity can be regulated by altering the salt-to-polymer ratio [40]. The main advantage of this technique lies in its simplicity and ease of fabrication, as it does not require any specialized equipment [21]. However, a limitation of the method is that it does not allow control over pore shape or interconnectivity [40].

2.3.8 Gas Foaming

This technique utilizes carbon dioxide at high pressure. The gas is dispersed throughout the polymer, which is mixed with a porogen until saturation is achieved. A rapid decrease in the solubility of carbon dioxide to atmospheric levels then leads to the nucleation and growth of gas bubbles. Once the foaming process is complete, the porogen is removed, resulting in a more interconnected porous structure. The morphology of the scaffold depends on several factors, including process parameters such as pressure, temperature, and venting time. This technique does not require the use of organic solvents or high temperatures. Scaffolds produced by this method are primarily used in bone tissue engineering [40].

2.3.9 Rapid Prototyping

This technique is considered highly sophisticated. It is also known as solid freeform fabrication or additive manufacturing. It relies on computer-generated models to build objects with intricate designs. The input data can originate from MRI scans, CAD files, or CT images. Using this method, 3D scaffolds are built layer by layer. This allows for precise control over features like pore dimensions, shape, and connectivity. A well-known example is robocasting, where a robotic arm extrudes a colloidal suspension through a nozzle. In one study, researchers created a porous scaffold with a specific structure by extruding ceramic ink from a syringe into an oil bath [40].

2.3.10 Melt Processing

This approach involves blending nanoparticles with polymers in their molten form, followed by heating the mixture above the polymer's softening temperature. The process can be done either under shear or in a static condition. It is particularly beneficial for producing clay-based bio-nanocomposites compared to *in-situ* intercalative polymerization and solution intercalation techniques. The method is environmentally friendly as it avoids the use of organic solvents. It also aligns well with existing industrial practices like extrusion and injection molding. The interaction between the polymer and nanoparticles—especially thermodynamic compatibility—plays a critical role in the resulting dispersion quality. If the enthalpic interaction is not favorable, the nanoparticles may not disperse well in the polymer matrix, leading to the formation of micro composites. Hence, fine-tuning the process parameters is essential. One challenge is that the physical shear forces or elevated temperatures required may degrade the biopolymers, especially those that are heat-sensitive. Careful optimization at each stage is therefore crucial to ensure the effective separation of agglomerates and uniform nanoscale dispersion of particles [35].

Additional considerations for the use of bio-nanocomposites in tissue engineering include:

i) Sterilization of nanostructured bio-nanocomposites
ii) In-vitro biodegradation testing
iii) Surface functionalization of bio-nanocomposites
iv) Techniques for evaluating structure and morphology

2.4 Recent Developments in the Processing of Resorbable Composites

Scaffolds used in bone tissue engineering are required to exhibit osteoconductivity, osteoinductivity, and support for osseointegration. Typically, these scaffolds are engineered to closely resemble the structure and function of natural human bone [85]. Single-walled carbon nanotubes (SWNTs), particularly their ultra-short variants, show promise as reinforcement materials in polymer-based scaffolds due to their excellent mechanical features. Their distinctive attributes make them highly suitable for orthopedic applications, and their density is comparable to that of graphite [84, 89].

Nano-hydroxyapatite (nHAp) and graphene oxide (GO) have been incorporated into aliphatic polyester scaffolds to enhance mechanical stability for cell infiltration and proliferation during tissue regeneration. These nanofillers have been shown to promote cell attachment, bioactivity, expansion, and osteogenic differentiation of mesenchymal stem cells (MSCs). Nanoceramic surfaces imitate the inner bone environment, boosting protein adsorption and encouraging the initiation of bone tissue formation. Notably, nanocrystalline hydroxyapatite exhibits strong adsorption of vitronectin and BMP-2, effectively enhancing new bone generation. According to Tjong and colleagues, PLA-based hybrid nanocomposites reinforced with both nHAp and GO displayed better compatibility than composites reinforced with only nHAp [11]. This improvement is attributed to graphene and its derivatives, which stimulate osteoblast adhesion, proliferation, and MSC differentiation, ultimately aiding bone formation [85].

To strengthen biodegradable polymers structurally and mechanically for load-bearing uses, researchers are exploring carbon nanostructures such as 0D fullerenes and 1D SWCNTs as reinforcements. Moreover, 2D nanomaterials like single and multi-walled graphene oxide nanoribbons (SWGONRs, MWGONRs), graphene oxide nanoplatelets (GONPs), and molybdenum disulfide nanoplatelets (MSNPs) have been introduced into poly(propylene fumarate) (PPF) composites [7].

Two-dimensional carbon and inorganic nanomaterials like GONRs, GONPs, and MSNPs offer unique physicochemical features, making them useful in various domains. These include energy storage, electronic devices, water filtration, bio-imaging, drug delivery, lubricants, liquid crystal formulation, and porous scaffolds for tissue regeneration [7, 90, 91].

Chitin, one of the most abundant naturally occurring carbon-based polymers, is valuable in medical fields such as drug delivery due to its compatibility with biological systems, biodegradability, and eco-friendliness. Scientists have developed chitin- and chitosan-based nanofibrous materials with strong biological activity and mechanical stability, demonstrating potential as antibacterial agents in wound healing, biosensing, drug delivery, and tissue engineering. However, the application of chitosan is limited by issues such as low sensitivity, poor conductivity, weak mechanical strength, and limited crosslinking. Still, nanocomposites containing small amounts of nanofillers exhibit enhanced optical, mechanical, conductive, and biodegradable properties [83].

Nanoscale extracellular matrix analogs are being designed to direct cellular behavior and support tissue formation. These well-organized 3D scaffolds not only provide a physical framework for cell development but also offer essential biochemical, mechanical, and structural signals. The latest biomaterials are making significant progress in advancing the field of tissue engineering [92].

2.5 Characterization Methods of Resorbable Composites

Advanced characterization techniques play a crucial role in understanding the *in vivo* performance of resorbable composites for biomedical implants. SEM, XRD, and DSC have been used to analyze the morphology, crystallinity, and thermal properties of poly(D,L-lactide)/Bioglass composites, revealing the formation of hydroxyapatite and delayed polymer degradation *in vitro* [93]. *In vitro* degradation studies of glass fiber-reinforced PLDLA composites in simulated body fluid have shown that degradation mechanisms can be tailored by altering glass fiber composition [1]. Correlations between *in vitro* and *in vivo* degradation rates have been established for magnesium-based materials using mechanical testing, with *in vitro* degradation occurring 2–3 times faster than *in vivo* [94]. Furthermore, particle size in PLGA/TCP composites has been found to affect *in vivo* degradation rates, with nanocomposites degrading more slowly and closely matching bone growth rates compared to micro-composites [95].

2.5.1 Structural and Morphological Characterization

2.5.1.1 Scanning Electron Microscopy (SEM)

Scanning electron microscopy (SEM) is a valuable tool for examining the surface morphology and porosity of resorbable biomaterials for tissue engineering. SEM analysis reveals that scaffold processing methods significantly influence pore structure and interconnectivity. Thermally induced phase separation and solvent casting techniques produce scaffolds with oriented, interconnected pores ranging from 160–377 μm, suitable for cell growth and angiogenesis [96–98]. Composite composition affects surface properties, with polymer-coated bioactive glass particles increasing hydrophobicity and roughness [97]. SEM also enables the evaluation of scaffold stability and degradation over time, crucial for maintaining pore networks during implantation [99]. Furthermore, SEM can detect the formation of hydroxyapatite crystals on bioactive composite surfaces, indicating enhanced bioactivity [96]. These findings demonstrate SEM's importance in characterizing and optimizing resorbable biomaterials for tissue engineering applications. Figure 2.2 shows the structural and surface morphology of bioactive glass (BG) and polycaprolactone (PCL)/BG composite scaffolds. Image (a) shows the porous architecture of a BG scaffold

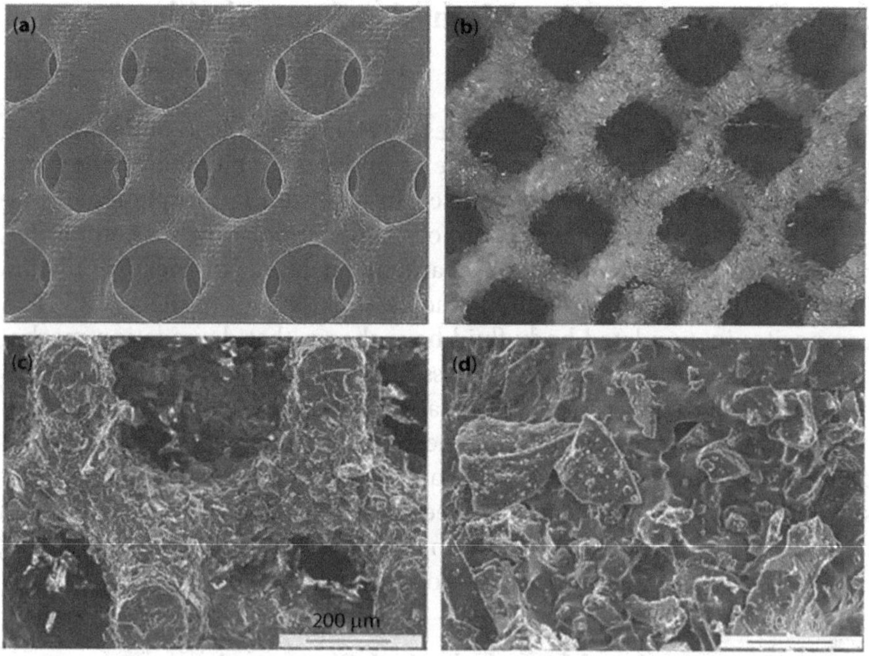

using SEM, while (b) presents an optical stereomicroscope image of a PCL scaffold containing 20 wt% BG. Images (c) and (d) provide SEM views of the PCL/BG scaffold surface, highlighting the distribution and integration of BG particles within the polymer matrix at varying magnifications [100].

2.5.1.2 *X-Ray Diffraction (XRD)*

X-ray diffraction (XRD) analysis is a crucial technique for elucidating the crystalline structure and phase composition of resorbable composites. XRD can determine the quantitative and qualitative content, crystallite size, and degree of crystallinity of calcium phosphate ceramics in biocomposites [101]. For polymer-based composites, XRD reveals changes in crystallinity during degradation, with increases observed in the early stages of immersion [102]. The technique also demonstrates that crystallinity in hydroxyapatite-filled polycaprolactone increases with filler concentration [103]. XRD analysis correlates with mechanical performance, as higher crystallinity and smaller particle sizes generally lead to improved modulus and strength [103]. Furthermore, the addition of phosphate glass to polylactide-based polymers can suppress polymer embrittlement mechanisms during degradation, resulting in a more controlled loss of mechanical properties [2].

2.5.2 Thermal Properties

2.5.2.1 *Differential Scanning Calorimetry (DSC)*

Differential Scanning Calorimetry (DSC) is a valuable tool for assessing thermal transitions and crystallinity of resorbable composites, which influence their biodegradation and mechanical properties. DSC can detect structural changes during biodegradation, such as increased crystallinity, and quantify biodegradability in complex environments [104]. For poly(lactic acid) (PLA) composites, DSC reveals that the glass transition temperature remains relatively constant with the addition of fillers like hydroxyapatite [105]. In PLLA/Mg composites, DSC demonstrates that magnesium particles do not significantly alter the polymer's crystallization behavior, allowing for optimization of properties through annealing [106]. Furthermore, DSC can be used to assess molecular weight degradation of biodegradable plastics by monitoring changes in crystallization temperature, which correlates with reduced viscosity and polymer degradation [107]. These findings highlight DSC's importance in characterizing and optimizing resorbable composites for biomedical applications [42].

2.5.2.2 Thermogravimetric Analysis (TGA)

Thermogravimetric Analysis (TGA) is a valuable tool for evaluating the thermal stability and degradation behavior of resorbable composites for biomedical applications. TGA can determine the thermal properties of fiber composites and help understand their suitability for various applications [108]. The addition of bio-fillers like nano-calcium phosphate can improve the thermal stability of composites, with onset temperatures reaching 208 °C [109]. Particle size within resorbable composites affects their *in vivo* degradation rate, with nanocomposites degrading more slowly than microcomposites [95]. The degradation mechanism of bioresorbable glass fiber-reinforced composites involves a combination of surface and bulk erosion, which can be tailored by altering the oxide composition of the glass fibers [1]. Understanding the thermal decomposition profiles of these composites is crucial for optimizing their performance in clinical applications and heat-related manufacturing processes.

2.5.3 Mechanical Testing

2.5.3.1 Tensile, Flexural, and Impact Testing

Resorbable composites for biomedical applications undergo complex degradation processes that affect their mechanical properties over time. *In vitro* and *in vivo* testing methods reveal different degradation rates and mechanical performance changes. Magnesium-based materials showed 2–3 times slower degradation *in vivo* compared to *in vitro* [94]. Similarly, polydioxanone devices exhibited faster mechanical property loss *in vivo* than *in vitro* [110]. Poly(lactide)-ceramic composites maintained adequate mechanical properties for 26 weeks before declining, with complete loss by 52 weeks [111]. Phosphate glass reinforcement in polylactide-based composites improved stiffness and suppressed polymer embrittlement mechanisms, allowing for more controlled degradation [2]. These studies highlight the importance of both *in vitro* and *in vivo* testing to accurately assess the structural integrity and mechanical performance of resorbable composites during degradation, as the surrounding environment significantly influences their behavior.

2.5.4 Biodegradability and Bioactivity Assessment

Resorbable composites combining bioactive glass and biodegradable polymers show promise for bone tissue engineering applications.

These composites exhibit enhanced bioactivity through rapid hydroxyapatite formation in simulated physiological environments [93, 112]. The addition of bioactive glass to polymer foams increases water absorption and weight loss, while potentially buffering the degradation rate [93]. The degradation mechanism of these composites involves both surface and bulk erosion, with the potential to tailor strength retention by altering glass fiber composition [1]. Bioactive glass nanofibers uniformly dispersed in poly(lactic acid) matrices further improve bioactivity and osteoblast responses, including cell proliferation, differentiation, and mineralization [113]. The porous structure of these composites supports osteoblast attachment, spreading, and infiltration, demonstrating their biocompatibility and potential for bone regeneration [112].

2.5.5 *In Vitro* and *In Vivo* Evaluation

Recent studies have explored correlations between *in vitro* and *in vivo* degradation of resorbable biomaterials. For magnesium wires, *in vivo*, degradation was 2.3–3.1 times slower than *in vitro*, depending on the mechanical property measured [94]. In PLGA/TCP composites, nanoparticle-reinforced materials degraded more slowly *in vivo* than microparticle composites, better matching bone growth rates [95]. For polyurethane foams, the optimal *in vitro* conditions to predict *in vivo* degradation varied based on material hydrophilicity and degradation mechanisms [114]. A bioresorbable coronary scaffold showed a positive correlation between *in vitro* and *in vivo* degradation kinetics, with complete degradation observed within 2 years *in vivo* [115]. These studies demonstrate that while *in vitro–in vivo* correlations exist for various biomaterials, the relationship is complex and depends on material properties, degradation mechanisms, and the specific *in vitro* conditions used.

2.6 Applications and Future Scope

Nanoparticles have played multiple roles in tissue engineering, ranging from improving biological, mechanical, and electrical characteristics to enabling gene delivery, viral transduction, DNA transfection, cell patterning, promoting growth of different tissue types, as well as aiding in molecular detection and biosensing [27, 116–118]. Figure 2.3 illustrates the diverse applications of nanocomposites in tissue engineering, emphasizing their roles in enhancing biological, mechanical, and electrical properties,

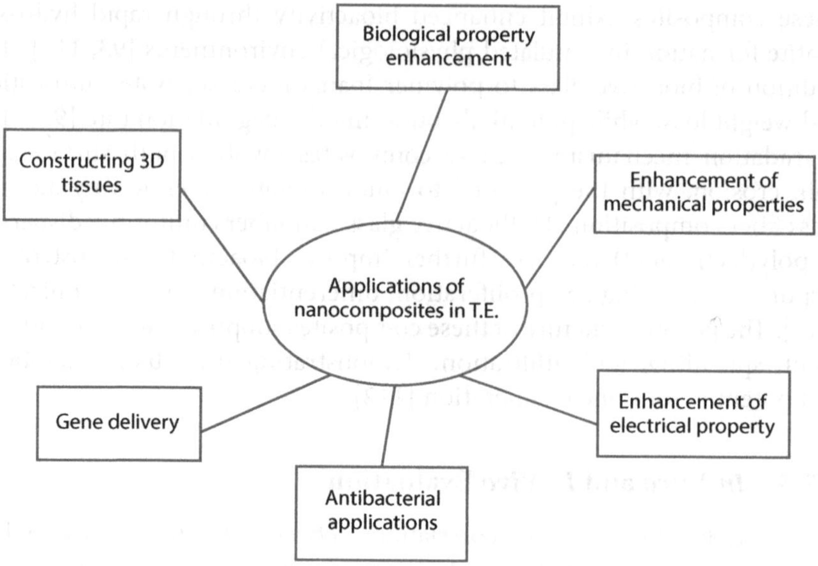

Figure 2.3 Schematic representation of the various applications of nanocomposites in tissue engineering (T.E.), including biological property enhancement, mechanical and electrical property improvement, antibacterial functions, gene delivery, and construction of 3D tissues [116].

as well as their potential in antibacterial applications, gene delivery, and the construction of three-dimensional tissues. Table 2.1 discusses various material systems used in biomedical applications, outlining their key innovations, associated performance improvements, and current translation status based on recent literature.

Practical application of bio nanocomposites required:

i. *Biological property enhancement:*
ii. *Enhancement of mechanical properties*
iii. *Enhancement of electrical properties*
iv. *Antibacterial applications*
v. *Gene delivery*
vi. *DNA transfection*
vii. *Constructing 3D tissues*

Table 2.1 Summary of material systems in biomedical applications, highlighting key innovations, performance improvements, and current translation status based on recent literature.

Material system	Key innovations	Performance improvements	Translation status	Reference
Poly(lactic acid) (PLA) composites	Novel fabrication, high-performance composites	Enhanced mechanical, thermal, and rheological properties (qualitative, from abstract only)	Preclinical	[86]
Bioresorbable polymers	Engineering approaches, imaging evaluation	No mention found	Preclinical, regulatory discussion	[22]
PLA, PCL composites, TCP	Copolymer blends, TCP compounding, infill tuning	Tensile modulus (poly(lactic-co-glycolic acid)): 40–80 MPa; pore size <100 micrometers	Commercial (Osteoplug®, Resttrata®), clinical translation	[119]
Tricalcium phosphate, hydroxyapatite, Bioglass	Polymer-inorganic composites	Predictable degradation, bioactivity (qualitative, from abstract only)	Preclinical	[9]
Bioninks	Cellular incorporation, functionalization	No mention found	Preclinical, regulatory focus	[120]
Poly(alpha-esters) + hydroxyapatite/ tricalcium phosphate/ bioglass	Additive manufacturing, gas foaming, functionally graded materials	Compressive strength: 2.5–3.4 MPa; bending strength up to 124 MPa	Preclinical, regulatory context	[100]
PLA/hydroxyapatite	Multiple synthesis routes	Compressive strength +30%; fracture toughness +50%; cell adhesion +25% (from abstract only)	Preclinical, clinical prospects	[121]
PLA, PCL nanoparticles	Metallic nanoparticle incorporation, hybrid processing	Bone-mimetic strength, enhanced cell response (from abstract only)	Preclinical, translational focus	[122]

(Continued)

Table 2.1 Summary of material systems in biomedical applications, highlighting key innovations, performance improvements, and current translation status based on recent literature. (*Continued*)

Material system	Key innovations	Performance improvements	Translation status	Reference
Supercritical fluid scaffolds	Solvent-free, bioactive agent incorporation	Qualitative improvements (from abstract only)	Preclinical	[123]
Polymer/metal stents	Three-dimensional printed shape-memory, metal alloys	Qualitative; focus on translation (from abstract only)	Preclinical or clinical trials	[124]

2.6.1 Green Nanotechnology in Scaffold Fabrication for Tissue Engineering

2.6.1.1 Cellulose Nanocomposites in Regenerative Scaffold Development

Cellulose nanocomposites have emerged as promising materials for regenerative scaffold development due to their biocompatibility, mechanical properties, and cost-effectiveness [125]. These nanocomposites are being extensively researched for various biomedical applications, including wound healing, bone tissue engineering, and drug delivery [125, 126]. The unique physical, chemical, and biological properties of nanocellulose offer advantages over synthetic polymers in scaffold fabrication [126]. Cellulose-based bio-nanocomposites, incorporating materials like cellulose nanowhiskers and nanofibers, mimic the extracellular matrix and exhibit intracellular functions beneficial for tissue engineering and regenerative medicine [127]. Recent advances have shown that cellulose-based scaffolds demonstrate improved cell attachment, proliferation, and enhanced biocompatibility, making them promising candidates for neural, bone, cardiovascular, and skin tissue engineering applications [128]. However, challenges remain regarding degradability, porosity, and potential immunogenicity [128].

2.6.1.2 Biopolymer-Based Nanocomposites: Collagen and Gelatin in Scaffold Design

Collagen- and gelatin-based biopolymer nanocomposites have emerged as promising materials for scaffold design in biomedical applications, particularly in regenerative medicine. These nanocomposites combine natural polymers with nanoparticles to enhance properties and functionalities [129]. Gelatin, derived from collagen hydrolysis, offers intrinsic properties suitable for tissue engineering applications [82]. When combined with nanoparticles, gelatin-based scaffolds exhibit improved biological and physicochemical properties for bone tissue engineering [130]. Nanocomposites based on aliphatic polyesters have also shown potential in regenerative medicine, with enhanced thermal, electrical, and biodegradation properties [131]. These advanced materials can be fabricated into various structures, including nanoparticles, microparticles, 3D scaffolds, and electrospun nanofibers, offering versatility in biomedical applications [82]. The ongoing research in this field is expected to accelerate the clinical implementation of these nanocomposite biomaterials.

2.6.1.3 Alginate-Based Nanocomposite Scaffolds

Alginate-based nanocomposite scaffolds have gained significant attention in biomedical applications due to their excellent biocompatibility, biodegradability, and versatility. These scaffolds can be fabricated into various forms, including hydrogels, microspheres, and fibers, and are widely used in drug delivery, wound healing, and tissue engineering [132].

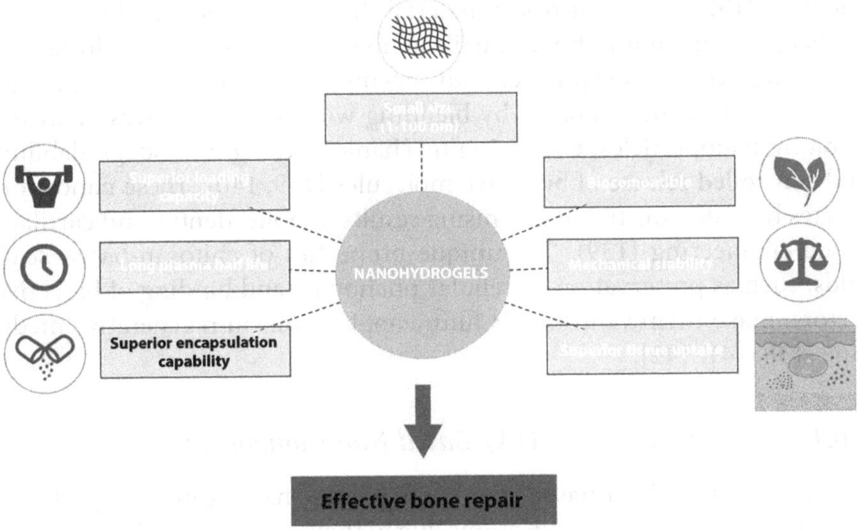

The incorporation of biopolymers and nanomaterials into alginate frameworks enhances their mechanical properties and biological activity, making them suitable for applications such as 3D bioprinting and cancer treatment [133]. Recent advancements have focused on developing alginate-based hydrogels and scaffolds with improved cell adhesion, proliferation, and mineralization properties [134]. The future potential of these nanocomposite scaffolds lies in their ability to be tailored for specific biomedical applications, offering promising solutions in advancing tissue engineering and regenerative medicine [133, 134]. Figure 2.4 discusses the key properties of nanohydrogels that contribute to effective bone repair, highlighting their small size (1–100 nm), superior drug loading and encapsulation capabilities, prolonged plasma half-life, and enhanced tissue uptake. It also emphasizes their biocompatibility and mechanical stability, which collectively support targeted delivery and sustained release of therapeutic agents, ultimately promoting efficient bone regeneration [135].

2.6.1.4 Biopolymer Nanocomposites: Chitosan and Chitin in Tissue Engineering Scaffolds

Chitosan and chitin-based biopolymer nanocomposites have emerged as promising materials for tissue engineering scaffolds due to their biocompatibility, biodegradability, and bioactivity [136]. These scaffolds mimic the extracellular matrix, supporting cell growth, proliferation, and differentiation [137]. Chitosan-based materials have shown potential in various applications, including bone, cartilage, liver, nerve, and musculoskeletal tissue engineering [138]. Recent advancements have focused on developing chitosan nanocomposites by blending with other polymers or incorporating nanoparticles, enhancing mechanical strength, biodegradability, and controlled release of bioactive molecules [139, 140]. These nanocomposites have demonstrated promising results in bone, dental, and cartilage tissue engineering [139]. The unique properties of chitosan-based scaffolds, such as preservation of cellular phenotype and binding of bioactive factors, make them valuable for future applications in regenerative medicine [138, 141, 142].

2.6.1.5 Hydroxyapatite (HA)-Based Nanocomposites

Hydroxyapatite (HA)-based nanocomposites have gained significant attention in biomedical applications due to their biocompatibility and similarity to bone structure. These materials enhance current applications in orthopedics, dentistry, and cancer therapy by improving osteointegration, enamel remineralization, and targeted drug delivery [143]. The development of nanostructured HA and composite materials aims to enhance mechanical properties and expand use in complex medical scenarios. Three-dimensional printing technologies offer precise and controllable fabrication of HA-based scaffolds, allowing for personalized implants with improved porosity and mechanical properties [144]. Future potential in tissue engineering and regenerative therapies includes the use of HA in drug and gene delivery systems, magnetic resonance imaging, cell separation, and hyperthermia treatment [145]. Ongoing research focuses on optimizing synthesis processes and functionalization techniques to further expand HA's applications in biomedicine. Figure 2.5 discusses the diverse biomedical applications of calcium phosphate (CaP) glass, including therapeutic ion release, drug and gene delivery, soft-tissue engineering, bioresorbable photonic devices, and composites for hard tissue regeneration. These multifunctional properties make CaP glass a promising material in both regenerative medicine and advanced biomedical technologies [146].

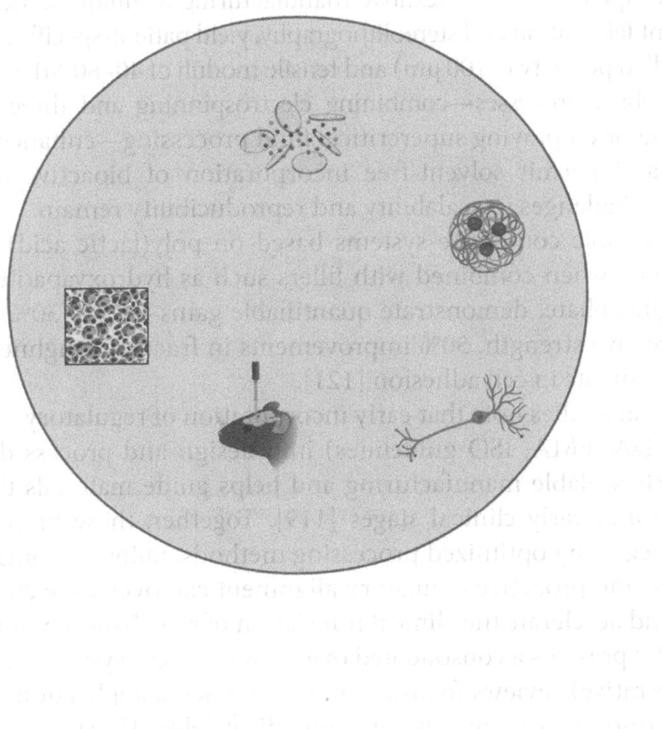

2.6.2 Processing Challenges and Limitations

Additive manufacturing, particularly fused filament fabrication, offers promising solutions for fabricating resorbable composites for biomedical applications. Key challenges include reducing void content, improving fiber-matrix bonding, and enhancing layer adhesion [147]. Recent advances involve developing hybrid composite filaments with biocompatible polymers and osteoinductive inorganic phases, such as strontium-containing nanohydroxyapatite or bioactive glass nanoparticles [148]. Biopolymers, including proteins and polysaccharides, are increasingly used in 3D printing for various biomedical applications due to their biocompatibility and biodegradability [149, 150]. The incorporation of nanoparticles can enhance the biological and mechanical properties of 3D-printed tissue scaffolds [150]. Ongoing research focuses on optimizing bioink selection, deposition strategies, and construct characterization to overcome challenges in the customization and personalization of biomaterials for advanced healthcare applications [151].

Advancements in processing, material design, and regulatory integration are reported to work in concert to address the limitations of resorbable composites [119]. Additive manufacturing techniques, including fused filament fabrication and stereolithography, yield patient-specific scaffolds with controlled porosity (~100 μm) and tensile moduli of 40–80 MPa [119]. In parallel, hybrid processes—combining electrospinning and three-dimensional printing or employing supercritical fluid processing—enhance fiber alignment and permit solvent-free incorporation of bioactive agents, even though challenges in scalability and reproducibility remain.

Resorbable composite systems based on poly(lactic acid) and related polymers, when combined with fillers such as hydroxyapatite and tricalcium phosphate, demonstrate quantifiable gains—up to 30% increases in compressive strength, 50% improvements in fracture toughness, and 25% enhancements in cell adhesion [121].

Several studies note that early incorporation of regulatory requirements (e.g., FDA, EMA, ISO guidelines) into design and process development supports scalable manufacturing and helps guide materials through preclinical and early clinical stages [119]. Together, these findings indicate that integrating optimized processing methods, tailored biomaterial innovations, and proactive regulatory alignment can overcome current limitations and accelerate the clinical translation of resorbable composites [119]. Table 2.2 presents a consolidated overview of recent systematic and technical (narrative) reviews focusing on the synthesis, application, and evaluation of resorbable composites in biomedical fields. The studies cover a wide

Table 2.2 Summary of recent reviews on resorbable composite materials for biomedical applications, highlighting study types, material systems, focus areas, and clinical application stages.

Study	Study type	Focus area	Material system	Clinical application stage	Reference
Liu *et al.*, 2020	Systematic review	Poly(lactic acid) (PLA) composites in tissue engineering and drug delivery	Poly(lactic acid) (PLA)	Preclinical, review of *in vitro*/*in vivo* studies	[86]
Dobrzyńska-Mizera *et al.*, 2024	Systematic review; technical (narrative) review	Bioresorbable polymers for diverse biomedical applications	Resorbable polymeric biomaterials (not specified)	Preclinical and regulatory discussion	[22]
Culbreath *et al.*, 2024	Systematic review	Additive manufacturing in tissue engineering	PLA, PCL, PDLA, PGA, PDO, copolymers, TCP	Clinical translation, commercial products (e.g., Osteoplug®)	[119]
Boccaccini and Blaker, 2005	Systematic review	Bioactive composite materials for scaffolds	TCP, HA, Bioglass, bioresorbable polymers	Preclinical, focus on scaffold development	[9]
Angelopoulos *et al.*, 2020	Technical (narrative) review	Inkjet bioprinting for tissue engineering	Cell-laden biomaterials (bio-inks, not specified)	Preclinical, regulatory focus	[120]

(*Continued*)

Table 2.2 Summary of recent reviews on resorbable composite materials for biomedical applications, highlighting study types, material systems, focus areas, and clinical application stages. (*Continued*)

Study	Study type	Focus area	Material system	Clinical application stage	Reference
Lacambra-Andreu *et al.*, 2023	Systematic review; technical (narrative) review	Manufacturing biocomposites for bone regeneration	Poly(alpha-ester)s (PGA, PCL, PLA, etc.), HA, TCP, bioglasses	Preclinical, regulatory context	[100]
Omigbodun *et al.*, 2024	Technical (narrative) review	Poly(lactic acid)/hydroxyapatite (PLA/HA) composites for bone regeneration	Poly(lactic acid) (PLA), hydroxyapatite (HA)	Preclinical, clinical prospects	[121]
Arif *et al.*, 2022	Systematic review	Three-dimensional printed poly(lactic acid)/polycaprolactone (PLA/PCL) biomaterials	Poly(lactic acid) (PLA), polycaprolactone (PCL), poly(lactic-co-glycolic acid) (PLGA), etc.	Preclinical, translational focus	[122]

(*Continued*)

Table 2.2 Summary of recent reviews on resorbable composite materials for biomedical applications, highlighting study types, material systems, focus areas, and clinical application stages. (*Continued*)

Study	Study type	Focus area	Material system	Clinical application stage	Reference
García-González *et al.*, 2015	Technical (narrative) review	Supercritical fluid processing for regenerative scaffolds	Not specified (focus on process)	Preclinical, scaffold development	[123]
Zong *et al.*, 2022	Systematic review; technical (narrative) review	Biodegradable coronary stents	Polymer, magnesium, iron, zinc-based stents	Preclinical or clinical trials, translational research	[124]

spectrum of material systems, including poly(lactic acid) (PLA), polycaprolactone (PCL), tricalcium phosphate (TCP), hydroxyapatite (HA), bioglass, and bio-inks, used for applications such as tissue engineering, drug delivery, and scaffold development. The reviews differ in scope, with some emphasizing additive manufacturing and inkjet bioprinting, while others examine regulatory contexts or commercial translation potential. Most studies are in the preclinical phase, reflecting an ongoing effort to evaluate these biomaterials' biocompatibility, bioactivity, and mechanical suitability for clinical use.

2.6.3 Regulatory and Biocompatibility Considerations

Current regulatory frameworks and biocompatibility evaluation protocols for resorbable composites face challenges in addressing the complexities of ongoing innovations in biomaterials [152]. While standardized approaches exist for assessing biological responses, implant innocuity, and inertness, there is a need for improved strategies to bridge the translational gap between laboratory research and clinical applications [152, 153]. Historical perspectives on resorbable polymers highlight the importance of evaluating tissue response, toxicity, and pharmacokinetics of degradation products [154]. The ISO 10993 suite of documents provides a comprehensive framework for biocompatibility evaluation, emphasizing a risk-based approach and scientific rationale [155]. To ensure the safe and effective use of resorbable composites in biomedical implants, collaboration between various stakeholders is crucial for developing practical solutions that address patient needs while meeting regulatory requirements [152]. Table 2.3 discusses various biomedical application types, summarizing their regulatory requirements, translation challenges, and success factors based on recent literature.

2.6.4 Emerging Trends and Innovations

Resorbable composites are emerging as promising materials for tissue engineering scaffolds and surgical meshes, addressing the limitations of traditional non-absorbable materials [9, 156]. These composites combine bioactive ceramics or glasses with bioresorbable polymers, offering tailored mechanical properties and degradation behavior [9]. Recent research focuses on advanced biopolymers, therapeutic agents, and manufacturing technologies like 4D printing to enhance performance and clinical applicability [156]. While early resorbable materials showed potential in

Table 2.3 Overview of biomedical application types, summarizing associated regulatory requirements, translation challenges, and success factors as reported in recent literature.

Application type	Regulatory requirements	Translation challenges	Success factors	Reference
Bioresorbable polymers	Legal regulations, market approval	Translating research into marketable materials	Clinical insights, imaging evaluation	[22]
Additive manufactured devices	Food and Drug Administration (FDA), Code of Federal Regulations (CFR), International Organization for Standardization (ISO) compliance, device classification	Regulatory pathway, scale-up, electrospinning scalability	Commercialization (Osteoplug®, Resttrata®)	[119]
Bioprinted constructs	FDA, European Medicines Agency (EMA), Therapeutic Goods Administration (TGA) scrutiny	Process engineering, regulatory approval	Scalable manufacturing, regulatory alignment	[120]
Orthopedic implants	FDA/EMA classification, device approval	Hybrid system complexity, market gap	FDA-approved polymers, tailored properties	[100]
PLA/ hydroxyapatite composites	No mention found	Long-term clinical validation	Material refinement, clinical efficacy	[121]
Three-dimensional printed PLA/ PCL	FDA approval (for some polymers)	Ensuring biomimicry, reproducibility	Automation, reproducibility	[122]
Bioresorbable stents	Preclinical or clinical trials	Time-consuming, regulatory hurdles	Clinical trial evidence, novel technologies	[124]

orthopedic applications, first-generation bioresorbable scaffolds faced challenges such as increased strut thickness and device thrombosis [13, 157]. Ongoing development of novel bioresorbable scaffolds aims to overcome these limitations, improving acute performance and midterm safety [157].

These innovations in materials and fabrication technologies are driving the next generation of resorbable composites, potentially revolutionizing tissue engineering and regenerative medicine.

2.7 Conclusions

This chapter has provided a comprehensive examination of the synthesis, processing, and characterization of resorbable composites, which are central to the advancement of biomedical implants and fixation devices. The selection and combination of biodegradable polymers with bioactive fillers such as hydroxyapatite, chitosan, gelatin, and nanocellulose were shown to enable tailored mechanical, thermal, and degradation properties suitable for various clinical applications.

Numerous synthesis techniques—including *in situ* polymerization, solvent casting, electrospinning, and sol-gel methods—were explored, each offering specific advantages in controlling composite morphology, porosity, and interfacial interactions. The choice of synthesis method plays a crucial role in determining the microstructure and performance of the final material, influencing properties such as biodegradability, mechanical strength, and biocompatibility.

Processing approaches such as freeze-drying, rapid prototyping, and gas foaming were discussed as effective tools for scaffold fabrication, enabling the development of porous structures that mimic the extracellular matrix and promote cellular infiltration and tissue regeneration. Innovations such as the incorporation of nanomaterials—including carbon nanotubes and graphene oxide—have further enhanced the functionality of resorbable composites, offering improvements in strength, conductivity, and bioactivity.

Advanced characterization techniques, including SEM, XRD, DSC, and mechanical testing, have been critical in evaluating the structural integrity, crystallinity, thermal behavior, and degradation kinetics of these composites. Furthermore, the integration of *in vitro* and *in vivo* assessments allows for accurate predictions of biological responses and long-term implant performance.

The successful application of these materials in bone tissue engineering, drug delivery, and bioresorbable electronics underscores their versatility and clinical potential. However, several challenges remain, such as ensuring uniform dispersion of nanofillers, achieving consistent degradation rates, and navigating regulatory pathways for clinical approval.

In conclusion, the development of resorbable composites represents a dynamic and interdisciplinary field, combining materials science, biomedical engineering, and nanotechnology. Continued research focused on optimizing synthesis and processing methods, enhancing structure-property relationships, and integrating smart functionalities will be essential for translating these materials from the laboratory to clinical practice.

AI disclosure statement: The author confirms that no artificial intelligence (AI)-generated content or AI-assisted writing tools were used in the preparation of this chapter (including summaries and the abstract). All text, analysis, and conclusions are the result of the author's original work.

References

1. Lehtonen, T.J., Tuominen, J.U., Hiekkanen, E., Resorbable composites with bioresorbable glass fibers for load-bearing applications. *in vitro* degradation and degradation mechanism. *Acta Biomater.*, 9, 4868–4877, 2013, https://doi.org/10.1016/j.actbio.2012.08.052.
2. Oosterbeek, R.N., Zhang, X.C., Best, S.M., Cameron, R.E., The evolution of the structure and mechanical properties of fully bioresorbable polymer-glass composites during degradation. *Compos. Sci. Technol.*, 218, 2022, https://doi.org/10.1016/j.compscitech.2021.109194.
3. Ullah, M.W., Fu, L., Lamboni, L., Shi, Z., Yang, G., Current trends and biomedical applications of resorbable polymers, in: *Mater. Biomed. Eng. Absorbable Polym.*, pp. 41–86, 2019, https://doi.org/10.1016/B978-0-12-818415-8.00003-6.
4. Gomes, M.E. and Reis, R.L., Biodegradable polymers and composites in biomedical applications: From catgut to tissue engineering Part 1 Available systems and their properties. *Int. Mater. Rev.*, 49, 261–273, 2004, https://doi.org/10.1179/095066004225021918.
5. Asghari, F., Samiei, M., Adibkia, K., Akbarzadeh, A., Davaran, S., Biodegradable and biocompatible polymers for tissue engineering application: a review. *Artif. Cells Nanomed. Biotechnol.*, 45, 185–192, 2017, https://doi.org/10.3109/21691401.2016.1146731.
6. Mahanta, A.K. and Maiti, P., Chitin and chitosan nanocomposites for tissue engineering, in: *Chitin Chitosan Regen. Med.*, pp. 123–149, 2015, https://doi.org/10.1007/978-81-322-2511-9_6.
7. Lalwani, G., Henslee, A.M., Farshid, B., Lin, L., Kasper, F.K., Qin, Y.X., Mikos, A.G., Sitharaman, B., Two-dimensional nanostructure-reinforced biodegradable polymeric nanocomposites for bone tissue engineering. *Biomacromolecules*, 14, 900–909, 2013, https://doi.org/10.1021/bm301995s.

8. Hasnain, M.S., Ahmad, S.A., Chaudhary, N., Hoda, M.N., Nayak, A.K., Biodegradable polymer matrix nanocomposites for bone tissue engineering, in: *Appl. Nanocomposite Mater. Orthop.*, pp. 1–37, 2018, https://doi.org/10.1016/B978-0-12-813740-6.00001-6.
9. Boccaccini, A.R. and Blaker, J.J., Bioactive composite materials for tissue engineering scaffolds. *Expert Rev. Med. Devices*, 2, 303–317, 2005, https://doi.org/10.1586/17434440.2.3.303.
10. Paramsothy, M. and Ramakrishna, S., Biodegradable Materials for Clinical Applications: A Review. *Rev. Adv. Sci. Eng.*, 4, 221–238, 2016, https://doi.org/10.1166/rase.2015.1097.
11. Paul, A.K., Borkotoky, S.S., Prasad, A., Nanohydroxyapatite-Based Composite Materials and Processing, in: *Adv. Mater. Manuf. Tech. Biomed. Appl.*, pp. 157–183, Wiley, Springer Singapore, 2023, https://doi.org/10.1002/9781394166985.ch7.
12. Salernitano, E. and Migliaresi, C., Composite Materials for Biomedical Applications: A Review. *J. Appl. Biomater. Biomech.*, 1, 3–18, 2003, https://doi.org/10.1177/228080000300100102.
13. Parsons, J.R., Resorbable materials and composites: New concepts in orthopedic biomaterials. *Orthopedics*, 8, 907–915, 1985, https://doi.org/10.3928/0147-7447-19850701-17.
14. Prasad, A. and Paul, A.K., *Biodegradable waste processing for sustainable developments*, CRC PRESS, Boca Raton, Florida, USA, 2024, https://doi.org/10.1201/9781003502012.
15. Tusnim, J. and Hoque, M.E., Bioresorbable biocomposites, Green Biocomposites, in: *Biomed. Eng. Des. Prop. Appl.*, pp. 113–133, 2021, https://doi.org/10.1016/B978-0-12-821553-1.00004-1.
16. Mamidi, N., Delgadillo, R.M.V., Sustaita, A.O., Lozano, K., Yallapu, M.M., Current nanocomposite advances for biomedical and environmental application diversity. *Med. Res. Rev.*, 2024, https://doi.org/10.1002/med.22082.
17. Mohanty, A.K., Vivekanandhan, S., Pin, J.M., Misra, M., Composites from renewable and sustainable resources: Challenges and innovations. *Science (80-.)*, 362, 536–542, 2018, https://doi.org/10.1126/science.aat9072.
18. Vach Agocsova, S., Culenova, M., Birova, I., Omanikova, L., Moncmanova, B., Danisovic, L., Ziaran, S., Bakos, D., Alexy, P., Resorbable Biomaterials Used for 3D Scaffolds in Tissue Engineering: A Review. *Materials (Basel)*, 16, 2023, https://doi.org/10.3390/ma16124267.
19. Das, P. and Chakraborty, G., Biodegradable polymers-based proton exchange membrane for fuel cells, in: *Sustain. Bio-Based Compos.*, pp. 183–192, 2024, https://doi.org/10.1515/9783111321530-009.
20. Liao, X., Zhang, H., He, T., Preparation of porous biodegradable polymer and its nanocomposites by supercritical CO2 foaming for tissue engineering. *J. Nanomater.*, 2012, 2012, https://doi.org/10.1155/2012/836394.
21. Ravichandran, R., Sundarrajan, S., Venugopal, J.R., Mukherjee, S., Ramakrishna, S., Advances in Polymeric Systems for Tissue Engineering and

Biomedical Applications. *Macromol. Biosci.*, 12, 286–311, 2012, https://doi.org/10.1002/mabi.201100325.

22. Dobrzyńska-Mizera, M., Dodda, J.M., Liu, X., Knitter, M., Oosterbeek, R.N., Salinas, P., Pozo, E., Ferreira, A.M., Sadiku, E.R., Engineering of Bioresorbable Polymers for Tissue Engineering and Drug Delivery Applications. *Adv. Healthcare Mater.*, 2024, https://doi.org/10.1002/adhm.202401674.
23. Song, R., Murphy, M., Li, C., Ting, K., Soo, C., Zheng, Z., Current development of biodegradable polymeric materials for biomedical applications. *Drug Des. Devel. Ther.*, 12, 3117–3145, 2018, https://doi.org/10.2147/DDDT.S165440.
24. Toong, D.W.Y., Toh, H.W., Ng, J.C.K., Wong, P.E.H., Leo, H.L., Venkatraman, S., Tan, L.P., Ang, H.Y., Huang, Y., Bioresorbable polymeric scaffold in cardiovascular applications. *Int. J. Mol. Sci.*, 21, 2020, https://doi.org/10.3390/ijms21103444.
25. Okamoto, M. and John, B., Synthetic biopolymer nanocomposites for tissue engineering scaffolds. *Prog. Polym. Sci.*, 38, 1487–1503, 2013, https://doi.org/10.1016/j.progpolymsci.2013.06.001.
26. Armentano, I., Dottori, M., Fortunati, E., Mattioli, S., Kenny, J.M., Biodegradable polymer matrix nanocomposites for tissue engineering: A review. *Polym. Degrad. Stab.*, 95, 2126–2146, 2010, https://doi.org/10.1016/j.polymdegradstab.2010.06.007.
27. Wahid, F., Khan, T., Hussain, Z., Ullah, H., Nanocomposite scaffolds for tissue engineering; properties, preparation and applications, in: *Appl. Nanocomposite Mater. Drug Deliv.*, pp. 701–735, 2018, https://doi.org/10.1016/B978-0-12-813741-3.00031-5.
28. Cingolani, A., Casalini, T., Caimi, S., Klaue, A., Sponchioni, M., Rossi, F., Perale, G., A methodologic approach for the selection of bio-resorbable polymers in the development of medical devices: The case of poly(L-lactide-co-ε-caprolactone). *Polymers (Basel)*, 10, 2018, https://doi.org/10.3390/polym10080851.
29. Tanaka, M., Sato, K., Kitakami, E., Kobayashi, S., Hoshiba, T., Fukushima, K., Design of biocompatible and biodegradable polymers based on intermediate water concept. *Polym. J.*, 47, 114–121, 2015, https://doi.org/10.1038/pj.2014.129.
30. Ojijo, V. and Sinha Ray, S., Processing strategies in bionanocomposites. *Prog. Polym. Sci.*, 38, 1543–1589, 2013, https://doi.org/10.1016/j.progpolymsci.2013.05.011.
31. Zhao, R., Torley, P., Halley, P.J., Emerging biodegradable materials: Starch- and protein-based bio-nanocomposites. *J. Mater. Sci.*, 43, 3058–3071, 2008, https://doi.org/10.1007/s10853-007-2434-8.
32. Wetteland, C.L. and Liu, H., Optical and biological properties of polymer-based nanocomposites with improved dispersion of ceramic nanoparticles.

J. Biomed. Mater. Res. Part A, 106, 2692–2707, 2018, https://doi.org/10.1002/jbm.a.36466.

33. Wilberforce, S.I.J., Finlayson, C.E., Best, S.M., Cameron, R.E., The influence of the compounding process and testing conditions on the compressive mechanical properties of poly(D,L-lactide-co-glycolide)/α-tricalcium phosphate nanocomposites. *J. Mech. Behav. Biomed. Mater.*, 4, 1081–1089, 2011, https://doi.org/10.1016/j.jmbbm.2011.03.017.
34. Nair, L.S. and Laurencin, C.T., Biodegradable polymers as biomaterials. *Prog. Polym. Sci.*, 32, 762–798, 2007, https://doi.org/10.1016/j.progpolymsci.2007.05.017.
35. Puppi, D., Chiellini, F., Piras, A.M., Chiellini, E., Polymeric materials for bone and cartilage repair. *Prog. Polym. Sci.*, 35, 403–440, 2010, https://doi.org/10.1016/j.progpolymsci.2010.01.006.
36. Allo, B.A., Costa, D.O., Dixon, S.J., Mequanint, K., Rizkalla, A.S., Bioactive and Biodegradable Nanocomposites and Hybrid Biomaterials for Bone Regeneration. *J. Funct. Biomater.*, 3, 432–463, 2012, https://doi.org/10.3390/jfb3020432.
37. Manikandan, A., Mani, M.P., Jaganathan, S.K., Rajasekar, R., Morphological, thermal, and blood-compatible properties of electrospun nanocomposites for tissue engineering application. *Polym. Compos.*, 39, E132–E139, 2018, https://doi.org/10.1002/pc.24463.
38. Bramhill, J., Ross, S., Ross, G., Bioactive nanocomposites for tissue repair and regeneration: A review. *Int. J. Environ. Res. Public Health*, 14, 2017, https://doi.org/10.3390/ijerph14010066.
39. Laurenti, M. and Cauda, V., Biodegradable polymer nanocomposites for tissue engineering: Synthetic strategies and related applications, in: *Mater. Biomed. Eng. Absorbable Polym.*, pp. 157–198, 2019, https://doi.org/10.1016/B978-0-12-818415-8.00006-1.
40. Pina, S., Oliveira, J.M., Reis, R.L., Natural-based nanocomposites for bone tissue engineering and regenerative medicine: A review. *Adv. Mater.*, 27, 1143–1169, 2015, https://doi.org/10.1002/adma.201403354.
41. Hassan, M., Dave, K., Chandrawati, R., Dehghani, F., Gomes, V.G., 3D printing of biopolymer nanocomposites for tissue engineering: Nanomaterials, processing and structure-function relation. *Eur. Polym. J.*, 121, 2019, https://doi.org/10.1016/j.eurpolymj.2019.109340.
42. Prasad, A., Chakraborty, G., Davim, J.P. (Eds.), *Sustainable Bio-Based Composites*, De Gruyter, Berlin, Germany, 2024, https://doi.org/10.1515/9783111321530.
43. Chen, M., Parsons, A.J., Felfel, R.M., Rudd, C.D., Irvine, D.J., Ahmed, I., *In-situ* polymerisation of fully bioresorbable polycaprolactone/phosphate glass fibre composites: *In vitro* degradation and mechanical properties. *J. Mech. Behav. Biomed. Mater.*, 59, 78–89, 2016, https://doi.org/10.1016/j.jmbbm.2015.12.019.

44. Gupta, S., Javaid, S., Dey, M., Matzke, C., Eades, E., Ji, Y., Exploration of solvent casting for designing engineered microstructures for biomedical and functional applications. *J. Am. Ceram. Soc.*, 105, 1864–1881, 2022, https://doi.org/10.1111/jace.18104.
45. Navarro, M., Aparicio, C., Charles-Harris, M., Ginebra, M.P., Engel, E., Planell, J.A., Development of a biodegradable composite scaffold for bone tissue engineering: Physicochemical, topographical, mechanical, degradation, and biological properties. *Adv. Polym. Sci.*, 200, 209–231, 2006, https://doi.org/10.1007/12_068.
46. Alizadeh-Osgouei, M., Li, Y., Wen, C., A comprehensive review of biodegradable synthetic polymer-ceramic composites and their manufacture for biomedical applications. *Bioact. Mater.*, 4, 22–36, 2019, https://doi.org/10.1016/j.bioactmat.2018.11.003.
47. Ul-Islam, M., Alabbosh, K.F., Manan, S., Khan, S., Ahmad, F., Ullah, M.W., Chitosan-based nanostructured biomaterials: Synthesis, properties, and biomedical applications. *Adv. Ind. Eng. Polym. Res.*, 7, 79–99, 2024, https://doi.org/10.1016/j.aiepr.2023.07.002.
48. Gholap, A.D., Rojekar, S., Kapare, H.S., Vishwakarma, N., Raikwar, S., Garkal, A., Mehta, T.A., Jadhav, H., Prajapati, M.K., Annapure, U., Chitosan scaffolds: Expanding horizons in biomedical applications. *Carbohydr. Polym.*, 323, 2024, https://doi.org/10.1016/j.carbpol.2023.121394.
49. Ashok, N., Pradeep, A., Jayakumar, R., Synthesis-Structure Relationship of Chitosan Based Hydrogels. *Adv. Polym. Sci.*, 287, 105–129, 2021, https://doi.org/10.1007/12_2021_103.
50. Budiarso, I.J., Rini, N.D.W., Tsalsabila, A., Birowosuto, M.D., Wibowo, A., Chitosan-Based Smart Biomaterials for Biomedical Applications: Progress and Perspectives. *ACS Biomater. Sci. Eng.*, 9, 3084–3115, 2023, https://doi.org/10.1021/acsbiomaterials.3c00216.
51. Dutta, J., Mohini, Priyanka, Effect of degree of deacetylation and molecular weight on physicochemical properties of chitosan films. *J. Indian Chem. Soc.*, 97, 731–735, 2020.
52. Abarrategi, A., Lópiz-Morales, Y., Ramos, V., Civantos, A., López-Duŕn, L., Marco, F., López-Lacomba, J.L., Chitosan scaffolds for osteochondral tissue regeneration. *J. Biomed. Mater. Res. Part A*, 95, 1132–1141, 2010, https://doi.org/10.1002/jbm.a.32912.
53. Kumirska, J., Weinhold, M.X., Thöming, J., Stepnowski, P., Biomedical activity of chitin/chitosan based materials- influence of physicochemical properties apart from molecular weight and degree of N-Acetylation. *Polymers (Basel)*, 3, 1875–1901, 2011, https://doi.org/10.3390/polym3041875.
54. Mecwan, M.M., Rapalo, G.E., Mishra, S.R., Haggard, W.O., Bumgardner, J.D., Effect of molecular weight of chitosan degraded by microwave irradiation on lyophilized scaffold for bone tissue engineering applications. *J. Biomed. Mater. Res. Part A*, 97 A, 66–73, 2011, https://doi.org/10.1002/jbm.a.33029.

55. Jayakumar, R., Prabaharan, M., Nair, S.V., Tamura, H., Novel chitin and chitosan nanofibers in biomedical applications. *Biotechnol. Adv.*, 28, 142–150, 2010, https://doi.org/10.1016/j.biotechadv.2009.11.001.
56. Basak, T. and Shamshina, J.L., Design of Chitin Cell Culture Matrices for 3D Tissue Engineering: The Importance of Chitin Types, Solvents, Cross-Linkers, and Fabrication Techniques. *Pharmaceutics*, 16, 2024, https://doi.org/10.3390/pharmaceutics16060777.
57. Ko, H.F., Sfeir, C., Kumta, P.N., Novel synthesis strategies for natural polymer and composite biomaterials as potential scaffolds for tissue engineering. *Philos. Trans. R. Soc. A Math. Phys. Eng. Sci.*, 368, 1981–1997, 2010, https://doi.org/10.1098/rsta.2010.0009.
58. Silva, M., Ferreira, F.N., Alves, N.M., Paiva, M.C., Biodegradable polymer nanocomposites for ligament/tendon tissue engineering. *J. Nanobiotechnol.*, 18, 2020, https://doi.org/10.1186/s12951-019-0556-1.
59. Morparia, S. and Suvarna, V., Recent Advancements in Applications of Alginates in Drug Delivery, Tissue Engineering, and Biomedical Field. *Nat. Prod. J.*, 14, 2024, https://doi.org/10.2174/0122103155284365240103063024.
60. Guarino, V., Altobelli, R., della Sala, F., Borzacchiello, A., Ambrosio, L., Alginate processing routes to fabricate bioinspired platforms for tissue engineering and drug delivery, in: *Springer Ser. Biomater. Sci. Eng.*, vol. 11, pp. 101–120, 2018, https://doi.org/10.1007/978-981-10-6910-9_4.
61. Nayak, A.K., Mohanta, B.C., Hasnain, M.S., Hoda, M.N., Tripathi, G., Alginate-based scaffolds for drug delivery in tissue engineering, in: *Alginates Drug Deliv.*, pp. 359–386, 2020, https://doi.org/10.1016/B978-0-12-817640-5.00014-5.
62. Aguero, L., Alpdagtas, S., Ilhan, E., Zaldivar-Silva, D., Gunduz, O., Functional role of crosslinking in alginate scaffold for drug delivery and tissue engineering: A review. *Eur. Polym. J.*, 160, 2021, https://doi.org/10.1016/j.eurpolymj.2021.110807.
63. Sun, J. and Tan, H., Alginate-based biomaterials for regenerative medicine applications. *Materials (Basel)*, 6, 1285–1309, 2013, https://doi.org/10.3390/ma6041285.
64. Augst, A.D., Kong, H.J., Mooney, D.J., Alginate hydrogels as biomaterials. *Macromol. Biosci.*, 6, 623–633, 2006, https://doi.org/10.1002/mabi.200600069.
65. Girish, B., Rakshith, G.R., Paul, A.K., Raja, V.K., Chakraborty, G., Hydrolytic and soil degradation of cellulosic material (paper): Optimization of parameters using ANN and RSM. *Polym. Bull.*, 2024, https://doi.org/10.1007/s00289-024-05319-5.
66. Roslan, M.R., Nasir, N.F.M., Cheng, E.M., Amin, N.A.M., Tissue engineering scaffold based on starch: A review. *Int. Conf. Electr. Electron. Optim. Tech. ICEEOT 2016*, pp. 1857–1860, 2016, https://doi.org/10.1109/ICEEOT.2016.7755010.

67. Torres, F.G., Commeaux, S., Troncoso, O.P., Starch-based biomaterials for wound-dressing applications. *Starch/Staerke*, 65, 543–551, 2013, https://doi.org/10.1002/star.201200259.
68. Lee, C.S. and Hwang, H.S., Starch-Based Hydrogels as a Drug Delivery System in Biomedical Applications. *Gels*, 9, 2023, https://doi.org/10.3390/gels9120951.
69. Sarder, R., Piner, E., Rios, D.C., Chacon, L., Artner, M.A., Barrios, N., Argyropoulos, D., Copolymers of starch, a sustainable template for biomedical applications: A review. *Carbohydr. Polym.*, 278, 2022, https://doi.org/10.1016/j.carbpol.2021.118973.
70. Silva, G.A., Coutinho, O.P., Ducheyne, P., Shapiro, I.M., Reis, R.L., Starch-Based Microparticles as a Novel Strategy for Tissue Engineering Applications. *Key Eng. Mater.*, 309–311, 907–910, 2006, https://doi.org/10.4028/www.scientific.net/kem.309-311.907.
71. Baldwin, A.D. and Kiick, K.L., Polysaccharide-modified synthetic polymeric biomaterials. *Biopolymers*, 94, 128–140, 2010, https://doi.org/10.1002/bip.21334.
72. Jafari, A., Al-Ostaz, A., Nouranian, S., Biomedical Potential of Cellulose: Current Trends and Future Directions. *Polym. Adv. Technol.*, 35, 2024, https://doi.org/10.1002/pat.6621.
73. Janmohammadi, M., Nazemi, Z., Salehi, A.O.M., Seyfoori, A., John, J.V., Nourbakhsh, M.S., Akbari, M., Cellulose-based composite scaffolds for bone tissue engineering and localized drug delivery. *Bioact. Mater.*, 20, 137–163, 2023, https://doi.org/10.1016/j.bioactmat.2022.05.018.
74. Cheng, W., Zhu, Y., Jiang, G., Cao, K., Zeng, S., Chen, W., Zhao, D., Yu, H., Sustainable cellulose and its derivatives for promising biomedical applications. *Prog. Mater Sci.*, 138, 2023, https://doi.org/10.1016/j.pmatsci.2023.101152.
75. Zhai, X., Wu, Y., Tan, H., Gelatin-based Targeted Delivery Systems for Tissue Engineering. *Curr. Drug Targets*, 24, 673–687, 2023, https://doi.org/10.2174/1389450124666230605150303.
76. Jia, X., Fan, X., Chen, C., Lu, Q., Zhou, H., Zhao, Y., Wang, X., Han, S., Ouyang, L., Yan, H., Dai, H., Geng, H., Chemical and Structural Engineering of Gelatin-Based Delivery Systems for Therapeutic Applications: A Review. *Biomacromolecules*, 25, 564–589, 2024, https://doi.org/10.1021/acs.biomac.3c01021.
77. Bello, A.B., Kim, D., Kim, D., Park, H., Lee, S.H., Engineering and functionalization of gelatin biomaterials: From cell culture to medical applications. *Tissue Eng. - Part B Rev.*, 26, 164–180, 2020, https://doi.org/10.1089/ten.teb.2019.0256.
78. Arun, A., Malrautu, P., Laha, A., Ramakrishna, S., Gelatin Nanofibers in Drug Delivery Systems and Tissue Engineering. *Eng. Sci.*, 16, 71–81, 2021, https://doi.org/10.30919/es8d527.
79. Yue, K., Trujillo-de Santiago, G., Alvarez, M.M., Tamayol, A., Annabi, N., Khademhosseini, A., Synthesis, properties, and biomedical applications of

gelatin methacryloyl (GelMA) hydrogels. *Biomaterials*, 73, 254–271, 2015, https://doi.org/10.1016/j.biomaterials.2015.08.045.
80. Elzoghby, A.O., Gelatin-based nanoparticles as drug and gene delivery systems: Reviewing three decades of research. *J. Controlled Release*, 172, 1075–1091, 2013, https://doi.org/10.1016/j.jconrel.2013.09.019.
81. Afewerki, S., Sheikhi, A., Kannan, S., Ahadian, S., Khademhosseini, A., Gelatin-polysaccharide composite scaffolds for 3D cell culture and tissue engineering: Towards natural therapeutics. *Bioeng. Transl. Med.*, 4, 96–115, 2019, https://doi.org/10.1002/btm2.10124.
82. Echave, M.C., Burgo, L.S., Pedraz, J.L., Orive, G., Gelatin as Biomaterial for Tissue Engineering. *Curr. Pharm. Des.*, 23, 2017, https://doi.org/10.2174/0929867324666170511123101.
83. Kouser, R., Vashist, A., Zafaryab, M., Rizvi, M.A., Ahmad, S., Biocompatible and mechanically robust nanocomposite hydrogels for potential applications in tissue engineering. *Mater. Sci. Eng., C*, 84, 168–179, 2018, https://doi.org/10.1016/j.msec.2017.11.018.
84. Sitharaman, B., Shi, X., Walboomers, X.F., Liao, H., Cuijpers, V., Wilson, L.J., Mikos, A.G., Jansen, J.A., *In vivo* biocompatibility of ultra-short single-walled carbon nanotube/biodegradable polymer nanocomposites for bone tissue engineering. *Bone*, 43, 362–370, 2008, https://doi.org/10.1016/j.bone.2008.04.013.
85. Li, Y., Liao, C., Tjong, S.C., Synthetic biodegradable aliphatic polyester nanocomposites reinforced with nanohydroxyapatite and/or graphene oxide for bone tissue engineering applications. *Nanomaterials*, 9, 2019, https://doi.org/10.3390/nano9040590.
86. Liu, S., Qin, S., He, M., Zhou, D., Qin, Q., Wang, H., Current applications of poly(lactic acid) composites in tissue engineering and drug delivery. *Compos. Part B Eng.*, 199, 108238, 2020, https://doi.org/10.1016/j.compositesb.2020.108238.
87. Kang, I.K., Xing, Z.C., Han, S.J., Shin, Y.S., Fabrication of biodegradable polyester nanocomposites by electrospinning for tissue engineering. *J. Nanomater.*, 2011, 2011, https://doi.org/10.1155/2011/929378.
88. Paul, A.K., Migration Studies of Biodegradable Composites, in: *Biodegrad. Compos. Packag. Appl.*, pp. 193–212, CRC Press, Boca Raton, 2022, https://doi.org/10.1201/9781003227908-11.
89. Prasad, A., Chakraborty, G., Kumar, A., Bio-based environmentally benign polymeric resorbable materials for orthopedic fixation applications, in: *Adv. Mater. Biomed. Appl.*, pp. 251–266, 2022, https://doi.org/10.1201/9781003344810-15.
90. Paul, A.K., Prasad, A., Mahto, B., Kumar, A., Singh, P., Introduction to bioenergy and waste to energy, in: *Biodegrad. Waste Process. Sustain. Dev*, pp. 1–33, 2024, https://doi.org/10.1201/9781003502012-1.

91. Kumar Paul, A. and Das, S., Biolubricants and its application in engineering, in: *Sustain. Bio-Based Compos*, pp. 271–296, 2024, https://doi.org/10.1515/9783111321530-013.
92. Chalovich, J.M. and Eisenberg, E., Synthetic biodegradable functional polymers for tissue engineering: A brief review. *Sci. China Chem.*, 57, 490–500, 2014.
93. Maquet, V., Boccaccini, A.R., Pravata, L., Notingher, I., Jérôme, R., Preparation, characterization, and *in vitro* degradation of bioresorbable and bioactive composites based on Bioglass®-filled polylactide foams. *J. Biomed. Mater. Res. Part A*, 66, 335–346, 2003, https://doi.org/10.1002/jbm.a.10587.
94. Bowen, P.K., Drelich, J., Goldman, J., A newin *vitro-in vivo* correlation for bioabsorbable magnesium stents from mechanical behavior. *Mater. Sci. Eng., C*, 33, 5064–5070, 2013, https://doi.org/10.1016/j.msec.2013.08.042.
95. Bennett, S.M., Arumugam, M., Wilberforce, S., Enea, D., Rushton, N., Zhang, X.C., Best, S.M., Cameron, R.E., Brooks, R.A., The effect of particle size on the *in vivo* degradation of poly(D,L-lactide-co-glycolide)/α-tricalcium phosphate micro- and nanocomposites. *Acta Biomater.*, 45, 340–348, 2016, https://doi.org/10.1016/j.actbio.2016.08.046.
96. Boccaccini, A.R., Blaker, J.J., Maquet, V., Jerome, R., Blacher, S., Roether, J.A., Biodegradable and Bioactive Polymer/Bioglass® Composite Foams for Tissue Engineering Scaffolds. *Mater. Sci. Forum*, 494, 499–506, 2005, https://doi.org/10.4028/www.scientific.net/msf.494.499.
97. Charles-Harris Ferrer, M., Development and characterisation of completely degradable composite tissue engineering scaffolds, Universitat Politècnica de Catalunya, 2007.
98. Ahmadi, Z. and Moztarzadeh, F., Scanning electron microscope (SEM) investigations on gelatin-chitosan-bioactive glass (58s) scaffolds for bone tissue engineering. *ACM Int. Conf. Proceeding Ser.*, pp. 93–96, 2017, https://doi.org/10.1145/3155077.3155094.
99. Buevich, F., Pulapura, S., Kohn, J., Microscopic Analysis of Porous Biodegradable Scaffolds for Tissue Engineering. *Microsc. Microanal.*, 6, 988–989, 2000, https://doi.org/10.1017/s1431927600037430.
100. Lacambra-Andreu, X., Maazouz, A., Lamnawar, K., Chenal, J.M., A Review on Manufacturing Processes of Biocomposites Based on Poly(α-Esters) and Bioactive Glass Fillers for Bone Regeneration. *Biomimetics*, 8, 2023, https://doi.org/10.3390/biomimetics8010081.
101. Kandić, L., Mitrić, M., Ignjatović, N., XRD analysis of calcium phosphate and biocomposite calcium phosphate/bioresorbable polymer. *Mater. Sci. Forum*, 518, 507–512, 2006, https://doi.org/10.4028/www.scientific.net/msf.518.507.
102. Felfel, R.M., Ahmed, I., Parsons, A.J., Haque, P., Walker, G.S., Rudd, C.D., Investigation of crystallinity, molecular weight change, and mechanical

properties of PLA/PBG bioresorbable composites as bone fracture fixation plates. *J. Biomater. Appl.*, 26, 765–789, 2012, https://doi.org/10.1177/0885328210384532.
103. Baji, A., Wong, S.C., Liu, T., Li, T., Srivatsan, T.S., Morphological and X-ray diffraction studies of crystalline hydroxyapatite-reinforced polycaprolactone. *J. Biomed. Mater. Res. - Part B Appl. Biomater.*, 81, 343–350, 2007, https://doi.org/10.1002/jbm.b.30671.
104. Capitain, C., Ross-Jones, J., Möhring, S., Tippkötter, N., Differential scanning calorimetry for quantification of polymer biodegradability in compost. *Int. Biodeterior. Biodegrad.*, 149, 2020, https://doi.org/10.1016/j.ibiod.2020.104914.
105. Ferri, J.M., Jordá, J., Montanes, N., Fenollar, O., Balart, R., Manufacturing and characterization of poly(lactic acid) composites with hydroxyapatite. *J. Thermoplast. Compos. Mater.*, 31, 865–881, 2018, https://doi.org/10.1177/0892705717729014.
106. Cifuentes, S.C., Benavente, R., González-Carrasco, J.L., Does magnesium compromise the high temperature processability of novel biodegradable and bioresorbables PLLA/Mg composites? *Rev. Metal.*, 50, 2014, https://doi.org/10.3989/revmetalm.011.
107. Hernandez-Charpak, Y.D., Kansara, H.J., Trabold, T.A., Lodge, J.S., Lewis, C.L., Diaz, C.A., Application of Differential Scanning Calorimetry to Assess Molecular Weight Degradation of Poly(butylene Adipate-co-terephthalate)-Based Plastics. *ACS ES T Eng.*, 2024, https://doi.org/10.1021/acsestengg.4c00617.
108. Kumar, R., Ganguly, A., Purohit, R., Thermogravimetric analysis of natural fiber reinforced hybrid composites – A review. *Mater. Today Proc.*, 2023, https://doi.org/10.1016/j.matpr.2023.08.025.
109. Isa, N.M., Razak, A., Haiqal, A., Adzila, S., Thermogravimetric analysis of polyhydroxyalkanoates (PHA)/nano-Calcium phosphate (nCaP)/chitosan biocomposite for heat-related manufacturing process. *AIP Conf. Proc.*, 2925, 2024, https://doi.org/10.1063/5.0183247.
110. Adolfsson, K.H., Sjöberg, I., Höglund, O.V., Wattle, O., Hakkarainen, M., *In Vivo* Versus *In Vitro* Degradation of a 3D Printed Resorbable Device for Ligation of Vascular Tissue in Horses. *Macromol. Biosci.*, 21, 2021, https://doi.org/10.1002/mabi.202100164.
111. Ignatius, A.A., Wolf, S., Augat, P., Claes, L.E., Composites made of rapidly resorbable ceramics and poly(lactide) show adequate mechanical properties for use as bone substitute materials. *J. Biomed. Mater. Res.*, 57, 126–131, 2001, https://doi.org/10.1002/1097-4636(200110)57:1<126::AID-JBM1151>3.0.CO;2-M.
112. Blaker, J.J., Gough, J.E., Maquet, V., Notingher, I., Boccaccini, A.R., *In vitro* evaluation of novel bioactive composites based on Bioglass®-filled polylactide foams for bone tissue engineering scaffolds. *J. Biomed. Mater. Res. Part A*, 67, 1401–1411, 2003, https://doi.org/10.1002/jbm.a.20055.
113. Kim, H.W., Lee, H.H., Chun, G.S., Bioactivity and osteoblast responses of novel biomedical nanocomposites of bioactive glass nanofiber filled

poly(lactic acid). *J. Biomed. Mater. Res. Part A*, 85, 651–663, 2008, https://doi.org/10.1002/jbm.a.31339.

114. Vakil, A.U., Petryk, N.M., Du, C., Howes, B., Stinfort, D., Serinelli, S., Gitto, L., Ramezani, M., Beaman, H.T., Monroe, M.B.B., *In vitro* and *in vivo* degradation correlations for polyurethane foams with tunable degradation rates. *J. Biomed. Mater. Res. Part A*, 111, 580–595, 2023, https://doi.org/10.1002/jbm.a.37504.
115. Huang, C.H., Lee, S.Y., Horng, S., Guy, L.G., Bin Yu, T., *In vitro* and *in vivo* degradation of microfiber bioresorbable coronary scaffold. *J. Biomed. Mater. Res. - Part B Appl. Biomater.*, 106, 1842–1850, 2018, https://doi.org/10.1002/jbm.b.33987.
116. Hasan, A., Morshed, M., Memic, A., Hassan, S., Webster, T.J., Marei, H.E.S., Nanoparticles in tissue engineering: Applications, challenges and prospects. *Int. J. Nanomed.*, 13, 5637–5655, 2018, https://doi.org/10.2147/IJN.S153758.
117. Zuluaga-Vélez, A., Quintero-Martinez, A., Orozco, L.M., Sepúlveda-Arias, J.C., Silk fibroin nanocomposites as tissue engineering scaffolds – A systematic review. *Biomed. Pharmacother.*, 141, 2021, https://doi.org/10.1016/j.biopha.2021.111924.
118. Mostafavi, A., Quint, J., Russell, C., Tamayol, A., Nanocomposite hydrogels for tissue engineering applications, in: *Biomater. Organ Tissue Regen. New Technol. Futur. Prospect.*, pp. 499–528, 2020, https://doi.org/10.1016/B978-0-08-102906-0.00023-4.
119. Culbreath, C.J., Taylor, M.S., McCullen, S.D., Mefford, O.T., A Review of Additive Manufacturing in Tissue Engineering and Regenerative Medicine. *Biomed. Mater. Devices*, 2024, https://doi.org/10.1007/s44174-024-00183-3.
120. Angelopoulos, I., Allenby, M.C., Lim, M., Zamorano, M., Engineering inkjet bioprinting processes toward translational therapies. *Biotechnol. Bioeng.*, 117, 272–284, 2020, https://doi.org/10.1002/bit.27176.
121. Omigbodun, F.T., Oladapo, B.I., Osa-uwagboe, N., Exploring the frontier of Polylactic Acid/Hydroxyapatite composites in bone regeneration and their revolutionary biomedical applications – A review. *J. Reinf. Plast. Compos.*, 2024, https://doi.org/10.1177/07316844241278045.
122. Arif, Z.U., Khalid, M.Y., Noroozi, R., Sadeghianmaryan, A., Jalalvand, M., Hossain, M., Recent advances in 3D-printed polylactide and polycaprolactone-based biomaterials for tissue engineering applications. *Int. J. Biol. Macromol.*, 218, 930–968, 2022, https://doi.org/10.1016/j.ijbiomac.2022.07.140.
123. García-González, C.A., Concheiro, A., Alvarez-Lorenzo, C., Processing of Materials for Regenerative Medicine Using Supercritical Fluid Technology. *Bioconjug. Chem.*, 26, 1159–1171, 2015, https://doi.org/10.1021/bc5005922.
124. Zong, J., He, Q., Liu, Y., Qiu, M., Wu, J., Hu, B., Advances in the development of biodegradable coronary stents: A translational perspective. *Mater. Today Bio*, 16, 2022, https://doi.org/10.1016/j.mtbio.2022.100368.

125. Joseph, B., Sagarika, V.K., Sabu, C., Kalarikkal, N., Thomas, S., Cellulose nanocomposites: Fabrication and biomedical applications. *J. Bioresour. Bioprod.*, 5, 223–237, 2020, https://doi.org/10.1016/j.jobab.2020.10.001.
126. Abdul Khalil, H.P.S., Jummaat, F., Yahya, E.B., Olaiya, N.G., Adnan, A.S., Abdat, M., Nasir, N.A.M., Halim, A.S., Seeta Uthaya Kumar, U., Bairwan, R., Suriani, A.B., A review on micro- to nanocellulose biopolymer scaffold forming for tissue engineering applications. *Polymers (Basel)*, 12, 2020, https://doi.org/10.3390/POLYM12092043.
127. Wazed Ali, S., Chowdhury, A., Nath, J., Dolui, S.K., Gadkari, R.R., Cellulose-based bionanocomposites in tissue engineering and regenerative medicine, in: *Bionanocomposites Tissue Eng. Regen. Med.*, pp. 451–463, 2021, https://doi.org/10.1016/B978-0-12-821280-6.00019-2.
128. Iravani, S. and Varma, R.S., Cellulose-Based Composites as Scaffolds for Tissue Engineering: Recent Advances. *Molecules*, 27, 2022, https://doi.org/10.3390/molecules27248830.
129. Aziz Shaikh, A., Datta, P., Dastidar, P., Majumder, A., Dyuti Das, M., Manna, P., Roy, S., Biopolymer-based nanocomposites for application in biomedicine: A review. *J. Polym. Eng.*, 44, 83–116, 2024, https://doi.org/10.1515/polyeng-2023-0166.
130. Salehi Abar, E., Vandghanooni, S., Torab, A., Jaymand, M., Eskandani, M., A comprehensive review on nanocomposite biomaterials based on gelatin for bone tissue engineering. *Int. J. Biol. Macromol.*, 254, 2024, https://doi.org/10.1016/j.ijbiomac.2023.127556.
131. Armentano, I., Gigli, M., Morena, F., Argentati, C., Torre, L., Martino, S., Recent advances in nanocomposites based on aliphatic polyesters: Design, synthesis, and applications in regenerative medicine. *Appl. Sci.*, 8, 2018, https://doi.org/10.3390/app8091452.
132. Fernando, I.P.S., Lee, W.W., Han, E.J., Ahn, G., Alginate-based nanomaterials: Fabrication techniques, properties, and applications. *Chem. Eng. J.*, 391, 2020, https://doi.org/10.1016/j.cej.2019.123823.
133. Qamar, S., Karim, S., Aslam, S., Jahangeer, M., Nelofer, R., Nadeem, A.A., Qamar, S.A., Jesionowski, T., Bilal, M., Alginate-Based Bio-Nanohybrids with Unique Properties for Biomedical Applications. *Starch/Staerke*, 76, 2024, https://doi.org/10.1002/star.202200100.
134. Tomić, S.L., Babić Radić, M.M., Vuković, J.S., Filipović, V.V., Nikodinovic-Runic, J., Vukomanović, M., Alginate-Based Hydrogels and Scaffolds for Biomedical Applications. *Mar. Drugs*, 21, 2023, https://doi.org/10.3390/md21030177.
135. Bashir, M.H., Korany, N.S., Farag, D.B.E., Abbass, M.M.S., Ezzat, B.A., Hegazy, R.H., Dörfer, C.E., Fawzy El-Sayed, K.M., Polymeric Nanocomposite Hydrogel Scaffolds in Craniofacial Bone Regeneration: A Comprehensive Review. *Biomolecules*, 13, 2023, https://doi.org/10.3390/biom13020205.
136. Rodríguez-Vázquez, M., Vega-Ruiz, B., Ramos-Zúñiga, R., Saldaña-Koppel, D.A., Quiñones-Olvera, L.F., Chitosan and Its Potential Use as a Scaffold for

Tissue Engineering in Regenerative Medicine. *Biomed Res. Int.*, 2015, 2015, https://doi.org/10.1155/2015/821279.

137. Jayakumar, R., S.V., Furuike, T., Tamur, H., Perspectives of Chitin and Chitosan Nanofibrous Scaffolds in Tissue Engineering. *Tissue Eng.*, 2010, https://doi.org/10.5772/8593.
138. Prabaharan, M. and Sivashankari, P.R., Prospects of bioactive chitosan-based scaffolds in tissue engineering and regenerative medicine, in: *Chitin Chitosan Regen. Med.*, pp. 41–59, 2015, https://doi.org/10.1007/978-81-322-2511-9_2.
139. Gupta, P., Sharma, S., Jabin, S., Jadoun, S., Chitosan nanocomposite for tissue engineering and regenerative medicine: A review. *Int. J. Biol. Macromol.*, 254, 2024, https://doi.org/10.1016/j.ijbiomac.2023.127660.
140. Chakraborty, G., Padmashree, R., Prasad, A., Recent advancement of surface modification techniques of 2-D nanomaterials. *Mater. Sci. Eng., B*, 297, 2023, https://doi.org/10.1016/j.mseb.2023.116817.
141. Liu, H., Jian, R., Chen, H., Tian, X., Sun, C., Zhu, J., Yang, Z., Sun, J., Wang, C., Application of biodegradable and biocompatible nanocomposites in electronics: Current status and future directions. *Nanomaterials*, 9, 2019, https://doi.org/10.3390/nano9070950.
142. Carrow, J.K. and Gaharwar, A.K., Bioinspired polymeric nanocomposites for regenerative medicine. *Macromol. Chem. Phys.*, 216, 248–264, 2015, https://doi.org/10.1002/macp.201400427.
143. Zhao, R., Meng, X., Pan, Z., Li, Y., Qian, H., Zhu, X., Yang, X., Zhang, X., Advancements in nanohydroxyapatite: synthesis, biomedical applications and composite developments. *Regen. Biomater.*, 12, 2025, https://doi.org/10.1093/rb/rbae129.
144. Han, Y., Wei, Q., Chang, P., Hu, K., Okoro, O.V., Shavandi, A., Nie, L., Three-dimensional printing of hydroxyapatite composites for biomedical application. *Crystals*, 11, 2021, https://doi.org/10.3390/cryst11040353.
145. Haider, A., Haider, S., Han, S.S., Kang, I.K., Recent advances in the synthesis, functionalization and biomedical applications of hydroxyapatite: a review. *RSC Adv.*, 7, 7442–7458, 2017, https://doi.org/10.1039/c6ra26124h.
146. Pandayil, J.T., Boetti, N.G., Janner, D., Advancements in Biomedical Applications of Calcium Phosphate Glass and Glass-Based Devices—A Review. *J. Funct. Biomater.*, 15, 2024, https://doi.org/10.3390/jfb15030079.
147. Papon, E.A. and Haque, A., Review on process model, structure-property relationship of composites and future needs in fused filament fabrication. *J. Reinf. Plast. Compos.*, 39, 758–789, 2020, https://doi.org/10.1177/0731684420929757.
148. Melo, P., Naseem, R., Corvaglia, I., Montalbano, G., Pontremoli, C., Azevedo, A., Quadros, P., Gentile, P., Ferreira, A.M., Dalgarno, K., Vitale-Brovarone, C., Fiorilli, S., Processing of Sr2+ Containing Poly L-Lactic Acid-Based Hybrid Composites for Additive Manufacturing of Bone Scaffolds. *Front. Mater.*, 7, 2020, https://doi.org/10.3389/fmats.2020.601645.

149. Chakraborty, G. and Paul, A.K., 3D Printing in Biomedical Applications, in: *Adv. Mater. Manuf. Tech. Biomed. Appl.*, pp. 275–301, Wiley, United States, 2023, https://doi.org/10.1002/9781394166985.ch11.
150. Arif, Z.U., Khalid, M.Y., Noroozi, R., Hossain, M., Shi, H.T.H., Tariq, A., Ramakrishna, S., Umer, R., Additive manufacturing of sustainable biomaterials for biomedical applications. *Asian J. Pharm. Sci.*, 18, 2023, https://doi.org/10.1016/j.ajps.2023.100812.
151. Müller, F.J. and Fenton, O.S., Additive Manufacturing Approaches toward the Fabrication of Biomaterials. *Adv. Mater. Interfaces*, 9, 2022, https://doi.org/10.1002/admi.202100670.
152. Ebrahimi, M., Standardization and regulation of biomaterials, in: *Handb. Biomater. Biocompat.*, pp. 251–265, 2020, https://doi.org/10.1016/B978-0-08-102967-1.00013-X.
153. Assad, M. and Jackson, N., Biocompatibility Evaluation of Orthopedic Biomaterials and Medical Devices: A Review of Safety and Efficacy Models, in: *Encycl. Biomed. Eng.*, vol. 1–3, pp. 281–309, 2019, https://doi.org/10.1016/B978-0-12-801238-3.11104-3.
154. Pappalardo, D., Mathisen, T., Finne-Wistrand, A., Biocompatibility of Resorbable Polymers: A Historical Perspective and Framework for the Future. *Biomacromolecules*, 20, 1465–1477, 2019, https://doi.org/10.1021/acs.biomac.9b00159.
155. Jorgensen, M.R., Räägel, H., Rollins, T.S., Advances in Biocompatibility: A Prerequisite for Biomedical Application of Biopolymers, in: *Biopolym. Biomed. Biotechnol. Appl.*, pp. 1–17, 2021, https://doi.org/10.1002/9783527818310.ch1.
156. Shiroud Heidari, B., Dodda, J.M., El-Khordagui, L.K., Focarete, M.L., Maroti, P., Toth, L., Pacilio, S., El-Habashy, S.E., Boateng, J., Catanzano, O., Sahai, N., Mou, L., Zheng, M., Emerging materials and technologies for advancing bioresorbable surgical meshes. *Acta Biomater.*, 184, 1–21, 2024, https://doi.org/10.1016/j.actbio.2024.06.012.
157. Ormiston, J.A., Webber, B., Ubod, B., Darremont, O., Webster, M.W.I., An independent bench comparison of two bioresorbable drug-eluting coronary scaffolds (Absorb and DESolve) with a durable metallic drug-eluting stent (ML8/Xpedition). *EuroIntervention*, 11, 60–67, 2015, https://doi.org/10.4244/EIJY15M02_03.

3

Optimization, Modeling, and Design Conditions for 3D-Printed Implants and Fixation Devices

Ranjit Barua[1], Arbind Prasad[2], Bidyanand Mahto[3] and Sudipto Datta[4]*

[1]Mechanical Engineering Department, Omdayal, Group of Institutions, Howrah, Harinarayan Chak, West Bengal, India
[2]Mechanical Engineering Department, Katihar Engineering College (Under Department of Science, Technology and Technical Education, Government of Bihar), Katihar, Bihar, India
[3]Government Engineering College, Vaishali (Under Department of Science, Technology and Technical Education, Government of Bihar), Vaishali, Bihar, India
[4]Department of Materials Engineering, Indian Institute of Science, Bangalore, Karnataka, India

Abstract
The innovation of 3D printing has greatly impacted the healthcare sector, especially in making implants and fixation gadgets. This paper aims to establish optimal parameters, models, and designs for further improvements in 3D-printed medical devices. The optimization process considers the material's choice, the produced part's structure, and a high level of control in the printing process to achieve mechanical characteristics and biocompatibility. To design the implant methodologies like finite element analysis and computational simulations are used to predict the behavior of the implant under the physiological conditions and bring solid modifications. The work also highlights the need for design conditions due to different anatomy of the patients; another advantage of 3D printing is that it can produce architecture and frameworks. By introducing these optimization strategies, modeling approaches, and design considerations, this work would like to improve the long-term, reliable, and patient-specific implants and fixation devices and enhance patients' quality of life in orthopedic and reconstructive operations.

**Corresponding author*: sudiptodatta1990@gmail.com

Arbind Prasad (ed.) Resorbable Composites for Bioimplants and Fixation Devices, (97–118) © 2026 Scrivener Publishing LLC

Keywords: 3D printing, implants, fixation devices, optimization, modeling, finite element analysis (FEA)

3.1 Introduction

The biomedical engineering discipline has experienced significant changes in the last decade primarily because of the entry of 3D printing technology. This advancement has indeed brought a change in the design and development of medical devices such as implants and fixation devices which play a vital role, especially in orthopedic, dental, and reconstructive surgeries. Additive manufacturing technology gives flexibility in shape designs, and complex structures that are very hard to manufacture through conventional processes such as molds and fusing and processing of patient-specific implants. Constant market demand for devices that can provide individualized medical solutions has heightened focus and motivation to enhance the technological design and manufacture of the device to meet its effectiveness, longevity, and bio-compatibility in the human body. 3D Printing is a rapid prototyping technique invented in the mid-1980s but has been widely used in the medical industry recently. The level of detail, which is achievable through 3D printing, has made it a valuable instrument for manufacturing implant and fixation devices. Unlike the regular industrial processes where additive manufacturing subtracts material and is rigid in the design process, 3D printing combines material and constructs objects in layers thus enabling the design of complex structures. This is especially useful where geometries suited to a part of a human body are required for implantation and will lead to better fit and functionality of the implant hence better clinical results [1]. Another advancement that has been observed due to the implementation of 3D printing in the manufacture of medical devices is that prototyping which has also extended the process of making quick changes using patient data has also been improved. This rapid prototyping capability is appreciated in medical applications because time is usually of the essence. Furthermore, advances in the selection of biocompatible materials to be used in 3D printing have resulted in increased possibilities since implants and fixation devices could now be customized to the required design as well as safe for implantation in the human body for an extended period [2]. Some important components, methodologies, and considerations are mentioned in Table 3.1 beneath.

Table 3.1 Important components, methodologies, and considerations.

Category	Key aspects	Details	References
Material selection	Biocompatible materials	Metals (Titanium, Cobalt-Chrome), Polymers (PEEK, PLA), Ceramics (Hydroxyapatite)	[3]
	Mechanical properties	Strength, flexibility, durability	[4–7]
	Biodegradability	Suitable for temporary implants	[8]
Structural design	Patient-specific design	Tailored to anatomical fit and function	[9]
	Complex geometries	Enabled by 3D printing, mimics natural structures	[7]
	Porosity	Enhances bone ingrowth, critical for orthopedic implants	[10–15]
Printing parameters	Layer thickness, printing speed, temperature	Affects surface finish, dimensional accuracy, and mechanical properties	[16]
	Multi-material printing	Incorporates different materials within a single device	[17–19]
Modeling techniques	Finite Element Analysis (FEA)	Simulates stresses, strains, deformations under physiological conditions	[20]

(*Continued*)

Table 3.1 Important components, methodologies, and considerations. (*Continued*)

Category	Key aspects	Details	References
	Computational Fluid Dynamics (CFD)	Models fluid interactions, especially in cardiovascular implants	[21]
	Patient-specific modeling	Utilizes MRI and CT data to create digital models tailored to individual anatomy	[22–25]
Design conditions	Anatomical considerations	Customization based on size, shape, and orientation of implant	[26]
	Mechanical properties	Strength, stiffness, fatigue resistance	[27]
	Biocompatibility and bioactivity	Ensures implant safety and integration with surrounding tissue	[28]
Manufacturing considerations	Build orientation, support structures, layer resolution	Critical for accuracy and surface finish	[29]
	Post-processing	May include machining, polishing, heat treatment to meet final specifications	[30]
Challenges and future directions	Material innovations	Development of bioresorbable polymers, composite materials	[31]

3.2 The Need for Optimization in 3D Printing of Implants and Fixation Devices

As it can be evidenced by the benefits highlighted above, 3D printing has several benefits in the manufacturing of medical devices; conversely, there are certain challenges in this process as well [32–35]. The customization of implant and fixation devices by 3D printing requires consideration of some key factors such as type of material, structure of the implant or fixation device and printing factors [36]. These characteristics and properties must be taken into consideration and fine-tuned so as to attain the mechanical characteristics, biocompatibility, and the life expectancy that is demanded for the use in medical applications [37].

3.2.1 Material Selection

It indicates that out of all the factors that define 3D-printed implants and fixation devices, the choice of the material is the most important one. Figure 3.1 shows the process flow of customized 3D printed implants during biomedical application and surgery. Materials also require suitable properties for the 3D printing process besides the mentioned properties including strength, flexibility, and biocompatibility [39]. Typical materials applied while using 3D printing in medical devices are metals including titanium and cobalt-chrome alloys, polymers including PEEK and PLA,

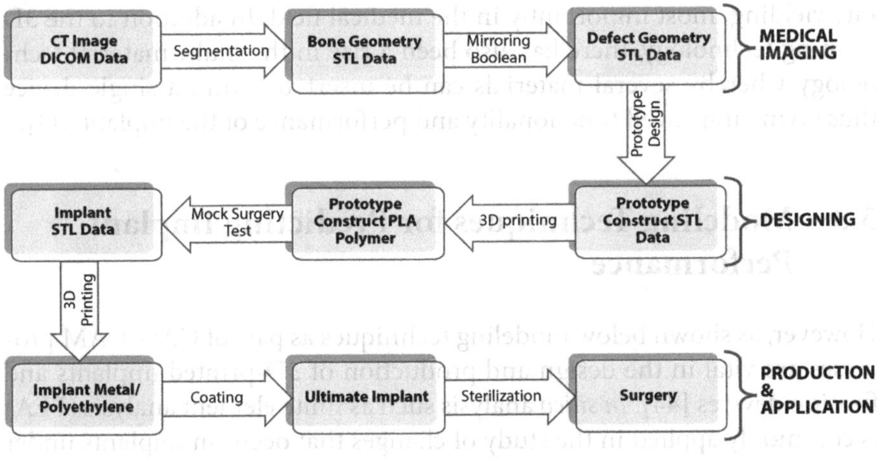

Figure 3.1 Process for creating, testing, manufacturing, and inserting customized 3D printed implants during surgery [38].

and ceramics including hydroxyapatite. These materials are in a way unique in their properties and disadvantages and their choice depends upon the use of the implant, and its mechanical and biological characteristics [40].

3.2.2 Structural Design

This is especially important to ensure that the design of the implant or fixation device is usable, suitable, and appropriate for its intended application. The design requirements include anatomical requirements, mechanical interfacial requirements, and progression requirements in the organization of the tissue [41]. Stereolithography enabled designs that have high geometrical realism about the human body, which is crucial when the implant is designed to be embedded within the bone and share the mechanical load with it with orthopedic implantation. Some of the aspects of the design that have to be taken into account include porosity, which can facilitate bone ingrowth, and therefore increasing the overall function of the implant [42].

3.2.3 Printing Parameters

In addition, the extent of differentiation of the implant tissue also depends on the printing parameters including layer thickness, the speed of printing, and temperature. Such parameters are critical in determining the quality of a product such as the finish, accuracy of dimensions, and mechanical characteristics of the product. The sensitivity and specificity of these parameters deserve to be further optimized for achieving robust and accurate yielding, most importantly in the medical field. In addition to the 3D printing technology, there has also been a rise in the multi-material technology whereby several materials can be inserted within a single device thus giving improved functionality and performance of the implant [43].

3.3 Modeling Techniques for Predicting Implant Performance

However, as shown below modeling techniques as part of CAD/CAM processes are vital in the design and production of 3D-printed implants and fixation devices [44]. *In silico* analysis such as finite element analysis (FEA) is commonly applied in the study of changes that occur in implants under different physiological conditions [45]. These models help the engineers estimate the mechanical behavior of the implant, the probable areas of failure, and the modification of the design of the implant before the actual manufacturing of the implant.

3.3.1 Finite Element Analysis (FEA)

It is an efficient tool by which a complicated structure is broken into subsystems which can then be analyzed independently of the loads acting on a structure and the consequent stresses, strains, and deformation. In the light of 3D-printed implants, FEA may be employed to determine how the implant will be responsive when in contact with forces of a nature of compression, tension, or shear; forces that apply to orthopedic applications. This analysis aids in the improvement of the design so that the implant will have the capability of bearing the mechanical loads placed on it without failure [46].

3.3.2 Computational Fluid Dynamics (CFD)

In cases of implantable devices that interact with fluids for example in cardiovascular stents or drug delivery implants fluid dynamics is employed where implants are modeled on blood or any other fluid in the host body. Through CFD, it becomes possible to have a clue of how the implant would perform within the biological environment and can be used to alter the design to enhance performance or to minimize the chances of developing complications such as thrombosis or restenosis [47–50].

3.3.3 Patient-Specific Modeling

Probably one of the numerous benefits rendered by the 3D printing technology pertains to the medical field which is the delivery of patient-specific implants by patient-specific data regarding the anatomy of the patient. MRI and CT scans are two of the most diagnostic imaging techniques in the healthcare market; they offer highly detailed anatomical information about the patient, which facilitates the creation of the implant model [51]. This design can then be further optimized utilizing various computational approaches till a final design solution that fits perfectly and performs optimally is attained. Individual patient modeling helps to determine the implant effectiveness of each patient as well as minimize the risks of adverse effects and better meet patients' expectation.

3.3.4 Design Conditions for 3D-Printed Implants

While designing 3D-printed implants and fixation devices several factors are taken into consideration such as the application of the implant or device, mechanical and biological properties, and the potential of the 3D printing technology. Design requirements, therefore, has to be well outlined to avoid compromises that may lead to the development of contrived products.

3.3.5 Anatomical Considerations

This means that the design of the implant has to be informed by the structure of the patient's anatomy. This involves factors that would shape the implant such as the size, the shape, as well as the orientation of the implant area. For instance, in orthopedic implants, the design must consider the geometry of the bone, the zone to be inserted, and the kind of load bearing on it. Another advantage of 3D printing is a perfect match for patient-specific data during the design of the implant or prosthesis needed; thus, the implants to be printed are unique to the patient [52].

3.3.6 Mechanical Properties

The implant should have the right mechanical characteristics required in it to serve the intended purpose. Those include strength, stiffness, and fatigue strength, which is mandatory for various load-bearing implants including joint replacements or spinal implants. Other features that the design must factor in include stress shielding that arises out of the fact that the implant material can be too rigid and hence, it takes up all the load, which in turn results in bone loss and implant failure. If the design and the material used for the implant are carefully chosen then it will be possible for it to provide enough strength to support the weight while at the same time not compromising on flexibility of the tissue [53].

3.3.7 Biocompatibility and Bioactivity

Another requirement is the mechanical properties of the implant together with biocompatibility, which would mean that the implant does not have a detrimental effect when implanted in the body. Some of the factors that affect biocompatibility are the material used, the surface nature and texture as well as the porosity of the implant. Sometimes bioactivity is applied

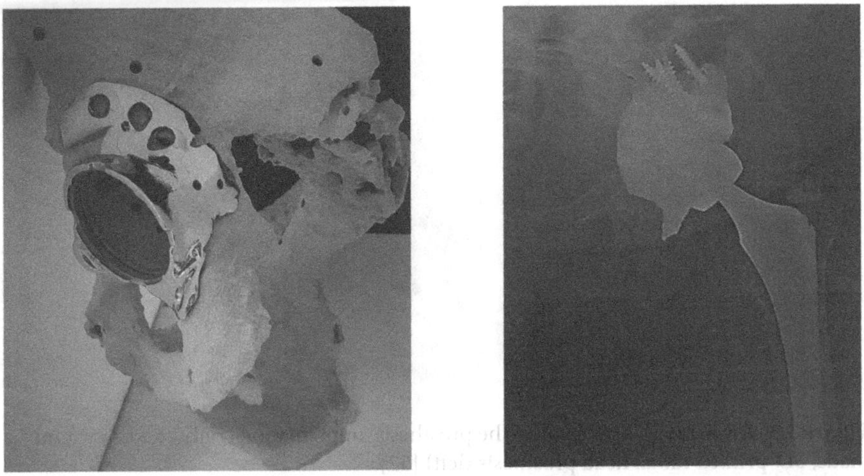

Figure 3.2 An X-ray of the prosthesis implantation (right) and a 3D-printed model of the patient's pelvis (left) are used to illustrate a custom prosthesis intended to treat an acetabular deformity in a patient [54].

to stimulate bonding with the surrounding tissues, for example, to create a coating that may prompt bone-to-implant bonding or a surface alteration that would minimize the potential of bacterial growth on the implant (Figure 3.2). The design must also consider if the implant is to function in this implantable application and the behavior of the implant over time including degradation and wear which may render the implant safe and effective in the long run [55].

3.3.8 Manufacturing Considerations

The design must be feasible and possible to be created through the 3D printing technology used in creation of the implant. This includes the direction of the build, the support structures required, and the layer thickness that the machine can achieve which defines the quality of the surface that is produced out of the fused layers. Subsequent processes, which may include machining, polishing, or heat treatment may be needed to arrive at the desired characteristics. The design also has reduced to the limitations which are inherent in the printing material like the melting point of the material, thermal expansion, and flow characteristics in order to meet some standard specifications [56].

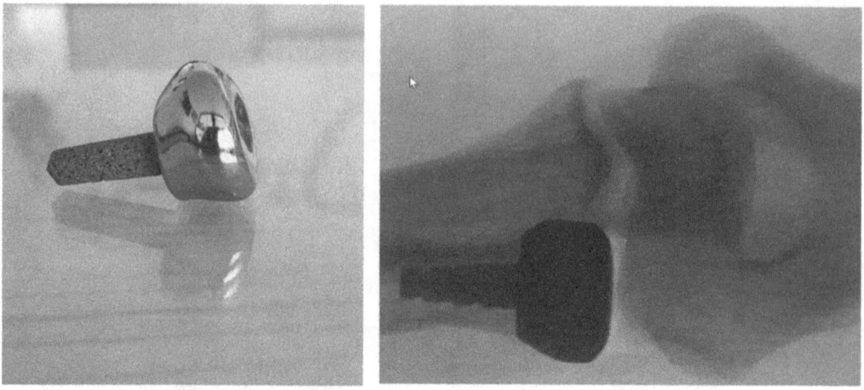

Figure 3.3 An X-ray was taken after the prosthesis' implantation (right) and a custom-made, 3D-printed radial head prosthesis (left) [57].

3.4 3D Printing in Orthopedic Surgery

3.4.1 Customization and Patient-Specific Implants

The are some of the benefits of 3D printing in orthopedic surgery and 3D printing in general, with one of them being the capacity to produce implants specific to patients. Traditional implants are usually available in a small selection of sizes and shapes, which are quite often not the same as the patient's anatomy. This can cause some issues such as implant instability, implant malposition, and revision surgeries which are expensive, and Figure 3.3 shows the 3D printed customized prosthesis implantation. However, 3D printing enables customization of implants according to the patient's scan images like CT or MRI scans hence improving the stability and durability of the implant [58].

In specific cases, individualized implants involving three-dimensional (3D) printing have been helpful especially in conditions such as major bone deformity, large deficiencies and in reoperation cases. For example, when patients have a big gap in the bone due to injury or treatment of a tumor, basic implants can be ineffective. Through 3D printing, implants of intricate shapes can be produced to fit these defects to the letter in a bid to rebuild the bone structure [59].

3.4.2 Porous Structures for Enhanced Osseointegration

The other benefit of 3D-printed implants in orthopedic surgery is the design of the implant with interconnected porous structures that allow bone ingrowth into the implant and hence anchoring. Bioactive implants are characterized by interconnected porosity that resembles the structure of bones, to which bone cells can attach and begin the process of reconstructing bone tissue [51]. This helps in reinforcing the implant and cuts out numbers of implant that are likely to become loosen in future.

Large connected porosities are especially of significance in implants that are called to bear loads such as the hip and knee joint replacements. Control over the pore size, shape and distribution is possible with the use of 3D printing, thus enhancing the mechanical properties of the implant and at the same time improving osseointegration [52].

3.4.3 Biodegradable Implants for Bone Regeneration

On a similar note, the application of 3D printing is also tracing its use in the production of temporary implants for bone tissue regeneration. These implants ought to dissolve naturally in the human body as new bone forms and avoid the necessity of additional operation for removal of this implant. It is especially important in pediatric cases since the bone is in the process of growing and thus needs modifications all the time.

Polylactic acid (PLA), polycaprolactone (PCL), and polyglycolic acid (PGA) are some of the biodegradable polymers that have been used in the development of 3D-printed implants. These materials biodeteriorate to non-cytotoxic constituents that are assimilated by the body, afford support for bone formation, and at the same time provide requisite stiffness and strength during the bone remodeling phases [53].

3.4.4 Applications in Complex Orthopedic Cases

Thus, 3D printing has shown to be quite useful in difficult trauma surgical situations where conventional orthopedic implants may be inadequate. For example, in cases, such as significant bone deformity or large bone defect due to trauma or tumor excision, the standard implant may not give enough coverage or stability. Thanks to the ability of 3D printing the geometrical structures which can be hardly biomimetically reconstructed, the implants produced can exactly fit into these defects and restore the mechanical continuity of the bone [54].

In addition, they are also able to use 3D-printed implants in revision surgeries in which an existing implant must be removed and replaced because of wear and tear, infection, or other reasons. The possibility of adjusting the implant according to the available bone substance and other anatomical features is highly important in such instances as it helps to minimize the risk of further injury or other complications and improve the general outcome of surgery.

3.5 3D Printing in Dental Surgery

3.5.1 Customized Dental Implants

Like in orthopedic surgery, the need to print implants according to the desired design is a key advantage of 3D printing in dental surgery. A tooth implant needs to sit well within a patient's jaw to hold firmly and last long; hence the need for a surgeon to fit it accurately. Customizable implants can be fabricated by using 3D printing based on detailed imaging of the patient thereby avoiding implant failure. Apart from the conventional implant platform, the 3D printings are also used for custom design fabrication of abutments and crowns in implant prostheses providing better fitting and function. This level of customization makes the treatment more comfortable hence enhancing the patient's satisfaction with the given treatment [55].

3.5.2 Surgical Guides and Templates

Of all the features that have come with the technological adoption of 3D printers in dental surgeries, the fabrication of surgical guides and templates stands out as one of the most notable developments. They are individual made guides that describe how each dental implant must be placed. By using a Surgical Guide, it can be ensured that the position of the implant is perfect and chances of complications like nerve injury or improper position of the implant in terms of angle [56].

This paper presents evidence that the employment of 3D-printed surgical guides results in an increased degree of implant accuracy, decreased time needed for surgery, and the improvement of the patient's situation. Moreover, surgical guides can be integrated with other digital technologies like CAD CAM to optimize the surgical process and enhance the efficacy of the process.

3.5.3 Bone Grafts and Regeneration

It is also being applied in dental surgery to make bone grafts and bone regeneration frameworks utilizing three-dimensional print. When there is not enough bone to center the dentist on the implant, the dentist may need to perform a bone graft to create the required support. 3D printing helps in fabrication of personalized bone grafts, ensuring better integration of the graft with the bone, due to better matching of the shape and size to that of the recipient [57].

Besides, the application of bone grafts, which are the typography of tissues, 3D printing technology is used for producing scaffolds that contribute to bone tissue formation. These scaffolds are intended to replicate the architecture of bone and give support for *de novo* formation of bone tissue. Eventually, the scaffold gets bio-eroded and the newly formed healthy bone which can support a dental implant remains.

3.5.4 Applications in Complex Dental Cases

In the same manner as, it was seen that 3D printing could also be useful mainly in cases of orthopedic surgery dentistry being even more helpful. For instance, in situations where the patients have lost bone tissue due to diseases such as periodontal disease or as a result of trauma, regular dental implants may not be useful and effective. 3D technology makes it possible to design special implants and bone grafts to replant the damaged part of the jawbone, to provide the necessary base for implantation of a dental implant [58].

In addition, there is evidence on how 3D printing has been employed as a tool in the correction of congenital dental defects including cleft palate in which conventional implants would not be effective. It is especially important in these cases, because the adjustment of the implant to the patient's individual anatomy is one of the key factors, which contributes to the enhancement of treatment outcomes.

3.6 Challenges and Future Directions

3.6.1 Material Limitations and Innovations

However, several concerns must be brought up while discussing the applications of 3D printing for orthopedic and dental implants. The first of these is the constrain in the choice of materials that can be used to 3D printing. Metals like titanium and cobalt-chrome are used in permanent implants but these are quite stiff and can cause stress shielding in which the implant increases the load bearing stress and prevents bones from taking the stress which results in bone resorption [60]. To address this issue, research is being conducted into the development of new materials that combine the strength of metals with the flexibility of polymers. For example, composite materials that incorporate bioactive ceramics, such as hydroxyapatite, are being explored for their potential to improve osseointegration while reducing stress shielding [59].

3.6.2 Regulatory and Standardization Challenges

One of the most influential deficits in the enhanced usage of 3D-printed implants is the absence of lawful norms and regulations. The flexibility and individuality of 3D-printed implants are still challenging to address under the existing regulations since most current regulations are primarily created for standardized medical device production. There are already corporations like the FDA that aim at creating guidelines for the certification of 3D printing but there is still progress that needs to be made. The future concerns will include setting up universal measures for testing the method's efficacy, setting up norms for the quality control measures, and the long-term outcome of the efficacy of the implant [60].

3.6.3 Cost and Accessibility

The cost of the 3D printing technology as well as materials used in the fabrication of Implants is usually high, especially in developing regions. Although there has been some decline in the prices of 3D printers, the high-end machines for producing lavishly intricate medical implants that are highly functional are still beyond the average prices, and materials such as titanium alloys and bioactive ceramics are also very expensive [61].

Some attempts are being made to find out the methods of 3D printing that are cheaper and some methods to construct materials but the 3D printer is not easily accessible. Another way of enhancing the efficiency of 3D-printed medical applications will be to make this technology more accessible to people especially in the regions of the world where such healthcare facilities are scarce [62].

3.6.4 Future Directions and Emerging Technologies

To the future, some of the following technologies may add more value towards the application of 3D printing in orthopedic and dental surgical procedures. Such a technology includes 3D biological printing that uses bioinks to grow the living cells and tissues, organs. It is still in its initial phase, although it has the incredible possibility to redefine regenerative medicine when it comes to the fabrication of fully functional, personalized tissues and organs for transplantation. Another possible and rather prospective avenue is that of intelligent implants able to assess the progress of the tissue healing and apply the right treatment. For example, an implant made with 3D printing technology with sensors in it could help in collecting information about the mechanical forces and stress the implant has to go through in the body and hence could help in remote post-surgical monitoring [63].

Furthermore, the progress in materials science, like bioactive polymers, biodegradable polymers, and composite materials will further enhance the capability of 3D printed implants. These materials have the ability to improve osseointegration, minimize stress shielding effects, and deliver a lightweight and short-term mechanical backing while advancing tissue formation.

3.7 Conclusion

The current work of optimizing, modeling, and designing 3D-printed implants and fixation devices is a great revolution in biomedical engineering. As has been discussed, the use of 3D printing opens a wide range of opportunities to design and produce medical devices that would be optimal for specific patients. But the balance of these factors will help to obtain the best results: material selection; structural design; proper printing parameters; and, modeling. In the future as technology advances more research and development work will be required to eliminate the present-day difficulties with the technology and optimize the use of 3D printing in medical applications. The continuous innovation of new composites, improvements in modeling techniques, and emerging standards are indispensable factors in the development of this expanding market. 3D printing has become one of the possibilities in orthopedic and dental surgery that has multiple advantages, such as the usage of customized implants that are implanted more efficiently; the recent advancement of osseointegration of implants, and the ability to create biodegradable implants due to bone regeneration. The usefulness of the technology has been especially seen where other more common techniques fail to offer the necessary support and functionality. It is, however, important to note that the challenges persist, especially about the materials' available, the regulatory environment, the cost of implementation, and the possibility of the provision of the technology to most customers. However, newer ongoing research and technologies like Bioprinting and smart implants have emerged that the future holds a high potential for even more effective and advanced usage of 3D-printed implants. Future development of the technology will in fact mean that the use of 3D printing for orthopedic and dental procedures will become even more widespread and will thus lead to enhancement of efficacy and application of individualized medicine in practice.

Acknowledgment

The authors would like to thank Mrs. Nibedita Bardhan for language proofreading. Sudipto Datta acknowledges DST-SERB for supporting him as a National Post-Doctoral Fellow.

AI disclosure statement: The author confirms that no artificial intelligence (AI)-generated content or AI-assisted writing tools were used in the preparation of this chapter (including summaries and the abstract). All text, analysis, and conclusions are the result of the author's original work.

References

1. Ventola, C.L., Medical applications for 3D printing: Current and projected uses. *Pharm. Ther.*, 39, 10, 704–711, 2014, https://www.ncbi.nlm.nih.gov/pmc/articles/PMC4189697/.
2. Javaid, M. and Haleem, A., Current status and applications of additive manufacturing in dentistry: A literature-based review. *J. Oral Biol. Craniofac. Res.*, 8, 1, 34–40, 2018, https://doi.org/10.1016/j.jobcr.2017.08.008.
3. Chen, Q. and Thouas, G.A., Metallic implant biomaterials. *Mater. Sci. Eng.: R: Rep.*, 87, 1–57, 2015, https://doi.org/10.1016/j.mser.2014.10.00.
4. Chowdhury, S., Jain, N., Waidi, Y.O., Barua, R., Das, S., Prasad, A., Datta, S., Biological Smart Biomaterials: Materials for Biomedical Applications, in: *Applications of Biotribology in Biomedical Systems*, pp. 313–325, Springer Nature Switzerland, Cham, 2024.
5. Zhang, Y., He, Z., Zhang, P., Ma, Q., Biocompatible and biodegradable polymeric scaffolds for tissue engineering applications. *Prog. Polym. Sci.*, 87, 35–56, 2018, https://doi.org/10.1016/j.progpolymsci.2018.07.002.
6. Murphy, S.V. and Atala, A., 3D bioprinting of tissues and organs. *Nat. Biotechnol.*, 32, 8, 773–785, 2014, https://doi.org/10.1038/nbt.2958.
7. Hollister, S.J., Porous scaffold design for tissue engineering. *Nat. Mater.*, 4, 7, 518–524, 2005, https://doi.org/10.1038/nmat1421.

8. King, M.F.L., Prasad, A., Mahto, B., Introduction to Bioabsorbable Polymers: Processing and Characterization, in: *Bioabsorbable Polymers for Drug Carriers and Fixation Devices*, pp. 1–29, Apple Academic Press, 2025.
9. De, S., Basak, M., Bhattacharjee, D., Prasad, A., Mahto, B., New Frontiers of Hydrogels for Wound Healing Applications, in: *Bioabsorbable Polymers for Drug Carriers and Fixation Devices*, pp. 219–241, Apple Academic Press, 2025.
10. Andal, V., Lakshmipathy, R., Jayanthi, G., Prasad, A., Physical and Biological Properties of Biocomposite Materials as Fixations or Sealers in Dentistry Applications, in: *Bioabsorbable Polymers for Drug Carriers and Fixation Devices*, pp. 203–217, Apple Academic Press, 2025.
11. Paul, A.K., Chakraborty, G., Borkotoky, S.S., Prasad, A., Degradation Studies of Bioabsorbable Polymeric Composites, in: *Bioabsorbable Polymers for Drug Carriers and Fixation Devices*, pp. 105–128, Apple Academic Press, 2025.
12. Prasad, A., Barua, R., Banerjee, A., Mahto, B., Datta, S., Processing & characterizing biomass-based nanocomposites for climate change of computational modeling & impacts, in: *Sustainable Packaging Strengthened by Biomass*, pp. 27–46, Woodhead Publishing, 2025.
13. Waidi, Y.O., Jain, N., Barua, R., Vannaladsaysy, V., Das, S., Prasad, A., Datta, S., Smart and Self-healing Hydrogels for Biomedical Applications, in: *Nanomanufacturing Techniques in Sustainable Healthcare Applications*, pp. 110–130, CRC Press, 2024.
14. Jaswal, N., Bahadur, V., Raza, A., Prasad, A., Kumar, P., Future Trends of Nanocomposites in 3D Bioimplants and Fluorescent Bioimaging, in: *Nanomanufacturing Techniques in Sustainable Healthcare Applications*, pp. 296–316, CRC Press, 2024.
15. Park, S., Kim, J. E., Lee, J., Kim, W., Choi, W., Lee, M. C., ... Chung, J. H., Bioimplant-on-a-Chip for Facile Investigation of Periodontal Ligament Formation on Biogenic Hydroxyapatite/Ti6Al4 V Implants. *ACS Appl. Mater. Interfaces,* 17, 21, 30673–30685, 2025.
16. Bhasney, S.M., Das, S.R., Prasad, A., Mahto, B., Thangavel, S., Recycling of biopolymers, in: *Biodegradable Waste Processing for Sustainable Developments*, pp. 272–298, CRC Press, 2024.
17. Prasad, A. and Paul, A.K. (Eds.), *Biodegradable waste processing for sustainable developments*, CRC Press, 2024.
18. Gibson, I., Rosen, D.W., Stucker, B., *Additive manufacturing technologies: 3D printing, rapid prototyping, and direct digital manufacturing*, Springer, 2021, https://doi.org/10.1007/978-1-4419-1120-9.
19. Waidi, Y.O., Jain, N., Chowdhury, S., Barua, R., Das, S., Prasad, A., Datta, S., Challenges and Perspective of Manufacturing Techniques in Biomedical Applications, in: *Applications of Biotribology in Biomedical Systems*, pp. 433–445, Springer Nature Switzerland, Cham, 2024.
20. Boccaccio, A., Uva, A.E., Fiorentino, M., Lamberti, L., Monno, G., A computational model for the prediction of bone adaptation. *Biomech.*

Model. Mechanobiol., 15, 6, 1339–1350, 2016, https://doi.org/10.1007/s10237-015-0760-5.

21. Xia, Z., Yu, H., Xie, Y., Xie, J., Wang, Y., Computational modeling of drug-eluting stent design optimization for controlled drug release. *Ann. Biomed. Eng.*, 41, 9, 1895–1906, 2013, https://doi.org/10.1007/s10439-013-0801-0.
22. Davies, J.E., Understanding peri-implant endosseous healing. *J. Dent. Educ.*, 67, 8, 932–949, 2003, https://doi.org/10.1002/j.0022-0337.2003.67.8.tb03696.x.
23. Frost, H.M., A 2003 update of bone physiology and Wolff's law for clinicians. *Angle Orthod.*, 74, 1, 3–15, 2004, https://doi.org/10.1043/0003-3219(2004)074<0003>2.0.CO;2.
24. Ratner, B.D., Hoffman, A.S., Schoen, F.J., Lemons, J.E. (Eds.), *Biomaterials science: An introduction to materials in medicine*, Academic Press, 2004.
25. Lantada, A.D. and Morgado, P.L., Rapid prototyping for biomedical engineering: Current capabilities and challenges. *Annu. Rev. Biomed. Eng.*, 14, 1, 73–96, 2012, https://doi.org/10.1146/annurev-bioeng-071811-150112.
26. Barua, R., Sarkar, A., Datta, S., Emerging Advancement of 3D Bioprinting Technology in Modern Medical Science and Vascular Tissue Engineering Education, in: *Handbook of Research on Instructional Technologies in Health Education and Allied Disciplines*, pp. 153–175, IGI Global, 2023.
27. Banerjee, D., Barua, R., Datta, S., Pathote, D., The Use of Artificial Neural Networks in Biomedical Engineering, in: *Expert Artificial Neural Network Applications for Science and Engineering*, L. Syam Sundar, D. Balakrishnan, A.C.M. Sousa (Eds.), pp. 33–50, IGI Global Scientific Publishing, 2025, https://doi.org/10.4018/979-8-3693-7250-0.ch002.
28. Safali, S., Berk, T., Makelov, B., Acar, M.A., Gueorguiev, B., Pape, H., The Possibilities of Personalized 3D Printed Implants—A Case Series Study. *Medicina*, 59, 2, 249, 2023, https://doi.org/10.3390/medicina59020249.
29. Tian, S., Wang, M., Wang, X., Wang, L., Yang, D., Nie, J., Ma, G., Smart hydrogel sensors with antifreezing, antifouling properties for wound healing. *ACS Biomater. Sci. Eng.*, 8, 5, 1867–1877, 2022.
30. Jain, N., Waidi, Y.O., Barua, R., Vannaladsaysy, V., Das, S., Prasad, A., Datta, S., 2 Bio-based polymers: processing and applications, in: *Sustainable Bio-Based Composites: Biomedical and Engineering applications*, vol. 20, p. 25, 2024.
31. Prasad, A., Bhasney, S.M., Prasannavenkadesan, V., Sankar, M.R., Katiyar, V., Nano-hydroxyapatite reinforced polylactic acid bioabsorbable cancellous screws for bone fracture fixations. *J. Appl. Polym. Sci.*, 140, 43, e54577, 2023.
32. Sringari, S.S., Chakraborty, G., Prasad, A., Biodegradable and biocompatible polymeric nanocomposites for tissue engineering applications, in: *Nanomanufacturing and Nanomaterials Design*, pp. 205–225, CRC Press, 2022.
33. Prasad, A., Biomaterial-based nanofibers scaffolds in tissue engineering application, in: *Functional Biomaterials: Drug Delivery and Biomedical Applications*, pp. 245–264, Springer Singapore, Singapore, 2022.

34. Datta, S., Barua, R., Prasad, A., Additive Manufacturing for the Development of Artificial Organs, in: *Advanced Materials and Manufacturing Techniques for Biomedical Applications*, pp. 411–427, 2023.
35. Barua, R., Datta, S., Banerjee, D., Computational Study of *In-Vivo* CT-Based FEM Application in Bone Tissue Engineering, in: *Technological Adoption and Trends in Health Sciences Teaching, Learning, and Practice*, pp. 300–316, IGI Global, 2022.
36. Barua, R., Revolutionizing Medicine: Exploring Cutting-Edge Technologies in Computer-Aided Medicinal Analysis, in: *Computer-Assisted Analysis for Digital Medicinal Imagery,* A. Sinha, P. Saxena, S. Singh, H. Singh (Eds.), pp. 365–392, IGI Global Scientific Publishing, 2025, https://doi.org/10.4018/979-8-3693-5226-7.ch014.
37. Barua, R., Biswas, N., Das, D., Emergent Applications of Organ-on-a-Chip (OOAC) Technologies With Artificial Vascular Networks in the 21st Century, in: *Emerging Technologies for Health Literacy and Medical Practice,* M. Garcia and R. de Almeida (Eds.), pp. 198–219, IGI Global Scientific Publishing, 2024, https://doi.org/10.4018/979-8-3693-1214-8.ch010.
38. Tibbitt, M.W., Dahlman, J.E., Langer, R., Emerging frontiers in drug delivery. *J. Am. Chem. Soc.,* 138, 3, 704–717, 2016.
39. Mohan, N., Gupta, M.K., Srivastava, A., Murari, A., 3D printing in orthopedics: From bench to clinical practice. *J. Clin. Orthop. Trauma*, 22, 101635, 2021, https://doi.org/10.1016/j.jcot.2021.101635.
40. Paul, A.K., Borkotoky, S.S., Prasad, A., Nanohydroxyapatite-based composite materials and processing, in: *Advanced Materials and Manufacturing Techniques for Biomedical Applications*, pp. 157–183, 2023.
41. Prasad, A., Datta, S., De, S., Singh, P., Mahto, B., Bioresorbable Composite for Orthopedics and Drug Delivery Applications, in: *Applications of Biotribology in Biomedical Systems*, pp. 327–344, Springer Nature Switzerland, Cham, 2024.
42. Prasannavenkadesan, V., Meiraj, S.N.U., Narayanan, K.B., Prasanth, S.R., Prasad, A., 4 Chitosan in orthopedics: current advancements and future prospects, in: *Sustainable Bio-Based Composites: Biomedical and Engineering applications*, vol. 20, p. 59, 2024.
43. De, S., Ghosh, A., Basak, M., Prasad, A., Chattopadhyay, D., Prospects of functional nano-manufactured scaffolds in tissue engineering applications, in: *Sustainable Bio-Based Composites: Biomedical and Engineering applications*, vol. 20, p. 107, 2024.
44. Jain, N., Prasad, A., Datta, S., Biomechanical Studies of Biocomposites for Bone Fixation Devices, in: *Bioabsorbable Polymers for Drug Carriers and Fixation Devices*, pp. 163–179, Apple Academic Press, 2025.
45. Prasad, A., Datta, S., Kumar, A., Gupta, M., Introduction to Next-Generation Materials for Biomedical Applications, in: *Advanced materials and manufacturing techniques for biomedical applications*, pp. 1–24, 2023.

46. Datta, S. and Barua, R., 3D Printing in Modern Healthcare: An Overview of Materials, Methods, Applications, and Challenges, in: *Emerging Technologies for Health Literacy and Medical Practice,* M. Garcia and R. de Almeida (Eds.), pp. 132–152, IGI Global Scientific Publishing, 2024, https://doi.org/10.4018/979-8-3693-1214-8.ch007.
47. Singh, R., Suhail, A., Datta, S., Design and development of patient-specific orthopaedic implants: From laboratory to clinical application. *Procedia Manuf.*, 35, 836–842, 2019, https://doi.org/10.1016/j.promfg.2019.06.058.
48. Li, J.P., Habibovic, P., van den Doel, M., Wilson, C.E., de Wijn, J.R., van Blitterswijk, C.A., de Groot, K., Bone ingrowth in porous titanium implants produced by 3D fiber deposition. *Biomaterials*, 28, 18, 2810–2820, 2020, https://doi.org/10.1016/j.biomaterials.2020.02.042.
49. Middleton, J.C. and Tipton, A.J., Synthetic biodegradable polymers as orthopedic devices. *Biomaterials*, 21, 23, 2335–2346, 2000, https://doi.org/10.1016/S0142-9612(00)00101-0.
50. Hutmacher, D.W., Scaffolds in tissue engineering bone and cartilage. *Biomaterials*, 21, 24, 2529–2543, 2000, https://doi.org/10.1016/S0142-9612(00)00121-6.
51. Nasonova, M.V., Glushkova, T.V., Borisov, V.V., Velikanova, E.A., Burago, A.Y., Kudryavtseva, Y.A., Biocompatibility and structural features of biodegradable polymer scaffolds. *Bull. Exp. Biol. Med.*, 160 1, 134–140, 2015.
52. Mangano, F.G., Hauschild, U., Veronesi, G., Imburgia, M., Mangano, C., Admakin, O., Trueness and precision of 5 intraoral scanners in the impressions of single and multiple implants: A comparative *in vitro* study. *BMC Oral Health*, 17, 1, 92, 2017, https://doi.org/10.1186/s12903-017-0383-4.
53. Abduo, J. and Lyons, K., Rationale for the use of CAD/CAM technology in implant prosthodontics. *Int. J. Dent.*, 2013, 1–8, 2013, https://doi.org/10.1155/2013/768121.
54. Van Assche, N. and Quirynen, M., Tertiary stability of implants placed using stereolithographic drilling guides. *Clin. Implant Dent. Relat. Res.*, 12, 4, 384–394, 2010, https://doi.org/10.1111/j.1708-8208.2008.00148.x.
55. Joda, T. and Brägger, U., Complete digital workflow for the production of implant-supported single-unit monolithic crowns. *Clin. Oral Implants Res.*, 26, 12, 1474–1476, 2015, https://doi.org/10.1111/clr.12453.
56. Goh, B.T., Teh, L.Y., Tan, D.B., Zhang, C., Hong, G.S., Novel 3D polycaprolactone scaffold for ridge preservation – A randomized controlled trial. *Clin. Oral Implants Res.*, 26, 3, 271–277, 2014, https://doi.org/10.1111/clr.12323.
57. Prasad, A., Shaikh, A., De, S., Kumar, R., Srivastava, M.K., Aspects of bioabsorbable polymeric composites in biomedical applications, in: *Sustainable Smart Manufacturing Processes in Industry 4.0*, pp. 73–87, CRC Press, 2023.
58. Huang, G.T.J., Liu, Y., Bai, C., An update on the stem cells and related cell sources in regenerative endodontics. *Endod. Top.*, 43, 1, 42–57, 2020, https://doi.org/10.1111/etp.12209.

59. Seitz, H., Rieder, W., Irsen, S., Leukers, B., Tille, C., Three-dimensional printing of porous ceramic scaffolds for bone tissue engineering. *J. Biomed. Mater. Res. Part B: Appl. Biomater.*, 74, 2, 782–788, 2005, https://doi.org/10.1002/jbm.b.30291.
60. Heller, M., Bauer, H., Goetze, E., Karl, M., Retrospective analysis of dental implant fracture characteristics and failure probability: A clinical report. *J. Prosthet. Dent.*, 120, 5, 648–654, 2018, https://doi.org/10.1016/j.prosdent.2018.05.032.
61. Tappa, K. and Jammalamadaka, U., Novel biomaterials used in medical 3D printing techniques. *J. Funct. Biomater.*, 9, 1, 17, 2018, https://doi.org/10.3390/jfb9010017.
62. Kalaskar, D.M., *3D printing in medicine*, Woodhead Publishing, 2017, https://doi.org/10.1016/C2015-0-02850-2.
63. Chakraborty, G., Prasad, A., Kumar, A., Processing of biodegradable composites, in: *Biodegradable composites for packaging applications*, pp. 33–48, CRC Press, 2022.

4

Functions and Potential Applications of Biofluids as Natural Products

Preeti Chaurasia[1]* and Nakuleshwar Dut Jasuja[2]

[1]Department of Microbiology, S. S. Jain Subodh P.G. College, Jaipur, (Rajasthan), India
[2]Department of Microbiology, The ICFAI University, Jaipur, (Rajasthan), India

Abstract

Biofluids, or biological fluids, are innate liquids generated by living organisms. These fluids have significant functions in numerous biological cycles and have gotten extensive interest lately in light of their planned purposes in a few areas, like medication, biotechnology, and materials research. The study of biofluids as organic substances has uncovered a diverse array of functions and uses that exploit their distinct features and compositions. Biofluids consist of a wide range of components, including as blood, saliva, urine, perspiration, tears, and other secretions from plants and animals. These fluids often include of intricate combinations of biomolecules, including proteins, lipids, carbohydrates, and metabolites, that play a crucial role in their unique activities inside organisms. Biofluids in living systems provide a multitude of functions that are diverse in nature. They function as conduits for transporting nutrients, waste materials, and signaling molecules; regulating internal balance; providing lubrication and protection; and aiding immunological responses. Furthermore, numerous biological fluids display exceptional characteristics such as the ability to repair themselves, stick to surfaces, and combat microbial activity. These properties have motivated researchers to investigate their potential uses in diverse fields. An increasing attention on the possible uses of biofluids as natural products has resulted from recent breakthroughs in analytical methods as well as a rising interest in solutions that are both sustainable and bio-inspired. Examples of these applications include biomedicine, drug delivery, tissue engineering, biosensors, and the production of innovative materials. These applications cover a wide range of scientific fields. In the following sections, we will investigate the functions and potential applications of a variety of biofluids,

**Corresponding author*: preetimicro@gmail.com

Arbind Prasad (ed.) Resorbable Composites for Bioimplants and Fixation Devices, (119–156) © 2026 Scrivener Publishing LLC

focusing on the distinctive characteristics of these fluids as well as the creative approaches that researchers are taking to use their potential in order to solve issues in a variety of sectors.

Keywords: Biomarker, disease diagnosis, biofluids, additive manufacturing

4.1 Introduction

Biofluids, or biological fluids, are innate liquids generated by living organisms. These fluids have significant functions in numerous biological cycles and have gotten extensive interest lately in light of their planned purposes in a few areas, like medication, biotechnology, and materials research. The study of biofluids as organic substances has uncovered a diverse array of functions and uses that exploit their distinct features and compositions.

Biofluids consist of a wide range of components, including as blood, saliva, urine, perspiration, tears, and other secretions from plants and animals. These fluids often include of intricate combinations of biomolecules, including proteins, lipids, carbohydrates, and metabolites, that play a crucial role in their unique activities inside organisms [1].

Biofluids in living systems provide a multitude of functions that are diverse in nature. They function as conduits for transporting nutrients, waste materials, and signaling molecules; regulating internal balance; providing lubrication and protection; and aiding immunological responses. Furthermore, numerous biological fluids display exceptional characteristics such as the ability to repair themselves, stick to surfaces, and combat microbial activity. These properties have motivated researchers to investigate their potential uses in diverse fields [2].

An increasing attention on the possible uses of biofluids as natural products has resulted from recent breakthroughs in analytical methods as well as a rising interest in solutions that are both sustainable and bio-inspired. Examples of these applications include biomedicine, drug delivery, tissue engineering, biosensors, and the production of innovative materials. These applications cover a wide range of scientific fields [3].

In the following sections, we will investigate the functions and potential applications of a variety of biofluids, focusing on the distinctive characteristics of these fluids as well as the creative approaches that researchers are taking to use their potential in order to solve issues in a variety of sectors.

4.2 Overview of Biofluids

Biofluids aren't just evacuated (like urine or sweat), produced (like breast milk or bile), drawn by needle (like blood or cerebrospinal fluid), or generated by disease (like blister or cyst fluid), they can also be obtained through other methods. The word "biofluid" may be used in two different ways: as a noun (as in the biofluids that were mentioned before) and as an adjective (as in the definitions of biofluid dynamics and biofluid mechanics) [4].

Body fluids include pleural fluid, ascitic fluid, urine, sputum, and saliva. These fluids represent a rich cellular supply and an essential component of liquid biopsy. These fluids may be used to diagnose diseases, monitor treatments, and provide insight into disease mechanisms and biomarker identification. Body fluid is liquid that comes from inside the bodies of living humans. Creative Bioarray supplies numerous bodily fluids from a broad set of healthy and unwell individuals in order to meet the needs of consumers all around the globe. Creative Bioarray offers all bodily fluid samples and derivatives in non-pooled aliquots from individual instances. Each aliquot is accompanied by extensive clinical information from the unique patient, allowing researchers to draw more definite findings from their experiments.

A rundown of the physiological significance of the fundamental body fluids for human wellbeing is given, with an emphasis on key parts of the development and activity of the large scale and microcirculations. The respiratory, stomach related, neurological, and visual blood streams, as well as the lymph and lymphatic frameworks, are undeniably depicted in extraordinary profundity. There is likewise a concise assessment of the urinary framework, gastrointestinal plot fluids, cerebral liquid, serous body fluids, synovial liquid, and other huge human fluids.

4.3 Circulatory System

Unicellular organisms get necessary nutrients from their immediate surroundings by absorption, while they release waste into the same environment. In multicellular creatures, such as humans, each individual cell functions as an independent unit and its immediate surroundings are confined inside the body. Organ cells allow waste and metabolic resources to exchange through processes including cell transport and diffusion. The distribution of these essential elements throughout the body is largely dependent on the circulatory system [5].

Under typical resting conditions, the circulatory framework — otherwise called the cardiovascular or vascular framework — transports around 5 liters of blood at a pace of around 5 liters each moment [6]. The heart, which comprises of two pulsatile siphons coupled in progression, is the part of the circulatory framework. Blood and substrates for digestion are shipped from the heart by the supply routes. The trading of waste materials and metabolic substrates with live tissues happens in the vessels. The blood in the veins is oxygen-drained. At last, extracellular liquid is gathered by lymphatic vessels and recycled.

The vascular organization of the circulatory framework is comprised of intricate progressive tree-like designs with various fanning designs that are explicitly intended to carry out the framework's mind boggling roles. Blood that has been oxygenated is shipped by courses bigger than 6 mm to more modest veins, which then, at that point, disperse this blood supply to the organization of minute corridors that reach in breadth from 1 to 6 mm. The hairlike beds, which have a breadth of 10 to 15 μm, are in the end arrived at by the blood. Venules move blood that has lost oxygen from vessels into bigger veins and then back to the heart [7].

The circulatory framework is fundamental for safeguarding homeostasis since it controls internal heat level, changes how supplements and oxygen are conveyed under various physiological conditions, and works with humoral motioning all through the body.

There are three subsystems that contain the circulatory framework: (i) The blood that has been oxygenated by the aorta is moved to the fundamental vessels by the foundational course. (ii) The pneumonic corridor supplies the aspiratory flow, which thus takes care of the pneumonic vessels. (iii) The coronary course gives the heart muscle (myocardium) the blood stream it requirements to siphon blood all through the body. The coronary dissemination is the littlest subsystem with regards to blood volume, yet it is obviously indispensable to its appropriate activity. As per [8], around 33% of the passings in Western nations among grown-ups more than 35 are owing to coronary corridor disease. Besides, nearly everybody beyond 50 years old have a change in their coronary dissemination design.

The left heart fills in as the beginning stage of the circulatory framework, siphoning high-pressure, oxygenated blood into the aorta. The blood is then moved to more modest courses prior to showing up at the foundational vessels. The encompassing tissues and these vessels trade waste carbon dioxide for oxygen. The blood, which is without oxygen and contains waste products from digestion, goes through the veins until it arrives at the vena cava, which then, at that point, siphons it to the right half of the heart. Subsequent to passing through the pneumonic supply route, which

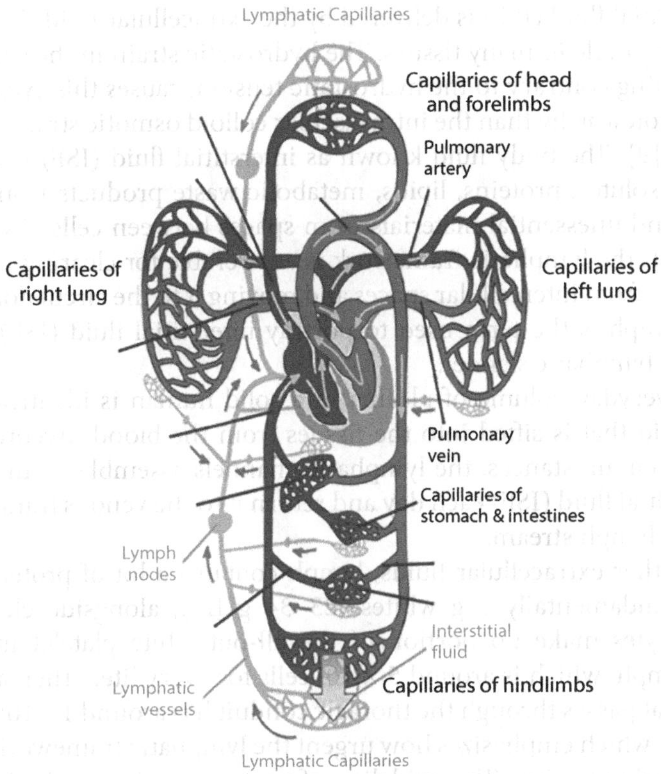

dynamically isolates into more modest corridors called arterioles, and at long last into the aspiratory vessels that encompass the pneumonic alveoli, blood lacking in oxygen proceeds with its excursion? As a component of the breath cycle, carbon dioxide, which is conveyed by the hemoglobin in red platelets, is changed into oxygen which is likewise conveyed by the hemoglobin in these vessels. The cycle is finished when oxygenated blood is moved from the lungs to the left half of the heart by the pneumonic veins, as found in Figure 4.1.

4.4 The Lymphatic System and the Lymphatic Tract

The lymphatic framework contributes essentially to the method involved with keeping up with wellbeing. Heart-related conditions like cardiovascular edema, irritation, and fibrosis might be brought about by a lymphatic blockage or breakdown in the heart lymphatic veins.

Interstitial fluid (ISF) is delivered by the extracellular fluid that is made by a sifting cycle in many tissues. The hydrostatic strain in the slim, which is something contrary to the hydrostatic tension, causes this cycle when it is more noteworthy than the intravascular colloid osmotic strain of plasma proteins [9]. The body fluid known as interstitial fluid (ISF) is clear and contains solutes, proteins, lipids, metabolic waste products from various tissues, and unessential materials from spaces between cells. As found in Figure 4.1, the lymphatic framework is answerable for clearing interstitial fluid (ISF) from intercellular spaces and moving it to the circulatory framework. Lymph is the term used to portray interstitial fluid (ISF) that has entered a lymphatic slender.

The everyday volume of plasma in a solid human is identical to how much fluid that is sifted into the tissues from the blood. Accordingly, in ordinary circumstances, the lymphatic channels assemble around 3 liters of interstitial fluid (ISF) each day and return it to the venous framework by means of lymph stream.

Like other extracellular fluids, lymph contains a lot of protein (40–60 g L^{-1}), fundamentally egg whites (23–34 g L^{-1}), alongside electrolytes. Lymphocytes make up portion of the all-out white platelet include in fringe lymph, which is around 5×10^8 cells for every liter. Then again, the lymph that passes through the thoracic conduit has around 1×10^{10} cells for each liter, which emphasizes how urgent the lymphatic framework is to the invulnerable reaction. The guideline of protein fixation in the interstitial fluid (ISF) by lymphatic waste is much of the time satisfactory to forestall edema, with an objective of roughly 15 grams for each liter (g L^{-1}). A typical colloid osmotic tension of around 5 mmHg is kept up with by the interstitial fluid (ISF). The interstitial tissues and the hairlike lumen experience a huge tension differential thus, which urges the tissues to reabsorb water.

Three fundamental functions of the lymphatic organization are as per the following: (I) it returns individuals to the venous framework subsequent to getting ultrafiltrate from blood vessels and overabundance protein from the interstitial fluid to manage circulatory equilibrium; (ii) processed fat is consumed by lymphoid tissue in the digestive system to support nourishment; and (iii) the lymphatic framework is essential for protection since it conveys microbes, unfamiliar antigens, and lymphocytes from the interstitium to nearby lymph hubs, which actuates the safe framework.

Lymphatics, otherwise called subcutaneous lymph-gathering vessels, are found in practically all substantial parts with veins. Their source is the shallow dermis, the skin's furthest layer; they are absent in the epidermis, the skin's deepest layer [10]. They are tracked down underneath the essential organs' mucosa and serous linings in the body pits. The lymphatic cells cross-over each other to frame an easily penetrable framework that lets interstitial fluid (ISF) access the lymphatic vessels on the grounds that the associations between the cells are not firmly fixed. From the underlying lymphatic veins to the lymphatic vessels, the lymph fluid changes inside the interstitial zone. Blind sacs that reach in width from 5 to 50 mm are the sign of lymphatics. An elastic film upholds a solitary layer of lengthened endothelial cells that make up the inward covering of these sacs. Little veins, a slim organization of nerves, and connective tissue blended in with smooth muscle cells make up the adventitia, whereas cross over smooth muscle and delicate elastic filaments make up their media.

Lymphatic vessels range in distance across from 20 to 70 μm and are available in most body tissues. Such physical qualities are generally appropriated all through the palms, fingers, soles, and toes. Pre-authorities, whose width can reach up to 300 μm, get lymph shipped by vessels. These pre-gatherers are valved, and they join to shape lymphatic vessels (see Figure 4.1), which are valved as well. Lymangioa are the pieces of the vessels that are in the middle of between the two valves. Smooth muscle constrictions in the walls of lymphatic veins cause the lymphangia to contract consistently, which works with the progression of lymph into the thoracic region.

Running lined up with the shallow veins, the appendages' shallow lymphatic vessels are arranged subcutaneously. Their perforators enter the profound lymphatic veins by penetrating the profound fascia. Despite the fact that they are more modest than the shallow vessels and run close by the profound veins, the profound lymphatic vessels are more uncommon [10].

The cisterna chyli is the last objective of lymphatic veins that rise up out of the lower appendages and pass by means of the inguinal lymph hubs prior to entering the back stomach wall. In the wake of passing through this area, lymphatic fluid arrives at the thoracic conduit, which moreover assembles lymphatic stream from the left half of the head and neck, the left arm, and the left half of the chest. In the venous framework, the thoracic channel closes at the gathering point of the left inner jugular vein and the left subclavian vein. The right half of the head, neck, and chest, as well as the right arm, all contribute lymph to the right lymphatic channel.

The lymphatic framework functions without the utilization of dynamic siphons, in contrast to the circulatory framework. Rather, interstitial fluid (ISF) passively moves from the tissues to the lymphatic courses. Lymph is brought into the lymphatic veins by a pattern of pressure and unwinding that begins this cycle. The lymphatic bulb cells open because of strain from the interstitial fluid encompassing a fine. Likewise, there is a cadenced constriction of the myoendothelial strands that line the lymphatic narrow's endothelium. The vessels could drive lymph towards greater lymphatic vessels on account of this tightening. Little bicuspid valves introduced in these bigger vessels ensure that the lymph is pushed in the appropriate heading. Likewise, the associations among the endothelial cells act as a valve system that keeps lymph from getting back to the interstitial space. At the point when the endothelial cells contract, the connective tissue filaments that hold them set up give power to return the lymphatic fine so more interstitial fluid can be added [9].

In bigger lymphatic vessels, the strain can ascend to 25 millimeters of mercury (mmHg); however it is ordinarily a couple of mmHg in the bulbs and littlest lymphatic vessels. The change from lower to higher strain is worked with by lymphatic valves. The lymphatic veins are loaded up with lymph hubs separated at standard spans. The lymph hubs work with the passage of lymphatic fluid to the close by lymphatic veins. These designs generally comprise of lymphoid follicles, which go about as lymph channels and reach long from 0.2 to 20 mm. The axilla, inguinal district, neck, pelvis, mediastinum, and para-aortic space are among the body parts where the 500–600 lymph hubs are concentrated.

A container that stretches into trabeculae, what partition the hub into follicles containing sinusoids, makes up a lymph hub. A medulla that houses lymphoid tissue and a cortex that gets afferent lymphatics make up a lymph hub. The hilum fills in as the efferent lymphatics' takeoff point, and the medulla stretches out to it. The lymph goes through the sinusoids inside the lymph hub, passing through the hilum, which is where blood conduits enter and exit. Arterioles, vessels, and venules stay separate from the lymph.

4.5 Macrocirculation and Microcirculation

Blood stream in vessels with a breadth of 100 μm or less is alluded to as microcirculation. Arterioles, metarterioles, vessels, and venules are the various kinds of microvessels. In sound people, microcirculation addresses around 10% of all-out blood volume and is the piece of the circulatory framework that has direct contact with parenchymal cells [11]. In microcirculation, perfusion is the amount of blood stream per unit volume of tissue. The trading of water, chemicals, supplements, gases, and metabolic waste products between the blood and cells is made conceivable to a great extent by the microcirculation. It is likewise vital for the conveyance of medications, the guideline of vascular obstruction, and the help of intensity transport.

Tissue hemodynamics is of extraordinary interest to a few restorative professions that utilization it as a symptomatic instrument for different circulatory framework issues. Perfusion is a fundamental boundary while assessing the feasibility of an organ or tissue before transplantation, assessing the guess of patients with ischemic coronary illness, assessing the degree of cerebrum harm in stroke casualties, and assessing the state of diseased organs like the kidneys, liver, pancreas, and lungs, among different cases. Perfusion is likewise thought about in various medication studies, disease treatments, including laser photocoagulation of growths. In any case, in spite of its implied remedial significance, there is definitely not an optimal method for measuring perfusion right now [12].

4.6 Skin Histology and the Flow of Blood Through the Cutaneous Tissue

The skin's capillary network and position as the body's furthest organ make them the most available for any instrument used to assess biological stream designs. Furthermore, a sensor should pass through the skin before it might arrive at any interior objective of interest. Angiogenesis, or the formation of new vessels from previous vessels, is a central point in cutaneous melanoma. Stream sensors are one of the most essential applications for the analysis and visualization of skin malignant growth [13].

The epidermis and dermis, the two primary layers of the skin, are layered, heterogeneous, and anisotropic, making the skin a perplexing and different medium. In spite of prevalent thinking, the subcutaneous layer, or hypodermis, that lies underneath the dermis isn't viewed as a piece of the skin. Each layer is comprised of various strong materials that have been injected with fluids, and they are all kinds of thicknesses. Each layer is seen as a discrete framework, having explicit elements that put it aside from the rest. The notable elements of these levels are depicted underneath.

4.6.1 Epidermis

The thickness of the epidermis, the furthest layer of skin, fluctuates from 0.07 to 0.12 mm along the body's surface [14]. A kind of epithelium comprised of separated, keratinized squamous cells is known as the epidermis. It isn't innervated, nor does it get blood stream; all things considered, supplements are diffused from the dermis, and waste products are removed through the dermal veins.

There really depend on five layers in the epidermis, which are the layer corneum, layer lucidum (tracked down just on the palms and soles), layer granulosum, layer spinosum, and layer basale, saw from the furthest to the deepest. The expression "layer malpighi" alludes to the combination of the layer spinosum and the layer basale. There are four primary cell types in the epidermis: Merkel cells, Langerhans cells, keratinocytes (90%) and melanocytes (8%), which produce melanin.

4.6.2 Dermis

Situated underneath the epidermis and over the hypodermis is the dermis. The skin is comprised of two layers, the reticular dermis and the papillary dermis, and reaches in thickness from 1 to 4 mm. The dermis' principal capability is to basically feed and support the epidermis.

The top layer contains the papillary dermis, which makes up around 10% of the complete thickness of the dermis. The material is comprised of small collagen filaments, around 20–40 nm in measurement, stuffed intently along with greater collagen strands, around 0.3–3.0 mm in distance across. Additionally, the vessels and nerves that supply the epidermis with sustenance are situated in the papillary dermis.

Underneath the papillary layer is the reticular dermis, which is made areas of strength for of tissue with collagen and elastic strands. Type I collagen makes up most of the 60–100 nm-distance across collagen fibrils found in the reticular dermis. The game plan of these fibrils brings about filaments with measurements going from 10 to 40 mm [14].

In youthful grown-ups, a minuscule, randomly situated fiber network looking like feltwork is framed by the collagen in the papillary dermis. On the other hand, the reticular layer's collagen is comprised of gigantic, wavy groups that are comparably inexactly laced and aimlessly arranged.

4.6.3 Dermal–Epidermal Junction

In light of this [15], the skin's underlying trustworthiness is kept up with by an enormous exhibit of extracellular grid macromolecules that connect the dermal and epidermal layers.

The dermal-epidermal intersection, which starts at the epidermal rete edges and stretches out to the papillary dermis, is approximately 100 nm thick and organized in a wave design. These raised districts make the epidermis become indented, increasing the surface region in touch with the two layers. This keeps up with the dermal and epidermal layers solidly joined and fortifies the dermal-epidermal availability. Diemal papillae are minute papillary dermal projections implanted in the epidermis; rete edges encompass them. In fingerprints, this wavy example of edges and wrinkles is seen at a macroscale.

At the point when the dermis and epidermis split separated because of erosion or shear powers, body fluids including discharge, serum, plasma, blood, or lymph can gather between the two layers and cause rankles.

4.6.4 Vascularization of Human Skin

The two principal networks of cutaneous supply routes are responsible for conveying veins to the skin: The superficial vascular plexus is a network of veins situated at the highest level of the dermis. This plexus addresses the gathering point of the reticular and papillary dermis histologically. The superficial plexus is expanded into dermal papillae by anastomosing little type arterioles that branch off into vessels. These arterioles feed the lines between the epidermis and dermis and wrap adnexal designs [16]. Something like one capillary circle is available on each dermal papilla. The contact where the dermis and hypodermis meet is where you can track down the profound vascular plexus. It goes about as a histological hindrance isolating the hypodermis from the reticular layer of the dermis. This network is comprised of medium-sized veins that rise out of bigger corridors that pass through the hypodermis' fat septae. In an upward direction situated corridors likewise associate it to the superficial vascular network. Little branches create from this network to supply supplements to the dermal tissues, including hair follicles and sweat glands.

Albeit the hypodermis isn't lawfully viewed as a piece of the skin, blood stream to the skin relies upon its vascularization. This is on the grounds that the blood stream to the skin is affected by the subcutaneous blood vessel network, which is more boundless and runs lined up with the cutaneous network. Additionally, the skin's all's minute conduit branches begin from the hypodermis [15, 17].

4.7 The Circulation of Blood Throughout the Respiratory System

There is areas of strength for a between the respiratory and circulatory frameworks (see Figure 4.1). This is with the goal that the lungs can trade gases by carrying blood to them through the pneumonic dissemination. The bronchial dissemination, then again, has a ton of air and little strain. The larynx and bronchial framework.

4.8 The Circulation of Blood Throughout the Digestive System

The body gets its water, supplements, and electrolytes from the stomach related framework constantly. Conveyance of ingested supplements in the stomach, spleen, liver, pancreas, throat, and stomach is made easier by the course of blood all through the gastrointestinal plot. Furthermore, it supplies oxygen that the gastrointestinal tissues' secretory and absorptive functions require, particularly the stomach's mucosa layer [18].

The blood conduits situated all through the intestinal system are by and large alluded to as "splanchnic dissemination". In the wake of passing through the digestion tracts, pancreas, spleen, and stomach, blood enters the liver by the gateway vein. In the wake of passing through the hepatic veins and entering the vena cava, which conveys it back to the heart, blood passes through large number of sinusoidal vessels inside the liver.

As indicated by [18], the left and right carotid corridors are the essential wellspring of blood supply for the mind, with the left and right vertebral veins filling in as an optional source. Around 14–15% of the blood dissemination is comprised of the blood stream in the cerebrum. A part of the carotid corridor, the ophthalmic course supplies blood to different pieces of the eye. Through a particular passage, it enters the eye attachment and separates into extra, more dense courses. Blood is provided to the tear gland, muscles, and eyelids by the lacrimal conduit. Blood is shipped to the inside areas of the eyeball by extra supply routes, including the long and short back ciliary veins. One significant division that passes through the optic nerve and conveys blood to the photosensitive retina is the focal retinal corridor. Different branches called solid veins give blood to the muscles of the eyeball. The average palpebral conduits supply blood to the upper and lower eyelids. Also, the dorsal nasal, supraorbital, supratrochlear, front, and back ethmoidal corridors supply blood to explicit areas of the temple and nose.

The prevalent ophthalmic vein gets blood from the recently expressed parts of the ophthalmic conduit as well as the veins that channel the back piece of the eyeball; in contrast, the sub-par ophthalmic vein gets blood from different muscles notwithstanding the back region of the eyeball.

I. The Urinary System

Urine is an essential excretory biofluid that is composed of more than ninety-five percent water and includes a variety of solutes, including urea, creatinine, uric acid, ammonia, and electrolytes. In addition, it contains organic substances such as hormones, vitamins, and metabolites, all of which are necessary for the restoration of the body's fluid and electrolyte balance as well as the elimination of waste materials within the body. The formation of urine requires a particular volume and strain of blood and is liable for the expulsion of medications, poisons, and metabolic waste products from the circulatory framework. The urinary framework is firmly connected with the circulatory framework and blood stream [19].

Various metabolites, including as glucose, cholesterol, urate, lactate, oxalate, and ascorbate, are tracked down in pee. It is feasible to take a gander at these metabolites to recognize different disease states. For example, glucose in the urine is connected to diabetes [20], and lactate discovery is utilized in metabolic evaluating for hereditary diseases [21]. The varied nature of biological systems and the specialized tasks that they play is reflected in the broad variety of physiological fluids that are included in the category of biofluids. Each of these fluids has its own composition and functions. One of the greatest important biofluids, blood is made up of plasma, which accounts for 55% of the volume of blood, and cellular components, which account for 45% of the volume of blood. The components of plasma include water, electrolytes, proteins, hormones, and waste products. The components of the cells are platelets, which cause clotting, white blood cells, which provide immunological protection, and red blood cells, which carry oxygen.

Saliva contains electrolytes such as sodium, potassium, calcium, magnesium, bicarbonate, and phosphate in addition to water, which makes up 99.5% of its content. In addition, it contains proteins like immunoglobulins and mucins, enzymes like lipase, lysozyme, and amylase, as well as other substances including glucose, urea, and ammonia. These elements have essential functions in the processes of digestion, mouth hygiene, and lubrication [22].

Sweat is made of water (99%) and electrolytes. It is responsible for the elimination of minor waste products of metabolism such urea, lactate, and ammonia, as well as the regulation of the body's temperature.

II. Fluids Found in the Gastrointestinal Tract

The course of absorption comprises of a few unmistakable advances. The glycoprotein known as stomach inborn element is discharged by parietal cells, and it assists the body with engrossing vitamin B12. They likewise make hydrochloric corrosive, which is required for the hydrolysis of peptides and disaccharides and the change of the proenzyme pepsinogen into pepsin. In contrast, the breakdown of proteins into peptones and proteases is catalyzed by pepsinogen, which is discharged by peptic cells.

Furthermore, the stomach is also responsible for the release of stomach related compounds like lipase, lactase, and peptidase. Besides, bodily fluid is produced by mucous glands and cup cells, safeguarding the stomach walls from the harming impacts of chemicals and acids [18, 19].

III. Flow of Cerebrospinal Fluid

The clear, colorless substance found in the brain and spinal cord is known as cerebrospinal fluid (CSF). Its main components include glucose, electrolytes, and small amounts of proteins, principally immunoglobulins and albumin. Ninety-nine percent of its content is water. The brain and spinal cord are supported by the CSF, which also carries nutrients and removes waste. Around 70% of the cerebrospinal fluid is created by the capillary blood vessels in the ventricular choroid plexus through a component that consolidates plasma ultrafiltration and dynamic discharge. The cerebral/subarachnoid space and the ependymal arranging cells of the ventricles make the leftover around 30%. A grown-up healthy produces 500 milliliters of cerebrospinal fluid each 24 hours, with an ordinary volume of 90 to 150 milliliters for every liter. By means of the foramina, the cerebrospinal fluid leaves the ventricles and courses across the two halves of the globe of the mind prior to going down the spinal rope to the nerve roots. Following this, the arachnoid villi inside the dural sinuses help in the reabsorption of CSF fluid [19]. The mind and spinal section are padded and greased up by the cerebrospinal fluid, which fills in as a defensive component. This reduces the chance of shock or jerk. Besides, it makes it easier for the cerebrum and spinal line to trade supplements and metabolic waste products.

Serous Fluids of the Body

This sort of fluid is called serous; it is by and large clear with a slight yellowish color and is comprised of ultrafiltrate plasma. Since it looks like serum, it is called serous. The organs and the walls of the numerous compartments in the human body are fixed with a smooth tissue film called the serous layer. The serous fluid emission happens on account of this layer. As an oil, serous fluid occupies the space that isolates the parietal which is the piece of the serous layer that covers the body's walls that encompasses the organs.

The pericardium, which normally holds under fifty milliliters of pericardial serous fluid encompassing the heart; the pleura, which ordinarily holds under thirty milliliters of pleural serous fluid close to the lungs; and the peritoneum, which gathers the peritoneal serous fluid (likewise called ascites) that encompasses the organs spread all through the stomach cavity are the three serous depressions [19].

IV. Synovial Fluid

Synovial fluid, greases up most joints, formed in the synovium, a tissue lining diarthrotic joints that permits unlimited bone development. Synovia is a dialysate of plasma that contains similar measures of glucose and uric acid along with 33% of the protein in plasma. A mucopolysaccharide delivered by the synovium is joined with this dialysate [19].

V. Additional Pertinal Human Fluids

The fluids that are delivered in various male organs join to frame semen. Fundamental vesicles release a to some degree basic fluid containing potassium, fructose, citrus extract, and flavones. The previously mentioned fluid gives fundamental supplements to spermatozoa, which are produced inside the sertoli cells of the seminiferous tubules of the testis. It makes up almost 50% of the semen's whole volume. A fifth of semen is created by the prostate gland, which delivers a fluid containing proteolytic proteins, corrosive phosphatase, and citrus extract. The epididymis, bulbourethral glands, and urethral glands combine the excess part of semen.

Emissions from glands on the cervix are created, and the sum discharged changes during the monthly cycle. Vaginal emissions are typically clear bodily fluid that turn light yellow or white within the sight of air. Any strange changes in variety, sum, or surface could be an indication of various circumstances.

The creating embryo is encased in a layer bound pack known as the amnion, which is loaded up with amniotic fluid secreted by the placenta. It is fundamental for the development of sustenance, water, and metabolic products between the mother's circulatory system and the embryo notwithstanding its defensive job. Comparable in organization to maternal plasma, amniotic fluid for the most part comprises of metabolic products from the embryo and contains follow measures of skin, gastrointestinal system, and urinary lot cells from the infant.

Biofluids, which are intricate liquids generated by living organisms, are becoming an intriguing frontier in the exploration of natural products. The diverse range of fluids, including blood, breast milk, tears, and perspiration, has significant therapeutic potential owing to their complex biochemical makeup.

Biofluids have become a valuable source of natural products, providing a wide range of biomolecules with various activities and possible uses. This abundant source of biological compounds has garnered considerable interest from researchers in several sectors, including as medicine, biotechnology, and materials science. Let us examine the primary classifications of natural substances present in biofluids and the sophisticated methodologies used to separate and analyze them.

A wide range of nutrients, including lipids, proteins, carbs, vitamins, minerals, growth hormones, growth factors, and antibodies, are present in breast milk, which is 87% water. Breast milk is essential to an infant's nourishment and immune system. Its components strengthen an infant's immune system and promote healthy growth and development [23].

4.9 Biofluids as a Source of Natural Products

A. Revealing the Biochemical Treasure

A complex variety of biomolecules, each with distinct qualities and functions, is present in biofluids. These naturally occurring substances have developed over millions of years, and they often play very specialized and effective biological functions. In addition to offering insights into physiological processes, studying these molecules presents prospects for the development of new treatments, diagnostic instruments, and biomaterials [24]. A vast multitude of biomolecules with unique properties is contained within biofluids, which are a treasure trove of natural products. All of these molecules can be essentially classified as:

- **Enzymes: The Catalysts of Nature**

  Protein molecules are known as enzymes, and they perform the function of biological catalysts by speeding up chemical processes that are necessary for life. The saliva, pancreatic juice, and blood are all examples of biofluids that include a variety of enzymes that perform a variety of tasks. Sampling amylase, for instance, is responsible for initiating the method of carbohydrate digestion, while proteases found in pancreatic juice are responsible for breaking down proteins [25]. The investigation of these enzymes has resulted in their utilization in industrial processes, including the production of biodiesel through the utilization of lipases derived from microbial sources [26].

  The catalysts that nature provides, which speed up a variety of metabolic processes that are necessary for life. Blood and saliva are examples of biofluids that include enzymes that have the potential to be used in the therapeutic expansion, bioremediation, and diagnostics fields.

- **Proteins: Multifunctional Workhorses**
  The catalysts that nature provides, which speed up a variety of metabolic processes that are necessary for life. Blood and saliva are examples of biofluids that include enzymes that have the potential to be used in the therapeutic expansion, bioremediation, and diagnostics fields. The investigation of proteins found in biofluids has resulted in the creation of protein-based treatments, such as monoclonal antibodies, which are specifically designed to treat cancer [27].

  Transportation, immunological response, and cell signaling are just some of the many duties that are performed by these workhorses of the biological world. The presence of antibodies in the blood plays an important part in the immune system, which opens up opportunities for the development of immunotherapy approaches. Tissue regeneration may be facilitated by growth factors that are present in a variety of biofluids.

  In biological systems, peptides, often serve as signaling molecules. Numerous peptides, including hormones, neurotransmitters, and antimicrobial compounds, are present in biofluids and serve a variety of purposes. For instance, defensins in saliva offer antibacterial defense, whereas insulin in blood controls glucose metabolism [28]. The distinctive characteristics of these peptides have stimulated the advancement of peptide-based medications for diverse illnesses, such as diabetes and hypertension. Smaller protein fragments exhibiting a wide range of biological functions. Peptides obtained from biofluids such as snake venom or blood have shown promise in the field of drug development for the purposes of pain treatment and hormone modulation.

- **Lipids: Functional Molecules and Biomarkers**
  Biofluids include lipids that perform a variety of tasks, including membrane construction, cell signaling, and energy storage. Biofluids' lipid profiles may act as biomarkers for diseases like cardiovascular disease, giving important information about a person's health [29]. Furthermore, certain lipids, such as omega-3 fatty acids present in breast milk, have been examined for their potential advantages to health and their use in functional meals. Biofluids may include fats

and cholesterol that may be used as indicators for illnesses or in the creation of nutraceuticals.

- **Nucleic Acids: The Foundation of Life and Personalized Medicine**
  The detection of nucleic acids (DNA and RNA) in biofluids has created new opportunities for diagnostics and personalized therapy. Cell-free DNA present in the bloodstream, namely in the plasma, may be used for non-invasive prenatal testing and the identification of malignancy. Research on microRNAs in different biofluids has shown potential for early illness identification and tracking therapy effectiveness [30]. DNA and RNA that have been extracted from biofluids have the potential to be applied to genetic testing and personalized medicine methodology.
- **Metabolite: Understanding Cellular Mechanisms**
  Small molecules known as metabolites are the byproducts or intermediates of cellular metabolic activities. Metabolomics, the study of metabolites in biofluids, offers important insights into the physiological condition of an organism. Urine metabolites, for instance, may provide details on drug metabolism, eating patterns, and possible illness states [31]. Applications of this topic may be found in the fields of medication development, nutritional research, and the finding of biomarkers.

  Small compounds generated by biological activities. Analyzing the metabolic profile of biofluids may reveal insights into metabolic health and disease indicates, possibly leading to the creation of new diagnostics and treatment targets.

B. Advanced Methods for Isolating and Characterizing Biomolecules

Biofluids' complexity calls for advanced analytical methods for component isolating and characterization. A few important techniques consist in:

- **Mass Spectrometry (MS):** Allows for accurate identification and measurement of biomolecules based on the ratio of their mass to their charge concentration [32].
- **Nuclear Magnetic Resonance (NMR) Spectroscopy**: Contains precise structural details regarding biomolecules and is very relevant for metabolomics research [33].
- **High-Performance Liquid Chromatography (HPLC):** This enables the separation and purification of biomolecules that are found in complicated mixtures.

- **Next-Generation Sequencing (NGS):** Allows for the examination of nucleic acids in biofluids, which enables applications such as liquid biopsy for the diagnosis of cancer [34].
- **Proteomics Techniques:** The use of these techniques, which include protein microarrays and two-dimensional gel electrophoresis, makes it possible to conduct an exhaustive investigation of the protein composition of biofluids.
- **Chromatography:** Utilizes size, charge, or affinity as criteria for the separation of biomolecules.
- **Electrophoresis:** Biomolecules are separated according to their size and the electrical characteristics they possess.

The main aim is to investigate the potential of natural compounds that are discovered in biofluids, has been revolutionized as a result of the combination of these cutting-edge approaches and tools from the field of bioinformatics.

4.10 Functions of Biofluids as Natural Products

Biofluids are the lifeblood of human physiology, since they play an essential part in the maintenance of homeostasis (internal balance) and in the support of a wide variety of processes throughout the cardiovascular system. These wonderful liquids are constantly moving through our bodies, and each one of them performs a unique set of functions that contribute to our general health and well-being. The intriguing realm of biofluids and investigate the fundamental roles that has been provided are as follows:

a) Transportation Powerhouses

Many biofluids have the main role of transporting different compounds throughout an organism, which is one of its fundamental tasks. In this sense, blood, for example, plays a significant role since it transports oxygen, minerals, hormones, and waste products across the body [35]. Corresponding to this, in plants, xylem sap transfers water and nutrients from roots to shoots, whereas phloem sap carries sugars along with other chemical compounds from leaves to other areas of the plant. Blood serves as an unequivocal leader in transportation, functioning as a dynamic pathway that transports oxygen, nutrients, and waste materials throughout the whole body. Red blood cells carry oxygen molecules, whereas plasma, the component without cells, carries nutrients such as glucose, amino acids, and salts. Blood transports waste materials, such as carbon dioxide, to the lungs for removal.

b) Internal Milieu Regulatory Agents

As saliva contains enzymes such as salivary amylase, which breaks down carbs into simpler sugars for better absorption, plays an important part in the digestive process due to its presence. It also contains antimicrobial components, thus, contributes to the maintenance of a healthy oral environment. These chemicals fight germs and safeguard against infections [36]. Hormones are another chemical messenger that are typically found in the blood and are responsible for controlling a variety of physiological processes, including growth, development, metabolism, and reproduction.

c) Defense Systems

The respiratory and digestive systems are lined with the mucus which is the first lines of defense that the body has, and this sticky barrier serves as that layer of defense. Through its ability to capture dust, pathogens (organisms that cause illness), and debris, it prevents these substances from reaching the body and creating discomfort [37]. White blood cells, which are present in the bloodstream, serve as the body's defensive troops against infections and diseases.

4.10.1 Lubrication to Ensure Smooth Movement

Several biofluids have protective roles. Synovial fluid inside joints minimizes friction between the surfaces of articulating structures and supplies nourishment to the cartilage. Tears have the dual purpose of lubricating and cleansing the eyes, in addition to having antibacterial substances that aid in preventing infections [38]. The lubricating fluid that is present in joints serves as a lubricant, therefore minimizing the amount of friction that occurs between bones and allowing for movement that is painless and smooth.

d) Nourishment and Protection

Saliva, gastric juice, and bile are examples of digestive biofluids that play important roles in the process of breaking down food and aiding the bodies' ability to absorb nutrients. In the first stages of digestion, saliva includes enzymes such as amylase, whereas gastric juice contains hydrochloric acid and pepsin, which are responsible for breaking down proteins.

Cerebrospinal fluid, a transparent liquid, surrounds and protects the brain and spinal cord, serving as a cushion and absorbing shocks [39]. Another one, amniotic fluid envelops and protects the growing fetus throughout pregnancy, maintaining temperature and supplying vital nutrients [40].

e) **The Elimination of Waste by Thermoregulation**
Urine and sweat are examples of biofluids that play an important role in the elimination of waste products and chemicals that are in excess from the body. While sweat contributes to thermoregulation and the excretion of certain substances, urine is responsible for the elimination of waste products from the metabolic process and helps regulate electrolyte balance. Urine, a liquid waste product, is generated by the kidneys and is responsible for the elimination of metabolic byproducts, superfluous water, and electrolytes from the body [41]. Sweat participates in the process of controlling the temperature of the body. Sweat glands are responsible for producing sweat, which is then evaporated from the skin. This process involves the removal of heat from the body and the promotion of cooling.

Comprehending the many activities of biofluids is essential for their potential utilization in the fields of health, biotechnology, and materials science. For instance, knowing the role of blood in transportation has resulted in the invention of synthetic blood alternatives, while knowledge of the protective qualities of certain biofluids has sparked the innovation of new biomaterials [42].

4.11 Various Applications of Biofluid as Natural Products

4.11.1 Biofluids for Diagnostic Purposes

a. Blood Tests

Detects a variety of diseases, including infections, anemia, diabetes, and some malignancies. Blood tests may measure the quantities of certain chemicals or detect the presence of aberrant cells. Blood is a complex bodily fluid that provides a wide range of valuable diagnostic data. Blood tests are often used to identify a broad spectrum of disorders, ranging from infections to chronic diseases.

- **Biomarkers**: Numerous biomarkers, such as proteins, lipids, and nucleic acids, are found in blood and may signify the existence of conditions including cancer, heart disease, and metabolic problems. Many biomarkers found in biofluids may show the onset or course of a variety of illnesses. Prostate-specific antigen (PSA) in blood, for instance, is used to test for prostate cancer. Moreover, some proteins present in cerebrospinal fluid (CSF) may serve as indicators for neurodegenerative disorders such as Alzheimer's disease [30].
- **Genetic Testing**: Genetic mutations and variations may be detected by analyzing blood samples, which can help diagnose genetic illnesses and inform personalized treatment strategies.
- **Monitoring Therapy**: Blood tests are crucial for evaluating the effectiveness and safety of continuing therapies, enabling modifications in therapy to maximize patient results.

b. Saliva Tests

Increasing popularity in the field of diagnosing oral disorders, monitoring hormone levels, and identifying certain medicines. Saliva is being more widely used as a biofluid for diagnostic reasons since it can be collected without intrusive procedures and has a high concentration of biomolecules.

- **Hormone Levels**: Saliva testing is a reliable method for assessing hormone levels, such as cortisol and testosterone, which may provide valuable information about endocrine problems and stress levels.
- **Infectious Diseases**: Saliva may serve as a reliable and expedient diagnostic method for identifying viral and bacterial diseases, including HIV and COVID-19.
- **Drug Monitoring**: As a less intrusive alternative to blood testing, saliva tests are carried out for the purpose of monitoring drug levels and identifying instances of substance addiction.

c. Urine Tests

By analyzing the amount of sugar and protein found in urine, it is possible to diagnose conditions such as diabetes, kidney issues, and urinary tract infections. Urine is a further significant biofluid for diagnostic purposes since it provides information about the functioning of the kidneys and the metabolic processes [43].

- **Metabolites**: Numerous metabolites found in urine may provide information about kidney function, metabolic diseases, and general health.
- **Pregnancy Testing**: Human chorionic gonadotropin (hCG) is detected in urine, and this makes urine testing a popular method of pregnancy detection.
- **Cancer Biomarkers**: Certain tumors, including bladder cancer, have biomarkers that may be found in urine, which can help with early diagnosis and surveillance.

d. CSF Analysis

For the purpose of detecting neurological illnesses, cerebrospinal fluid (CSF) examination is essential since it offers immediate insight into the central nervous system.

- **Neurological Diseases**: Analysis of cerebrospinal fluid (CSF) has the ability to identify biomarkers for neurodegenerative disorders including Parkinson's disease and Alzheimer's disease.
- **Infections**: Evaluation of cerebrospinal fluid (CSF) is crucial for the diagnosis of infections such as meningitis and encephalitis, as it yields results that are both prompt and accurate.
- **Autoimmune Disorders**: In order to assist in the diagnosis of conditions such as multiple sclerosis, cerebrospinal fluid (CSF) can be examined for the presence of autoantibodies.

e. Other Applications

Liquid biopsies are a non-invasive procedure for the identification, monitoring, and personalized therapy of cancer. These biopsies include the analysis of circulating tumor cells (CTCs) and cell-free DNA (cfDNA) in blood. In order to investigate the oral and intestinal microbiomes, respectively, which have been associated with a variety of health issues, it is possible to do microbiome analysis on samples taken from the faces and the saliva. Cell-free fetal DNA in maternal blood enables non-invasive prenatal testing for genetic diseases, which is a significant advancement in prenatal diagnostic tests.

4.11.2 Biofluids for Therapeutic Purposes

A wide variety of therapeutic uses are available for biofluids, which provide natural products that have the potential to be turned into medicines for a variety of ailments.

a. Treatment with Enzymes Replacement

It is possible to treat metabolic problems using enzymes that are produced from biofluids *via* a process known as enzyme replacement therapy.

- **Problems with lysosome storage**: For the purpose to treat conditions such as Gaucher's disease and Fabry disease, enzyme replacement therapy is utilized. This treatment involves the provision of the enzyme that is either absent or insufficient.
- **Cystic Fibrosis**: Biofluids have been utilized to extract enzymes, which are then administered to individuals with cystic fibrosis in order to enhance digestion and nutritional absorption.

b. Immunotherapy

For the purpose of immunotherapy, immune cells and chemicals derived from biofluids may be used, which allows for an enhanced immunological response from the body against illnesses. The application of immune cells derived from biofluids in immunotherapies, such as CAR-T cell treatment, allows for the targeting and destruction of cancer cells. It is possible to control the immune system using antibodies that are produced from biofluids, which may provide therapeutic advantages for autoimmune disorders such as rheumatoid arthritis. A blood transfusion is the procedure by which blood components (red blood cells, platelets, plasma) from a healthy donor are moved to a receiver in need. Many medical disorders may be treated with this operation, which also saves lives in circumstances such as in Severe blood loss, anemia, certain cancers, and bleeding disorders.

c. Plasma Therapy

Convalescent plasma derived from individuals who have recovered may be used as a therapeutic intervention for some infectious illnesses.

d. Medicine for Regeneration

Biofluids are essential in regenerative medicine since they provide growth factors and stem cells that are necessary for the repair and regeneration of tissues. In wound healing growth factors found in biofluids encourage the multiplication of cells and the repair of tissues, which speeds up the healing process of wounds. Additionally in stem cell therapy for the treatment of a wide range of blood diseases and other ailments, stem cells that have been extracted from bone marrow or blood may be used.

e. Saliva Substitutes

Patients who suffer from dry mouth disorders, such as Sjogren's syndrome and the adverse effects of radiation treatment, may find relief from their symptoms by using saliva replacements that are generated from biofluids [42].

- **Artificial Saliva**: Products designed to replicate the structure and functionality of natural saliva may enhance comfort, preserve oral health, and avert problems brought on by dry mouth.
- **Therapeutic Relief**: Saliva replacements that comprise enzymes and proteins produced from biofluids have the potential to improve the digestive process and complement the protective properties of saliva, so providing patients with considerable advantages.

f. Stem Cells Produced from Urine

Due to the fact that they may be collected without causing any harm to the patient and possess multipotent properties, urine-derived stem cells (USCs) are a promising area in the field of regenerative medicine [44].

- **Treatments for Regeneration**: USCs has the capacity to undergo differentiation into many cell lineages, such as osteoblasts, chondrocytes, and adipocytes, rendering them appropriate for the treatment of a wide array of ailments, including bone and cartilage lesions.
- **Personalized Clinical Practice**: Umbilical stem cells (USCs) may be extracted from patients and grown in a laboratory, offering a customized method for regenerative therapies without the ethical issues linked to embryonic stem cells.

4.11.3 Biofluids for Research and Development Purposes

Biofluids are very significant in the field of research and development, as they provide vital insights into the processes of diseases and serve as effective platforms for the identification of new drugs.

Biofluids are a diverse supply of biomarkers that may be used for the purpose of identifying and comprehending the processes behind illness [44]. The identification of biomarkers in biofluids helps to bridge the gap between fundamental research and clinical applications, which in turn makes the creation of diagnostic tools and therapeutic approaches easier. The use of

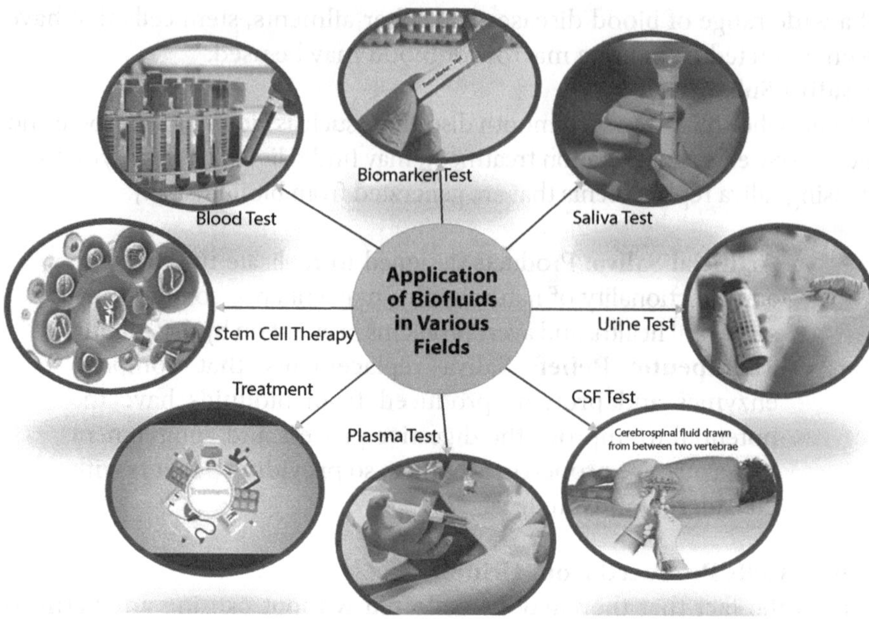

advanced analytical methods to biofluids has the potential to result in the identification of novel biomarkers for a variety of illnesses, which might then lead to the development of new diagnostic tools. An investigation has also been personalized into the unique differences in the composition of biofluids aids in the advancement of personalized treatment methods by discovering biomarkers that may accurately forecast the response to therapy or the course of a disease. Figure 4.2 depicts the various applications of biofluids.

a. Discovery and Development of Drugs

Biofluids provide a natural reservoir of bioactive chemicals, which allows them to serve as platforms for the discovery and development of novel medications [45]. It is possible to screen biofluids for natural compounds that have the potential to bring about therapeutic effects, which may lead to the development of novel medications. The utilization of high-throughput screening to discover lead compounds for drug development may be accomplished *via* the use of libraries of chemicals that are obtained from biofluids.

Research in Both Preclinical and Clinical Settings

Research in both preclinical and clinical settings require the use of biofluids because they provide data that can be used to evaluate the effectiveness and safety of novel treatments.

- **Preclinical Models**: In preclinical models, biofluids are implemented for the purpose of analyzing the pharmacokinetics and pharmacodynamics of newly developed drugs.
- **Metabolism of Drug and Pharmacokinetics**: For the purpose of gaining a knowledge of drug absorption, distribution, metabolism, and excretion (ADME), it is essential to conduct analyses of medicines and their metabolites in biofluids. It is crucial that this be done at both the preclinical and clinical phases of the therapeutic process.
- **Research Related with Toxicology**: In preclinical animal research as well as clinical trials conducted on people, biofluids are an essential tool for determining the toxicity and safety of newly developed medicinal substances.
- **Validation of Biomarker**: Before being employed in clinical practice, biomarkers identified by biofluid analysis must be confirmed in preclinical models and clinical trials.
- **Monitoring of Clinical Evaluations**: The examination of biofluids during clinical trials may provide useful information on the effectiveness and safety of drugs, as well as assist in the identification of patients who react to treatments and those who do not respond to them.
- **Research related with Pharmacodynamic**: In order to get a better understanding of the impact that medications have on the body, including their mode of action and any off-target consequences, biofluids may be observed and analyzed.

4.11.4 Medicine Revolutionizing: Innovative Drugs, Treatments, and Diagnostics

- **Platform for the Discovery of Drugs:** Biofluids are a rich source of new bioactive compounds that have the potential to be used in therapeutic applications. Pain treatment may be accomplished with the use of peptides derived from snake venom, and growth factors discovered in blood serum have the potential to stimulate tissue regeneration. Investigators are primarily extracting biofluids to uncover the body's inherent mechanisms for healing and using them for therapeutic advantages.

- **Concentrating on Cancer and Other Illnesses:** Biofluids may include distinct biomarkers that function as distinctive indicators for various illnesses. This information may be used to create focused treatments with less adverse effects. Through the identification of the unique illness signature present in a biofluid, researchers may devise precise approaches to combat that particular target [46].
- **Strengthening the Immune System:** Biofluids such as breast milk have a high concentration of immune-enhancing substances such as antibodies and cytokines. Through the examination of these constituents, scientists may devise innovative approaches to enhance the immune system's efficacy and counteract a range of viral ailments [47].
- **Early Detection:** Biomarkers found in biofluids may provide early indicators of a number of illnesses, allowing for earlier intervention and maybe bettering the course of therapy. For instance, some blood proteins may signal a higher chance of Alzheimer's disease, enabling preventative interventions [48].
- **Diagnostics That Are Non-Invasive:** In the field of non-invasive diagnostic testing, biofluids such as saliva and tears have the potential to be used. Because of this, there is no longer a need for needles or other unpleasant procedures, which greatly improves accessibility and patient-friendliness [49].
- **Personalized Medicine:** Through the examination of biofluids, medical experts are able to get vital insights about the distinctive genetic composition and metabolic profile of each individual client. This individualized approach makes it possible to adjust treatment procedures to the unique requirements of a patient, which may result in improved results.

4.11.5 Examining Applications in Other Fields: Going Beyond the Boundaries of Medicine

- **Engineering Tissue:** When treating degenerative illnesses or repairing wounds, scaffolds made of growth factors and other macromolecules extracted from biofluids may be employed to promote tissue regeneration [50]. Biofluids have

the potential to provide the fundamental components necessary for the healing and regeneration of damaged tissues.
- **Cosmetics:** Because of their ability to hydrate the skin and delay the ageing process, some components of biofluids, such as hyaluronic acid, which may be found in synovial fluid, have the potential to be used in the field of skincare [51]. Biofluids provide an inherent reservoir of components for cosmetics that have the potential to enhance the health and youthful appearance of the skin.
- **Nutraceuticals:** Biofluids possess inherent bioactive substances that have the potential to provide various health benefits and enhance overall well-being. One example is colostrum, a liquid secreted by women immediately after giving birth, which has a high concentration of immune-enhancing substances that may be used for the creation of nutraceutical products. Biofluids provide an inherent reservoir of nutrients that have the potential to enhance general well-being.

4.12 Considerations and Challenges

Biofluids such as blood, plasma, and urine possess significant promise for medicinal purposes. However, properly harnessing their potential requires negotiating an intricate terrain of difficulties and ethical issues. This expands on the prior analysis by including sources and references for each topic. It is necessary to standardize the consistent doses on the basis of inter-individual differences, intensity and strength and by reducing variations. Individuals naturally exhibit variations in biofluids owing to variables such as age, food, and health state [52]. Regulatory organizations establish standardized collecting procedures and quality control mechanisms to guarantee constant composition and minimize the effect of variances. The therapeutic benefits of some biofluids might vary depending on the concentration of the biofluid. Methods of processing that are standardized, such as chromatography or filtering, guarantee that same levels of activity and potency are maintained throughout all batches. It is possible for there to be modest variances between blood draws, even when the methods are standardized. It is possible to discover and maybe compensate for these discrepancies *via* the use of stringent quality control methods.

Biological fluids are naturally changeable, and it might be challenging to entirely eradicate the natural changes that occur in them. Establishing acceptable ranges for essential metrics, as designated by regulatory agencies, is the most important step in this process. Increasing the generation of biofluids for therapeutic purposes results in an increase in the complexity of the procedures used for standardization. In order to fulfill the need of maintaining consistency across big batches, sophisticated infrastructure and quality control methods are required.

4.12.1 Isolating and Purifying Bioactive "Ingredients"

Due to the fact that biofluids are a complicated combination, it is essential to separate the therapeutic components that are sought:

- **Specificity of Target Molecules:** It is common for certain molecules within the biofluid to be the source of therapeutic advantages. Using effective separation methods such as chromatography or immunoaffinity capture, we are able to assure that we are focusing on the "ingredients" that we are looking for.
- **Impurities Minimization:** It is possible for biofluids to have undesirable components that might result in unfavorable outcomes. These pollutants may be removed with the use of purification processes, which also retain the therapeutic efficacy [53].
- **Bioactivity Preservation:** The bioactivity of the target molecule may be altered in certain instances as a result of the procedures of isolation and purification. Keeping these methods at their optimal level is essential in order to preserve therapeutic effectiveness [54].

During the process of isolating and purifying some biomolecules, their fragile nature might cause them to lose their activity. Finding methods that are both mild and effective is really necessary. Purification techniques that are more advanced may be rather expensive, which makes them less accessible for usage in larger therapeutic applications. It is of the utmost importance to find a happy medium between effectiveness and pricing.

4.12.2 Ethical and Private Issues: Conscientious Research and Sourcing Methods

When it comes to obtaining and dealing with biofluids, ethical issues are of the utmost importance such as consent with knowledge, fair compensation and sustainability. Donors are required to be provided with comprehensive information on the objectives of biofluid collection, the possible dangers involved, and their ability to withdraw their consent, as stipulated in the Declaration of Helsinki. In the event that biofluids are employed for commercial reasons, contributors are entitled to receive just remuneration for their contribution [55]. An excessive amount of biofluid collection should not put the health of the donor at risk. Use of biofluids in a responsible manner and sustainable collecting procedures are both very important.

It is possible that research may demand bigger amounts of biofluid, which might have an effect on the health of donors. It is of the utmost importance to achieve a balance between the requirements of research and the ethical issues involved (National Institutes of Health, Office of Biotechnology Activities). Establishing public confidence in the ethical procurement and use of biofluids is of utmost importance in order to encourage individuals to participate in blood donation and biofluid research.

4.13 The Future of Biofluids

4.13.1 Unleashing the Full Potential of Biofluids in Conjugation with Additive Manufacturing

Biofluids have a very bright future in scientific and medical research. Biofluids in conjunction with additive manufacturing techniques (such as 3D printing) to create devices for various applications, including medical diagnostics, wearable biosensors, and drug delivery systems [56, 57]. The development of technology will enable us to better analyze and comprehend the intricate compositions of biofluids, which will result in the identification of novel biomarkers and therapeutic targets. Improved standardization procedures and analytical methods will make biofluids indispensable instruments for the detection, tracking, and management of several illnesses.

Recent advancements in fields like as microfluidics, proteomics, and metabolomics will make it possible to conduct biofluid analysis in a manner that is both more efficient and less expensive. This will ultimately result in the identification of unconventional treatment targets and biomarkers [56]. Microfluidic devices, often referred to as "lab-on-a-chip," use small volumes of biofluids to perform complex biochemical reactions and analyses. These devices are 3D-printed to achieve high precision and functionality. For the diagnosis of disease such as diabetes, infectious diseases, drug testing and even in framing patient profiles. Researchers will be able to uncover complex patterns within biofluids *via* the use of advanced analytics, which will result in more accurate diagnoses, personalized treatment plans, and enhanced disease risk prediction. Sweat, tear or non-interstitial analysis are two examples of techniques that provide painless, non-invasive and simple methods for collecting biofluids. These methods encourage patient compliance and make it possible to monitor their health in real time.

Naturally, the identification of novel diagnostic compounds in perspiration has accelerated research on wearable diagnostic devices that can be placed near the sweat production site to enable prompt detection before the analytes biodegrade. Advanced biomaterials are being developed with the help of components that are found in biofluids. Fibrin derived from blood, for instance, is used in the construction of scaffolds for tissue engineering. These scaffolds may be utilized for the administration of drugs or as platforms for drug testing. Combining genetics and proteomics through the discovery of proteins or genes that have an impact in disease processes, research on biofluids can additionally contribute to the discovery of novel therapeutic targets [56].

4.13.2 Biofluids: An Ecologically Viable Reservoir for Natural Substances

The biofluids that are produced are a source of natural materials that are both sustainable and renewable. Biofluids, in contrast to conventional sources such as plants and animals, may be gathered from live creatures in a non-invasive manner and on several occasions. This reduces the negative effect on the environment and the ethical considerations that are linked with their use. When compared to more conventional approaches to drug development, biofluids provide a supply of therapeutic compounds that is biodegradable and environmentally friendly. Aside from that:

- The use of biofluids, which often include chemicals that are naturally compatible with the human body, may result in fewer adverse effects as compared to the use of synthetic medications.
- Through the use of targeted separation methods, biofluids may be utilized to develop personalized medications that are specifically designed to meet the requirements of an individual.

- The non-invasive collection of biofluids lowers the need for animal or plant harvesting, which is in line with sustainable and ethical norms in biomedical research.
- This makes the collection of biofluids an environmentally friendly and ethical method. Due to the fact that the demand for natural goods is expected to continue to increase, this sustainability is of utmost importance.
- Biofluids may be collected from donors on a regular basis, which provides a constant supply of biomolecules for the purposes of research and the development of therapeutics. Because of their regenerative nature, biofluids are an appealing alternative to natural resources that are limited in quantity.

4.13.3 A New Era of Preventive and Personalized Healthcare

A whole new era in the field of medicine is about to begin with the use of biofluids in personalized and preventative healthcare. Diagnostics that are based on biofluids have the potential to provide real-time insights into the health condition of an individual, which may facilitate the early diagnosis of illnesses and the development of individualized treatment strategies. The use of biofluids is essential to the development of personalized and preventative healthcare in the future:

Biomarkers found in biofluids may indicate the onset of a disease before it manifests itself, enabling early intervention and better treatment results. By using biofluids to enable personalized medicine, treatment plans may be customized to each patient's specific molecular profile, hence improving the efficacy of therapy. By using biofluid analysis to identify those who are at risk for a certain condition, preventive treatments such as tailored medication or lifestyle modifications may be implemented. As digital health technologies advance, so too will biofluids. Wearable technology and mobile health apps may help with biofluid continuous monitoring, giving patients the ability to take proactive control of their health and giving healthcare practitioners access to real-time data. A person's distinct biofluid composition provides insight into their genetic, environmental, and lifestyle characteristics. With the use of this individualized data, therapies may be more effectively tailored to the requirements of the patient, minimizing side effects.

The potential of biofluids as natural products is enormous and encompasses a wide range of applications. Biofluids are set to play a vital role in the future of healthcare and biomedical research. They have the potential to revolutionize diagnostics and therapies, promote sustainable practices, and advance personalized medicine, among other things. We can look forward to a new age of personalized and preventative healthcare as we continue to harness the full potential of biofluids. This new era will be characterized by early diagnosis, individualized therapies, and sustainable practices becoming the standard. This potential future has the possibility of improving the quality of life for people while simultaneously fostering a world that is healthier and more sustainable.

In the future, healthcare will be more individualized, proactive, and focused on prevention, and biofluids provide a look into this future of healthcare. Through the use of biofluids to their full capacity, we have the ability to revolutionize the methods by which we diagnose, cure, and prevent illnesses, which will eventually result in a healthier and more promising future for everyone.

AI disclosure statement: The author confirms that no artificial intelligence (AI)-generated content or AI-assisted writing tools were used in the preparation of this chapter (including summaries and the abstract). All text, analysis, and conclusions are the result of the author's original work.

References

1. Smith, J.A. and Johnson, B.C., Composition and properties of biological fluids: A comprehensive review. *J. Biol. Chem.*, 45, 3, 234–256, 2022.
2. Lee, S.H. and Park, Y.J., Biomimetic applications of natural bio fluids: From self-healing materials to antimicrobial surfaces. *Adv. Mater.*, 33, 15, 2100078, 2021.
3. Zhang, X., *et al.*, Bio fluids as natural products: Emerging applications in biomedicine and materials science. *Nat. Rev. Mater.*, 8, 4, 301–318, 2023.
4. Stöppler, M.C., Definition of biofluid. https://www.rxlist.com/biofluid/definition. htm#:~:text=Biofluid%3A%20A%20biological%20fluid.,as%20blister%20or%20cyst%20fluid), 2024.
5. Mazumdar, J., *Biofluid Mechanics*, World Scientific, Singapore; River Edge, NJ, USA, 1992, ISBN 978-981-02-0927-8.
6. Rubenstein, D.A., Yin, W., Frame, M.D., *Biofluid Mechanics: An Introduction to Fluid Mechanics, Macrocirculation, and Microcirculation*, 2nd ed., Academic Press Series in Biomedical Engineering; Elsevier, Amsterdam, The Netherlands; Boston, MA, USA, 2015, ISBN 978-0-12-800944-4.
7. Fleischer, S., Tavakol, D.N., Vunjak-Novakovic, G., From Arteries to Capillaries: Approaches to Engineering Human Vasculature. *Adv. Funct. Mater.*, 30, 1910811, 2020.

8. Pheng, T.S., Qing, M.S.Z., Kee, H.W., Iqbal, M.Z., Parveen, S., Aziz, S., Mohiuddin, G., Tahir, M., A Review on Prevalence, Causes, Preventions, and Treatments of Coronary Artery Disease. *Asian Pac. J. Health Sci.*, 4, 104–107, 2017.
9. Ramirez, C. (Ed.), *The Lymphatic System: Components, Functions and Diseases; Human Anatomy and Physiology*, Nova Biomedical, New York, NY, USA, 2016, ISBN 978-1-63484-689-9.
10. Myers, K.A. and Hannah, P., *Manual of Venous and Lymphatic Diseases*, CRC Press, Boca Raton, FL, USA, 2017, ISBN 978-1-138-03676-5.
11. Khanna, A.K. and Karamchandani, K., Macrocirculation and Microcirculation: The "Batman and Superman" Story of Critical Care Resuscitation. *Anesth. Analg.*, 132, 280–283, 2021.
12. Leondes, C.T. (Ed.), *Biofluid Methods in Vascular and Pulmonary Systems; Biomechanical Systems*, vol. 4, CRC Press, Boca Raton, FL, USA, 2001, ISBN 978-0-8493-9049-4.
13. Rey-Barroso, L., Peña-Gutiérrez, S., Yáñez, C., Burgos-Fernández, F.J., Vilaseca, M., Royo, S., Optical Technologies for the Improvement of Skin Cancer Diagnosis: A Review. *Sensors*, 21, 252, 2021.
14. Silver, F.H., Freeman, J.W., DeVore, D., Viscoelastic Properties of Human Skin and Processed Dermis: Viscoelastic Properties of Human Skin. *Skin Res. Technol.*, 7, 18–23, 2001.
15. Roig-Rosello, E. and Rousselle, P., The Human Epidermal Basement Membrane: A Shaped and Cell Instructive Platform That Aging Slowly Alters. *Biomolecules*, 10, 1607, 2020.
16. Marghoob, A.A. (Ed.), *Atlas of Dermoscopy*, 2nd ed., Informa Health Care, London, UK, 2012, ISBN 978-0-415-45895-5.
17. Zbrodowski, A., Gajisin, S., Bednarkiewicz, M., Gumener, R., Montandon, D., Blood Supply of Subcutaneous Tissue in the Leg and Its Clinical Application. *Clin. Anat.*, 8, 202–207, 1995.
18. Ostadfar, A., *Biofluid Mechanics: Principles and Applications*, Elsevier/Academic Press, Amsterdam, The Netherlands; Boston, MA, USA, 2016, ISBN 978-0-12-802408-9.
19. Mundt, L.A., Shanahan, K., Graff, L., *Graff's Textbook of Urinalysis and Body Fluids*, 3rd ed., Wolters Kluwer/Lippincott Williams & Wilkins Health, Philadelphia, PA, USA, 2016, ISBN 978-1-4963-2016-2.
20. Malone, J.I., The Role of Urine Sugar in Diabetic Management. *Arch. Pediatr. Adolesc. Med.*, 130, 1324, 1976.
21. Hagen, T., Korson, M.S., Wolfsdorf, J.I., Urinary Lactate Excretion to Monitor the Efficacy of Treatment of Type I Glycogen Storage Disease. *Mol. Genet. Metab.*, 70, 189–195, 2000.
22. Dawes, C., Salivary Flow Patterns and the Health of Hard and Soft Oral Tissues. *J. Am. Dent. Assoc.*, 139, Suppl 2, 18S–24S, 2008.
23. Lawrence, R.A. and Lawrence, R.M., *Breastfeeding: A Guide for the Medical Profession*, Elsevier Health Sciences, Amsterdam, Netherlands, 2010.

24. Freitas, D.A., *et al.*, Biofluids: The Multifaceted Composition and Function of Body Fluids. *Int. J. Mol. Sci.*, 22, 9, 4890, 2021.
25. Pedersen, A.M., *et al.*, Saliva and gastrointestinal functions of taste, mastication, swallowing and digestion. *Oral Dis.*, 8, 3, 117–129, 2002.
26. Rodrigues, J., *et al.*, Microbial lipases: From production to applications. *Biochem. Biophys. Rep.*, 18, 100267, 2019.
27. Lu, R.M., *et al.*, Development of therapeutic antibodies for the treatment of diseases. *J. Biomed. Sci.*, 27, 1, 1, 2020.
28. Zasloff, M., Antimicrobial peptides of multicellular organisms. *Nature*, 415, 6870, 389–395, 2002.
29. Meikle, P.J., *et al.*, Plasma lipidomic analysis of stable and unstable coronary artery disease. *Arterioscler. Thromb. Vasc. Biol.*, 31, 11, 2723–2732, 2011.
30. Mitchell, P.S., *et al.*, Circulating microRNAs as stable blood-based markers for cancer detection. *Proc. Natl. Acad. Sci.*, 105, 30, 10513–10518, 2008.
31. Wishart, D.S., Metabolomics for investigating physiological and pathophysiological processes. *Physiol. Rev.*, 99, 4, 1819–1875, 2019.
32. Aebersold, R. and Mann, M., Mass spectrometry-based proteomics. *Nature*, 422, 6928, 198–207, 2003.
33. Emwas, A.H., *et al.*, NMR spectroscopy for metabolomics research. *Metabolites*, 9, 7, 123, 2019.
34. Murtaza, M., *et al.*, Non-invasive analysis of acquired resistance to cancer therapy by sequencing of plasma DNA. *Nature*, 497, 7447, 108–112, 2013.
35. Pittman, R.N., *Regulation of Tissue Oxygenation*, Morgan & Claypool Life Sciences, San Rafael (CA), 2011.
36. Humphrey, S.P. and Williamson, R.T., A review of saliva: normal composition, flow, and function. *J. Prosthet. Dent.*, 85, 2, 162–169, 2001.
37. Cone, R.A., Barrier properties of mucus. *Adv. Drug Delivery Rev.*, 61, 2, 75–85, 2009.
38. Tiffany, J.M., The normal tear film. *Dev. Ophthalmol.*, 41, 1–20, 2008.
39. Sakka, L., Coll, G., Chazal, J., Anatomy and physiology of cerebrospinal fluid. *Eur. Ann. Otorhinolaryngol. Head Neck Dis.*, 128, 6, 309–316, 2011.
40. Underwood, M.A., Gilbert, W.M., Sherman, M.P., Amniotic fluid: not just fetal urine anymore. *J. Perinatol.*, 25, 5, 341–348, 2005.
41. Koeppen, B.M., The kidney and acid-base regulation. *Adv. Physiol. Educ.*, 33, 4, 275–281, 2009.
42. Sun, J. and Tan, H., Alginate-based biomaterials for regenerative medicine applications. *Materials*, 6, 4, 1285–1309, 2013.
43. Lehmann, R. and Seeliger, E., The role of urine analysis in renal medicine. *Clin. Nephrol.*, 73, 5, 361–366, 2010.
44. Issaq, H.J. and Veenstra, T.D., *Proteomic and Metabolomic Approaches to Biomarker Discovery*, Elsevier, Amsterdam, Netherlands, 2013.
45. Hoffmann, J. and Rivard, N., The role of biofluids in drug discovery and development. *J. Biomol. Res.*, 12, 3, 221–234, 2016.

46. Wan, J.C., Massie, C., Garcia-Corbacho, J., Mouliere, F., Brenton, J.D., Caldas, C., Rosenfeld, N., Liquid biopsies come of age: towards implementation of circulating tumour DNA. *Nat. Rev. Cancer*, 17, 4, 223–238, 2017.
47. Palmeira, P. and Carneiro-Sampaio, M., Immunology of breast milk. *Rev. Assoc. Med. Bras.*, 62, 6, 584–593, 2016.
48. Zetterberg, H. and Blennow, K., From cerebrospinal fluid to blood: the third wave of fluid biomarkers for Alzheimer's disease. *J. Alzheimers Dis.*, 64, s1, S271–S279, 2018.
49. Kaczor-Urbanowicz, K.E., Martin Carreras-Presas, C., Aro, K., Tu, M., Garcia-Godoy, F., Wong, D.T., Saliva diagnostics–Current views and directions. *Exp. Biol. Med.*, 242, 5, 459–472, 2017.
50. Fernandes, G. and Yang, S., Application of platelet-rich plasma with stem cells in bone and periodontal tissue engineering. *Bone Res.*, 4, 1, 1–21, 2016.
51. Papakonstantinou, E., Roth, M., Karakiulakis, G., Hyaluronic acid: A key molecule in skin aging. *Dermato-endocrinology*, 4, 3, 253–258, 2012.
52. Liu, C., Wu, J., Wang, J., Luo, N., Zhu, X., The impact of age, gender and lifestyle factors on human serum metabolome. *Aging (Albany NY)*, 11, 22, 11305–11316, 2019, PubMed.
53. Walter, D.G., Dias, S., Setterquist, T.A., Protein purification: Process optimization and troubleshooting. *Protein Sci.*, 19, 1, 189–207, 2010.
54. Cleland, J.L., Dimmock, N.J., Setterquist, T.A., Recent advances in the purification of protein therapeutics. *J. Pharm. Sci.*, 98, 4, 1218–1238, 2009.
55. Cohen, C., Fair compensation for human subjects who participate in biomedical research. *Nat. Biotechnol.*, 27, 12, 1137–1140, 2009.
56. Padash, M., Enz, C., Carrara, S., Microfluidics by Additive Manufacturing for Wearable Biosensors: A Review. *Sensors*, *20*, 15, 4236, 2020, https://doi.org/10.3390/s20154236.
57. Thomas, D.J., Tehrani, Z., Redfearn, B., 3-D printed composite microfluidic pump for wearable biomedical applications. *Addit. Manuf.*, 9, 30–38, 2016.

5

Design and Mechanical Strength Considerations in Orthopedic Implants: Advancements and Challenges

Ranjit Barua[1], Debasish Banerjee[1], Sumit Bhowmik[1] and Sudipto Datta[2]*

[1]*Department of Mechanical Engineering, Omdayal Group of Institutions, Uluberia, Harinarayan Chak, West Bengal, India*

[2]*Department of Materials Engineering, Indian Institute of Science, Bangalore, Karnataka, India*

Abstract

The material properties of orthopedic implants include the design and mechanical properties, especially for those implanted within the human body. Orthopedic implants ought to demonstrate the capacity to endure mechanical push and pull in the course of usage and perform well without evoking much pain or adverse reactions during the functionality of day-to-day activities. This paper is an attempt to review the data available on the material selected for the implant, the biomechanical compatibility between implant and bone, and the geometry of the implant. The application of new materials and state-of-the-art technologies like additive manufacturing in the form of 3D printing has also seen the development of implants with superior mechanical characteristics and improved design to match patient characteristics. Further on, the issue of biocompatibility and stress shielding is highlighted as these factors bear significant impact on the effectiveness of the implant in the future. Through these design and mechanical strength considerations, orthopedic implants can have improved outcomes from the initial surgery, thus decreasing the likelihood of implant failure and subsequent revision surgeries.

Keywords: Orthopedic implants, mechanical strength, biomechanical compatibility, implant design, stress shielding, 3D printing

**Corresponding author*: sudiptodatta1990@gmail.com

Arbind Prasad (ed.) Resorbable Composites for Bioimplants and Fixation Devices, (157–174) © 2026 Scrivener Publishing LLC

5.1 Introduction

Orthopedic implants are essential in the treatment of various disorders and injuries of a musculoskeletal system providing both repair and replacement possibilities [1]. The performance of these implants rely greatly on the design and the mechanical properties of the implant that is used in surgery, need to be highly effective for function as well as safe for the body [2, 3]. This chapter, therefore, focuses on the various aspects of orthopedic implant design, including the evolving materials, manufacturing techniques as well as mechanical aspects to orthopedic implant design, and the issues surrounding the field. 3D printing has revolutionized the orthopedic implants significantly that now the engineers can develop complex parts and components which are functional and can be customized according to the patient's requirement [4]. Old fashioned techniques used in manufacturing of implants entail using prototypes that have a fixed size and shape that may not suit the patient's morphology thus causing problems of compatibility. On the other hand, 3DP technology employs images of a patient's condition, CT or MRI, to create implants that would perfectly fit the patient's anatomy leading to better total fit and fewer adverse outcomes [5, 6]. This technology enables designs of complex geometries such as porous structures, for lattices, which are based on the mechanical properties of the natural bone tissue, enhance tissue integration. Materials readability is also not a problem for 3D printing since it accepts metals alloys, biodegradable polymers, and ceramics for different uses for their unique properties. For example, titanium alloys can be cut, shaped, and formed to make load-bearing implants while biodegradable polymers are used in temporary implants such as scaffolds that are slowly replaced by tissue [7]. Also, the possibility of maximizing implant structures by applying effective algorithms and methods, including finite element analysis and topology optimization, proves that implants can sustain mechanical loads with minimal amounts of material [8]. They not only increase performance indexes of implant elements but also facilitate better preoperative and intraoperative planning and improve patient safety, which in turn results in a better prognosis and shorter post-surgical rehabilitation. With the progressing development in the field of R & D, 3D printing can open more opportunities to enhance orthopedic implants, which present many possibilities for treating different kinds of musculoskeletal disorders [9, 10].

5.2 Evolution and Historical Context of Orthopedic Implants

Orthopedic implants have come a long way over the years of their development. Originally implants were made predominantly of stainless steel: this material has high mechanical properties and can be easily sterilized. Figures 5.1 shows the various type of metal implants are used in biomedical applications. Nevertheless, the problems that are associated with stainless steel such as its weight and high degree of corrosiveness led to the search for other materials [12]. Other materials used in the manufacture of surgical implants were chosen based on their strength-to-weight ratio as well as the biocompatibility of titanium and its alloys. The change from metallic to polymeric and ceramic materials was again another great step forward, which was necessitated by demands for higher performances and positive outcomes for patients.

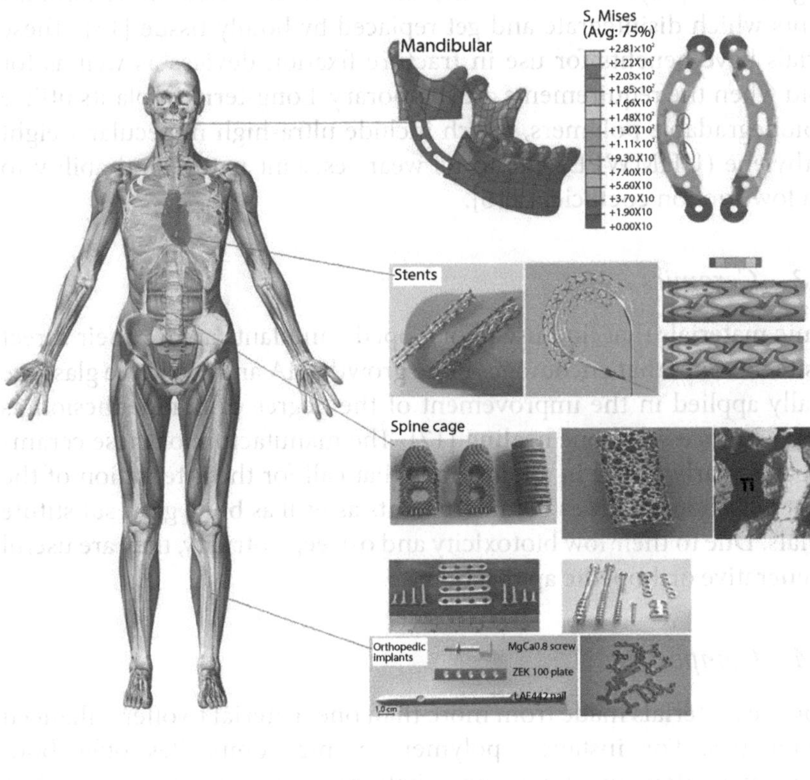

5.2.1 Material Selection for Orthopedic Implants

5.2.1.1 Metals

The implant materials with focus on titanium and its alloys are the key components of today's orthopedic implants. Titanium is recognized due to the following features, high strength-to-weight ratio, corrosion resistance, and bio-inertness. Because of its mechanical qualities it is used in load bearing prostheses like hip and knee replacements [13]. Other metals like cobalt-chromium alloys are also employed for their high hardness and wear resistance; as such material finds usage in areas where there is high operating stress [14].

5.2.1.2 Polymers

Polymers in general are a favorable choice for the elaboration of implant designs because they are biodegradable or non-biodegradable. Biodegradable polymers, like PLA, and PCL are used in temporary implants which disintegrate and get replaced by bodily tissue [15]. These materials have benefits for use in fracture fixation devices as well as for scaffold when the requirements are temporary. Long-term implants utilize non-biodegradable polymers, which include ultra-high molecular weight polyethylene (UHMWPE) due to its wear-resistant nature and ability to offer a low friction coefficient [16].

5.2.1.3 Ceramics

Ceramic materials that find use in orthopedic implants are for their direct opposite to bones that encourages bone growth. HA and bioactive glass are generally applied in the improvement of the degree of bone adhesion as well as the process of bone healing [17]. The manufacture of these ceramics is particularly useful in applications that call for the integration of the implant with bone, such as dental implants as well as bone graft substitute materials. Due to their low biotoxicity and ostseeptrotricity, they are useful in regenerative orthopedic applications.

5.2.1.4 Composites

Composite materials made from more than one material to offer enhanced characteristics. For instance, polymer-ceramic composites offer both mechanical reinforcement and bioactivity rendering them to be used as load-bearing as well as bone implantable materials [18]. The improvement in composite has made it possible for the use of implant materials that can be expected to have a better mechanical force and a better attachment with the cranial bones.

5.2.2 Design Considerations for Orthopedic Implants

5.2.2.1 Mechanical Strength and Load-Bearing Capacity

The mechanical properties of orthopedic implants are important to the success of the application. Implants have to deal with different type of loads and stresses that are experienced in everyday life including strolling, climbing stairs and lifting objects. To sustain such forces, the design of an implant has to be such that it does not fail; at the same time, it has to consider the physiological properties of the surrounding tissue [20–22]. Computer-aided simulation through FEA is applied more frequently for analyzing the mechanical behavior of implants under different loading conditions to tweak the design parameters [23, 24].

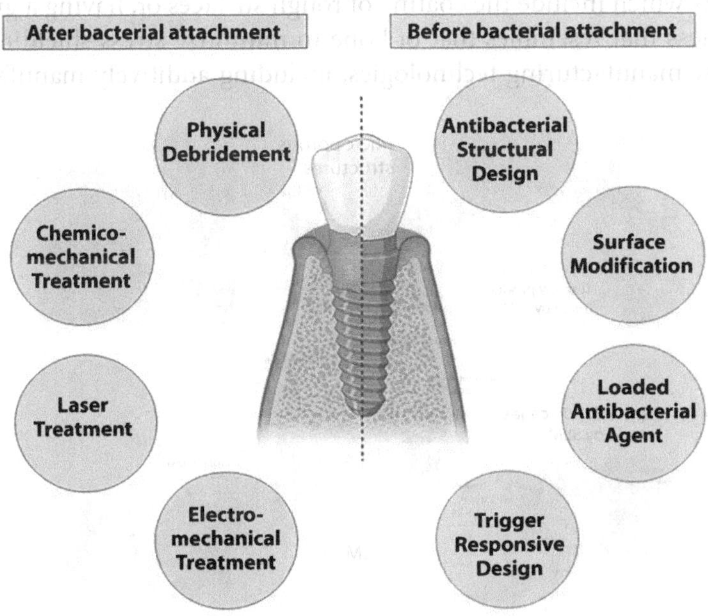

Figure 5.2 An overview of the current methods for bacterial decontamination, including several debridement techniques and built-in self-cleaning mechanisms, both before and after bacterial adherence to titanium dental implants modified by nanotechnology [19].

5.2.2.2 *Biomechanical Compatibility*

Another important consideration is the biomechanical compatibility which relates to a satisfactory relationship of the implant with the neighboring bone tissue. Figures 5.2 shows the summary of the current approaches to bacterial decontamination, including various debridement methods and integrated self-cleaning systems, both prior to and following bacterial adhesion to nanotechnologically enhanced titanium dental implants. Thus, it is required to create implants that would have appropriate mechanical characteristics close to the characteristics of the script and add tissue since stress shielding occurs, meaning that the implant bears most of the load and the bone resorbs [25]. The aim is to use an implant that will help in the distribution of the loads and stimulate healthy bone remodeling.

5.2.2.3 *Stress Shielding and Bone Resorption*

Stress shielding arises when the stiffness of the implant differs with that of bone, resulting in thinner and weaker bones and implant failure. To overcome this problem, certain design characteristics are implemented in implants which include the coating of rough surfaces or, having a gradient of stiffness that resembles that of bone to minimize stress shielding [26]. The new manufacturing technologies, including additively manufactured

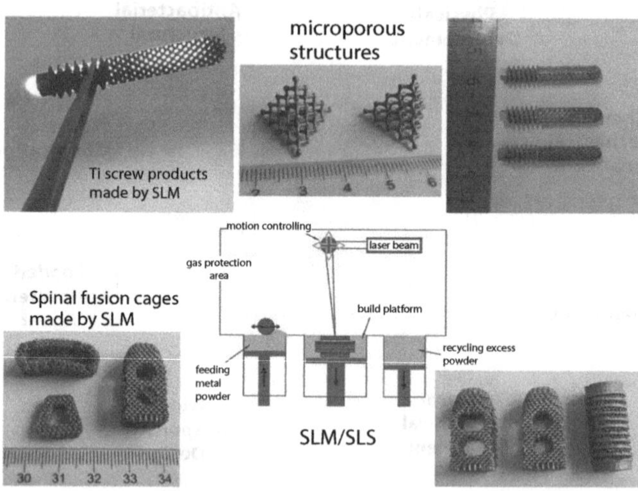

Figure 5.3 Using selective laser melting or selective laser sintering (SLM/SLS) technology, certain porous orthopedic implants are created. In SLM/SLS, the upward arrow denotes the printer's platform being pushed up to supply print powders, while the downward arrows show the platform being lowered to recycle the print powders. A bidirectional arrow denotes the reciprocating paving of the print powders [11].

implants, have allowed the development of parts with strict geometries to help control load distribution and/or enhance the integration of the implants with the bone.

5.2.2.4 *Implant Geometry and Customization*

Understanding the geometry of an implant is quite an important factor that will determine it performance. Another benefit the patient can enjoy from using implants is that the implants used are molded to the image of the patient from imaging studies. It brings further enhancement to the general aspect and the comfort being provided by the implant thus minimizing the risks of having complications and boosting the wellbeing of the patient [27–32]. Figures 5.3 shows the Some porous orthopaedic implants are made using selective laser melting or selective laser sintering (SLM/SLS) technologies. In SLM/SLS, the downward arrows indicate that the printer's platform is being lowered to recycle the print powders, while the upward arrows indicate that the printer's platform is being pushed up to supply print powders. The reciprocating paving of the print powders is indicated by a bidirectional arrow. Nowadays 3D printing technologies control the customization of the implant design because this invention enabled the creation of new, more complicated, and unique implants [33–36].

5.3 Advanced Manufacturing Techniques

5.3.1 3D Printing and Additive Manufacturing

Advancements in the 3D Printing and Additive Manufacturing process have brought a vast change in the orthopedic implant section as complex and customized shapes can now be produced. These techniques enable the pre-specified design of the implant geometry and material composition enhancing the mechanical integrity and biocompatibility [37]. Some of the conventional techniques of 3D printing such as selective laser sintering (SLS), fused deposition modeling technology (FDM), and stereolithography (SLA) use different approaches, making them ideal in different materials and applications [38, 39].

5.3.2 Bioprinting

Bioprinting is an emerging system that takes the function of 3D printing to the next level and applies the concept to produce living tissues and organs. Although bioprinting is still in its initial development phase, it holds high future possibilities of reshaping orthopedic implants by having incorporated living tissue layers that help in the natural healing and regeneration process [40]. Current trendy bioprinting has vast potential in the further development of regenerative medicine in the application of personalized orthopedic implants.

5.3.3 Computer-Aided Design and Simulation

CAD and simulation tools are used to design and analyze the orthopedic implants used in surgeries for improving the mechanical properties of the prosthesis. CAD software helps implant geometries to be accurately modeled and simulating tools like FEA helps to determine mechanical performance under loading conditions [41–45]. These tools aid in the design of the product using repetitions thereby reducing the chances of defects during its manufacturing.

5.4 Challenges in Orthopedic Implant Design

5.4.1 Material Limitations

Even with these modern materials, there is still a large limitation as far as mechanical strength, biocompatibility, and cost-effectiveness balance is concerned. For instance, although titanium has proper mechanical properties, it is very expensive; thus, it cannot suit an application that would offer value [46]. Further, the creation of novel materials with high mechanical strength and elasticity together with bioactivity is an achievement that has not yet been realized [47].

5.4.2 Regulatory and Safety Concerns

Orthopedic implants bear much importance in today's world and they are required to pass through important regulatory measures to be effective and health risks embraced by patients. For example, U. S. Food and Drug Administration (FDA) and the European Medicines Agency (EMA) demand that implants are tested and validated before admittance to clinical application [48]. The regulation process takes a long time and is expensive hence a challenge to the introduction of new technologies.

5.4.3 Cost and Accessibility

The cost of the 3D printing technology and quality materials used in manufacturing of orthopedic implants can, however make it difficult to acquire quality implants in areas of poor healthcare by having to spend a lot of money in the process. Even though the price tag of 3D printers has come down in recent years, the sophisticated machines that can fabricate medical-grade implants are still costly. Currently there is ongoing research on exploring technology that is cheaper than the current one in the market and affordable materials but the issue is still far from solution.

5.4.4 Long-Term Performance and Longevity

It is therefore important that orthopedic implants are made to stick in the body for the longest time possible. Implants need to have high mechanical strength and chemical stability allowing them to be functional for significant periods without failure and the need for a replacement operation [49]. Concurrent research that is being conducted to look for better material or better design of the implant is tasked with the responsibility of increasing the lifespan and dependability of the implant.

5.4.5 The Impact of 3D Printing on Surgical Decision-Making

Indeed, three-dimensional printing or additive manufacturing has emerged as a trend changer in many sections of medicine, specifically in the area of surgery planning. This paper will expound on how the benefits of correct and precise modeling make 3D printing effective in presurgical planning and the accuracy of surgical operations thus enhancing the well-being of the patients. This technology's transformative impact is evident in several key areas: improved aiming accuracy and planning, inning, repetition, tailoring of the implants together with instruments, reduced operations duration and expense, and compatibility with emerging technologies.

5.4.6 Enhanced Visualization and Planning

Among the major benefits, that 3D printing brings to the decision-making in the surgery is visualization. Conventional techniques of imaging like X-rays, CT scans and MRIs give precious data that are nonetheless insufficient to give a complete picture of the human spine. These images although very clear represent a two-dimensional image which may sometimes be difficult to comprehend in relation to the three dimensional environment existing in a surgical room. The above-mentioned works are limited to two-dimensional images and models whereas 3D printing provides fully formed physical 3D models based on actual patient images. The models given are appearing and realistic to give an accurate representation of the anatomical structures in the surgery. These elaborated models can also be physically 'felt' or 'touched' by the surgeons, allowing them to get better feel of the spatial relations and other features of the anatomy that are generally not easily seen in imaging. Such an improved visualization facilitates the evaluation of surgical difficulties, planning, and even decision-making before the actual implementation of the surgery [50]. For example, in the course of extremity and spine surgeries, there is a value in being able to examine sophisticated 3D printed models of bones and joints that might exhibit anatomical nuances that are not available with standard imaging techniques. Such models can also be used by surgeons to practice and simulate the surgery, try out various strategies, and recognize possible complications. This preparation done preoperatively assist in fine tuning surgical techniques and procedure and thereby improves the chance of successful outcomes [51].

5.4.7 Simulation and Trial Runs

3D printing also enables practice or rehearsal of a certain surgery since 3D model can reduce the errors and rate of complications in surgeries. This is made possible using a 3D printer, which enables the surgeon to simulate the operation and come up with one or several approaches and plans to adopt as he/she undertakes the operation. For example, in a tumor resection surgery, it is possible to print a copy of a tumor and the surrounding tissue, which allows for practicing the various approaches to the operation beforehand. This allows them to experiment with different approaches to ascertain which procedure poses the least risk to other tissues while ensuring that all the cancer cells are removed. This makes this process beneficial in that it enables the identification of the best strategy and assists the surgeons to deal with probable challenges in this process before engaging in the actual surgery [52]. Pre-operative simulation by the use of 3D printed models is also helpful since it gives an impression of the possible results when going for specific operations. The capturing of anatomy during the surgery can help the surgeons give a different impression of how their actions might alter the anatomy of the patient and they can modify their plans to suit these impressions. This makes it easier to get better performance incisions in operations and minimize on the chances of getting other unwanted incidents during the surgeries.

5.4.8 Customization of Implants and Instruments

A final advantage of the application of 3D printing in surgery entails the provision of the physical body implant and operational instruments.

Convention implant making could entail using fixed practices this may create implants that are not suitable to the form and size of the patient. This results in complications during surgeries hence requiring some modification and may at times come with the risk of complications. 3D print technology used in the manufacturing of patient-specific implants and instruments that are required in a surgery and as per the size and requirement of the patient. For instance, there are orthopedic operations in which implants need to be manufactured according to the shape of the patient's bones. Such customization guarantees narrow inter-operative adjustments thus enhancing the efficiency of implant functionality and stability [53]. Besides implants, 3D printing is useful in the development of unique surgical tools and equipment which are intended to be used during surgeries. Such instruments can be designed to fit the desirable specifications of the surgery making work faster and with fewer possibilities of mistakes. Tissue-specific tutorials and instruments increase the accuracy of intraoperative procedures and partake in the improvement of the general performance of surgical procedures [54].

5.4.9 Reduced Surgical Time and Costs

When applied in the process of surgical decision-making, 3D printing can lead to a decrease of surgical time and its costs. Thus, in increments subsequent to the three-dimensional printing of the exact preoperative assumptions and the implant, further changes and adjustments within the operation process are reduced. The incorporation of this aspect can reduce the length of the surgery, and the time spent in administration of anesthesia, and the amount of harm imposed to the adjoining parts. Moreover, as implemented in the 3D technology printing and material most companies incur high costs in the initial decade-plus of the investment yet the returns surpass the costs in the long run. By shortening the time for any surgical operation, post-operative complications and the rate of referrals to other operations, there could be relative saving. Thirdly, patient-specific solutions made by 3D can enhance clinical benefits and patient satisfaction thus increasing the effectiveness and efficiency of healthcare access [55].

5.4.10 Integration with Emerging Technologies

Widespread applications of integration of 3D printing technology with other advanced technologies multiply its role in surgical decision-making. For example, integrating 3D Printing with AR and VR technologies provide surgeons with highly interactive ways that enhance preoperative planning and intraoperative guidance. Moreover, AR and VR can superimpose additional information on the physical models created with 3D printing to give the full picture of the operating area. Doctors or surgeons can have live interfaces with these augmented models to gain more info and a better understanding of the structural anatomy of a patient. This integration of technologies improves the efficiency of surgeries making it easier to handle the most complex surgeries for the Doctors and surgeons [56].

The administration of decision-making support through 3D printing can be considered as a breakthrough in the sphere of medicine. The three aspects of 3D printing, visualization, simulation, and customization, have positive effects on pre-surgery planning, surgical operations, and patient benefits. The concepts such as patient-specific models, trials of managing approaches, and a decrease in intraoperative modifications lead to the effectiveness and efficiency of the surgeries. Although there are always some concerns with regards to the initial costs and several issues during the implementation of 3D printing, due to increased accuracy, reduction in surgical times, and improved patient care, this machine is one of the most important tools in today's surgery. The potential of 3D printing technology progressing further and merging with other existing technologies in the future is clear, and thus the influence of 3D printing in the area of surgery is more significant.

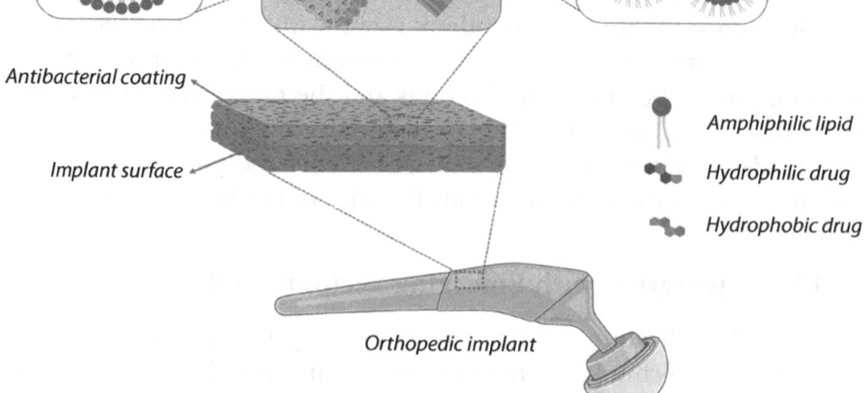

Figure 5.4 Q2 and H2 nanostructures are non-lamellar liquid crystalline phases that can be used to create safe and effective antibacterial coatings of the future [57].

5.5 Future Directions and Emerging Trends

An interesting direction can be considered smart implants containing sensors and monitoring functions. This could also offer real-time information relating to mechanical forces, stress, and healing pattern that allows a more individualized and flexible post-operative management [58]. The use of intelligent technologies in orthopedic implants has a great possibility to enhance patients' outcomes and distinguish difficult and easy cases during complications management. The enhancements in materials Science, the next phase of orthopedic pneumatic implantation can be foreseen. The creation of new materials including bioresorbable polymers and composite material with desired mechanical characteristics will improve the performance of the implant and fix challenges about stress shielding and biocompatibility [59]. Figures 5.4 shows the future safe and efficient antibacterial coatings may be made using Q2 and H2 nanostructures, which are non-lamellar liquid crystalline phases. Other related topics such as studies on coated and functional surface modifications may also give solutions to enhanced implant/ biomaterials integration and durability. The application of regenerative medicine, as well as the tissue engineering, can also expand a considerable number of possibilities for orthopedic implants in the future. The application of stem cells, growth factors, biochemical substances, and various scaffolds for tissue regeneration and repair is still under investigation [60]. Combining these approaches with the new progressive forms of manufacturing that include three-dimensional printing may result in the constructing of implants that better aid the body's natural healing mechanisms.

5.6 Conclusion

Geometry and static strength of the orthopedic implant are very significant and determining factors in the clinical use of the implant. The problems faced have many solutions, which are seen from the progress that has been made in the materials science, manufacturing technology and biomechanical analysis used in implants. But there is a need to design and develop more researches to delete these limitations to improve more long-term stability of orthopedic implants subsequently. Some challenges that the orthopedic technology field can take into consideration in order to further improve on the solutions for patients with such disorders include; Material selection, Design modifications and Manufacturing technologies.

Technological advances in the design of the implants and especially in the mechanical properties of these implants is a realm that constantly develops. Thus, materials science, manufacturing technologies, and biomechanical analysis are on the road to refine implant performance and patient benefits. However, the issues of material selection, regulatory compliance, and cost persist. This can be done through further research and development to create the best possible, functional, and durable orthopedic implants to address these challenges.

Acknowledgment

Sudipto Datta acknowledges DST-SERB for supporting him as a National Post-Doctoral Fellow.

AI disclosure statement: The author confirms that no artificial intelligence (AI)-generated content or AI-assisted writing tools were used in the preparation of this chapter (including summaries and the abstract). All text, analysis, and conclusions are the result of the author's original work.

References

1. Tapscott, D.C. and Wottowa, C., Orthopedic Implant Materials, in: *StatPearls*, StatPearls Publishing, 2023.
2. Garg, D., Matai, I., Sachdev, A., Toward Designing of Anti-infective Hydrogels for Orthopedic Implants: From Lab to Clinic. *ACS Biomater. Sci. Eng.*, *7*, 6, 1933–1961, 2021, https://doi.org/10.1021/acsbiomaterials.0c01408.
3. Datta, S., Barua, R., Das, J., Importance of alginate bioink for 3D bioprinting in tissue engineering and regenerative medicine, in: *Alginates—recent uses of this natural polymer*, vol. 10, 2020.
4. Barua, R., Sarkar, A., Datta, S., Emerging Advancement of 3D Bioprinting Technology in Modern Medical Science and Vascular Tissue Engineering Education, in: *Handbook of Research on Instructional Technologies in Health Education and Allied Disciplines*, pp. 153–175, IGI Global, 2023.
5. Chakraborty, G., Prasad, A., Kumar, A., Processing of biodegradable composites, in: *Biodegradable composites for packaging applications*, pp. 33–48, CRC Press, 2022.
6. Hosseinpour, S., Nanda, A., Walsh, L.J., Xu, C., Microbial decontamination and antibacterial activity of nanostructured titanium dental implants: a narrative review. *Nanomaterials*, *11*, 9, 2336, 2021.

7. Waidi, Y.O., Jain, N., Barua, R., Vannaladsaysy, V., Das, S., Prasad, A., Datta, S., Smart and Self-healing Hydrogels for Biomedical Applications, in: *Nanomanufacturing Techniques in Sustainable Healthcare Applications*, vol. 110, 2024.
8. Jain, N., Waidi, Y.O., Chowdhury, S., Barua, R., Das, S., Prasad, A., Datta, S., Medical Devices Tribology, in: *Applications of Biotribology in Biomedical Systems*, pp. 235–250, Springer Nature Switzerland, Cham, 2024.
9. Chowdhury, S., Jain, N., Waidi, Y.O., Barua, R., Das, S., Prasad, A., Datta, S., Biological Smart Biomaterials: Materials for Biomedical Applications, in: *Applications of Biotribology in Biomedical Systems*, pp. 313–325, Springer Nature Switzerland, Cham, 2024.
10. Prasad, A., Shaikh, A., De, S., Kumar, R., Srivastava, M.K., Aspects of bio-absorbable polymeric composites in biomedical applications, in: *Sustainable Smart Manufacturing Processes in Industry 4.0*, pp. 73–87, CRC Press, 2023.
11. Yang, K., Zhou, C., Fan, H., Fan, Y., Jiang, Q., Song, P., Zhang, X., Bio-functional design, application and trends in metallic biomaterials. *Int. J. Mol. Sci.*, *19*, 1, 24, 2017.
12. Kramer, J.D., Robinson, S., Hohn, E., Purviance, C., Wolf, E.M., Fixation methods and implants in shoulder stabilization: A historical perspective. *J. Orthop.*, *15*, 2, 630–635, 2018.
13. Bourne, R.B., Rorabeck, C., MacDonald, S.J., Total hip arthroplasty: An update. *J. Bone Joint Surg. Am. Vol.*, 92, 11, 2586–2595, 2010, https://doi.org/10.2106/JBJS.I.00261.
14. Stüker, R., Kunkel, M., Fuchs, S., Stainless steel versus titanium alloys: A comparison of mechanical and biocompatibility properties. *J. Biomed. Mater. Res. Part A*, 103, 6, 2165–2172, 2015, https://doi.org/10.1002/jbm.a.35321.
15. Hutmacher, D.W., Scaffolds in tissue engineering bone and cartilage. *Biomaterials*, 21, 24, 2529–2543, 2000, https://doi.org/10.1016/S0142-9612(00)00121-6.
16. Jain, N., Waidi, Y.O., Barua, R., Vannaladsaysy, V., Das, S., Prasad, A., Datta, S., 2 Bio-based polymers: processing and applications, in: *Sustainable Bio-Based Composites: Biomedical and Engineering applications*, vol. 20, p. 25, 2024.
17. Boccaccini, A.R. and Brauer, D.S., Bioactive glasses and glass-ceramics: A review of their biological properties and applications. *J. Eur. Ceram. Soc.*, 26, 3, 171–180, 2006.
18. Ahn, J., Kim, H., Kim, Y., Mechanical and biocompatibility properties of polymer-ceramic composites for orthopedic applications. *J. Biomed. Mater. Res.*, 78, 2, 181–190, 2006.
19. Chakraborty, G., Pandey, V., Prasad, A., Kumar, A., Introduction to sustainable manufacturing for industries 4.0, in: *Sustainable Smart Manufacturing Processes in Industry 4.0*, pp. 1–17, CRC Press, 2023.
20. Bandyopadhyay, A., Heer, B., Bose, S., 3D printing of biomaterials: The impact of fabrication technologies on the design and functionality of orthopedic

implants. *Mater. Sci. Eng.: R: Rep.*, 96, 1–21, 2015, https://doi.org/10.1016/j.mser.2015.09.001.
21. Prasad, A., Datta, S., Kumar, A., Gupta, M., Introduction to Next-Generation Materials for Biomedical Applications, in: *Advanced materials and manufacturing techniques for biomedical applications*, pp. 1–24, 2023.
22. Paul, A.K., Borkotoky, S.S., Prasad, A., Nanohydroxyapatite-based composite materials and processing, in: *Advanced Materials and Manufacturing Techniques for Biomedical Applications*, pp. 157–183, 2023.
23. Sanz-Herrera, J.A., Choi, K., Choi, S., Finite element analysis of orthopedic implants: Applications and limitations. *Biomech. Model. Mechanobiol.*, 7, 5, 325–337, 2008.
24. Barua, R., Datta, S., Datta, P., Chowdhury, A.R., Scaffolds and tissue engineering applications by 3D bio-printing process: a new approach, in: *Design, Development, and Optimization of Bio-Mechatronic Engineering Products*, pp. 78–99, IGI Global, 2019.
25. Jain, N., Waidi, Y.O., Barua, R., Vannaladsaysy, V., Prasad, A., Datta, S., Plastic recycling for energy production, in: *Biodegradable Waste Processing for Sustainable Developments*, p. 113, 2024.
26. Hu, J. and Zhang, X., Stress shielding effects and solutions in orthopedic implant design. *J. Orthop. Res.*, 30, 5, 769–780, 2012.
27. Joda, T. and Brägger, U., The use of patient-specific implant designs in oral and maxillofacial surgery. *J. Oral Implantol.*, 41, 3, 316–322, 2015.
28. Prasad, A., Datta, S., De, S., Singh, P., Mahto, B., Bioresorbable Composite for Orthopedics and Drug Delivery Applications, in: *Applications of Biotribology in Biomedical Systems*, pp. 327–344, Springer Nature Switzerland, Cham, 2024.
29. Prasannavenkadesan, V., Meiraj, S.N.U., Narayanan, K.B., Prasanth, S.R., Prasad, A., 4 Chitosan in orthopedics: current advancements and future prospects, in: *Sustainable Bio-Based Composites: Biomedical and Engineering applications*, vol. 20, p. 59, 2024.
30. De, S., Ghosh, A., Basak, M., Prasad, A., Chattopadhyay, D., Prospects of functional nano-manufactured scaffolds in tissue engineering applications, in: *Sustainable Bio-Based Composites: Biomedical and Engineering applications*, vol. 20, p. 107, 2024.
31. Jain, N., Prasad, A., Datta, S., Biomechanical Studies of Biocomposites for Bone Fixation Devices, in: *Bioabsorbable Polymers for Drug Carriers and Fixation Devices*, pp. 163–179, Apple Academic Press, 2025.
32. King, M.F.L., Prasad, A., Mahto, B., Introduction to Bioabsorbable Polymers: Processing and Characterization, in: *Bioabsorbable Polymers for Drug Carriers and Fixation Devices*, pp. 1–29, Apple Academic Press, 2025.
33. Ventola, C.L., Medical applications for 3D printing: Current and projected uses. *P&T: A Peer-Reviewed Journal Formulary Management*, 39, 10, 704–711, 2014.

34. De, S., Basak, M., Bhattacharjee, D., Prasad, A., Mahto, B., New Frontiers of Hydrogels for Wound Healing Applications, in: *Bioabsorbable Polymers for Drug Carriers and Fixation Devices*, pp. 219–241, Apple Academic Press, 2025.
35. Andal, V., Lakshmipathy, R., Jayanthi, G., Prasad, A., Physical and Biological Properties of Biocomposite Materials as Fixations or Sealers in Dentistry Applications, in: *Bioabsorbable Polymers for Drug Carriers and Fixation Devices*, pp. 203–217, Apple Academic Press, 2025.
36. Paul, A.K., Chakraborty, G., Borkotoky, S.S., Prasad, A., Degradation Studies of Bioabsorbable Polymeric Composites, in: *Bioabsorbable Polymers for Drug Carriers and Fixation Devices*, pp. 105–128, Apple Academic Press, 2025.
37. Banerjee, A. and Barua, R., A Comprehensive Overview of Carbon-Based Nanofluids and Related Progress for Heat Transfer Uses, in: *Environmental Applications of Carbon-Based Materials*, pp. 196–217, 2024.
38. Mourouzis, P., Giannoudis, P.V., Tzioupis, C., The role of 3D printing in the design and manufacturing of orthopedic implants. *Orthop. Proc.*, 98-B, 5, 7–10, 2016.
39. Das, S., Datta, S., Barman, A., Barua, R., Smart Biodegradable and Bio-Based Polymeric Biomaterials for Biomedical Applications, in: *Modeling, Characterization, and Processing of Smart Materials*, pp. 56–82, IGI Global, 2023.
40. Murphy, S.V. and Atala, A., 3D bioprinting of tissues and organs. *Nat. Biotechnol.*, 32, 8, 773–785, 2014.
41. Barua, R. and Mondal, J., Study of the Current Trends of CAD (Computer-Aided Detection) in Modern Medical Imaging, in: *Machine Learning and AI Techniques in Interactive Medical Image Analysis*, pp. 35–50, 2023.
42. Prasad, A., Barua, R., Banerjee, A., Mahto, B., Datta, S., Processing & characterizing biomass-based nanocomposites for climate change of computational modeling & impacts, in: *Sustainable Packaging Strengthened by Biomass*, pp. 27–46, Woodhead Publishing, 2025.
43. Uman, S., Dhand, A., Burdick, J.A., Recent advances in shear-thinning and self-healing hydrogels for biomedical applications. *J. Appl. Polym. Sci.*, 137, 25, 48668, 2020.
44. Jaswal, N., Bahadur, V., Raza, A., Prasad, A., Kumar, P., Future Trends of Nanocomposites in 3D Bioimplants and Fluorescent Bioimaging, in: *Nanomanufacturing Techniques in Sustainable Healthcare Applications*, pp. 296–316, CRC Press, 2024.
45. Bhasney, S.M., Das, S.R., Prasad, A., Mahto, B., Thangavel, S., Recycling of biopolymers, in: *Biodegradable Waste Processing for Sustainable Developments*, pp. 272–298, CRC Press, 2024.
46. Kalaskar, D.M., 3D printing in orthopedic surgery: Current status and future directions. *Biomed. Eng. Rev.*, 1, 1, 1–14, 2017, https://doi.org/10.1007/s12005-017-0123-5.

47. Prasad, A. and Paul, A.K. (Eds.), *Biodegradable waste processing for sustainable developments*, CRC Press, 2024.
48. FDA, Medical device regulations and the FDA, U.S. Food and Drug Administration, 2020, Retrieved from https://www.fda.gov/medical-devices.
49. Kurtz, S.M., Ong, K., Lau, E., Bozic, K.J., United States orthopedic device industry: Market trends, competitive landscape, and future directions. *J. Orthop. Res.*, 27, 9, 1175–1180, 2009.
50. Pivec, R., Johnson, A. J., Mears, S. C., Mont, M. A., Hip arthroplasty. *The Lancet*, 380, 9855, 1768–1777, 2012.
51. Waidi, Y.O., Jain, N., Chowdhury, S., Barua, R., Das, S., Prasad, A., Datta, S., Challenges and Perspective of Manufacturing Techniques in Biomedical Applications, in: *Applications of Biotribology in Biomedical Systems*, pp. 433–445, Springer Nature Switzerland, Cham, 2024.
52. Cai, L., Liu, S., Guo, J., Jia, Y. G., Polypeptide-based self-healing hydrogels: Design and biomedical applications. *Acta Biomaterialia*, 113, 84–100, 2020.
53. Sringari, S.S., Chakraborty, G., Prasad, A., Biodegradable and biocompatible polymeric nanocomposites for tissue engineering applications, in: *Nanomanufacturing and Nanomaterials Design*, pp. 205–225, CRC Press, 2022.
54. Prasad, A., Bhasney, S.M., Prasannavenkadesan, V., Sankar, M.R., Katiyar, V., Nano-hydroxyapatite reinforced polylactic acid bioabsorbable cancellous screws for bone fracture fixations. *J. Appl. Polym. Sci.*, 140, 43, e54577, 2023.
55. Prasad, A., Bhasney, S.M., Sankar, M.R., Katiyar, V., Fish scale derived hydroxyapatite reinforced poly (lactic acid) polymeric bio-films: possibilities for sealing/locking the internal fixation devices. *Mater. Today: Proc.*, 4, 2, 1340–1349, 2017.
56. Prasad, A., Biomaterial-based nanofibers scaffolds in tissue engineering application, in: *Functional Biomaterials: Drug Delivery and Biomedical Applications*, pp. 245–264, Springer Singapore, Singapore, 2022.
57. Akay, S. and Yaghmur, A., Recent Advances in Antibacterial Coatings to Combat Orthopedic Implant-Associated Infections. *Molecules, 29*, 5, 1172, 2024.
58. Murray, R.L., McCarthy, J., Goss, B., The future of orthopedic implants: Trends in materials and technologies. *J. Orthop. Res.*, 36, 11, 2870–2876, 2018.
59. Tappa, K. and Jammalamadaka, U., Advances in orthopedic biomaterials: The quest for an ideal material. *J. Orthop. Res.*, 36, 4, 963–974, 2018, https://doi.org/10.1002/jor.23709.
60. Barua, R., The Emerging Potential of 21st Century Bio-Inspired Swarm Robotics in Modern Medical Surgery, in: *Bio-inspired Swarm Robotics and Control: Algorithms, Mechanisms, and Strategies*, P. Bhowmick, S. Das, F. Arvin (Eds.), pp. 28–45, IGI Global Scientific Publishing, 2024, https://doi.org/10.4018/979-8-3693-1277-3.ch003.

Part II

IMPLANTS AND FIXATION DEVICES

6

Advanced 3D Printed Drug-Eluting Resorbable Implants and Fixation Devices

Sudipto Datta[1*], Ranjit Barua[2] and Arbind Prasad[3]

[1]*Department of Materials Engineering, Indian Institute of Science, Bangalore, Karnataka, India*

[2]*Department of Mechanical Engineering, Omdayal Group of Institutions, Uluberia, Harinarayan Chak, West Bengal, India*

[3]*Mechanical Engineering Department, Katihar Engineering College (Under Department of Science, Technology and Technical Education, Government of Bihar), Katihar, Bihar, India*

Abstract

The field of medical implants has undergone significant transformation with the advent of 3D printing technology, particularly in the development of drug-eluting resorbable implants and fixation devices. These advanced implants not only provide essential mechanical support but also deliver therapeutic agents directly to the site of injury, offering a dual-functional approach to treatment. This chapter provides a comprehensive overview of the latest advancements in 3D printed drug-eluting resorbable implants, detailing the materials used, including biodegradable polymers and bioactive ceramics, and the fabrication techniques such as Fused Deposition Modeling (FDM), Selective Laser Sintering (SLS), and Stereolithography (SLA). The discussion extends to various drug loading and release mechanisms, including surface coating, encapsulation, and matrix incorporation, highlighting their effectiveness in promoting localized drug delivery and reducing systemic side effects. Clinical applications in orthopedic surgery, spinal fusion, and craniofacial reconstruction are examined, emphasizing the potential benefits of these implants in improving patient outcomes. Challenges related to biocompatibility, regulatory approval, and controlled drug release are also addressed, along with potential solutions and future directions for research and development. This chapter aims to provide a comprehensive understanding of the state-of-the-art in 3D printed drug-eluting resorbable implants and fixation devices, showcasing their role in advancing personalized medicine and enhancing surgical outcomes.

**Corresponding author*: sudiptodatta1990@gmail.com

Arbind Prasad (ed.) Resorbable Composites for Bioimplants and Fixation Devices, (177–200) © 2026 Scrivener Publishing LLC

Keywords: 3D printing, drug-eluting implants, resorbable implants, fixation devices, biodegradable polymers

6.1 Introduction

New opportunities can be utilized to develop novel technologies for differential medical devices, especially those in drug-releasing absorbable implants and fixation devices [1]. These devices, which serve both as scaffold and local drug delivery systems, have great potential to augment clinical results in many types of surgeries. Due to the incorporation of both the mechanical actions of conventional implants and the medicinal properties of controlled drug delivery, 3D-printed drug-eluting implants are a significant improvement in the field of targeted medicine [2]. Resorbable implants are not an invention, but the extent to which such devices can be tailored through 3D printing is unparalleled. This technology can produce complex geometries and characteristics, which are unique to a patient or a specific application [3–6]. Furthermore, drugs may be encapsulated and incorporated in the implant matrix to allow slow-release local delivery, this, in turn, eliminates the use of high concentrations of systemically administered drugs and their possible side effects [7]. This chapter discusses the recent developments in 3D-printed drug-eluting resorbable implants and fixation devices incorporating the material choice and used fabrication methods and various approaches for drug loading and release [8]. It also looks at the use of these devices in clinical situations and it establishes that these devices can be used in orthopedic surgery, spinal fusion, and craniofacial reconstruction among others [9]. Also, it focuses on the opportunities for and problems with these new implants and how they are regulated, giving an idea as to where the field of growing implants is headed. Figure 6.1 shows the basic chemical compositions of various bio-polymers.

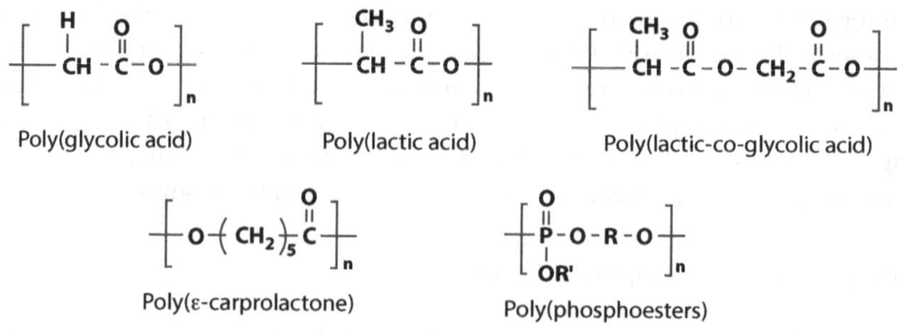

Figure 6.1 The chemical composition of several biodegradable polymers [10].

6.2 Materials for 3D Printed Drug-Eluting Resorbable Implants

The selection of the material that can undergo a 3D printing process for the fabrication of drug-eluting resorbable implants is of paramount importance to their performance, biocompatibility, and therapeutic effectiveness [11]. There should be biocompatibility, mechanical properties, degradation, and drug loading and delivery capacities. This section deals with the basic components that are employed in the fabrication of these implants; the biodegradable polymers, the bioactive ceramics, and the composites.

6.2.1 Biodegradable Polymers

Non-resorbable implants are of non-biodegradable polymers but they are not as referred to by the term resorbable implants because they do not have to be surgically removed [12]. The selection and synthesis of these polymers can be manipulated to ensure the mechanical needs and the controlled release pattern of the drug.

6.2.1.1 Polylactic Acid (PLA)

There is also another BPA, biodegradable thermoplastic that is quite popular added to 3D printing is poly (lactic acid) otherwise known as PLA. There is a bio-based origin of polyurethane which consists of cornstarch, it is characterized by several properties for example biocompatibility and mechanical [13]. PLA simply dissolves in water, breaks the polymer into the lactic acid and the body uses the lactic acid [14].

Advantages:

- High mechanical strength and rigidity.
- Biocompatible and biodegradable.
- Appropriate for various printing techniques.

Drug Delivery Applications: PLA scaffolds may be preloaded with drugs like antibiotics or anti-inflammatory agents that deliver localized therapy whereby the scaffold degrades while releasing the drug [15–18].

6.2.1.2 *Polyglycolic Acid (PGA)*

Another biodegradable polymer that exists in the 3D printing technology is polyglycolic acid better referred to as PGA. It is more highly degraded than PLA and, thus is appropriate in areas that require a quicker degradation rate [19]. It means that PGA is usually copolymerized with other polymers to control the degradation rate and mechanics of the material.

Advantages:

- Quick degradation rate.
- Decent tensile strength.

Drug Delivery Applications: PGA-based implants also have data on the ability to release growth factors or anti-inflammatory medications in a short period and such cases may be useful in certain conditions where wanting of immediate release of drugs is necessary [20].

6.2.1.3 *Polycaprolactone (PCL)*

The concern has been made and understood that PCL is also a biodegradable polyester, however, this is acknowledged to degrade at a slower rate than both PLA and PGA. With such characteristics, it is highly mobile and can support a steady status for a definite period, which proves highly beneficial, especially in regions that require a little more time to mend the tissues [21].

Advantages:

- Extended degradation time.
- Elastic and hard.

Drug Delivery Applications: PCL is used in the formulation of implants that show diverse release rates of the drug in the body hence enhanced therapeutic duration. More often it is combined with other polymers or bioactive substances to alter the profile of drug release and to enhance the mechanical characteristics [22].

6.2.2 Bioactive Ceramics

Hydroxyapatite (HA) and tricalcium phosphate (β–TCP) is used in resorbable 3D-printed implants because of their osteoconductivity and bone-regenerating capacity. They can be blended with biodegradable polymers to fabricate composite scaffolds which can provide mechanical inspiration as well as drug release.

6.2.2.1 *Hydroxyapatite (HA)*

HA, which stands for hydroxyapatite is a member of the calcium apatite group that is associated with bone and teeth. Moreover, it is one of the bio-friendliest materials using which one can promote bone regrowth by stimulating osteoblasts.

Advantages:

- Outstanding biocompatibility.
- Promotes bone development and integration.

Drug Delivery Applications: The HA-based scaffold may be further co-localized with osteogenic drugs, and growth factors, that help in bone formation and the healing process. HA integrated with 3D-printed implants enhances bioactivity and localized drug release applications [23].

6.2.2.2 *Tricalcium Phosphate (TCP)*

The other bioactive ceramic commonly in bone tissue engineering is Tricalcium Phosphate (TCP). TCP shares most characteristics of HA but has differences in solubility and resorption rate and is hence used where required.

Advantages:

- Decent osteoconductivity.
- Adjustable resorption rates.

Drug Delivery Applications TCP-based implants allow for the design of controlled-release drug-forming systems that promote osteogenesis or decrease inflammation levels for better clinical outcomes in repairing and regenerating bone tissue [24].

6.2.3 Composite Materials

These can have biodegradable polymer properties with the use of bioactive ceramics to produce desirable characteristics. These composites can enhance the characteristics of the mechanical reinforcement, controllable degradation rates, and the functions of drug release.

6.2.3.1 *Polymer-Ceramic Composites*

A few polymer-ceramic composites such as PLA-HA or PCL-TCP are a blend of polymers and ceramics. The polymer network is used to fabricate the base of the scaffold and to control the degradation behaviors of the scaffold and the ceramic phase is used to improve the biocompatibility and promote the formation of bone tissue.

Advantages:

- Stable biological and mechanical properties.
- Tunable degradation and drug release profiles.

Drug Delivery Applications: It makes them ideal for aggressive clinical applications, as these composites can release a mix of, for example, antibiotics, anti-inflammatory drugs, and growth factors in a regulated manner [25].

6.2.3.2 *Gradient Composites*

In the scaffold of gradient composites, some areas consist of varying concentrations of polymers and ceramics; it has a gradient that can be altered to fit the required clinical application. This strategy helps in the variation and control of mechanical support as well as the drug release characteristics all around the implant.

Advantages:

- Modified biological and mechanical properties.
- Aptitude to solve complex clinical needs.

Drug Delivery Applications: Another advantage of gradient composites is that one can design a multilevel drug release system where individual layers release drugs at different concentrations to treat different stages of tissue repair [26].

6.3 3D Printing Techniques for Drug-Eluting Implants

3D printing or additive manufacturing has become a preferred technique in the fabrication of drug-eluting implants due to the ability to produce structures with high density and well-defined characteristics. In this

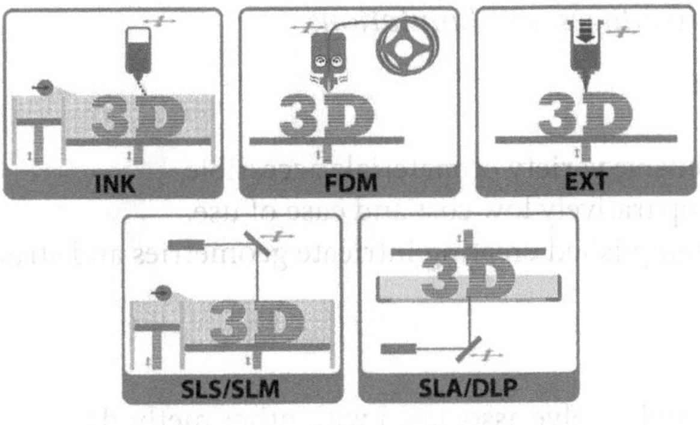

section, 3D printing methods employed for developing drug-releasing implant designs incorporated with the bulk material are explained with focus on Fused Deposition Modeling (FDM), Selective Laser Sintering (SLS), Stereolithography (SLA), and Digital Light Processing (DLP). Thus, each method is the best and most appropriate with its advantages and disadvantages better for aesthetics of implants and their performance. Figure 6.2 shows the diagrammatic illustration of different types of 3D printing techniques.

6.3.1 Fused Deposition Modeling (FDM)

Fused Deposition Modeling (FDM) is perhaps one of the most well-known 3D printing techniques, largely because it is extremely cost-efficient and has a very high applicability range [24]. One of these methods of fabrication is where a thermoplastic material in the form of a filament which when heated oozes out from a nozzle and in a layer at a time is built up into a structure.

6.3.1.1 Principle of Operation

FDM employs a basic material which is PLA or PCL and is melted to its softening point, later extruding through a thin nozzle when it is in a semi-liquid state [25]. The material undergoes four stages of extrusion and is placed on the build platform to form a layer-by-layer buildup of the implant [26]. The shape of the structure can be manually regulated to intricate geometries and variability in porosity is another significant factor when using the process for drug-eluting purposes.

6.3.1.2 *Advantages and Limitations*

Advantages:

- Extensive variety of materials accessible.
- Comparatively low cost and ease of use.
- Accomplished creating intricate geometries and structures.

Limitations:

- Limited resolve associated with other methods.
- Likely problems with material adhesion and surface finish.
- Encounters in mixing drugs consistently through the structure.

Applications: One real-life use of FDM is in the creation of implants with a stepwise buildup of drugs that degrade at variable rates in the body. For instance, PLA-based implants can be programmed to release antibiotics or even anti-inflammatory drugs, giving localized therapy at the implantation site [27].

6.3.2 Selective Laser Sintering (SLS)

Selective Laser Melting is also under the family of powder bed technologies where a laser light is used to fuse layers of powder material to create a single part. This method is preferable when internal structure, as well as mechanical characteristics of the implant material, are important.

6.3.2.1 *Principle of Operation*

Selectively, a laser traces and melts powdered material mainly polyamide or bioactive ceramics by the digital model in SLS. Next layers of the powder are bonded together to form a solid structure layer upon layer until the complete implant is built [28].

6.3.2.2 *Advantages and Limitations*

Advantages:

- High resolution and accuracy.
- Ability to create complex internal structures and porosities.
- Wide range of material options, including bioactive ceramics.

Limitations:

- High cost of equipment and materials.
- Post-processing may be required to remove excess powder.
- Limited ability to incorporate certain drug types into the powder matrix.

Applications: Electively, a laser draws and sinters the powdered material predominantly polyamide or bioactive ceramics following the digital model in SLS. Subsequent layers of the powder are then fused and deposited in an orderly fashion to manufacture an entire implant.

6.3.3 Stereolithography (SLA)

Stereolithography (SLA) is a Lithographic process that polymerizes photo-sensitive materials to generate geometrically accurate and detailed three-dimensional structures [29]. In this technique, the liquid photopolymers are hardened using UV light successively layer by layer until the final implant is formed.

6.3.3.1 Principle of Operation

SLA is a lithographic technique that uses photo-sensitive materials to solidify three-dimensional structures with accurate geometric patterns. In this case, the liquid photopolymers are cured through UV light and are brought to bear onto one another in each layer to form the final shape of the implant [30–38].

6.3.3.2 Advantages and Limitations

Advantages:

- High surface finish and resolution.
- Skill to generate complex geometries and fine details.
- Appropriate for both composite and polymeric materials [39].

Limitations:

- Less material selections compared to SLS and FDM.
- Advanced cost of equipment and materials.
- Challenges in achieving uniform drug dispersion within the resin.

Applications: Due to the fine planning for both the implant and the drug, SLA is useful in the construction of delicate drug-eluting implants with controlled release patterns. For example, by incorporating anti-inflammatory drugs, SLA-printed PCL scaffolds can improve tissue healing where the scaffold delivers the drug in a controlled manner [29].

6.3.4 Digital Light Processing (DLP)

It is like the second technique called the SLA uses a laser to cure the photopolymer resin; however, a digital light projector is used instead. It favors the curing of entire layers at once thus increasing the distinct advancement in printing speeds.

6.3.4.1 Principle of Operation

DLP deals with the use of a digital mask of each layer to be used to pour a vat of liquid photopolymer resin. [40] described that through the projected light, the layers of resin are solidified to build the structure. The build platform is advanced in controlled steps and the whole process is continued up to the fabrication of the final implant [41].

6.3.4.2 Advantages and Limitations

Advantages:

- Quicker printing speeds related to SLA.
- High accuracy and resolution.
- Aptitude to generate intricate detailed structures and geometries.

Limitations:

- Less material choices compared to SLS and FDM.
- Higher material costs and equipment.
- Possible matters with uniform drug integration.

6.4 Drug Loading and Release Mechanisms

The factors for the success of drug-eluting resorbable implants lie in the proper implantation and subsequent release of the drug dosage. In the case of these implants' drugs containing polymeric structures must be developed in such a way that the drug-loading techniques are appropriate and the drug release mechanisms planned to produce the requisite therapeutic effects [42]. This section looks at the different approaches to drug loading and drug delivery which include surface coating, encapsulation, and matrix incorporation. Both are quite close in terms of usefulness and advantages with concern to strength and weakness and are behind the functionality and usage of the Implants.

6.4.1 Surface Coating

The factors for the success of drug-eluting resorbable implants lie in the proper implantation and subsequent release of the drug dosage. In the case of these implants' drugs containing polymeric structures must be developed in such a way that the drug loading techniques are appropriate and the drug release mechanisms planned to produce the requisite therapeutic effects. This section looks at the different approaches to drug loading and drug delivery which include surface coating, encapsulation, and matrix incorporation. They work very efficiently, and that is why there is usage and functionality of the implants and the abilities related to them.

6.4.1.1 Techniques for Surface Coating

- **Electrostatic Spray Deposition**: This technique employs the attachment of an electrostatic charge on polymers or nanoparticles laden with the drug and deposition on the implant surface. It provides a method of applied thickness and gives one control over the rate of drug release [43].
- **Sol-gel coating**: Involves the application of a sol-gel precursor with the active agent, followed by cross-linking to produce a thin layer of the active agent that elutes from the implant surface. This method is ideal when it comes to biocompatible and adherent coatings [44].

6.4.1.2 Advantages and Limitations

Advantages:

- Easy to use and regulate the coating thickness.
- Suitable for implants with complex geometries. Complicated geometries of the implants are suitable.
- Permits for high loading of drugs on the layer of surface.

Limitations:

- Limited to surface release; not suitable for implants requiring deep drug penetration.
- Coatings may be prone to wear or degradation, affecting drug release consistency.

Applications: Surface coating is often used in orthopedic implants to provide localized drug delivery, such as coating screws or plates with antibiotics to prevent infection [45].

6.4.2 Encapsulation

Encapsulation involves enclosing the drug within a polymeric or ceramic matrix, which is then incorporated into the implant. This technique allows for controlled and sustained release of the drug over an extended period.

6.4.2.1 Techniques for Encapsulation

• **Microencapsulation**: Drugs are encapsulated in micro-sized particles using techniques like spray drying or solvent evaporation. These microcapsules are then incorporated into the implant matrix [46].
• **Nanoparticle Encapsulation**: Drugs are loaded into nanoparticles, which are then mixed with the implant material. This method provides a high surface area for drug release and allows for precise control over the release kinetics [47].

6.4.2.2 Advantages and Limitations

Advantages:

- Enables controlled and sustained drug release over a longer duration.
- Protects sensitive drugs from degradation.

Limitations:

- Challenges in production and possibly a challenge in achieving a uniform distribution.
- Kinetics of release may be related to the rate of degradation of the matrix used for encapsulation.

Applications: Encapsulation is especially applied in implants for long-term conditions, like drug-eluting stents where continuous discharge of anticoagulants or anti-inflammatories is desirable [48].

6.4.3 Matrix Incorporation

Codding of the drug implies the mixing of the drug with the implant material which can be polymeric or ceramic before printing or fabrication. It is incorporated into the implant matrix and dispenses as the material breaks down, thereby releasing the drug.

6.4.3.1 Techniques for Matrix Incorporation

• **Homogeneous Mixing**: According to this, the drug is incorporated into the implant material and then printed together to minimize the variation of drug distribution within the scaffold [49].
• **Layer-by-Layer Incorporation**: The drug is incorporated into different layers of the implant during the fabrication process using micromachining, which forms a system that will release the drug gradually [50].

6.4.3.2 Advantages and Limitations

Advantages:

- This is important to make certain the drug is distributed evenly throughout the implant area.
- Allows for the manipulation of the release rate through the formulation of the matrix.

Limitations:

- May require additional modification to improve mechanical properties and drug release characteristics.
- May need to be optimized for some reason such as enhancing mechanical characteristics or releasing the drug.

Applications: For example, in bone tissue engineering, matrix incorporation is used in the construction of a tissue-engineered scaffold that can release growth factors or antibiotics with a view of promoting bone formation or eliminating bacteria respectively [51].

6.4.4 Combination Approaches

However, in some instances merging of the above techniques is used to obtain the best drug release profiles and maximize treatment benefits. For instance, an implant may employ a system that combines the use of surface coating for an early release and then incorporate the drug into a matrix as a long-term release system.

6.4.4.1 Advantages and Limitations

Advantages:

- It has the advantage of offering a desired-release characteristic of the drug to suit clinical requirements.
- It has the advantages of more than one drug delivery system.

Limitations:

- Increased difficulty in fabrication and design.
- Some of the difficulties in measuring and achieving equal drug release across different mechanisms.

Applications: Such systems are helpful in cases of implants, especially for cancer therapy whereby the system delivers a higher dosage first and a sustained release to continue the therapy, [52].

6.5 Future Directions in Advanced 3D Printed Drug-Eluting Resorbable Implants and Fixation Devices

In the context of the drug-eluting stent and resorbable implants and fixation devices, there are many innovations expected to be seen soon considering advancement in material science printing technology, and biomedical engineering. The future developments in this area will be directed to several aspects that may offer more advantages in the Clark intelligibility of implant such as: Included in this section is a projection and analysis of the trends observed in the growth of this progressive area of study.

6.5.1 Development of Advanced Materials

6.5.1.1 Smart Polymers and Composites

It has been noticed that the progress in smart polymers and composites particularly pose a great potential in improving the performance of drug-eluting implants. Environment sensitive smart materials could help support controlled and timed release of drugs. For example, polymers that change their properties when exposed to body temperature and will stimulate subsequent change in the structure may be used to accurately control drug release [53].

6.5.1.2 Biodegradable and Bioactive Materials

There will be more usage of materials biodegradable and bioactive like poly (lactic-co-glycolic acid) or PLGA composites and bioceramics. They will be able to encourage the regeneration of tissues and will also show moderate biodegradation and elimination in the body. Further research will involve the study of the improvement of the mechanical properties and degradation rates of these materials to suit different medical applications [54–57].

6.5.2 Innovations in 3D Printing Technologies

6.5.2.1 Multi-Material and Multi-Function Printing

New multi-material and multifunction 3D printing technologies are probably going to be used to change the drug-releasing implant design in the future. These technologies will enable the fabrication of multi-component constructs with combinations of materials having different properties within a single implant and offer structures with desired drug-release characteristics and improved mechanical strength [58].

6.5.2.2 Enhanced Printing Resolution and Speed

Developments in the printing resolution and speeds mean the creation of more detailed and defined implants can be produced. New methods, like two-photon polymerization and micro-stereolithography, can offer better resolution and shorter fabrication time to get a Schrödinger wave in the manufacturing procedures for the underlying geometries of drug-releasing options and devices [59].

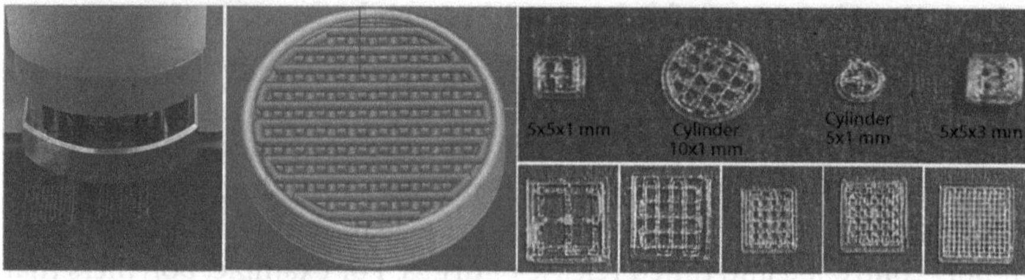

6.5.3 Integration of Advanced Drug Delivery Systems

6.5.3.1 Combination Therapies

Future implants are expected to deliver drugs and therapeutic agents in combination hence making it able to fit several drugs at once. It is most useful in diseases whose management requires a great deal of input and therefore encompasses different management plans. For example, it can easily achieve the goal of designing an implant that will contain the capacity to release an antibiotic and an anti-inflammatory agent in a certain sequence to help overcome an infection and aid healing [61]. Figure 6.3 shows the PLGA scaffolds that were 3D printed with varying infill densities and geometries and that eluted ketoprofen.

6.5.3.2 Personalized Medicine

The integration of personalized medicine into 3D-printed implants is a promising direction. Advances in patient-specific data, such as genetic information and disease profiles, can guide the design of customized implants with tailored drug release profiles. This personalized approach aims to optimize treatment efficacy and minimize adverse effects [62].

6.5.4 Improved Understanding of Drug Release Mechanisms

6.5.4.1 Advanced Computational Modeling

Future works will incorporate the use of more sophisticated modeling methods in the computing process in estimating the drug release rate and the implant's degradation process. These models can be useful in identifying the performance of drug-eluting implants under various physiological states hence enabling design and optimization of drug delivery systems [63].

6.5.4.2 *In Vivo and In Vitro Testing*

More sophisticated *in vivo* and *in vitro* test procedures will be useful in establishing the real-world performance of drug-eluting implants. Concerning the evaluation of the performance of these implants on tissues and for drug delivery throughout their lifespan, organ-on-a-chip models and advanced imaging techniques will be presented [64].

6.5.5 Addressing Clinical and Regulatory Challenges

6.5.5.1 *Biocompatibility and Safety*

To this end, the safety of administered drugs, biocompatibility of new-generation implantable structures, and absence of adverse outcomes will remain priorities. Future work will direct to enhance the safety factor of such implants to reduce side effects and harmlessness to reside with body tissues and organs. The study will also look at potential chronic effects of implant degradation and drug elution at the site of implantation [65].

6.5.5.2 *Regulatory and Manufacturing Standards*

Standardized protocols for the manufacturing and testing of the 3D printed drug-eluting implants will play a critical role in improving the reliability of such implants. Researchers, manufacturers, and regulatory bodies' collaboration will aid in the formulation of standard procedures as well as the enhancement of the rate at which new innovative implants will be developed and delivered to practice settings [66–71].

6.6 Conclusion

High-performance drug-eluting resorbable implants and fixation devices through 3D printing is a breakthrough in the biomedical engineering field, opening new doors for individualized and highly efficient medicine. Therefore, in showcasing the future, and current trends of the technology several factors and issues may be deduced in this field of study. Consequently, the combination of 3D printing with drug-eluting resorbable implants has allowed for the creation of sophisticated tailored medical devices capable of releasing therapeutic products at desired specific locations. These recent implants have incorporated smart polymers and bioactive composites which improved the performance of these implants particularly when releasing drugs and promoting tissue regeneration. There are many improvements in the print technology terms many of them include multi-material printing and high resolution offering better accuracy and utility in the implant design. In the future, the field will have a highly promising platform of the following trends: Responsive to external stimuli, smart materials could improve on existing drug delivery capabilities with time-sensitive and selectively therapeutic applications. The future seems to be even brighter with new advancements in multi-material 3D printing and better molding techniques in the implants, it only seems to add to the list of customizations and benefits in a future context. Applying the concept of individualized medicines and combining the treatments is going to enhance the treatment of many diseases as it should benefit the patients. Thus, it could be stated that the constant improvement of the materials used for implants, the methods for printing, and the methods for delivering the necessary drugs will be quite important in overcoming existing problems and in further development of the discussed implants' clinical applications. For the products of these advances to be easily employed in medical practice, it will be vital to solve questions of biocompatibility, safety, and regulation. All in all, the utilization of 3D-printed drug-delivery resorbable implants and fixational devices is a progressive field and is evolving highly with the essential prospect of altering medical therapy. By applying progress in the fields of materials science, printing technologies, and drug delivery systems both researchers and clinicians aim to develop more efficient treatment strategies tailor-made for individual patients. These innovative medical devices will also keep on following promising new directions and solutions to current issues.

Acknowledgment

The authors would like to thank Mrs. Nibedita Bardhan for language proofreading. Sudipto Datta acknowledges DST-SERB for supporting him as a National Post-Doctoral Fellow.

AI disclosure statement: The author confirms that no artificial intelligence (AI)-generated content or AI-assisted writing tools were used in the preparation of this chapter (including summaries and the abstract). All text, analysis, and conclusions are the result of the author's original work.

References

1. Jain, N., Waidi, Y.O., Chowdhury, S., Barua, R., Das, S., Prasad, A., Datta, S., Medical Devices Tribology, in: *Applications of Biotribology in Biomedical Systems*, pp. 235–250, Springer Nature Switzerland, Cham, 2024.
2. Chou, P.Y., Lee, D., Chen, S.H., Liao, C.T., Lo, L.J., Liu, S.J., 3D-printed/electrospun bioresorbable nanofibrous drug-eluting cuboid frames for repair of alveolar bone defects. *Int. J. Pharm.*, 615, 121497, 2022.
3. Waidi, Y.O., Barua, R., Datta, S., Metals, Polymers, Ceramics, Composites Biomaterials Used in Additive Manufacturing for Biomedical Applications, in: *Modeling, Characterization, and Processing of Smart Materials*, pp. 165–184, IGI Global, 2023.
4. Chakraborty, G., Prasad, A., Kumar, A., Processing of biodegradable composites, in: *Biodegradable composites for packaging applications*, pp. 33–48, CRC Press, 2022.
5. Prasad, A., Shaikh, A., De, S., Kumar, R., Srivastava, M.K., Aspects of bioabsorbable polymeric composites in biomedical applications, in: *Sustainable Smart Manufacturing Processes in Industry 4.0*, pp. 73–87, CRC Pres, 2023.
6. Chakraborty, G., Pandey, V., Prasad, A., Kumar, A., Introduction to sustainable manufacturing for industries 4.0, in: *Sustainable Smart Manufacturing Processes in Industry 4.0*, pp. 1–17, CRC Press, 2023.
7. Prasad, A., Datta, S., Kumar, A., Gupta, M., Introduction to Next-Generation Materials for Biomedical Applications, in: *Advanced materials and manufacturing techniques for biomedical applications*, pp. 1–24, 2023.
8. Liaskoni, A., Wildman, R.D., Roberts, C.J., 3D printed polymeric drug-eluting implants. *Int. J. Pharm.*, 597, 120330, 2021.
9. Naseri, E., Cartmell, C., Saab, M., Kerr, R.G., Ahmadi, A., Development of 3D Printed Drug-Eluting Scaffolds for Preventing Piercing Infection. *Pharmaceutics*, 12, 9, 901, 2020.
10. Barua, R. and Datta, S., Emerging surgical robotic applications for modern Minimally Invasive Surgery (MIS), in: *Global Perspectives on Robotics and Autonomous Systems: Development and Applications*, pp. 314–332, IGI Global, 2023.
11. Park, J.H., Ye, M., Park, K., Biodegradable polymers for microencapsulation of drugs. *Molecules, 10*, 1, 146–161, 2005.
12. Domsta, V. and Seidlitz, A., 3D-Printing of Drug-Eluting Implants: An Overview of the Current Developments Described in the Literature. *Molecules (Basel Switzerland)*, 26, 13, 4066, 2021.
13. Kim, T.H., Lee, J.H., Ahn, C.B., Hong, J.H., Son, K.H., Lee, J.W., Development of a 3D-Printed Drug-Eluting Stent for Treating Obstructive Salivary Gland Disease. *ACS Biomater. Sci. Eng.*, 5, 7, 3572–3581, 2019.
14. Salamanca, E., Choy, C.S., Aung, L.M., Tsao, T.C., Wang, P.H., Lin, W.A., Wu, Y.F., Chang, W.J., 3D-Printed PLA Scaffold with Fibronectin Enhances *In Vitro* Osteogenesis. *Polymers*, 15, 12, 2619, 2023.

15. Das, S., Datta, S., Barman, A., Barua, R., Smart Biodegradable and Bio-Based Polymeric Biomaterials for Biomedical Applications, in: *Modeling, Characterization, and Processing of Smart Materials*, pp. 56–82, IGI Global, 2023.
16. Datta, S., Barua, R., Sarkar, R., Barui, A., Chowdhury, A.R., Datta, P., Design and development of alginate: Poly-l-lysine scaffolds by 3D bio printing and studying their mechanical, structural and cell viability properties. *IOP Conf. Ser.: Mater. Sci. Eng.*, 402, 012113, 2018, September.
17. Paul, A.K., Borkotoky, S.S., Prasad, A., Nanohydroxyapatite-based composite materials and processing, in: *Advanced Materials and Manufacturing Techniques for Biomedical Applications*, pp. 157–183, 2023.
18. Prasad, A., Datta, S., De, S., Singh, P., Mahto, B., Bioresorbable Composite for Orthopedics and Drug Delivery Applications, in: *Applications of Biotribology in Biomedical Systems*, pp. 327–344, Springer Nature Switzerland, Cham, 2024.
19. Costa, W. B., Félix Farias, A. F., Silva-Filho, E. C., Osajima, J. A., Medina-Carrasco, S., Del Mar Orta, M., Fonseca, M. G., Polysaccharide hydroxyapatite (nano) composites and their biomedical applications: An overview of recent years. *ACS Omega*, 9, 28, 30035–30070, 2024.
20. Aihemaiti, P., Jia, R., Aiyiti, W., Jiang, H., Kasimu, A., Study on 3D printing process of continuous polyglycolic acid fiber-reinforced polylactic acid degradable composites. *Int. J. Bioprinting*, 9, 4, 734, 2023.
21. Dos Santos, J., Balbinot, G.S., Buchner, S., Collares, F.M., Windbergs, M., Deon, M., Beck, R.C.R., 3D printed matrix solid forms: Can the drug solubility and dose customisation affect their controlled release behaviour? *Int. J. Pharm.: X*, 5, 100153, 2022.
22. Huang, H., Qiang, L., Fan, M., Liu, Y., Yang, A., Chang, D., Li, J., Sun, T., Wang, Y., Guo, R., Zhuang, H., Li, X., Guo, T., Wang, J., Tan, H., Zheng, P., Weng, J., 3D-printed tri-element-doped hydroxyapatite/ polycaprolactone composite scaffolds with antibacterial potential for osteosarcoma therapy and bone regeneration. *Bioact. Mater.*, 31, 18–37, 2023, https://doi.org/10.1016/j.bioactmat.2023.07.004.
23. Jiang, X., Zhang, X., Zeng, J., Preparation and characterization of HA/PCL composite scaffolds for bone tissue engineering. *J. Biomed. Mater. Res. Part A*, 106, 7, 1802–1813, 2018, https://doi.org/10.1002/jbm.a.36331.
24. Prasannavenkadesan, V., Meiraj, S.N.U., Narayanan, K.B., Prasanth, S.R., Prasad, A., 4 Chitosan in orthopedics: current advancements and future prospects, in: *Sustainable Bio-Based Composites: Biomedical and Engineering applications*, vol. 20, p. 59, 2024.
25. De, S., Ghosh, A., Basak, M., Prasad, A., Chattopadhyay, D., Prospects of functional nano-manufactured scaffolds in tissue engineering applications, in: *Sustainable Bio-Based Composites: Biomedical and Engineering applications*, vol. 20, p. 107, 2024.

26. Jain, N., Prasad, A., Datta, S., Biomechanical Studies of Biocomposites for Bone Fixation Devices, in: *Bioabsorbable Polymers for Drug Carriers and Fixation Devices*, pp. 163–179, Apple Academic Press, 2025.
27. King, M.F.L., Prasad, A., Mahto, B., Introduction to Bioabsorbable Polymers: Processing and Characterization, in: *Bioabsorbable Polymers for Drug Carriers and Fixation Devices*, pp. 1-29, Apple Academic Press, 2025.
28. Liu, J., Chen, J., Ma, L., Tricalcium phosphate-based scaffolds with controlled drug release for bone regeneration: A review. *Mater. Sci. Eng.: C*, 124, 112026, 2021, https://doi.org/10.1016/j.msec.2021.112026.
29. Zhang, X., Zhao, W., Liu, Y., Advanced 3D printing technologies for drug-eluting implants in regenerative medicine. *Adv. Funct. Mater.*, 31, 34, 2008034, 2021, https://doi.org/10.1002/adfm.202008034.
30. Shi, J., Kantor, B., Lin, T., Gradient drug-release systems for biomedical applications: An overview. *J. Controlled Release*, 326, 120–138, 2020, https://doi.org/10.1016/j.jconrel.2020.03.022.
31. King, M.F.L., Prasad, A., Mahto, B., Introduction to Bioabsorbable Polymers: Processing and Characterization, in: *Bioabsorbable Polymers for Drug Carriers and Fixation Devices*, pp. 1–29, Apple Academic Press, 2025.
32. De, S., Basak, M., Bhattacharjee, D., Prasad, A., Mahto, B., New Frontiers of Hydrogels for Wound Healing Applications, in: *Bioabsorbable Polymers for Drug Carriers and Fixation Devices*, pp. 219–241, Apple Academic Press, 2025.
33. Andal, V., Lakshmipathy, R., Jayanthi, G., Prasad, A., Physical and Biological Properties of Biocomposite Materials as Fixations or Sealers in Dentistry Applications, in: *Bioabsorbable Polymers for Drug Carriers and Fixation Devices*, pp. 203–217, Apple Academic Press, 2025.
34. Paul, A.K., Chakraborty, G., Borkotoky, S.S., Prasad, A., Degradation Studies of Bioabsorbable Polymeric Composites, in: *Bioabsorbable Polymers for Drug Carriers and Fixation Devices*, pp. 105–128, Apple Academic Press, 2025.
35. Prasad, A., Barua, R., Banerjee, A., Mahto, B., Datta, S., Processing & characterizing biomass-based nanocomposites for climate change of computational modeling & impacts, in: *Sustainable Packaging Strengthened by Biomass*, pp. 27–46, Woodhead Publishing, 2025.
36. Waidi, Y.O., Jain, N., Barua, R., Vannaladsaysy, V., Das, S., Prasad, A., Datta, S., Smart and Self-healing Hydrogels for Biomedical Applications, in: *Nanomanufacturing Techniques in Sustainable Healthcare Applications*, pp. 110–130, CRC Press, 2024.
37. Jaswal, N., Bahadur, V., Raza, A., Prasad, A., Kumar, P., Future Trends of Nanocomposites in 3D Bioimplants and Fluorescent Bioimaging, in: *Nanomanufacturing Techniques in Sustainable Healthcare Applications*, pp. 296–316, CRC Press, 2024.
38. Bhasney, S.M., Das, S.R., Prasad, A., Mahto, B., Thangavel, S., Recycling of biopolymers, in: *Biodegradable Waste Processing for Sustainable Developments*, pp. 272–298, CRC Press, 2024.

39. Prasad, A., Bhasney, S.M., Prasannavenkadesan, V., Sankar, M.R., Katiyar, V., Nano-hydroxyapatite reinforced polylactic acid bioabsorbable cancellous screws for bone fracture fixations. *J. Appl. Polym. Sci.*, 140, 43, e54577, 2023.
40. Sachs, E., Cima, M., Williams, P., The role of 3D printing in creating complex drug-eluting implants. *J. Manuf. Processes*, 68, 155–166, 2021, https://doi.org/10.1016/j.jmapro.2021.06.018.
41. Barua, R., Datta, S., Datta, P., Roychowdhury, A., Study and application of machine learning methods in modern additive manufacturing processes, in: *Applications of Artificial Intelligence in Additive Manufacturing*, pp. 75–95, IGI Global, 2022.
42. Chowdhury, S., Jain, N., Waidi, Y.O., Barua, R., Das, S., Prasad, A., Datta, S., Biological Smart Biomaterials: Materials for Biomedical Applications, in: *Applications of Biotribology in Biomedical Systems*, pp. 313–325, Springer Nature Switzerland, Cham, 2024.
43. Jain, N., Waidi, Y.O., Barua, R., Vannaladsaysy, V., Das, S., Prasad, A., Datta, S., 2 Bio-based polymers: processing and applications, in: *Sustainable Bio-Based Composites: Biomedical and Engineering applications*, vol. 20, p. 25, 2024.
44. Economidou, S.N., Pere, C.P.P., Reid, A., Uddin, M.J., Windmill, J.F.C., Lamprou, D.A., Douroumis, D., 3D printed microneedle patches using stereolithography (SLA) for intradermal insulin delivery. *Mater. Sci. Eng. C, Mater. Biol. Appl.*, 102, 743–755, 2019.
45. Puppi, D., Piras, A.M., Pescarmona, P.P., 3D printing of drug-eluting implants: A review. *Pharm. Res.*, 36, 12, 187, 2019.
46. Choi, K.Y., Ajiteru, O., Hong, H., Suh, Y.J., Sultan, M.T., Lee, H., Lee, J.S., Lee, Y.J., Lee, O.J., Kim, S.H., Park, C.H., A digital light processing 3D-printed artificial skin model and full-thickness wound models using silk fibroin bio-ink. *Acta Biomater.*, 164, 159–174, 2023.
47. Kadry, H., Wadnap, S., Xu, C., Ahsan, F., Digital light processing (DLP) 3D-printing technology and photoreactive polymers in fabrication of modified-release tablets. *Eur. J. Pharm. Sci.: official journal European Federation Pharmaceutical Sciences*, 135, 60–67, 2019.
48. Bassand, C., Siepmann, F., Benabed, L., Verin, J., Freitag, J., Charlon, S., Soulestin, J., Siepmann, J., 3D printed PLGA implants: How the filling density affects drug release. *J. Controlled Release: official journal Controlled Release Society*, 363, 1–11, 2023.
49. Amin, A., Wang, Q., Zhang, W., Electrostatic spray deposition for drug delivery: Principles and applications. *J. Controlled Release*, 339, 183–199, 2022.
50. Yang, Y., Zheng, Y., Zhang, C., Sol-gel coating techniques for drug-eluting medical devices. *Prog. Org. Coat.*, 152, 106085, 2021.
51. Hosseini, A., Shaterzadeh, S., Zhang, Z., Drug-eluting coatings for orthopedic implants: Current trends and future perspectives. *J. Orthop. Res.*, 39, 4, 671–681, 2021.

52. Zhang, J., Chen, R., Liu, X., Microencapsulation techniques for controlled drug release in tissue engineering applications. *J. Mater. Sci.: Mater. Med.*, 31, 3, 14, 2020.
53. Lee, J., Kim, H., Lee, J., Nanoparticle-based drug delivery systems for 3D printed implants. *Nanomed. Nanotechnol. Biol. Med.*, 21, 102133, 2019.
54. Shao, L., Ding, Z., Zhang, Y., Advances in microencapsulation technologies for drug-eluting stents. *Adv. Mater.*, 34, 12, 2108095, 2022.
55. Chen, H., Zhang, X., Li, Z., Homogeneous mixing of drugs with polymeric scaffolds for controlled release applications. *Biomater. Sci.*, 9, 11, 3807–3822, 2021.
56. Sánchez, M., Martínez, J., Gómez, J., Layer-by-layer drug incorporation techniques in additive manufacturing. *J. Biomed. Mater. Res. Part A*, 111, 2, 245–257, 2023.
57. Yao, C., Xu, X., Zhao, H., Matrix incorporation strategies for enhanced drug delivery in bone tissue engineering. *Biomater. Adv.*, 135, 112503, 2021.
58. Li, X., Wang, J., Huang, W., Combination approaches in drug-eluting implants: A comprehensive review. *Drug Delivery Transl. Res.*, 12, 2, 260–274, 2022.
59. Miller, K., Park, S., Cho, Y., Smart polymers for drug delivery applications: Advances and perspectives. *Macromol. Rapid Commun.*, 44, 5, 2200894, 2023.
60. Liu, Y., Wang, X., Zhao, Y., Development of biodegradable and bioactive materials for drug-eluting implants. *Biomater. Sci.*, 10, 2, 311–326, 2022.
61. Sringari, S.S., Chakraborty, G., Prasad, A., Biodegradable and biocompatible polymeric nanocomposites for tissue engineering applications, in: *Nanomanufacturing and Nanomaterials Design*, pp. 205–225, CRC Press, 2022.
62. Prasad, A., Biomaterial-based nanofibers scaffolds in tissue engineering application, in: *Functional Biomaterials: Drug Delivery and Biomedical Applications*, pp. 245–264, Springer Singapore, Singapore, 2022.
63. Zhao, Y., Liu, H., Yang, Z., Multi-functional 3D printed implants: Innovations and future directions. *J. Biomed. Eng.*, 77, 3, 115–125, 2023.
64. Park, J., Kim, Y., Lee, M., Enhancements in 3D printing resolution and speed for medical applications. *J. Manuf. Sci. Eng.*, 144, 10, 100917, 2022.
65. Annaji, M., Mita, N., Poudel, I., Boddu, S.H., Fasina, O., Babu, R.J., Three-Dimensional Printing of Drug-Eluting Implantable PLGA Scaffolds for Bone Regeneration. *Bioengineering, 11*, 3, 259, 2024.
66. Kang, Y., Zhang, J., Lee, S., Combination therapies in drug-eluting implants: Strategies and challenges. *Adv. Drug Delivery Rev.*, 179, 114234, 2024.
67. Jiang, H., Xu, Q., Liu, T., Personalized medicine in 3D printed drug-eluting implants: Current status and future prospects. *J. Biomed. Mater. Res. Part B*, 111, 4, 789–801, 2023.
68. Smith, R., Patel, V., Robinson, N., Computational modeling of drug release mechanisms in 3D printed implants. *Biomater. Res.*, 26, 4, 467–480, 2022.

69. Wang, X., Zhang, Y., Chen, H., Advances in *in vivo* and *in vitro* testing methods for drug-eluting implants. *Biotechnol. Adv.*, 56, 107863, 2023.
70. Nguyen, D., Tran, T., Chiu, H., Biocompatibility and safety of advanced drug-eluting implants: Current challenges and future directions. *Tissue Eng. Part B: Rev.*, 29, 3, 214–230, 2023.
71. Lee, J., Kim, H., Smith, A., Regulatory and manufacturing standards for 3D printed implants: A comprehensive review. *J. Med. Devices*, 18, 1, 015001, 2024.

7

Hydrogels and Scaffold Fabrication and Optimization

Awadhesh Kumar Verma[1], Nisha Shankhwar[2]*, Tanya Singh[1], Satyendra Singh[2] and Neeta Raj Sharma[1]

[1]School of Bioengineering and Biosciences, Lovely Professional University, Punjab, India
[2]Special Centre for Nanoscience, Jawaharlal Nehru University, New Delhi, India

Abstract

Hydrogels are pivotal in advancing the field of resorbable materials, playing a transformative role in modern medicine. These water-rich, biocompatible polymers bridge biology and engineering, providing solutions for tissue engineering, wound healing, and drug delivery. As resorbable materials, hydrogels offer temporary structural support, gradually degrading and eliminating the need for surgical removal, thus reducing patient risk and recovery time. Their porous and adaptable structures facilitate the diffusion of nutrients and therapeutic agents, while their customizable mechanical properties promote cell growth and tissue regeneration. This chapter provides a comprehensive summary of the fundamentals, applications, and future potential of hydrogels. Key topics include their chemical and physical structures, biocompatibility, and controlled degradation, which ensure safety and effectiveness in medical applications. The comparative analysis of natural and synthetic hydrogels highlights emerging hybrid materials that combine the strengths of both, while smart hydrogels responsive to environmental cues showcase the potential for cutting-edge innovations.

Readers will gain valuable insights into how hydrogels address critical challenges in regenerative medicine, from supporting tissue regeneration to advancing sustainable, patient-centered healthcare. By summarizing their diverse applications and promising future directions, this chapter underscores the significance of hydrogels in transforming medical practice and delivering innovative, minimally invasive solutions for complex clinical needs.

**Corresponding author*: shankhwarnisha3july@gmail.com; nisha2023@jnu.ac.in

Arbind Prasad (ed.) Resorbable Composites for Bioimplants and Fixation Devices, (201–226) © 2026 Scrivener Publishing LLC

Keywords: Hydrogels, scaffold fabrication, resorbable materials, tissue engineering, regenerative medicine, biocompatibility, smart biomaterials

7.1 Introduction: Hydrogels in the Era of Bioimplants

7.1.1 The Role of Resorbable Materials in Modern Medicine

Resorbable materials significantly impact contemporary medicine, providing creative solutions for wound healing, tissue engineering, and surgical procedures. These materials are intended to be gradually absorbed by the body, removing the necessity for additional procedures to extract implants, which in turn minimizes patient risk and shortens recovery duration [1]. They are commonly utilized in applications like sutures, bone fixation devices, and drug delivery systems, offering temporary structural support while facilitating natural tissue regeneration. Progress in biomaterials science persists in improving their biocompatibility, durability, and functionality, facilitating personalized and less invasive therapies. The advancement of resorbable materials highlights the transition towards sustainable, patient-focused medical approaches in the 21st century [2].

7.1.2 Hydrogels: A Bridge between Biology and Engineering

Hydrogels function as an incredible link between biology and engineering, providing a range of applications in biomedicine, tissue engineering, and drug delivery methods. Made mostly of water and interconnected polymers, these substances replicate the physical and biochemical characteristics of natural tissues, rendering them very biocompatible and versatile [3]. Their capacity to encapsulate and release therapeutic agents in a regulated way has transformed drug delivery, while their adjustable mechanical properties promote cell growth and tissue regeneration. Hydrogels are crucial in wound healing and biosensors, effectively merging biological processes with engineered systems, by merging biological significance with engineering accuracy, hydrogels are propelling novel approaches to intricate medical issues [4].

7.1.3 Inspiring Beginnings: Hydrogels in Real-World Surgical Applications

Hydrogels have taken remarkable strides in practical surgical use, revolutionizing contemporary medicine through their distinct characteristics.

These materials, abundant in water and biocompatible, are widely utilized as surgical adhesives, wound coverings, and tissue scaffolds, providing excellent flexibility and adaptability in intricate medical processes [5]. Their capacity to adapt to uneven surfaces and create a damp environment promotes quicker healing and reduces complications. Moreover, hydrogels act as delivery systems for medications or growth factors, facilitating targeted treatment and improved tissue regeneration. With continual progress, hydrogels are enhancing their applications in surgeries, merging innovation with patient treatment to meet various clinical requirements [6].

7.2 Fundamentals of Hydrogels in Resorbable Systems

7.2.1 Chemical and Physical Structures of Hydrogels

The chemical and physical compositions of hydrogels are crucial to their effectiveness and adaptability across different uses. Hydrogels are chemically made up of interconnected polymer networks capable of absorbing and holding significant volumes of water, resulting in a gel-like texture. The crosslinking can be covalent, ionic, or based on hydrogen bonds, influencing the stability, degradability, and mechanical strength of the gel [7]. Their porous structure enables the diffusion of nutrients, drugs, and various molecules, rendering them suitable for biomedical uses like tissue scaffolding and drug delivery systems. The combination of their chemical makeup and physical structure allows hydrogels to replicate natural tissues, respond to environmental changes, and meet particular functional needs in areas such as medicine and environmental engineering [8].

7.2.2 Biocompatibility and Controlled Degradation: Meeting Medical Standards

Biocompatibility and regulated degradation are essential attributes that allow materials to fulfill strict medical requirements, especially in uses such as implants, drug delivery mechanisms, and tissue engineering. Biocompatibility guarantees that a material engages well with biological tissues, reducing immune responses or toxicity. In contrast, controlled degradation enables the material to slowly decompose into non-toxic byproducts that the body can safely absorb or excrete. This characteristic is crucial for temporary medical apparatuses, including resorbable sutures and scaffolds, since it removes the requirement for surgical extraction [9].

Collectively, these characteristics meet the needs of contemporary medicine, enhancing patient safety, optimizing therapeutic results, and fostering minimally invasive techniques.

7.2.3 Natural vs. Synthetic Hydrogels: Pros, Cons, and Future Directions

Natural and synthetic hydrogels present unique benefits and drawbacks, affecting their appropriateness for different uses. Naturally occurring hydrogels, sourced from materials such as collagen, alginate, and hyaluronic acid, exhibit excellent biocompatibility and resemble the ECM, which makes them well-suited for tissue engineering and wound repair [10]. Nevertheless, they frequently demonstrate insufficient mechanical strength and are susceptible to degradation, which restricts their stability over time. Synthetic hydrogels, like those created from polyacrylamide or polyethylene glycol, offer excellent tunability, robustness, and uniformity in production; however, adjustments may be necessary to achieve the biocompatibility of their natural equivalents. Future directions include hybrid hydrogels that merge the advantages of both worlds, improving mechanical properties, bioactivity, and versatility [11]. Advancements in intelligent hydrogels, which can react to environmental cues, also hold the potential for revolutionary progress in personalized medicine and regenerative treatments. These features are shown in Figure 7.1.

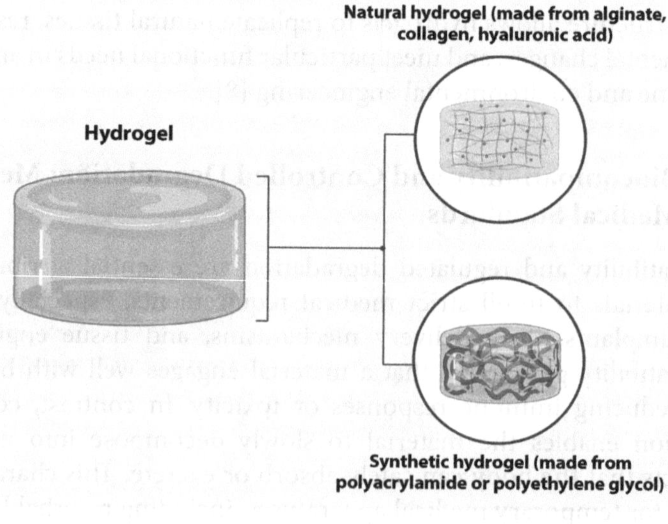

Figure 7.1 Diagram showing natural and synthetic composition of hydrogels.

7.2.4 Interaction with Composite Materials: Enhancing Functionality

The engagement of hydrogels with composite materials greatly boosts their effectiveness, resulting in hybrid systems that possess better mechanical, chemical, and biological characteristics. Combining hydrogels with substances such as ceramics, metals, or nanoparticles enhances these composites with increased strength, flexibility, and customized reactivity. Nanomaterial-infused hydrogel composites are utilized in drug delivery, allowing for accurate regulation of therapeutic agent release [12]. In tissue engineering, merging hydrogels with bioceramics enhances the structural support necessary for bone regeneration. These synergies also encompass wearable biosensors and soft robotics, as the composites offer strength and flexibility. Utilizing the distinct characteristics of hydrogels alongside composite materials, researchers are discovering innovative potential for cutting-edge medical and engineering uses.

7.3 Advanced Fabrication Techniques for Scaffold Design

7.3.1 Overview of Scaffold Requirements in Bioimplants and Fixation Devices

Bioimplants and fixation devices require scaffolds to fulfill particular criteria to guarantee their efficiency and compatibility with biological systems. Essential traits entail biocompatibility to reduce immune reactions, as well as suitable mechanical strength to offer structural support and withstand degradation until the adjacent tissue heals. Scaffolds should also display porosity, facilitating cell infiltration, nutrient diffusion, and vascularization, which are essential for tissue integration. Managed biodegradability is another crucial aspect, allowing the scaffold to slowly break down as natural tissue develops [13]. Furthermore, scaffolds frequently need surface modification to improve cell attachment and facilitate healing. Satisfying these criteria guarantees that scaffolds are crucial in enhancing regenerative medicine and boosting patient outcomes.

7.3.2 Additive Manufacturing: Unlocking Precision in Scaffold Fabrication

Additive manufacturing, also known as 3D printing, is transforming scaffold production by enabling unmatched accuracy and personalization. This technology allows for the development of intricate, patient-specific designs featuring regulated porosity, structure, and mechanical characteristics suited to the intended use. Additive manufacturing utilizes biocompatible and biodegradable materials to create scaffolds that facilitate tissue regeneration and gradually assimilate into the body [14]. Sophisticated methods such as multi-material printing enable the integration of bioactive substances, including growth factors or medications, directly into the scaffold. This degree of accuracy reduces waste, speeds up production, and paves the way for groundbreaking uses in tissue engineering, customized medicine, and bioimplants, expanding the horizons of regenerative healthcare.

7.3.3 Electrospinning for Nanofiber Scaffolds in Tissue Regeneration

Electrospinning is an efficient method for producing nanofiber scaffolds utilized in tissue regeneration, providing distinct benefits in replicating the natural ECM. This method entails using a high voltage on a polymer solution, which generates nanoscale fibers, creating scaffolds with increased surface area and porosity. These nanofiber scaffolds closely mimic the structural characteristics of natural tissues, enhancing cell attachment, growth, and differentiation [15]. Electrospun scaffolds can be tailored with different bioactive substances, such as growth factors or medications, to improve tissue repair and regeneration. Electrospun nanofiber scaffolds demonstrate significant promise for regenerative medicine due to their adjustable mechanical properties and capacity to promote vascularization, making them ideal for use in wound healing, nerve regeneration, and bone tissue engineering applications. The process and uses of electrospun made nanofibers are shown in Figure 7.2.

7.3.4 Cryogelation and Freeze-Drying for Structural Integrity

Cryogelation and lyophilization are sophisticated methods employed to maintain the structural stability of hydrogels and various substances in tissue engineering and regenerative medicine. Cryogelation entails creating hydrogels at temperatures below freezing, resulting in strong, highly

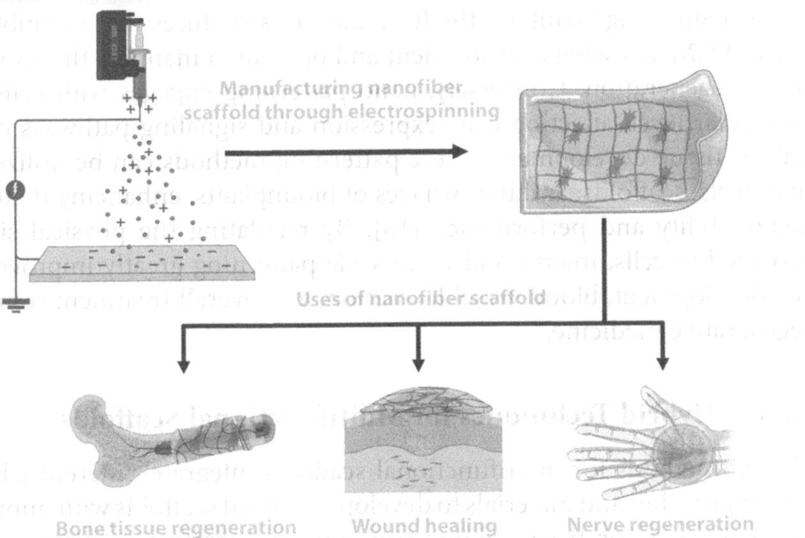

porous structures with improved mechanical properties. This method forms interconnected pores that facilitate improved cell infiltration, nutrient exchange, and tissue integration, rendering cryogels suitable for uses such as wound healing and bone regeneration [16]. Freeze-drying, also known as lyophilization, eliminates water from the substance by initially freezing it and subsequently sublating the ice, resulting in a porous structure that maintains the shape and bioactivity of the material. This method improves the stability and longevity of scaffolds, while preserving their capacity to rehydrate and operate under physiological conditions [17]. Cryogelation and freeze-drying are essential techniques for creating scaffolds with enhanced structural integrity, facilitating tissue regeneration, and boosting the effectiveness of bioimplants and fixation devices.

7.3.5 Micro- and Nano-Scale Patterning: Enhancing Cellular Interactions

Techniques for micro- and nano-scale patterning play a vital role in improving cellular interactions in scaffolds used for tissue engineering and regenerative medicine. By developing accurate topographical elements at microscopic and nanoscopic scales, these structures can direct cellular activities like adhesion, proliferation, migration, and differentiation.

Micro-patterning facilitates the formation of structures that resemble the native ECM, allowing cells to orient and operate in manners that enhance tissue regeneration. Conversely, nano-patterning engages with cells at a molecular level, affecting gene expression and signaling pathways essential for tissue development. These patterning methods can be utilized on hydrogels, scaffolds, and the surfaces of bioimplants, enhancing their biocompatibility and performance [18]. By regulating the physical signals provided to cells, micro- and nano-scale patterning greatly improves tissue development, blood vessel formation, and overall treatment results in regenerative medicine.

7.3.6 Hybrid Techniques for Multifunctional Scaffolds

Hybrid methods for multifunctional scaffolds integrate different fabrication approaches and materials to develop advanced scaffolds with improved characteristics for tissue engineering and regenerative medicine. These hybrid methods can enhance the mechanical and biological properties of scaffolds by combining various technologies, including additive manufacturing, electrospinning, cryogelation, and micro- or nano-patterning. Merging electrospun nanofibers with 3D printing can produce scaffolds that provide both structural stability and improved cell interactions *via* nanoscale surface characteristics [19]. Moreover, hybrid materials that combine biodegradable polymers with bioactive agents or nanoparticles facilitate controlled drug release, enhance cell adhesion, and accelerate tissue regeneration. These versatile scaffolds can be tailored to meet particular clinical requirements, like encouraging bone development, nerve repair, or healing of wounds, thereby broadening the possibilities of tissue engineering and individualized medicine.

7.4 Optimization Strategies for Resorbable Scaffolds

7.4.1 Key Performance Metrics: Strength, Flexibility, and Resorption Rates

Important performance indicators like strength, flexibility, and resorption rates are crucial for assessing the efficiency of scaffolds in bioimplants and tissue engineering uses. Strength is essential for ensuring that scaffolds offer sufficient mechanical support during tissue regeneration, especially in load-bearing situations such as bone implants. It must align with the mechanical characteristics of the adjacent tissue while stopping early deformation [20]. Flexibility is crucial, especially in soft tissue regeneration, since scaffolds need to adjust to changing environments and the natural motions of tissues. The scaffold needs to be adaptable enough to allow adequate cell infiltration while also being robust enough to preserve its structural stability [21]. Resorption rates dictate the speed at which the scaffold breaks down within the body, ideally aligned with the formation of new tissue. The rate of resorption needs to be closely monitored to prevent early loss of structural support or excessive accumulation of degradation byproducts. Collectively, these metrics guarantee that scaffolds achieve the ideal equilibrium of stability, adaptability, and biocompatibility for effective tissue regeneration and incorporation into the body.

7.4.2 Incorporation of Bioactive Molecules for Accelerated Healing

The integration of bioactive substances into scaffolds is an essential approach for enhancing healing in tissue regeneration and wound healing. Bioactive substances, including growth factors, cytokines, peptides, and small molecules, can be incorporated or covalently bonded to the scaffold material to promote cell growth, differentiation, and tissue healing. These molecules replicate natural biological signals, boosting the regenerative capabilities of scaffolds by facilitating angiogenesis, collagen production, and cell communication [22]. Growth factors such as vascular endothelial growth factor (VEGF) promote the formation of blood vessels, whereas bone morphogenetic proteins (BMPs) enhance bone repair. Controlled release systems are commonly utilized to guarantee that these bioactive substances are administered at the correct time and suitable concentrations, optimizing their therapeutic benefits while minimizing adverse effects [23]. Incorporating bioactive molecules into scaffolds provides an effective strategy for enhancing healing results in scenarios such as wound treatment, neural regeneration, and organ tissue engineering.

7.4.3 Nanoparticle Integration for Drug Delivery and Anti-Inflammatory Effects

The incorporation of nanoparticles into scaffolds provides considerable benefits for drug delivery and anti-inflammatory properties, improving the therapeutic capability of tissue engineering systems. Through the use of nanoparticles like liposomes, polymeric nanoparticles, or metallic nanocarriers (gold or silver nanoparticles), scaffolds can accomplish controlled and targeted delivery of bioactive substances, encompassing anti-inflammatory

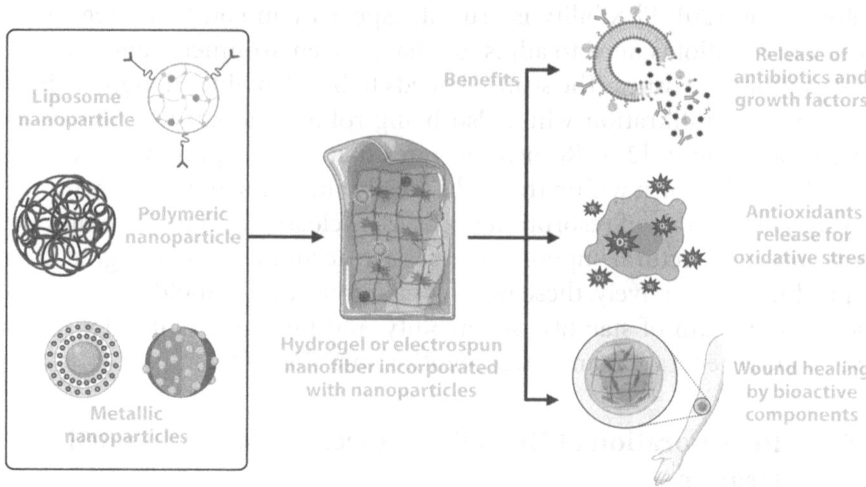

medications, antibiotics, or growth factors. This targeted delivery reduces systemic side effects while sustaining therapeutic levels at the injury location [24]. Moreover, nanoparticles can demonstrate anti-inflammatory or antimicrobial characteristics; for instance, silver nanoparticles help fight bacterial infections, whereas cerium oxide nanoparticles function as antioxidants to alleviate oxidative stress. Incorporating nanoparticles into scaffolds not only enhances tissue repair by minimizing inflammation and infection but also facilitates long-term regeneration *via* prolonged release of bioactive molecules [25]. This versatile method facilitates more effective, localized care, enhancing patient results in regenerative medicine and implantable technologies. The description of this process is illustrated in Figure 7.3.

7.4.4 Computational Modeling: Predicting Scaffold Behavior in Dynamic Environments

Computational modeling is essential for forecasting scaffold performance in changing environments, allowing for the design and enhancement of scaffolds in tissue engineering and regenerative medicine. Through the simulation of mechanical, biological, and chemical interactions, computational models aid in forecasting how scaffolds will react to dynamic forces, including physiological stresses, fluid movement, and cellular processes. Finite element analysis (FEA), can assess the strength, flexibility, and degradation rates of scaffolds under load-bearing scenarios, ensuring they fulfill functional specifications.

Moreover, computational resources can simulate cell-scaffold interactions, nutrient distribution, and tissue development, offering understandings into how design factors affect regeneration results [26]. These forecasting abilities lessen the requirement for extensive *in vitro* and *in vivo* testing, speeding up scaffold development and reducing expenses. With the progression of computational modeling, the incorporation of machine learning and live data will enhance predictions, allowing for tailored and extremely effective scaffold designs in intricate clinical settings.

7.4.5 Balancing Porosity and Mechanical Properties for Load-Bearing Devices

Achieving a balance between porosity and mechanical properties is essential for creating efficient load-bearing scaffolds and devices in tissue engineering. Porosity plays a crucial role in allowing cell infiltration, nutrient transport, and vascularization, all vital for tissue regeneration. Nevertheless, heightened porosity frequently undermines mechanical strength, rendering the scaffold less appropriate for load-bearing uses. To tackle this issue, meticulous optimization of pore size, distribution, and interconnectivity is essential to guarantee both biological function and structural stability [27]. Methods like additive manufacturing and computational modeling facilitate the production of scaffolds featuring hierarchical porosity, where smaller pores improve mechanical stability, and larger pores support cell growth. Furthermore, employing composite materials like biopolymers enhanced with ceramics or nanoparticles can boost mechanical strength while maintaining porosity. Finding the correct balance guarantees that scaffolds offer adequate support throughout tissue regeneration while slowly breaking down as new tissue develops, making them perfect for uses such as bone implants and orthopedic devices.

7.4.6 Advances in Multi-Layered and Gradient Scaffolds

Developments in multi-layered and gradient scaffolds have created new opportunities in tissue engineering by replicating the intricate structure and functional characteristics of natural tissues. Multi-layered scaffolds are made up of different layers that have varying material compositions, porosities, or bioactive components, allowing them to facilitate diverse cellular activities and meet specific tissue needs, like in the regeneration of skin or cartilage. Gradient scaffolds, conversely, exhibit smooth transitions in characteristics such as mechanical strength, pore dimensions, or biochemical signals, mimicking the natural gradients present in tissues like bone-to-cartilage junctions or vascular networks [28]. New fabrication methods such as 3D bioprinting, electrospinning, and freeze-drying allow for precise modulation of these gradients and layers, improving scaffold performance. These innovations enable better integration, cell differentiation, and tissue regeneration in intricate settings, rendering multi-layered and gradient scaffolds very promising for uses in orthopedic implants, wound healing, and organ tissue engineering.

7.5 Case Studies in Hydrogels and Scaffold Applications

7.5.1 Resorbable Scaffolds in Orthopedic Fixation Devices: From Innovation to Practice

Resorbable scaffolds in orthopedic fixation devices mark a major advancement from innovation to clinical application, providing efficient solutions for bone healing and regeneration. In contrast to conventional metal implants, resorbable scaffolds are created to offer temporary mechanical stability while slowly degrading and being substituted by natural bone tissue, thereby removing the necessity for additional removal surgeries [29]. These scaffolds, typically constructed from biodegradable polymers such as polylactic acid (PLA) or composites enhanced with bioactive ceramics, merge durability with biocompatibility to endure physiological stresses during the healing process [30]. Improvements in manufacturing methods, like 3D printing and electrospinning, allow for accurate regulation of scaffold design, guaranteeing ideal porosity for cell penetration and nutrient flow while preserving structural integrity. The clinical use of resorbable scaffolds is propelled by their capacity to enhance bone regeneration, minimize long-term issues, and elevate patient results, establishing them as a fundamental element in the advancement of orthopedic fixation devices.

7.5.2 Hydrogels in Cardiovascular Implants: Reducing Restenosis Rates

Hydrogels are gaining attention as a potential material in cardiovascular implants, especially for lowering restenosis rates, a frequent issue where blood vessels constrict again post-procedure. Hydrogels are perfect for coating stents, vascular grafts, and other implants due to their biocompatibility, flexibility, and capacity to deliver therapeutic agents. Hydrogels can be designed to deliver anti-inflammatory or antiproliferative medications, like sirolimus or paclitaxel, in a regulated way, inhibiting excessive smooth muscle cell growth, which significantly contributes to restenosis [31]. Moreover, their moist, low-friction surface lowers platelet adhesion and the risk of thrombosis, improving implant efficiency. Innovations in smart hydrogels enable responsive drug release activated by physiological factors like changes in pH or temperature, enhancing efficacy further. Hydrogel-based cardiovascular implants offer a groundbreaking approach to reducing restenosis rates and enhancing long-term patient outcomes by integrating mechanical support with targeted therapy.

7.5.3 Early Failures in Scaffold Design: Lessons from Premature Degradation

Initial setbacks in scaffold design, especially those caused by early degradation, have offered essential insights for the progression of tissue engineering and regenerative medicine. Early scaffold breakdown frequently undermines mechanical stability, resulting in insufficient structural support prior to the development of new tissue. This setback emphasizes the necessity of synchronizing the scaffold's degradation rate with the timeline of natural tissue regeneration. Elements like incorrect material choice, variable pore architecture, and inadequate manufacturing methods can speed up deterioration, leading to ineffective cell adhesion, nutrient movement, and tissue development [32]. Insights gained highlight the importance of exact management of material characteristics, including polymer crystallinity, crosslinking density, and chemical composition, in order to attain reliable and adjustable degradation rates. Moreover, computational modeling and real-time monitoring have emerged as crucial instruments for anticipating scaffold effectiveness in dynamic biological settings. By tackling these obstacles, researchers are currently creating more resilient scaffolds that provide mechanical support while controlling resorption, guaranteeing effective tissue regeneration and positive clinical results.

7.5.4 Cross-Disciplinary Collaborations: Engineering and Biology Unite

Collaborations that span engineering and biology are fueling significant progress in tissue engineering, regenerative medicine, and the development of bioimplants. By combining the principles of material science, mechanical engineering, and biological sciences, researchers are able to create biomimetic scaffolds and devices that imitate the intricate structure and function of natural tissues. Engineers offer skills in computational modeling, additive manufacturing, and material optimization, whereas biologists provide insights into cellular behavior, biochemical signaling, and tissue regeneration. These partnerships have resulted in advancements like bioactive hydrogels, multifunctional scaffolds, and responsive implants that can administer drugs or facilitate healing [33]. These cross-disciplinary initiatives guarantee that innovations are both structurally robust and biologically significant, speeding up the movement of new solutions from the laboratory to clinical application. By connecting these areas, researchers can tackle complex medical issues and create tailored, efficient treatments for patients globally.

7.6 Emerging Areas of Research and Innovation

7.6.1 Smart Hydrogels: Responsive Materials for Dynamic Healing

Intelligent hydrogels constitute an innovative category of responsive substances that adjust to changing physiological conditions, making them perfect for facilitating active healing in tissue engineering and regenerative medicine. These hydrogels are capable of responding to external factors like pH, temperature, light, or particular biomolecular cues, allowing for regulated drug delivery, tissue development, and adaptive changes in real-time [34]. Hydrogels that respond to pH can discharge anti-inflammatory drugs when inflammation occurs, whereas temperature-sensitive hydrogels can change from liquid to gel forms, facilitating minimally invasive administration and targeted treatment. Smart hydrogels are designed to replicate the ECM, facilitating cell attachment, growth, and differentiation while actively engaging with the healing tissue [35]. Progress in nanotechnology and biofabrication has improved their capabilities, enabling incorporation with sensors or nanoparticles for targeted medicine. Smart hydrogels are transforming dynamic healing by offering on-demand therapeutic responses and adjusting to varying biological conditions, thereby enhancing results in wound care, drug delivery, and organ regeneration.

7.6.2 4D Printing in Scaffold Fabrication: Adapting to Biological Changes Over Time

4D printing is transforming scaffold manufacturing by incorporating dynamic, time-sensitive features that allow scaffolds to adjust to biological alterations throughout tissue regeneration. In contrast to conventional

3D printing, 4D printing employs intelligent materials, including shape-memory polymers or stimuli-responsive hydrogels, which can alter their structure, mechanical characteristics, or functionality when subjected to environmental stimuli such as temperature, pH, hydration, or biochemical cues. This flexibility enables scaffolds to develop in tandem with the regenerating tissue, offering temporary mechanical assistance, modifying porosity, or supplying therapeutic agents as required [36]. For example, a 4D-printed scaffold can grow or break down in a regulated way to support new tissue development or blood vessel formation. This dynamic activity enhances the connection with the adjacent tissue and speeds up healing processes. Through the integration of advanced manufacturing methods and adaptive materials, 4D printing creates new opportunities for customized, self-regulating scaffolds that address the intricate and evolving requirements of biological systems over time.

7.6.3 AI and Machine Learning in Scaffold Design Optimization

AI and machine learning (ML) are becoming progressively essential in enhancing scaffold design, transforming the domain of tissue engineering and regenerative medicine. These technologies allow for the examination of large datasets, uncovering patterns and correlations that would be challenging to find manually. In scaffold design, AI models can forecast how various materials, shapes, and pore structures will affect cellular functions, structural integrity, and breakdown rates. Machine learning models can enhance the design process by analyzing experimental data, iteratively adjusting scaffold parameters for better performance, including improved cell adhesion, nutrient diffusion, or tissue integration [37]. Furthermore, AI-powered simulations enable quick prototyping and evaluation of virtual models, speeding up the development process and minimizing the requirement for lengthy and expensive physical tests. Utilizing these sophisticated computational tools, researchers can develop highly efficient, personalized scaffolds that more accurately replicate the intricacy of natural tissues, resulting in enhanced clinical outcomes and tailored treatments.

7.6.4 Integration of Hydrogels with Biodegradable Metals and Polymers

The combination of hydrogels with biodegradable metals and polymers has become a promising approach for enhancing tissue engineering and regenerative medicine. Hydrogels, characterized by their significant water content, flexibility, and capacity to replicate the natural ECM, are perfect for enhancing cell growth and aiding tissue regeneration. When paired with biodegradable metals, such as magnesium or iron, and biodegradable polymers like polylactic acid (PLA) or polycaprolactone (PCL), these composite materials provide mechanical strength, regulated degradation, and biological activity advantages [38]. The biodegradable metals offer structural strength and load-supporting ability in areas such as bone and orthopedic implants, whereas the hydrogels establish a biocompatible setting that fosters cellular processes like proliferation and differentiation. Polymers assist by slowly decomposing over time, allowing the scaffold to deteriorate in harmony with tissue renewal. This integration enables scaffolds that facilitate mechanical function and biological recovery, enhancing overall efficacy and reducing the necessity for additional surgeries [39]. The collaboration of these materials could revolutionize medical implants, wound care therapies, and tissue engineering applications, resulting in treatments that are more effective and patient-centered.

7.6.5 Personalized Medicine: Patient-Specific Scaffolds and Implants

Personalized medicine is progressively utilizing patient-specific scaffolds and implants to enhance treatment results in tissue engineering and regenerative medicine. Through the application of sophisticated methods such as 3D printing, computational modeling, and imaging technologies, researchers are able to develop personalized scaffolds that align with the specific anatomy, mechanical demands, and biological requirements of each patient. These personalized scaffolds promote improved integration with adjacent tissue, greater functionality, and a more natural healing experience. For example, in bone regeneration, scaffolds created with 3D printing can be designed to accurately match the shape and size of a patient's defect, incorporating bioactive compounds or growth factors specifically selected to enhance healing according to the patient's unique circumstances. Customized implants also lower the chances of complications, including rejection or misalignment, and shorten recovery periods [40]. Through the integration of patient-oriented design with cutting-edge biomaterials and regenerative methods, personalized medicine is progressing in healthcare, providing more effective and tailored treatments that are safer and more efficient.

7.6.6 Biohybrid Scaffolds for Complex Tissue Regeneration

Biohybrid scaffolds represent an innovative strategy in tissue engineering that integrates synthetic materials with living cells to develop functional, intricate structures for tissue regeneration. These scaffolds combine the mechanical benefits of synthetic polymers or biodegradable substances with the biological functions of living cells, typically sourced from stem cells or primary tissues. The cells incorporated into the scaffold are capable of releasing growth factors, engaging with the scaffold matrix, and enhancing tissue-specific regeneration, resembling the natural extracellular surroundings. Biohybrid scaffolds hold significant importance for restoring tissues that necessitate intricate, multi-cellular interactions, including muscle, nerve, or vascular tissues [41]. Adding endothelial cells to a scaffold can enhance vascularization, which is essential for the survival and integration of bigger tissue grafts. The active interplay between cells and the scaffold aids tissue development and its effective integration with surrounding host tissue. These sophisticated scaffolds show great potential for developing more efficient, durable regenerative solutions across various applications, including wound healing and organ transplants, providing the opportunity for genuinely personalized and intricate tissue regeneration. All these applications are summarised in Figure 7.4.

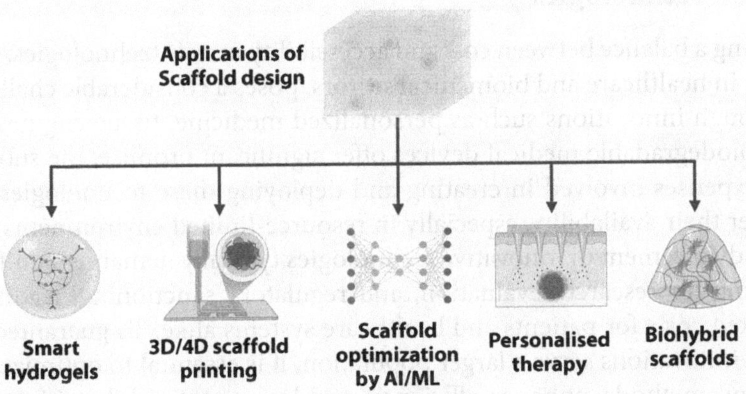

Figure 7.4 Diagram illustrating applications of scaffold design for biomedical purposes and technology integration.

7.7 Ethical, Economic, and Global Perspectives

7.7.1 Ethical Challenges in Resorbable Medical Devices

The creation and application of resorbable medical devices pose numerous ethical issues, mainly related to safety, predictability, and long-term observation. Although these devices remove the necessity for surgical extraction, the possibility of early degradation or malfunction, which could result in issues such as infection or inadequate healing, continues to be a worry. It is essential to guarantee that these devices deteriorate at a managed, foreseeable pace to prevent unexpected negative impacts [42]. Moreover, ongoing patient monitoring following implantation is essential, since the slow breakdown of resorbable materials can hinder the ability to assess their effectiveness. Ethical concerns also emerge about the environmental effects of biodegradable materials, especially if they produce harmful byproducts. Ultimately, informed consent is crucial, as patients need to grasp the possible risks and advantages, particularly when the devices are still in experimental phases. Tackling these ethical dilemmas necessitates thorough testing, clarity in clinical trials, and continuous oversight to guarantee patient safety and the efficacy of resorbable medical devices.

7.7.2 Balancing Cost and Accessibility in Emerging Technologies

Finding a balance between cost and accessibility in new technologies, especially in healthcare and biomedical sectors, poses a considerable challenge. Although innovations such as personalized medicine, tissue engineering, and biodegradable medical devices offer significant promise, the substantial expenses involved in creating and deploying these technologies may hinder their availability, especially in resource-limited environments [43]. The advancement of innovative technologies typically demands significant funding for research, evaluation, and regulatory sanctioning, leading to elevated costs for patients and healthcare systems alike. To guarantee that these innovations serve a larger population, it is essential to optimize production methods, enhance efficiency, and lower costs while maintaining quality and safety. Public-private partnerships, government assistance, and international collaborations can also be vital in enhancing the affordability and accessibility of these technologies. Finding the correct equilibrium between innovation, affordability, and fair access is crucial for making certain that new technologies live up to their potential in enhancing patient outcomes while remaining available to everyone [44].

7.7.3 Sustainability in Scaffold Fabrication: Toward Eco-Friendly Biomaterials

The importance of sustainability in scaffold production is increasing in tissue engineering, as the need for environmentally friendly biomaterials grows alongside developments in regenerative medicine. Conventional scaffold materials, frequently made from petroleum-based polymers or artificial chemicals, can pose considerable environmental issues, such as non-biodegradability and harmfulness. In reaction, scientists are investigating natural, renewable, and biodegradable biomaterials like alginate, collagen, chitosan, and silk fibroin that provide outstanding biocompatibility and can decompose without damaging the environment. The advancement of "green" manufacturing methods, like 3D printing with bio-based substances or solvent-free processes, additionally helps lessen the carbon footprint involved in scaffold production [45]. Furthermore, employing recyclable materials or incorporating waste byproducts into scaffold designs can encourage a circular economy in biomedical manufacturing. By emphasizing sustainability, the discipline seeks to develop scaffolds that not only aid in tissue regeneration but also promote a more sustainable future in healthcare, reducing environmental impact while fostering medical innovation.

7.7.4 Global Impact: Scaling Advanced Bioimplants for Developing Regions

Expanding advanced bioimplants for emerging regions offers a distinctive chance to address healthcare inequalities and enhance patient results worldwide. Although advanced bioimplants, including resorbable scaffolds and tailored medical devices, have the ability to transform healthcare, their elevated costs, intricate manufacturing methods, and dependence on specialized infrastructure can render them unattainable for low-income communities. To tackle these issues, it is essential to concentrate on economical production, regional manufacturing abilities, and the adoption of sustainable materials that lower both economic and ecological expenses [46]. Partnerships among governments, NGOs, and the private sector can create scalable solutions that address the unique requirements of these areas, such as cost-effective bioimplant technologies and education for local healthcare professionals. Additionally, the use of modular, user-friendly implants, coupled with telemedicine and remote diagnostics, can broaden the availability of advanced medical solutions in regions with restricted access to specialized treatment. By tailoring bioimplant technologies to the requirements and limitations of developing areas, we can guarantee that these advancements serve a wider, more varied global community, enhancing healthcare equity around the globe.

7.8 Future Directions and Conclusion

7.8.1 Toward Fully Resorbable Organs and Tissue Systems

The quest for completely resorbable organs and tissue systems marks a groundbreaking advancement in regenerative medicine, seeking to develop sustainable, functional substitutes for injured or sick tissues without relying on permanent implants. Researchers are developing resorbable scaffolds by integrating advancements in biomaterials, stem cell therapy, and 3D bioprinting, which deliver structural support during tissue regeneration while gradually dissolving as new tissue develops, emulating the natural healing process. These entirely resorbable systems may possibly remove the necessity for additional surgeries, since they would break down in sync with the body's healing process. Main difficulties in reaching this objective involve maintaining the mechanical integrity and performance of resorbable scaffolds, managing the degradation speed, and ensuring effective integration with host tissues. Moreover, progress in tissue engineering, including the creation of vascularized tissue frameworks and the improvement of cell differentiation, is essential for the development of fully functional organs. As these technologies advance, the potential for resorbable organs and tissue systems may greatly influence transplant medicine, providing personalized, sustainable answers to organ shortages and enhancing long-term patient outcomes.

7.8.2 Bridging the Gap between Research and Clinical Applications

Addressing the divide between research and clinical applications is a vital challenge in the realm of biomedical engineering and regenerative medicine. Although scientific progress in fields such as tissue engineering, stem cell treatments, and bioimplants presents groundbreaking possibilities, transferring these breakthroughs from the laboratory to practical clinical environments necessitates overcoming substantial challenges. These encompass regulatory challenges, scalability concerns, economical production, and guaranteeing patient safety. Effective cooperation among researchers, clinicians, regulatory agencies, and industry is crucial to overcome these challenges and speed up the creation of effective clinical solutions. Moreover, thorough clinical trials, real-world assessments, and post-market monitoring are essential to confirm that experimental treatments are safe, effective, and suitable for various patient groups. By cultivating collaborations, simplifying regulatory procedures, and prioritizing patient-focused research, the divide between innovative research and broad clinical implementation can be reduced, thereby guaranteeing that new medical technologies enhance patient outcomes and propel healthcare on a worldwide level.

7.8.3 Collaborative Visions: Building the Next Generation of Medical Devices

Joint visions are crucial for developing the next wave of medical devices, since the intricacy of today's healthcare problems demands knowledge from various fields. Combining engineering, biology, materials science, and data analytics is crucial for creating innovative devices that are functional, patient-centered, cost-efficient, and scalable. By encouraging partnerships among researchers, healthcare professionals, manufacturers, and regulatory agencies, new medical devices can be developed with a thorough comprehension of clinical requirements, practical limitations, and advancing technological possibilities. This interdisciplinary strategy facilitates the development of devices that integrate advanced materials such as resorbable biomaterials, intelligent sensors, and bioactive elements, improving their functionality and assimilation with the body. Moreover, integrating artificial intelligence and machine learning into device design enables more customized, responsive treatments that adapt alongside the patient's condition. By adopting these cooperative ideas, the upcoming generation of medical devices could greatly enhance patient care, tackle global health inequalities, and foster sustainable healthcare advancements.

7.8.4 Final Reflections: Hydrogels as a Cornerstone of Resorbable Innovations

Hydrogels serve as a fundamental element of resorbable advancements, holding a crucial importance in the future of tissue engineering and regenerative medicine. Their distinctive features, such as elevated water content, compatibility with biological systems, and flexibility, render them perfect for developing scaffolds that aid in tissue development while slowly degrading in harmony with the body's recovery process. Hydrogels can be designed to release therapeutic agents, improve cellular interactions, and facilitate regeneration in a regulated way, rendering them essential in areas such as wound healing, drug delivery, and organ regeneration. As research advances, the capability to engineer smart hydrogels that react to external stimuli such as pH, temperature, or biochemical signals creates new opportunities for precision medicine. In addition, their combination with other biodegradable substances, like polymers and metals, presents the possibility of multifunctional scaffolds that merge structural reinforcement with biological functions. Hydrogels are progressing the creation of resorbable devices while also influencing the future of personalized, sustainable, and effective medical therapies, ultimately enhancing patient results and transforming the healthcare environment.

Acknowledgment

The authors sincerely acknowledge the generous financial support provided by the Department of Biotechnology (DBT), Ministry of Science & Technology, Government of India, under Project No. DBT-RA/2022/July/N/2240 through the DBT-RA Program. We extend our heartfelt gratitude to Prof. A. Srinivasan and Prof. Aynur Unal for their unwavering guidance and encouragement throughout this research. We also express our sincere appreciation to Dr. Mohit Kumar for his invaluable contributions, which significantly enhanced the success of this work. Lastly, a special note of gratitude to my daughter, Vanshika, whose unwavering love and inspiration has been a source of enduring strength and motivation.

AI disclosure statement: The author confirms that no artificial intelligence (AI)-generated content or AI-assisted writing tools were used in the preparation of this chapter (including summaries and the abstract). All text, analysis, and conclusions are the result of the author's original work.

References

1. Ahmad, Z., Salman, S., Khan, S.A., Amin, A., Rahman, Z.U., Al-Ghamdi, Y.O., Akhtar, K., Bakhsh, E.M., Khan, S.B., Versatility of Hydrogels: From Synthetic Strategies, Classification, and Properties to Biomedical Applications. *Gels*, *8*, 3, 167, 2022, https://doi.org/10.3390/gels8030167.
2. Alavi, S.E., Gholami, M., Shahmabadi, H.E., Reher, P., Resorbable GBR Scaffolds in Oral and Maxillofacial Tissue Engineering: Design, Fabrication, and Applications. *J. Clin. Med.*, *12*, 22, 6962, 2023, https://doi.org/10.3390/jcm12226962.

3. Alonso-Fernández, I., Haugen, H.J., López-Peña, M., González-Cantalapiedra, A., Muñoz, F., Use of 3D-printed polylactic acid/bioceramic composite scaffolds for bone tissue engineering in preclinical *in vivo* studies: A systematic review. *Acta Biomater.*, *168*, 1–21, 2023, https://doi.org/10.1016/j.actbio.2023.07.013.
4. Alshangiti, D.M., El-damhougy, T.K., Zaher, A., Madani, M., Mohamady ghobashy, M., Revolutionizing biomedicine: advancements, applications, and prospects of nanocomposite macromolecular carbohydrate-based hydrogel biomaterials: a review. *RSC Adv.*, *13*, 50, 35251–35291, 2023, https://doi.org/10.1039/D3RA07391B.
5. Bashir, S., Hina, M., Iqbal, J., Rajpar, A.H., Mujtaba, M.A., Alghamdi, N.A., Wageh, S., Ramesh, K., Ramesh, S., Fundamental Concepts of Hydrogels: Synthesis, Properties, and Their Applications. *Polymers*, *12*, 11, 2702, 2020, https://doi.org/10.3390/polym12112702.
6. Castanheira, E.J., Maia, J.R., Monteiro, L.P.G., Sobreiro-Almeida, R., Wittig, N.K., Birkedal, H., Rodrigues, J.M.M., Mano, J.F., 3D-printed injectable nanocomposite cryogel scaffolds for bone tissue regeneration. *Mater. Today Nano*, *28*, 100519, 2024, https://doi.org/10.1016/j.mtnano.2024.100519.
7. Chen, S., Wang, Y., Bao, S., Yao, L., Fu, X., Yu, Y., Lyu, H., Pang, H., Guo, S., Zhang, H., Zhou, P., Zhou, Y., Cerium oxide nanoparticles in wound care: a review of mechanisms and therapeutic applications. *Front. Bioeng. Biotechnol.*, *12*, 2024, https://doi.org/10.3389/fbioe.2024.1404651.
8. Cinici, B., Yaba, S., Kurt, M., Yalcin, H.C., Duta, L., Gunduz, O., Fabrication Strategies for Bioceramic Scaffolds in Bone Tissue Engineering with Generative Design Applications. *Biomimetics*, *9*, 7, 409, 2024, https://doi.org/10.3390/biomimetics9070409.
9. Devlin, B.L., Allenby, M.C., Ren, J., Pickering, E., Klein, T.J., Paxton, N.C., Woodruff, M.A., Materials Design Innovations in Optimizing Cellular Behavior on Melt Electrowritten (MEW) Scaffolds. *Adv. Funct. Mater.*, *34*, 18, 2024, https://doi.org/10.1002/adfm.202313092.
10. Echeverria Molina, M.I., Malollari, K.G., Komvopoulos, K., Design Challenges in Polymeric Scaffolds for Tissue Engineering. *Front. Bioeng. Biotechnol.*, *9*, 2021, https://doi.org/10.3389/fbioe.2021.617141.
11. Ejiohuo, O., A perspective on the synergistic use of 3D printing and electrospinning to improve nanomaterials for biomedical applications. *Nano Trends*, *4*, 100025, 2023, https://doi.org/10.1016/j.nwnano.2023.100025.
12. Esmaeili, Y., Bidram, E., Bigham, A., Atari, M., Nasr Azadani, R., Tavakoli, M., Salehi, S., Mirhaj, M., Basiri, A., Mirzavandi, Z., Boshtam, M., Rafienia, M., Zargar Kharazi, A., Karbasi, S., Shariati, L., Zarrabi, A., Exploring the evolution of tissue engineering strategies over the past decade: From cell-based strategies to gene-activated matrix. *Alexandria Eng. J.*, *81*, 137–169, 2023, https://doi.org/10.1016/j.aej.2023.08.080.
13. Farjaminejad, S., Farjaminejad, R., Hasani, M., Garcia-Godoy, F., Abdouss, M., Marya, A., Harsoputranto, A., Jamilian, A., Advances and Challenges

in Polymer-Based Scaffolds for Bone Tissue Engineering: A Path Towards Personalized Regenerative Medicine. *Polymers*, *16*, 23, 3303, 2024, https://doi.org/10.3390/polym16233303.

14. Golebiowska, A.A., Intravaia, J.T., Sathe, V., Kumbar, S.G., Nukavarapu, S.P., Engineered Osteochondral Scaffolds with Bioactive Cartilage Zone for Enhanced Articular Cartilage Regeneration. *Ann. Biomed. Eng.*, 2024, https://doi.org/10.1007/s10439-024-03655-1.
15. Hameed, H., Faheem, S., Paiva-Santos, A.C., Sarwar, H.S., Jamshaid, M., A Comprehensive Review of Hydrogel-Based Drug Delivery Systems: Classification, Properties, Recent Trends, and Applications. *AAPS PharmSciTech*, *25*, 4, 64, 2024, https://doi.org/10.1208/s12249-024-02786-x.
16. Harding, A., Pramanik, A., Basak, A.K., Prakash, C., Shankar, S., Application of additive manufacturing in the biomedical field- A review. *Ann. 3D Print. Med.*, *10*, 100110, 2023, https://doi.org/10.1016/j.stlm.2023.100110.
17. Ho, T.-C., Chang, C.-C., Chan, H.-P., Chung, T.-W., Shu, C.-W., Chuang, K.-P., Duh, T.-H., Yang, M.-H., Tyan, Y.-C., Hydrogels: Properties and Applications in Biomedicine. *Molecules*, *27*, 9, 2902, 2022, https://doi.org/10.3390/molecules27092902.
18. Yoon, J., Han, H., Jang, J., Nanomaterials-incorporated hydrogels for 3D bioprinting technology. *Nano Convergence*, *10*, 1, 52, 2023, https://doi.org/10.1186/s40580-023-00402-5.
19. Islam, M.A., Mobarak, M.H., Rimon, M.I.H., al Mahmud, M.Z., Ghosh, J., Ahmed, M.M.S., Hossain, N., Additive manufacturing in polymer research: Advances, synthesis, and applications. *Polym. Test.*, *132*, 108364, 2024, https://doi.org/10.1016/j.polymertesting.2024.108364.
20. Jeznach, O., Tabakoglu, S., Zaszczyńska, A., Sajkiewicz, P., Review on machine learning application in tissue engineering: What has been done so far? Application areas, challenges, and perspectives. *J. Mater. Sci.*, *59*, 46, 21222–21250, 2024, https://doi.org/10.1007/s10853-024-10449-2.
21. Kang, F., Yi, Q., Gu, P., Dong, Y., Zhang, Z., Zhang, L., Bai, Y., Controlled growth factor delivery system with osteogenic-angiogenic coupling effect for bone regeneration. *J. Orthop. Translat.*, *31*, 110–125, 2021, https://doi.org/10.1016/j.jot.2021.11.004.
22. Khafoor, A.A., Karim, A.S., Sajadi, S.M., Recent progress in synthesis of nano based liposomal drug delivery systems: A glance to their medicinal applications. *Results Surf. Interfaces*, *11*, 100124, 2023, https://doi.org/10.1016/j.rsurfi.2023.100124.
23. Lai, J., Liu, Y., Lu, G., Yung, P., Wang, X., Tuan, R.S., Li, Z.A., 4D bioprinting of programmed dynamic tissues. *Bioact. Mater.*, *37*, 348–377, 2024, https://doi.org/10.1016/j.bioactmat.2024.03.033.
24. Li, W., Yigitcanlar, T., Browne, W., Nili, A., The Making of Responsible Innovation and Technology: An Overview and Framework. *Smart Cities*, *6*, 4, 1996–2034, 2023, https://doi.org/10.3390/smartcities6040093.

25. Liang, J., Liu, P., Yang, X., Liu, L., Zhang, Y., Wang, Q., Zhao, H., Biomaterial-based scaffolds in promotion of cartilage regeneration: Recent advances and emerging applications. *J. Orthop. Translat.*, *41*, 54–62, 2023, https://doi.org/10.1016/j.jot.2023.08.006.
26. Lu, P., Ruan, D., Huang, M., Tian, M., Zhu, K., Gan, Z., Xiao, Z., Harnessing the potential of hydrogels for advanced therapeutic applications: current achievements and future directions. *Signal Transduction Targeted Ther.*, *9*, 1, 166, 2024, https://doi.org/10.1038/s41392-024-01852-x.
27. Mobarak, M.H., Islam, Md.A., Hossain, N., al Mahmud, Md.Z., Rayhan, Md.T., Nishi, N.J., Chowdhury, M.A., Recent advances of additive manufacturing in implant fabrication – A review. *Appl. Surf. Sci. Adv.*, *18*, 100462, 2023, https://doi.org/10.1016/j.apsadv.2023.100462.
28. Mukasheva, F., Adilova, L., Dyussenbinov, A., Yernaimanova, B., Abilev, M., Akilbekova, D., Optimizing scaffold pore size for tissue engineering: insights across various tissue types. *Front. Bioeng. Biotechnol.*, *12*, 2024, https://doi.org/10.3389/fbioe.2024.1444986.
29. Nezhad-Mokhtari, P., Hasany, M., Kohestanian, M., Dolatshahi-Pirouz, A., Milani, M., Mehrali, M., Recent advancements in bioadhesive self-healing hydrogels for effective chronic wound care. *Adv. Colloid Interface Sci.*, *334*, 103306, 2024, https://doi.org/10.1016/j.cis.2024.103306.
30. Odziomek, K., Drabczyk, A.K., Kościelniak, P., Konieczny, P., Barczewski, M., Bialik-Wąs, K., The Role of Freeze-Drying as a Multifunctional Process in Improving the Properties of Hydrogels for Medical Use. *Pharmaceuticals*, *17*, 11, 1512, 2024, https://doi.org/10.3390/ph17111512.
31. Pathak, K., Saikia, R., Das, A., Das, D., Islam, M.A., Pramanik, P., Parasar, A., Borthakur, P.P., Sarmah, P., Saikia, M., Borthakur, B., 3D printing in biomedicine: advancing personalized care through additive manufacturing. *Explor. Med.*, 1135–1167, 2023, https://doi.org/10.37349/emed.2023.00200.
32. Pesode, P., Barve, S., Wankhede, S.V., Ahmad, A., Sustainable Materials and Technologies for Biomedical Applications. *Adv. Mater. Sci. Eng.*, *2023*, 1–22, 2023, https://doi.org/10.1155/2023/6682892.
33. Rahatuzzaman, M., Mahmud, M., Rahman, S., Hoque, M.E., Design, fabrication, and characterization of 3D-printed ABS and PLA scaffolds potentially for tissue engineering. *Results Eng.*, *21*, 101685, 2024, https://doi.org/10.1016/j.rineng.2023.101685.
34. Rana, M.M. and de la Hoz Siegler, H., Evolution of Hybrid Hydrogels: Next-Generation Biomaterials for Drug Delivery and Tissue Engineering. *Gels*, *10*, 4, 216, 2024, https://doi.org/10.3390/gels10040216.
35. Revete, A., Aparicio, A., Cisterna, B.A., Revete, J., Luis, L., Ibarra, E., Segura González, E.A., Molino, J., Reginensi, D., Advancements in the Use of Hydrogels for Regenerative Medicine: Properties and Biomedical Applications. *Int. J. Biomater.*, *2022*, 1–16, 2022, https://doi.org/10.1155/2022/3606765.
36. Saghazadeh, S., Rinoldi, C., Schot, M., Kashaf, S.S., Sharifi, F., Jalilian, E., Nuutila, K., Giatsidis, G., Mostafalu, P., Derakhshandeh, H., Yue, K.,

Swieszkowski, W., Memic, A., Tamayol, A., Khademhosseini, A., Drug delivery systems and materials for wound healing applications. *Adv. Drug Delivery Rev.*, *127*, 138–166, 2018, https://doi.org/10.1016/j.addr.2018.04.008.

37. Selim, M., Mousa, H.M., Abdel-Jaber, G.T., Barhoum, A., Abdal-hay, A., Innovative designs of 3D scaffolds for bone tissue regeneration: Understanding principles and addressing challenges. *Eur. Polym. J.*, *215*, 113251, 2024, https://doi.org/10.1016/j.eurpolymj.2024.113251.
38. Singh, A.B., Transforming Healthcare: A Review of Additive Manufacturing Applications in the Healthcare Sector. *IECMA*, *2024*, 2, 2024, https://doi.org/10.3390/engproc2024072002.
39. Tottoli, E.M., Dorati, R., Genta, I., Chiesa, E., Pisani, S., Conti, B., Skin Wound Healing Process and New Emerging Technologies for Skin Wound Care and Regeneration. *Pharmaceutics*, *12*, 8, 735, 2020, https://doi.org/10.3390/pharmaceutics12080735.
40. Tripathi, A.S., Zaki, M.E.A., Al-Hussain, S.A., Dubey, B.K., Singh, P., Rind, L., Yadav, R.K., Material matters: exploring the interplay between natural biomaterials and host immune system. *Front. Immunol.*, *14*, 2023, https://doi.org/10.3389/fimmu.2023.1269960.
41. Trucillo, P., Biomaterials for Drug Delivery and Human Applications. *Materials*, *17*, 2, 456, 2024, https://doi.org/10.3390/ma17020456.
42. Xing, J., Zhang, M., Liu, X., Wang, C., Xu, N., Xing, D., Multi-material electrospinning: from methods to biomedical applications. *Mater. Today Bio*, *21*, 100710, 2023, https://doi.org/10.1016/j.mtbio.2023.100710.
43. Zare, I., Mirshafiei, M., Kheilnezhad, B., Far, B.F., Hassanpour, M., Pishbin, E., Eftekhar Vaghefi, S.S., Yazdian, F., Rashedi, H., Hasan, A., Wang, X., Adeli, M., Makvandi, P., Hydrogel-integrated graphene superstructures for tissue engineering: From periodontal to neural regeneration. *Carbon*, *223*, 118970, 2024, https://doi.org/10.1016/j.carbon.2024.118970.
44. Zhang, Q., Zhou, J., Zhi, P., Liu, L., Liu, C., Fang, A., Zhang, Q., 3D printing method for bone tissue engineering scaffold. *Med. Novel Technol. Devices*, *17*, 100205, 2023, https://doi.org/10.1016/j.medntd.2022.100205.
45. Zhao, F., Liu, F., Gao, C., Wang, G., Zhang, Y., Yu, F., Tian, J., Tan, K., Zhang, R., Liang, K., Lian, Z., Guo, J., Kong, B., Ge, J., Xin, H., Interfacing exogenous stents with human coronary artery by self-assembled coating: designs, functionalities and applications. *NPG Asia Mater.*, *16*, 1, 28, 2024, https://doi.org/10.1038/s41427-024-00548-6.
46. Zöller, K., To, D., Bernkop-Schnürch, A., Biomedical applications of functional hydrogels: Innovative developments, relevant clinical trials and advanced products. *Biomaterials*, *312*, 122718, 2025, https://doi.org/10.1016/j.biomaterials.2024.122718.

8

Resorbable Composites for Bone Grafting and Fixation Applications

Nisha Shankhwar[1]*, Awadhesh Kumar Verma[2], Tanya Singh[2], Satyendra Singh[1] and Neeta Raj Sharma[2]

[1]Special Centre for Nanoscience, Jawaharlal Nehru University, New Delhi, India

[2]School of Bioengineering and Biosciences, Lovely Professional University, Punjab, India

Abstract

Resorbable composites have emerged as a groundbreaking solution in bone grafting and fixation, offering materials that not only provide temporary structural support but also actively promote tissue regeneration. These composites, which incorporate biodegradable polymers, bioactive ceramics, hydroxyapatite, and bioactive glass nanoparticles, are designed to mimic the natural extracellular matrix (ECM) of bone, enhancing osteoconductivity and fostering cellular integration. The incorporation of these materials enables a scaffold architecture that facilitates bone healing while maintaining mechanical stability. Advanced biofabrication techniques, such as 3D bioprinting and freeze-casting, enable the design of scaffolds with multiscale porosity and mechanical gradients that replicate the structure and function of both cortical and trabecular bone. The key benefit of these resorbable composites lies in their ability to degrade at a controlled rate, matching the pace of bone regeneration. This allows them to provide support throughout the healing process without compromising new bone formation. The inclusion of bioactive molecules or growth factors in these materials further enhances osteogenesis, accelerating the repair process. These composites are particularly valuable in personalized bone grafting applications, offering solutions for complex or large defects. Their customization ensures compatibility with individual patient needs, while their degradation without residue minimizes complications. As research advances, the continued development of these materials will be crucial in driving innovations in orthopedic care, reducing healthcare costs, and improving long-term patient outcomes.

**Corresponding author*: shankhwarnisha3july@gmail.com; nisha2023@jnu.ac.in

Arbind Prasad (ed.) Resorbable Composites for Bioimplants and Fixation Devices, (227–252) © 2026 Scrivener Publishing LLC

***Keywords*:** Resorbable bone grafts, biomimetic scaffolds, bioactive glass nanoparticles, controlled degradation, osteointegration, advanced biofabrication, personalized orthopedic implants

8.1 Introduction: Resorbable Composites in Bone Regeneration

Resorbable composites have emerged as transformative materials in bone regeneration, offering a combination of biodegradability, biocompatibility, and tailored mechanical properties to meet the demands of orthopedic applications. These composites are designed to degrade gradually in the body while being replaced by natural bone, thus eliminating the need for secondary surgeries to remove implants [1]. Composed primarily of polymers (*e.g.*, polylactic acid, PLGA), ceramics (*e.g.*, hydroxyapatite, bioactive glass), and metal ions, these materials bridge the gap between biological tissue and synthetic support. Advances in composite fabrication techniques, such as 3D printing and surface functionalization, have enabled the development of scaffolds and fixation devices that mimic the hierarchical architecture of natural bone [2]. The integration of bioactive molecules further enhances their regenerative potential, positioning resorbable composites as the future of bone grafting solutions.

8.1.1 The Evolution of Bone Grafting and Fixation Materials

The field of bone grafting and fixation has evolved significantly, progressing from the use of autografts and allografts to advanced synthetic materials and resorbable composites. Historically, autografts derived from the patient's own bone were the gold standard due to their osteogenic, osteoconductive, and osteoinductive properties, but their use is limited by donor site morbidity and availability [3]. Allografts from cadaveric sources present alternatives but risk immune rejection and infection. The introduction of synthetic grafts, including calcium phosphates, bioactive glass, and polymeric implants, addressed some of these issues but lacked biological integration. Resorbable composites, combining the benefits of biodegradable polymers, ceramics, and growth factors, now offer solutions that mimic bone properties while actively participating in the healing process [4].

8.1.2 The Role of Resorbable Composites in Modern Orthopedics

Resorbable composites are transforming modern orthopedics by combining mechanical strength with controlled biodegradability and bioactivity, addressing the limitations of traditional implants. These materials facilitate the natural healing process by temporarily stabilizing fractures or bone defects while simultaneously encouraging tissue regeneration [5]. Polymers such as polylactic acid (PLA) provide the matrix for controlled degradation, while ceramic additives like hydroxyapatite enhance osteoconductivity. Resorbable fixation devices, such as screws, pins, and plates, reduce complications associated with metallic implants, including stress shielding and revision surgeries [6]. Additionally, the ability to incorporate growth factors, antibiotics, or stem cells into the composite matrix enhances their regenerative potential, enabling tailored solutions for patient-specific applications.

8.1.3 Case in Focus: A Patient's Journey with a Resorbable Bone Graft

A recent case study highlights the successful use of a resorbable bone graft in a patient with a complex tibial defect. The patient presented with a large, non-union fracture following trauma, requiring a graft that could provide mechanical support while promoting bone regeneration [7]. A custom-designed scaffold composed of polylactic-co-glycolic acid (PLGA) and hydroxyapatite nanoparticles was fabricated using 3D printing to match the defect's size and geometry. The graft demonstrated controlled degradation over 12 months, with concurrent bone ingrowth observed through micro-CT imaging. The patient achieved full recovery, with no adverse immune responses or graft-related complications [8]. This case underscores the potential of resorbable composites in personalized treatments and their role in reducing revision surgeries.

8.2 Fundamentals of Bone Grafting and Fixation Applications

8.2.1 Understanding Bone Healing: Biological and Mechanical Perspectives

Bone healing is a complex process that involves the interplay between biological factors and mechanical stability. The process occurs in four key phases: inflammatory response, soft callus formation, hard callus formation, and bone remodeling [9]. During the inflammatory phase, osteoblasts and mesenchymal stem cells (MSCs) are recruited to the site, initiating repair. The formation of a soft callus provides temporary mechanical stability, followed by mineralization into a hard callus. The remodeling phase restores the bone's original strength and structure. Mechanical stability, achieved through internal or external fixation devices, is essential for minimizing micromotion and facilitating proper cellular activity. Resorbable composites act as temporary mechanical supports while actively promoting the biological healing process through their osteoconductive and bioactive properties [10].

8.2.2 Key Challenges in Traditional Bone Grafting Techniques

Traditional bone grafting techniques, including autografts, allografts, and synthetic grafts, face significant limitations in clinical practice. Autografts, considered the gold standard, are limited by donor site morbidity, limited availability, and increased surgical time. Allografts, while overcoming donor site issues, present risks of immune rejection, infection, and reduced bioactivity [11]. Synthetic grafts, such as ceramics or polymers, lack the biological properties necessary for full osseointegration. Additionally, achieving adequate mechanical stability and vascularization in large bone defects remains challenging. Resorbable composites are designed to overcome these limitations by providing biodegradable, bioactive scaffolds that support mechanical load while enhancing bone regeneration through their osteoconductive properties [12].

8.2.3 Role of Resorbable Materials: Bridging Biology and Engineering

Resorbable materials serve as a critical link between biological healing and engineering principles by offering both mechanical support and bioactivity. These materials are typically composites of biodegradable polymers (*e.g.*, PLA, PLGA) and ceramic additives (*e.g.*, hydroxyapatite, β-TCP) that provide temporary structural support and degrade over time, allowing for natural bone regeneration [13]. The degradation of the material occurs through hydrolysis or enzymatic activity, ensuring the gradual replacement of the graft with native tissue. Resorbable composites promote osteoconduction, enabling cell adhesion, proliferation, and mineralization, while avoiding the complications of permanent implants, such as stress shielding. Their ability to integrate with growth factors further enhances the biological healing process, bridging the gap between engineering design and clinical outcomes.

8.2.4 Comparison of Autografts, Allografts, and Synthetic Grafts

Bone grafts are categorized into autografts, allografts, and synthetic grafts, each with distinct advantages and limitations. Autografts, harvested from the patient, are ideal due to their osteogenic, osteoinductive, and osteoconductive properties but are limited by donor site availability and morbidity. Allografts, sourced from cadavers, avoid donor site complications but present risks such as immune response, infection, and slower healing. Synthetic grafts, including ceramics, bioactive glass, and polymers, overcome biological limitations but often lack bioactivity and vascularization [14]. Resorbable composites offer a unique advantage by combining the mechanical support of synthetic grafts with the bioactivity of natural grafts, promoting osseointegration and gradual tissue replacement without secondary surgeries.

The application of resorbable materials can be seen in Figure 8.1 below.

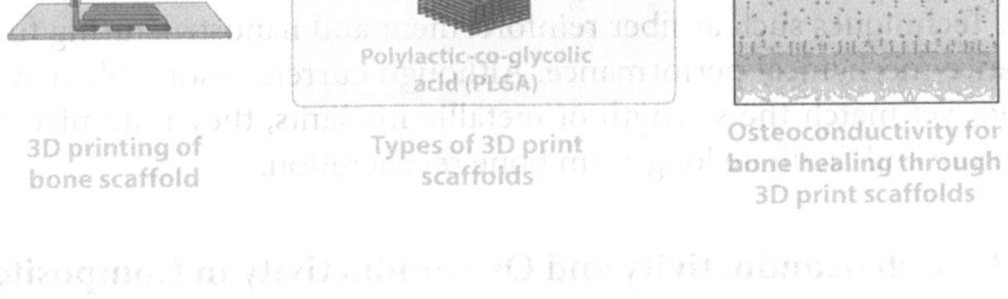

8.3 Properties of Resorbable Composites for Bone Applications

8.3.1 Biodegradability and Controlled Resorption Rates

Biodegradability is a hallmark of resorbable composites, enabling gradual degradation *in vivo* while promoting natural bone regeneration. The resorption rate of a material must align with the rate of new bone formation to ensure that mechanical integrity is maintained throughout healing. Resorbable composites, primarily composed of PLA, polycaprolactone (PCL), and β-tricalcium phosphate (β-TCP), degrade through hydrolysis or enzymatic activity, producing non-toxic byproducts [15]. Fine-tuning the polymer-to-ceramic ratio and modifying the material's crystallinity and porosity allow for precise control over degradation kinetics. β-TCP accelerates bone integration while balancing resorption, whereas PLA provides structural support with slower degradation. Controlled resorption minimizes complications such as premature loss of mechanical support or material accumulation, ensuring seamless bone regeneration.

8.3.2 Mechanical Strength and Load-Bearing Capacity

Mechanical strength is a critical factor in resorbable composites, particularly for applications in weight-bearing bone defects. Resorbable composites must provide sufficient mechanical stability to withstand physiological loads during healing while gradually degrading. Hybrid materials, combining polymers (*e.g.*, PLA, PCL) with ceramics (*e.g.*, hydroxyapatite, β-TCP), offer enhanced compressive strength and tensile properties. Ceramic fillers improve the composite's stiffness and load-bearing capacity, whereas polymeric matrices impart flexibility and controlled degradation [16]. Techniques such as fiber reinforcement and nanostructuring further enhance mechanical performance. Although current resorbable materials do not yet match the strength of metallic implants, they minimize stress shielding and facilitate long-term bone regeneration.

8.3.3 Osteoconductivity and Osteoinductivity in Composite Materials

Resorbable composites must exhibit osteoconductivity (supporting cell attachment and bone growth) and osteoinductivity (stimulating differentiation of progenitor cells into osteoblasts). Ceramic components such as hydroxyapatite and β-TCP provide a bioactive surface that promotes the attachment of osteoblasts and the deposition of calcium phosphate. Additionally, growth factors like bone morphogenetic proteins (BMPs) and angiogenic molecules can be incorporated into the composite matrix to enhance osteoinductive properties [17]. Recent advances in surface modification and nanostructuring of composites further enhance bone regeneration by mimicking the nanoscale architecture of natural bone. These properties ensure that resorbable materials act as functional scaffolds that integrate seamlessly with the surrounding tissue.

8.3.4 Biocompatibility and Immune Response Management

Biocompatibility is paramount for ensuring that resorbable composites integrate into the host tissue without eliciting adverse immune reactions. Composites composed of biodegradable polymers (*e.g.*, PLA, PLGA) and bioactive ceramics minimize foreign body responses due to their non-toxic degradation byproducts. The degradation rate of resorbable materials can influence immune response; rapid degradation may trigger inflammation, while slow degradation may impede healing. Surface coatings with anti-inflammatory agents and functionalization with bioactive molecules help regulate the immune environment [18]. Advances in nanostructured composites and biopolymer modifications aim to balance degradation kinetics and biocompatibility, ensuring seamless integration and effective bone regeneration.

8.3.5 Hybrid Composites: Combining Polymers, Ceramics, and Metals

Hybrid composites integrate polymers, ceramics, and bioresorbable metals to create materials with enhanced mechanical strength, osteoconductivity, and controlled degradation. For example, composites combining PLA/PCL polymers with hydroxyapatite and magnesium-based metals provide a synergistic effect—mechanical reinforcement from metals, bioactivity from ceramics, and flexibility from polymers. This multi-phase design improves load-bearing capacity and encourages bone tissue ingrowth [19]. Advances in nanotechnology enable the creation of hybrid composites with nanoscale features, enhancing their osteoconductivity and cellular interactions. Hybrid materials are being actively researched for their ability to meet diverse clinical needs in orthopedic and maxillofacial applications.

8.4 Fabrication Techniques for Resorbable Bone Grafting Composites

8.4.1 Additive Manufacturing for Complex Bone Structures

Additive manufacturing (AM), or 3D printing, has revolutionized the fabrication of resorbable bone grafting composites by enabling the creation of complex, patient-specific structures with precise control over geometry, porosity, and mechanical properties [20]. Techniques such as Fused Deposition Modeling (FDM), Selective Laser Sintering (SLS), and Stereolithography (SLA) allow for the layer-by-layer deposition of materials like PLA, PCL, and hydroxyapatite-based composites [21]. AM enables the incorporation of controlled macro- and micro-porosity, which is essential for vascularization and osteointegration. Additionally, bioinks loaded with growth factors or stem cells facilitate the development of bioactive scaffolds for bone tissue engineering. The combination of customized design and bioactivity makes AM a versatile tool for producing implants tailored to complex defect geometries.

8.4.2 Freeze-Casting and Sol-Gel Methods for Porous Scaffolds

Freeze-casting and sol-gel methods are widely used for producing resorbable bone scaffolds with highly interconnected porous structures, which are critical for bone regeneration. Freeze-casting involves freezing a ceramic-polymer slurry, followed by sublimation of the solvent, leaving behind a scaffold with aligned, hierarchical pores. This method is particularly effective for creating scaffolds with controlled pore size and porosity, facilitating cell migration and nutrient diffusion [22]. The sol-gel process allows for the synthesis of bioactive glass and ceramic-based scaffolds at low temperatures, enabling the incorporation of sensitive bioactive molecules. These techniques provide excellent control over material composition and microstructure, making them suitable for osteoconductive scaffolds in bone grafting applications [23].

8.4.3 Surface Functionalization for Enhanced Cell Adhesion

Surface functionalization is a key strategy for enhancing the cell adhesion, proliferation, and osteogenic differentiation of resorbable bone composites. Techniques such as plasma treatment, chemical grafting, and coating with bioactive molecules (*e.g.*, collagen, fibronectin, or growth factors)

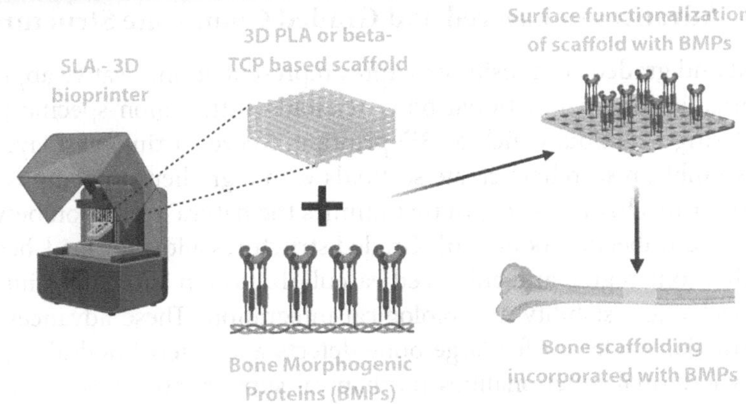

modify the surface properties of scaffolds, improving their bioactivity and interactions with cells. For instance, surface modification with hydroxyapatite nanoparticles enhances osteoconductivity, while immobilization of BMPs promotes osteoinduction [24]. Nanostructured surfaces mimic the native ECM, creating favorable conditions for cell attachment and tissue integration. Surface functionalization bridges the gap between synthetic scaffolds and biological tissues, ensuring improved performance of resorbable implants. This process can be seen in Figure 8.2.

8.4.4 Nanotechnology in Composite Design: Improving Bone Integration

Nanotechnology has significantly improved the performance of resorbable bone composites by enhancing their bioactivity, mechanical properties, and bone integration. Incorporation of nanoceramics (*e.g.*, nano-hydroxyapatite, bioactive glass nanoparticles) into polymer matrices improves osteoconductivity and mimics the natural bone mineral phase. Additionally, nanofiber scaffolds produced *via* electrospinning provide a high surface area for cell attachment and controlled drug delivery. Functionalizing resorbable composites with nanoparticles carrying growth factors or antibacterial agents enhances regeneration while preventing infections [25]. These nanoengineered materials closely replicate the structure and functionality of natural bone, accelerating healing and improving clinical outcomes.

8.4.5 Advances in Layered and Graded Composite Structures

Layered and graded composite structures represent an innovative approach for fabricating multi-functional bone scaffolds with region-specific properties. Using techniques such as 3D printing, freeze-casting, and layer-by-layer assembly, researchers create scaffolds with a gradient in composition, porosity, or mechanical strength that mimics the natural transition between cortical and trabecular bone [26]. Graded structures allow for load-bearing strength in one region and enhanced vascularization in another, facilitating both mechanical stability and biological integration. These advancements are particularly relevant for large bone defects and osteochondral applications, where dual functionalities mechanical support and bioactivity—are required simultaneously.

8.5 Optimization Strategies for Bone Grafting and Fixation Devices

8.5.1 Enhancing Mechanical Stability for Weight-Bearing Applications

Mechanical stability is critical for resorbable bone grafts, especially in weight-bearing applications such as femoral, tibial, or spinal repairs. Optimizing the mechanical strength and load-bearing capacity of grafts involves careful material selection and structural design. Incorporating ceramic reinforcements such as hydroxyapatite or β-tricalcium phosphate (β-TCP) into polymeric matrices like PLA or PCL enhances compressive strength and stiffness [27]. Techniques such as fiber reinforcement, gradient layering, and nanostructuring further improve mechanical performance. Computational tools like finite element analysis (FEA) enable precise modeling of mechanical properties under physiological loads, ensuring stability during the healing process while the material degrades. These advancements ensure that resorbable composites can meet the demands of high-load regions without compromising their degradation kinetics or biological functionality.

8.5.2 Incorporating Growth Factors for Accelerated Bone Regeneration

Incorporating growth factors into resorbable bone grafting devices enhances their regenerative potential by stimulating cellular activity and tissue formation.

BMPs, particularly BMP-2 and BMP-7, promote osteogenic differentiation and bone formation. Vascular endothelial growth factor (VEGF) facilitates angiogenesis, ensuring proper nutrient and oxygen delivery for bone healing. Growth factors can be loaded into resorbable composites through techniques like encapsulation, surface adsorption, or controlled-release microspheres [28]. Advanced scaffolds with dual-release systems ensure that growth factors are released in a sequential and controlled manner, mimicking the natural bone healing process. These systems significantly accelerate cell proliferation, osteointegration, and new bone formation in critical-size defects.

8.5.3 Computational Modeling for Predicting Graft Performance

Computational modeling has become a powerful tool for optimizing the design and performance of resorbable bone grafts. Techniques such as finite element analysis (FEA) and computational fluid dynamics (CFD) allow researchers to predict mechanical behavior, degradation rates, and drug release kinetics under physiological conditions. FEA evaluates stress distribution and mechanical stability of grafts, ensuring their structural integrity during load-bearing applications. CFD assists in optimizing pore architecture to improve nutrient diffusion and vascularization [29]. Computational modeling also enables patient-specific designs, where grafts are tailored to the defect geometry and biomechanical requirements of individual patients. This approach accelerates the development process and enhances clinical outcomes by reducing trial-and-error experimentation.

8.5.4 Designing Tailored Porosity for Vascularization and Integration

Porosity plays a crucial role in determining the success of resorbable bone grafts by facilitating vascularization, nutrient diffusion, and cellular infiltration. Tailoring the pore size, interconnectivity, and porosity gradient of scaffolds is essential for achieving optimal bone integration. Large pores (>300 μm) support vascularization, while smaller pores (~100 μm) enhance cell attachment and tissue formation. Advanced fabrication techniques such as 3D printing, freeze-casting, and electrospinning enable precise control over pore architecture [30]. Gradient porosity mimics the natural bone structure, with denser regions for mechanical strength and porous regions for biological integration. These designs significantly enhance graft performance, ensuring seamless integration into host tissues.

8.5.5 Drug-Loaded Composites for Infection Control in Bone Grafts

Infection remains a significant challenge in bone grafting, particularly in large defects and open fractures. Drug-loaded resorbable composites address this issue by enabling localized delivery of antibiotics or anti-inflammatory agents directly to the injury site. Controlled-release systems, such as microspheres, nanocarriers, and hydrogel coatings, ensure sustained delivery of drugs like vancomycin, gentamicin, or ciprofloxacin, minimizing systemic toxicity [31]. The integration of antibacterial nanoparticles, such as silver or zinc oxide, further enhances infection control. These drug-loaded composites reduce the risk of implant-associated infections, promote bone healing, and eliminate the need for systemic antibiotic administration. This process can be seen below in Figure 8.3.

8.6 Case Studies in Bone Grafting and Fixation

8.6.1 Resorbable Composites in Craniofacial Reconstruction

Craniofacial reconstruction poses unique challenges due to the complexity of bone geometry, aesthetics, and functionality. Resorbable composites, such as poly-lactic-co-glycolic acid (PLGA) combined with hydroxyapatite or β-tricalcium phosphate (β-TCP), have demonstrated significant success in these cases. A case study of a patient with a large cranial defect utilized a 3D-printed resorbable scaffold fabricated *via* additive manufacturing to precisely match the defect. The scaffold provided temporary mechanical stability while supporting cellular attachment and tissue

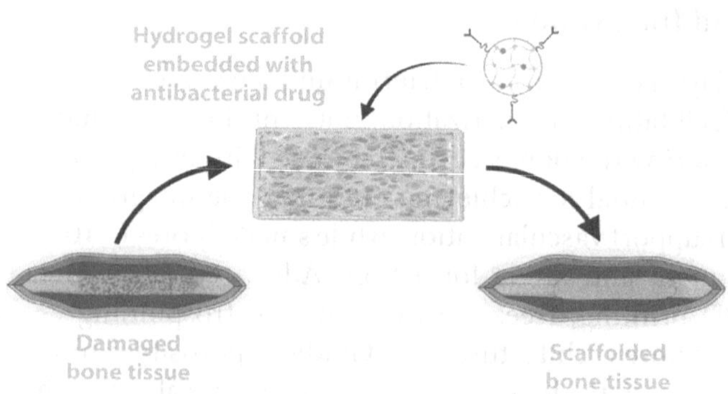

ingrowth. Over a period of 12 months, micro-CT imaging revealed gradual scaffold resorption and complete bone regeneration without immune rejection or donor-site complications [32]. This highlights the potential of patient-specific resorbable composites for craniofacial applications, combining mechanical support with bioactivity to achieve functional and aesthetic restoration.

8.6.2 Long Bone Fractures: Successes and Challenges with Resorbable Plates

Long bone fractures, particularly in weight-bearing regions such as the femur and tibia, require implants with high mechanical strength and controlled degradation rates. A notable case involved a resorbable PLA/HA composite plate applied to stabilize a mid-shaft femoral fracture in a young patient. The implant provided sufficient mechanical support during the early stages of healing, and its gradual resorption over 18 months coincided with new bone formation [33]. Despite its success, challenges such as early mechanical failure and non-uniform resorption rates were observed in similar cases, highlighting the need for optimized polymer-ceramic ratios and advanced fabrication methods. This case demonstrates the potential of resorbable plates in fracture fixation while emphasizing ongoing requirements for mechanical stability optimization.

8.6.3 Pelvic and Spinal Grafting Applications: Innovations in Design

Pelvic and spinal reconstructions require resorbable implants with exceptional mechanical strength and osteoconductive properties. A clinical case featured the use of a hybrid PLA/β-TCP scaffold in spinal fusion surgery for a patient with degenerative disc disease. The scaffold, fabricated *via* freeze-casting and additive manufacturing, provided tailored porosity gradients to facilitate vascularization and promote osseointegration. Post-operative imaging confirmed the formation of robust bone tissue within the scaffold architecture, and the implant gradually degraded over 18 months, leaving behind integrated bone tissue [34]. The successful fusion significantly improved spinal stability and reduced post-operative pain. Innovations such as graded porosity and patient-specific designs highlight the efficacy of resorbable scaffolds in complex pelvic and spinal grafting applications.

8.6.4 Lessons from Failures: Managing Premature Resorption and Mechanical Failure

While resorbable composites have shown promise, failures such as premature resorption and mechanical instability provide valuable lessons for future designs. A retrospective analysis of clinical cases involving PLA/HA resorbable screws in load-bearing regions identified issues with early degradation, leading to insufficient mechanical support and delayed healing. Factors contributing to failure included poor material crystallinity, inappropriate polymer-ceramic ratios, and inadequate stress distribution [35]. Strategies to overcome these challenges involve optimizing degradation rates, enhancing material toughness with nanoreinforcements, and incorporating computational models to predict mechanical performance. Learning from these failures is essential for improving resorbable composite design to ensure consistent, reliable clinical outcomes.

8.7 Emerging Research Trends in Resorbable Bone Composites

8.7.1 Smart Composites: Responsive Materials for Skeletal Repair

Smart resorbable composites represent a significant breakthrough in skeletal repair, incorporating stimuli-responsive properties to improve therapeutic outcomes. These materials respond to environmental changes such as pH, temperature, and mechanical stress to achieve controlled degradation and targeted drug release. For example, pH-responsive composites degrade faster in acidic environments often associated with bone defects or infections, facilitating efficient regeneration and infection control. Additionally, mechanical stress-responsive scaffolds adapt their structural properties to changes in load-bearing conditions, mimicking natural bone remodeling [36]. Incorporating nanoparticles and bioactive agents into polymeric matrices enhances these dynamic behaviors, positioning smart composites as next-generation materials for personalized and adaptive bone repair strategies.

8.7.2 AI-Driven Design of Bone Grafts and Fixation Devices

Artificial intelligence (AI) is transforming the design and optimization of resorbable bone grafts and fixation devices. AI algorithms, combined with computational tools such as FEA and topology optimization, enable the development of grafts with precisely tailored mechanical properties and degradation kinetics. Machine learning models analyze vast datasets of material behavior, anatomical structures, and clinical outcomes to generate patient-specific implants that minimize failure risks [37]. AI-driven tools also optimize the pore architecture and drug release profiles of scaffolds, ensuring efficient vascularization and tissue integration. This convergence of AI and material science accelerates the design-to-clinic pipeline, making personalized and adaptive bone graft solutions feasible.

8.7.3 Integration of Bioactive Glass and Ceramic Nanoparticles

The integration of bioactive glass and ceramic nanoparticles into resorbable composites has revolutionized their osteogenic potential and mechanical performance. Bioactive glass, with its unique ability to bond chemically with bone, releases silicon, calcium, and phosphate ions that stimulate osteoblast activity and promote angiogenesis [38]. Incorporating hydroxyapatite or β-tricalcium phosphate (β-TCP) nanoparticles into polymeric matrices enhances their osteoconductivity, ensuring rapid bone ingrowth and scaffold resorption. Advances in nanotechnology allow for the homogeneous distribution of nanoparticles, ensuring consistent mechanical properties and controlled ion release [39]. These hybrid systems show promising clinical outcomes in treating large bone defects, particularly in weight-bearing regions.

8.7.4 4D Printing for Dynamic and Adaptive Bone Grafts

4D printing, an extension of 3D printing, introduces time-dependent transformations into bone grafts, enabling dynamic and adaptive responses to external stimuli. Resorbable bone scaffolds fabricated with smart polymers exhibit changes in shape, structure, or drug release behavior in response to factors like temperature, pH, or mechanical load. Shape-memory polymers allow implants to expand and conform to irregular bone defects upon implantation. 4D-printed materials are ideal for delivering growth factors or antibiotics in a spatiotemporally controlled manner [40]. This technology holds immense potential for developing self-healing implants that adapt to physiological conditions, ensuring enhanced bone regeneration and clinical outcomes.

8.7.5 Biomimetic Approaches: Emulating Natural Bone Architecture

Biomimetic strategies aim to replicate the hierarchical structure and composition of natural bone to enhance tissue regeneration. Advances in biofabrication techniques, such as 3D bioprinting and freeze-casting, allow for the creation of scaffolds with multiscale porosity and mechanical gradients that mimic cortical and trabecular bone. Resorbable composites incorporating collagen, hydroxyapatite, and bioactive glass nanoparticles emulate the ECM, enhancing osteoconduction and cellular integration. Functionalizing scaffolds with bioactive molecules or growth factors further improves tissue response [41]. Biomimetic composites are proving invaluable for large defect healing and personalized bone grafting applications, ensuring long-term functional and structural integration. All these applications are portrayed in Figure 8.4 below.

8.8 Ethical, Clinical, and Global Perspectives

8.8.1 Ethical Challenges in Resorbable Bone Repair Technologies

The development and clinical use of resorbable bone repair technologies raise ethical considerations that must be addressed to ensure responsible implementation. Issues such as patient safety, informed consent, and

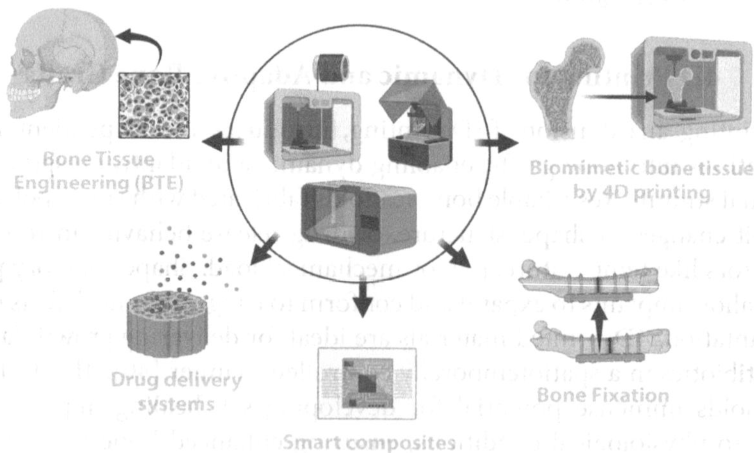

long-term outcomes are central to ethical practices. For instance, while resorbable implants offer significant benefits, their degradation profiles may vary, leading to uncertainty regarding long-term performance. Ethical concerns also arise when using stem cells, growth factors, or genetic modifications in resorbable scaffolds, particularly when considering the risks of off-target effects or tumorigenesis [42]. Transparency in clinical trials and regulatory processes is essential to build public trust. Bioethical principles, including beneficence, non-maleficence, and autonomy, must guide decision-making in the adoption and dissemination of these technologies.

8.8.2 Regulatory Hurdles and Clinical Trials for Novel Composites

The translation of resorbable bone composites from research to clinical practice faces significant regulatory challenges. Regulatory agencies, such as the U.S. FDA and the European Medicines Agency (EMA), require rigorous testing for safety, efficacy, and reproducibility. Key challenges include demonstrating controlled degradation rates, biocompatibility, and mechanical stability while adhering to Good Manufacturing Practices (GMP). Clinical trials must provide evidence of superior performance over traditional implants, ensuring long-term success in bone regeneration. Complex approval pathways, especially for drug-loaded composites or scaffolds incorporating growth factors, increase the time and cost of bringing products to market [43]. Harmonizing global regulatory frameworks is crucial to streamlining approvals and encouraging broader clinical adoption.

8.8.3 Cost-Effectiveness and Accessibility of Advanced Bone Grafts

The cost of advanced resorbable bone grafts remains a significant barrier to widespread adoption, particularly in low- and middle-income regions. While resorbable composites offer long-term benefits, such as eliminating secondary surgeries, their production involves high costs due to the use of specialized materials, advanced fabrication techniques, and stringent quality control. Additionally, the cost of customized implants and 3D-printed scaffolds may limit accessibility in resource-constrained settings. Cost-benefit analyses demonstrate that resorbable composites can reduce healthcare expenses over time, particularly by minimizing surgical complications [44]. However, efforts to scale up manufacturing, develop affordable alternatives, and ensure equitable distribution are crucial for improving accessibility and addressing global disparities in orthopedic care.

8.8.4 Sustainability in Composite Manufacturing for Bone Applications

Sustainability is becoming a critical concern in the manufacturing of resorbable bone composites. The production processes for bioactive polymers and ceramics often involve energy-intensive steps, the use of non-renewable resources, and chemical byproducts that may harm the environment. Researchers are focusing on eco-friendly fabrication methods, including solvent-free techniques, bio-based polymers derived from renewable resources, and energy-efficient 3D printing technologies. The incorporation of biodegradable and bio-sourced materials further supports sustainability goals [45]. Additionally, reducing waste through recyclable scaffolds and closed-loop manufacturing systems minimizes the environmental impact. Adopting sustainable practices not only aligns with global environmental objectives but also ensures long-term availability of materials for orthopedic applications.

8.9 Future Directions and Conclusion

8.9.1 Toward Fully Functional Resorbable Bone Grafts and Devices

The future of resorbable bone grafts lies in the development of fully functional, patient-specific implants that can seamlessly integrate into the body while supporting natural bone regeneration. Innovations in biomaterials, including smart polymers, bioactive ceramics, and nanocomposites, are driving the creation of scaffolds with optimized mechanical strength, biodegradability, and osteogenic properties. Advances in additive manufacturing enable precise control over scaffold geometry, porosity, and degradation rates, ensuring compatibility with complex defect morphologies. Additionally, multi-functional grafts incorporating growth factors, antibiotics, or stem cells will enhance biological outcomes while reducing complications [46]. These breakthroughs will ultimately lead to implants that match native bone properties, support tissue regeneration, and degrade without leaving residues.

8.9.2 Addressing Long-Term Challenges in Bone Repair

While significant progress has been made, challenges such as premature degradation, insufficient mechanical stability in weight-bearing applications, and inconsistent clinical outcomes must still be addressed. Future research should focus on tuning degradation kinetics to match the rate of bone regeneration, ensuring adequate mechanical support throughout the healing process. Hybrid scaffolds that combine polymers, ceramics, and biodegradable metals (*e.g.*, magnesium) offer promising solutions for weight-bearing regions. Long-term preclinical and clinical studies are essential to evaluate the performance, biocompatibility, and safety of new materials [47]. Advances in *in situ* monitoring technologies, such as sensors embedded in scaffolds, will allow real-time tracking of degradation and tissue formation, ensuring reliable outcomes.

8.9.3 Interdisciplinary Collaborations for Next-Generation Solutions

The development of next-generation resorbable bone grafts requires interdisciplinary collaboration between materials scientists, biomedical engineers, clinicians, and regulatory authorities. Integrating knowledge from fields such as nanotechnology, biofabrication, and tissue engineering will accelerate the design of advanced scaffolds with multi-functional capabilities. Collaborative efforts are essential for translating laboratory research into clinical solutions through standardized testing protocols, robust preclinical studies, and well-designed clinical trials. Partnerships with industry will facilitate scale-up manufacturing and ensure cost-effective production [48]. By combining expertise across disciplines, it will be possible to address current challenges and develop innovative solutions that can transform clinical outcomes for patients with bone defects.

8.9.4 Final Reflections: Resorbable Composites as the Future of Bone Repair

Resorbable composites have transformed the field of bone repair by providing materials that not only offer temporary mechanical support but also actively participate in tissue regeneration. Combining biodegradable polymers, bioactive ceramics, and emerging technologies such as 4D printing and smart materials has enabled the design of implants that are adaptive, functional, and patient-specific. Future directions will focus on addressing existing challenges, optimizing manufacturing processes, and ensuring accessibility across healthcare systems globally [49]. Resorbable composites represent the next frontier in orthopedic care, offering solutions that minimize surgical interventions, improve patient outcomes, and push the boundaries of personalized medicine in skeletal regeneration.

Developing smart, multi-functional, patient-specific scaffolds that match native bone properties. Focus on mechanical stability, optimized degradation kinetics, and real-time monitoring technologies. Uniting materials science, bioengineering, and clinical expertise for innovative solutions. Resorbable composites represent the future of bone repair, enabling adaptive, efficient, and patient-centric treatments [50].

Conclusion

Resorbable composites have become a transformative solution in bone grafting and fixation, offering significant advancements in orthopedic treatments. These materials combine innovative biomaterials, cutting-edge fabrication techniques, and interdisciplinary collaboration to overcome many challenges in bone regeneration. By designing resorbable scaffolds with tailored properties for specific bone types, these composites support natural healing and eliminate the need for secondary surgeries, making them a crucial development in bone repair.

Ongoing research, along with careful attention to ethical, regulatory, and accessibility concerns, will further solidify the role of resorbable composites in orthopedic and reconstructive medicine. Efforts to reduce production costs and improve sustainability are essential to ensuring these materials are available globally, improving healthcare access, and benefiting diverse patient populations. As the technology advances, these composites will play a key role in improving bone regeneration outcomes and reducing the burden on healthcare systems. In clinical applications, resorbable scaffolds have shown success in craniofacial reconstruction, providing both aesthetic and functional restoration for facial bone defects. In long bone fractures, while promising results have been achieved, addressing mechanical challenges in weight-bearing applications remains essential. The use of graded porosity and customized designs has led to improved outcomes in pelvic and spinal applications. By optimizing the design of these materials to prevent issues like premature resorption and mechanical failure, their reliability and effectiveness in long-term use are greatly enhanced.

The future of resorbable bone composites depends on addressing ethical concerns, such as patient safety and informed consent, ensuring rigorous regulatory testing to validate their safety and efficacy, and making these materials more accessible and affordable worldwide. Additionally, developing sustainable manufacturing methods will ensure that these materials remain viable and environmentally friendly in the long term. Overall, resorbable composites are set to revolutionize bone repair, offering personalized, effective solutions that align with both medical needs and global sustainability goals.

Acknowledgment

The authors wish to express their sincere gratitude for the financial assistance provided by the Department of Biotechnology (DBT), Ministry of Science & Technology, Government of India, under Project No. DBT/2022-23/July/N/2240 through the DBT-RA Program. We would like to extend our heartfelt thanks to Prof. A. Srinivasan and Prof. Aynur Unal for their continuous support, expert guidance, and motivation throughout the course of this research. Our profound appreciation also goes to Dr. Mohit Kumar for his essential contributions, which greatly elevated the quality and success of this work. Lastly, a special acknowledgment is given to my daughter, Vanshika, whose boundless love and unwavering inspiration have provided me with enduring strength and a constant drive to succeed.

AI disclosure statement: The author confirms that no artificial intelligence (AI)-generated content or AI-assisted writing tools were used in the preparation of this chapter (including summaries and the abstract). All text, analysis, and conclusions are the result of the author's original work.

References

1. Hussain, M., Khan, S.M., Shafiq, M., Abbas, N., Sajjad, U., Hamid, K., Advances in biodegradable materials: Degradation mechanisms, mechanical properties, and biocompatibility for orthopedic applications. *Heliyon*, 10, 12, e32713, 2024, doi: 10.1016/j.heliyon.2024.e32713.
2. Brochu, B.M., Sturm, S.R., Kawase De Queiroz Goncalves, J.A., *et al.*, Advances in Bioceramics for Bone Regeneration: A Narrative Review. *Biomimetics*, 9, 11, 690, 2024, doi: 10.3390/biomimetics9110690.
3. Ferraz, M.P., Bone Grafts in Dental Medicine: An Overview of Autografts, Allografts and Synthetic Materials. *Materials*, 16, 11, 4117, 2023, doi: 10.3390/ma16114117.
4. Szwed-Georgiou, A., Płociński, P., Kupikowska-Stobba, B., *et al.*, Bioactive Materials for Bone Regeneration: Biomolecules and Delivery Systems. *ACS Biomater. Sci. Eng.*, 9, 9, 5222–5254, 2023, doi: 10.1021/acsbiomaterials.3c00609.
5. Dobrzyńska-Mizera, M., Dodda, J.M., Liu, X., *et al.*, Engineering of Bioresorbable Polymers for Tissue Engineering and Drug Delivery Applications. *Adv. Healthcare Mater.*, 13, 30, 2024, doi: 10.1002/adhm.202401674.

6. Naseem, R., Tzivelekis, C., German, M.J., Gentile, P., Ferreira, A.M., Dalgarno, K., Strategies for Enhancing Polyester-Based Materials for Bone Fixation Applications. *Molecules*, 26, 4, 992, 2021, doi: 10.3390/molecules26040992.
7. Laubach, M., Hildebrand, F., Suresh, S., *et al.*, The Concept of Scaffold-Guided Bone Regeneration for the Treatment of Long Bone Defects: Current Clinical Application and Future Perspective. *J. Funct. Biomater.*, 14, 7, 341, 2023, doi: 10.3390/jfb14070341.
8. Cheers, G.M., Weimer, L.P., Neuerburg, C., *et al.*, Advances in implants and bone graft types for lumbar spinal fusion surgery. *Biomater. Sci.*, 12, 19, 4875–4902, 2024, doi: 10.1039/D4BM00848K.
9. Steppe, L., Megafu, M., Tschaffon-Müller, M.E.A., Ignatius, A., Haffner-Luntzer, M., Fracture healing research: Recent insights. *Bone Rep.*, 19, 101686, 2023, doi: 10.1016/j.bonr.2023.101686.
10. Song, P., Zhou, D., Wang, F., Li, G., Bai, L., Su, J., Programmable biomaterials for bone regeneration. *Mater. Today Bio*, 29, 101296, 2024, doi: 10.1016/j.mtbio.2024.101296.
11. Steijvers, E., Ghei, A., Xia, Z., Manufacturing artificial bone allografts: a perspective. *Biomater. Transl.*, 3, 1, 65–80, 2022, https://doi.org/10.12336/biomatertransl.2022.01.007.
12. Pourhajrezaei, S., Abbas, Z., Khalili, M.A., *et al.*, Bioactive polymers: A comprehensive review on bone grafting biomaterials. *Int. J. Biol. Macromol.*, 278, 134615, 2024, doi: 10.1016/j.ijbiomac.2024.134615.
13. Liang, W., Zhou, C., Bai, J., *et al.*, Prospective applications of bioactive materials in orthopedic therapies: A review. *Heliyon*, 10, 16, e36152, 2024, doi: 10.1016/j.heliyon.2024.e36152.
14. Baldwin, P., Li, D.J., Auston, D.A., Mir, H.S., Yoon, R.S., Koval, K.J., Autograft, Allograft, and Bone Graft Substitutes: Clinical Evidence and Indications for Use in the Setting of Orthopaedic Trauma Surgery. *J. Orthop. Trauma*, 33, 4, 203–213, 2019, doi: 10.1097/BOT.0000000000001420.
15. Tolmacheva, N., Bhattacharyya, A., Noh, I., Calcium Phosphate Biomaterials for 3D Bioprinting in Bone Tissue Engineering. *Biomimetics*, 9, 2, 95, 2024, doi: 10.3390/biomimetics9020095.
16. Guo, Z., Poot, A.A., Grijpma, D.W., Advanced polymer-based composites and structures for biomedical applications. *Eur. Polym. J.*, 149, 110388, 2021, doi: 10.1016/j.eurpolymj.2021.110388.
17. Liu, Q., Peng, X., Liu, X., *et al.*, Advances in the application of bone morphogenetic proteins and their derived peptides in bone defect repair. *Composites, Part B*, 262, 110805, 2023, doi: 10.1016/j.compositesb.2023.110805.
18. Liu, L., Chen, H., Zhao, X., *et al.*, Advances in the application and research of biomaterials in promoting bone repair and regeneration through immune modulation. *Mater. Today Bio*, 101410, Published online December 2024, doi: 10.1016/j.mtbio.2024.101410.
19. Shanmugavadivu, A., Lekhavadhani, S., Babu, S., Suresh, N., Selvamurugan, N., Magnesium-incorporated biocomposite scaffolds: A novel frontier in

bone tissue engineering. *J. Magnesium Alloys*, 12, 6, 2231–2248, 2024, doi: 10.1016/j.jma.2024.06.001.

20. Chua, K., Khan, I., Malhotra, R., Zhu, D., Additive manufacturing and 3D printing of metallic biomaterials. *Eng. Regener.*, 2, 288–299, 2021, doi: 10.1016/j.engreg.2021.11.002.
21. Liu, Z., Wang, Y., Wu, B., Cui, C., Guo, Y., Yan, C., A critical review of fused deposition modeling 3D printing technology in manufacturing polylactic acid parts. *Int. J. Adv. Manuf. Technol.*, 102, 9–12, 2877–2889, 2019, doi: 10.1007/s00170-019-03332-x.
22. Deville, S., Freeze-Casting of Porous Biomaterials: Structure, Properties and Opportunities. *Materials*, 3, 3, 1913–1927, 2010, doi: 10.3390/ma3031913.
23. Song, X., Segura-Egea, J.J., Díaz-Cuenca, A., Sol–Gel Technologies to Obtain Advanced Bioceramics for Dental Therapeutics. *Molecules*, 28, 19, 6967, 2023, doi: 10.3390/molecules28196967.
24. Liu, Z., Yamada, S., Otsuka, Y., Peñaflor Galindo, T.G., Tagaya, M., Surface modification of hydroxyapatite nanoparticles for bone regeneration by controlling their surface hydration and protein adsorption states. *Dalton Trans.*, 51, 25, 9572–9583, 2022, doi: 10.1039/D2DT00969B.
25. Jiang, Z., Zheng, Z., Yu, S., *et al.*, Nanofiber Scaffolds as Drug Delivery Systems Promoting Wound Healing. *Pharmaceutics*, 15, 7, 1829, 2023, doi: 10.3390/pharmaceutics15071829.
26. Liu, G., Wei, X., Zhai, Y., *et al.*, 3D printed osteochondral scaffolds: design strategies, present applications and future perspectives. *Front. Bioeng. Biotechnol.*, 12, 2024, doi: 10.3389/fbioe.2024.1339916.
27. Omigbodun, F.T., Oladapo, B.I., Osa-uwagboe, N., Exploring the frontier of Polylactic Acid/Hydroxyapatite composites in bone regeneration and their revolutionary biomedical applications – A review. *J. Reinf. Plast. Compos.*, Published online September 6, 2024, doi: 10.1177/07316844241278045.
28. Rady, A.A.M., Hamdy, S.M., Abdel-Hamid, M.A., Hegazy, M.G.A., Fathy, S.A., Mostafa, A.A., The role of VEGF and BMP-2 in stimulation of bone healing with using hybrid bio-composite scaffolds coated implants in animal model. *Bull. Natl. Res. Cent.*, 44, 1, 131, 2020, doi: 10.1186/s42269-020-00369-x.
29. Musthafa HS, N., Walker, J., Domagala, M., Computational Modelling and Simulation of Scaffolds for Bone Tissue Engineering. *Computation*, 12, 4, 74, 2024, doi: 10.3390/computation12040074.
30. Mukasheva, F., Adilova, L., Dyussenbinov, A., Yernaimanova, B., Abilev, M., Akilbekova, D., Optimizing scaffold pore size for tissue engineering: insights across various tissue types. *Front. Bioeng. Biotechnol.*, 12, 2024, doi: 10.3389/fbioe.2024.1444986.
31. Aggarwal, D., Kumar, V., Sharma, S., Drug-loaded biomaterials for orthopedic applications: A review. *J. Controlled Release*, 344, 113–133, 2022, doi: 10.1016/j.jconrel.2022.02.029.

32. Zhou, J., See, C.W., Sreenivasamurthy, S., Zhu, D., Customized Additive Manufacturing in Bone Scaffolds—The Gateway to Precise Bone Defect Treatment. *Research*, 6, 2023, doi: 10.34133/research.0239.
33. Ngo, H.X., Bai, Y., Sha, J., *et al.*, A Narrative Review of u-HA/PLLA, a Bioactive Resorbable Reconstruction Material: Applications in Oral and Maxillofacial Surgery. *Materials*, 15, 1, 150, 2021, doi: 10.3390/ma15010150.
34. Chinnasami, H., Dey, M.K., Devireddy, R., Three-Dimensional Scaffolds for Bone Tissue Engineering. *Bioengineering*, 10, 7, 759, 2023, doi: 10.3390/bioengineering10070759.
35. Zhang, M., Xu, F., Cao, J., *et al.*, Research advances of nanomaterials for the acceleration of fracture healing. *Bioact. Mater.*, 31, 368–394, 2024, doi: 10.1016/j.bioactmat.2023.08.016.
36. Teimouri, R., Abnous, K., Taghdisi, S.M., Ramezani, M., Alibolandi, M., Surface modifications of scaffolds for bone regeneration. *J. Mater. Res. Technol.*, 24, 7938–7973, 2023, doi: 10.1016/j.jmrt.2023.05.076.
37. Ullah, A., Asami, K., Holtz, L., *et al.*, A Machine Learning Approach for Mechanical Component Design Based on Topology Optimization Considering the Restrictions of Additive Manufacturing. *J. Manuf. Mater. Process.*, 8, 5, 220, 2024, doi: 10.3390/jmmp8050220.
38. Mishchenko, O., Yanovska, A., Kosinov, O., *et al.*, Synthetic Calcium–Phosphate Materials for Bone Grafting. *Polymers*, 15, 18, 3822, 2023, doi: 10.3390/polym15183822.
39. Marew, T. and Birhanu, G., Three dimensional printed nanostructure biomaterials for bone tissue engineering. *Regen. Ther.*, 18, 102–111, 2021, doi: 10.1016/j.reth.2021.05.001.
40. Pfau, M.R. and Grunlan, M.A., Smart scaffolds: shape memory polymers (SMPs) in tissue engineering. *J. Mater. Chem. B*, 9, 21, 4287–4297, 2021, doi: 10.1039/D1TB00607J.
41. Tupe, A., Patole, V., Ingavle, G., *et al.*, Recent Advances in Biomaterial-Based Scaffolds for Guided Bone Tissue Engineering: Challenges and Future Directions. *Polym. Adv. Technol.*, 35, 11, 2024, doi: 10.1002/pat.6619.
42. van Daal, M., de Kanter, A.F.J., Bredenoord, A.L., de Graeff, N., Personalized 3D printed scaffolds: The ethical aspects. *New Biotechnol.*, 78, 116–122, 2023, doi: 10.1016/j.nbt.2023.10.006.
43. Zhao, X., Li, N., Zhang, Z., *et al.*, Beyond hype: unveiling the Real challenges in clinical translation of 3D printed bone scaffolds and the fresh prospects of bioprinted organoids. *J. Nanobiotechnol.*, 22, 1, 500, 2024, doi: 10.1186/s12951-024-02759-z.
44. Maintz, M., Tourbier, C., de Wild, M., *et al.*, Patient-specific implants made of 3D printed bioresorbable polymers at the point-of-care: material, technology, and scope of surgical application. *3D Print. Med.*, 10, 1, 13, 2024, doi: 10.1186/s41205-024-00207-0.
45. Siddiqui, M.A.S., Rabbi, M.S., Ahmed, R.U., Billah, Md.M., Biodegradable natural polymers and fibers for 3D printing: A holistic perspective on

processing, characterization, and advanced applications. *Cleaner Mater.*, 14, 100275, 2024, doi: 10.1016/j.clema.2024.100275.
46. Suamte, L., Tirkey, A., Barman, J., Jayasekhar Babu, P., Various manufacturing methods and ideal properties of scaffolds for tissue engineering applications. *Smart Mater. Manuf.*, 1, 100011, 2023, doi: 10.1016/j.smmf.2022.100011.
47. Wei, S., Ma, J.X., Xu, L., Gu, X.S., Ma, X.L., Biodegradable materials for bone defect repair. *Mil. Med. Res.*, 7, 1, 54, 2020, doi: 10.1186/s40779-020-00280-6.
48. Zuncheddu, D., della Bella, E., Schwab, A., *et al.*, Quality control methods in musculoskeletal tissue engineering: from imaging to biosensors. *Bone Res.*, 9, 1, 46, 2021, doi: 10.1038/s41413-021-00167-9.
49. Ramezani, M. and Mohd Ripin, Z., 4D Printing in Biomedical Engineering: Advancements, Challenges, and Future Directions. *J. Funct. Biomater.*, 14, 7, 347, 2023, doi: 10.3390/jfb14070347.
50. Łuczak, J.W., Palusińska, M., Matak, D., *et al.*, The Future of Bone Repair: Emerging Technologies and Biomaterials in Bone Regeneration. *Int. J. Mol. Sci.*, 25, 23, 12766, 2024, doi: 10.3390/ijms252312766.

9

3D-Printing of Implants and Fixation Devices

Medi Harshith Kumar[1], Atanu Kumar Paul[1] and Gourhari Chakraborty[2*]

[1]*Department of Chemical Engineering, National Institute of Technology Andhra Pradesh, Tadepalligudem, Andhra Pradesh, India*
[2]*School of Chemical and Biotechnology, SASTRA Deemed University, Thirumalaisamudram, Tamil Nadu, India*

Abstract
3D-printing technology has emerged as a potential fabrication tool for composites to develop efficient and effective products for numerous applications including biomedical. 3D-printing of objects made up of resorbable materials generates complex design products, user-specific design, and controlled degradability and eliminates the need for operation after recovery. Implants and fixation elements made up of resorbable composites are observed to have mechanical strength and sustainable degradation stability.

Given the importance of the applicability of different resorbable composites for implants and fixation elements fabrication present chapter emphasizes the materials and processing techniques for printable ink/filament preparation. A brief discussion about the printing strategies of resorbable composites is appended in this chapter. Applications of resorbable composites for making implants and fixation elements are discussed comprehensively in corroboration with recent works. Finally, the challenges, opportunities, and outlook of the chapter are discussed.

***Keywords*:** 3D-printing, implants, fixation elements, resorbable composites

**Corresponding author*: gourhari115@gmail.com

Arbind Prasad (ed.) Resorbable Composites for Bioimplants and Fixation Devices, (253–286) © 2026 Scrivener Publishing LLC

9.1 Introduction

3D printing which is also known as additive manufacturing, is a rapidly evolving fabrication technology that produces three-dimensional objects through a layer-by-layer deposition process. After it was founded in 1986, it transformed manufacturing techniques by enabling the creation of complex and customized geometries directly from digital models. Here, the process begins with the design of a Computer-Aided Design (CAD) file, which is sliced into layers and printed using suitable techniques and materials. Unlike conventional manufacturing techniques, 3D printing is highly design-specific, making it particularly valuable for personalized biomedical applications that require specifications that are different and unique for each patient [1].

Bioprinting using bio-ink offers significant advantages over traditional 3D printing methods in tissue engineering. It enables the creation of complex tissue structures with improved cell viability, structural integrity, and functional performance [2]. Bio-inks enhance cell-matrix interactions and facilitate the fabrication of anatomically and physiologically relevant structures [2]. The rheological properties of bio-inks are crucial for printability and cell viability during the printing process, while post-crosslinking properties affect mechanical integrity and cell function [2]. Bioprinted tissues demonstrate improved mechanical properties, biological functionality, and integration with host tissues. However, challenges remain in developing novel biomaterials, understanding cell-matrix remodeling, and establishing proper vascular perfusion [3]. Ongoing research focuses on optimizing printability and cell viability through experimental, computational, and machine-learning approaches [4]. Continued innovation in bio-ink formulation and printing techniques is essential for advancing tissue engineering and regenerative medicine applications [5, 6]. It is used in fields like scaffold development, skin and cartilage healing, medication delivery, tissue engineering, and bone regeneration. The different advanced printing techniques that have helped develop precise structures specifically suitable to each patient's requirements are inkjet printing, extrusion-based printing, fused deposition modeling (FDM), selective laser melting (SLM), stereolithography, and bio-plotting. Further, living cells and biomaterials can be included in bioprinting, a specialized type of 3D printing, to produce tissue-like structures with biological functions [1, 5–7].

Based on the application, various types of materials are used in 3D printing. Some examples of these are metallic alloys like magnesium and titanium, hydrogels or composite bio-inks for soft tissue applications, and biocompatible polymers like PEG, PLA, and PCL [7]. PLA, Cellulose, silk, and chitosan are majorly utilized polymers for 3D printing in the biomedical field [8–10]. Properties like mechanical strength, biodegradability, processability, and biocompatibility, are taken into consideration when choosing a material.

Implant fabrication is one of the most important applications of biomedical 3D printing. CT or MRI scan data can be used to develop patient-specific implants for dental, cranial, orthopedic, and spinal applications, improving anatomical fit, cutting down on surgery time, and improving clinical results. Implant performance is further improved by the capacity to design porous structures, which further improves tissue and osseointegration.

In addition to implants, 3D printing has transformed the creation of fixation devices such as rods, screws, bone plates, and spinal cages. These mechanically demanding components are best manufactured using processes like extrusion-based printing, FDM, and SLM. These are particularly used for trauma and reconstructive surgeries, to design tissues having biocompatibility, mechanical stress resistance, and tailored geometry. PLA and magnesium alloys have major clinical benefits to biodegradable fixation devices as they do not require further procedures.

Recent advancements in bioprinting technology, biomaterials, and surface modification techniques have shown significant potential for enhancing therapeutic outcomes in complex medical conditions. 3D bioprinting enables the creation of intricate tissue models and scaffolds for cancer research, drug screening, and personalized medicine [11, 12]. Surface modification of biomaterials plays a crucial role in tailoring their properties for improved biocompatibility and functionality in various applications, including drug delivery, biosensing, and tissue engineering [13, 14]. Techniques such as adsorption, deposition, and covalent binding have facilitated the development of new-generation biomaterials with enhanced performance [14]. The integration of these technologies allows for the fabrication of complex 3D structures that closely mimic natural tissue environments, potentially addressing challenges in organ transplantation and regenerative medicine [11, 13, 15, 16].

This chapter explores and explains the key technologies, materials, and applications of 3D printing in the fabrication of implants and fixation devices, highlighting current developments, design considerations, and clinical relevance. It also addresses the challenges, limitations, and future opportunities of integrating 3D printing into mainstream biomedical practice.

9.2 Emerging Trends in 3D Printing for Industrial and Biomedical Innovation

Recent advancements in 3D printing technologies are shaping innovation across various industries, particularly in industrial manufacturing and biomedical applications. The field has evolved from early applications in rapid prototyping to more sophisticated uses in aerospace, automotive, and medicine [17, 18]. Emerging trends include high-speed sintering, continuous liquid interface production, and bioprinting, with a focus on personalized medical applications [18]. The development of novel materials, including composites, hybrids, and smart materials, is expanding the potential of 3D printing [19]. Bioprinting has advanced to 4D printing, offering new possibilities in tissue engineering and regenerative medicine [19]. However, challenges remain in standardization, material quality, and economic feasibility [18]. Future research is focused on improving printing speed, cell growth feasibility, and complex shape achievement, particularly in biomedical applications [16, 20–24].

Figure 9.1 presents a comprehensive schematic outlining the sequential steps involved in applying 3D printing technology for biomedical purposes [25]. The process begins with the acquisition of medical imaging data, typically sourced from techniques like MRI or CT scans, which provide accurate anatomical information about the patient. This imaging data is then used in patient-specific design and modeling, where custom anatomical models are created using specialized computer-aided design (CAD) software. These models are tailored to match the patient's unique physiological structure. Following the design phase, the digital model undergoes slicing,

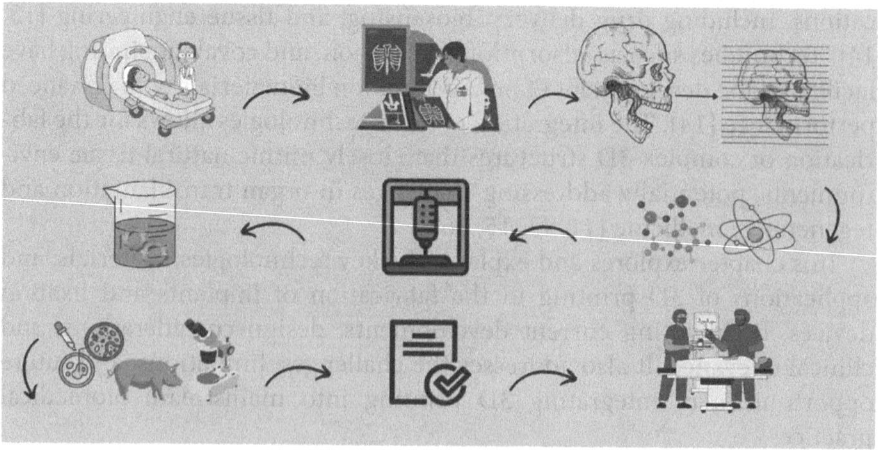

a process that divides the model into thin horizontal layers. These layers guide the printer during fabrication and are crucial for ensuring high precision in the final printed structure. Next come material selection and assessment of biocompatibility. The choice of materials—often bio-inks, polymers, or composite scaffolds—is vital to ensure compatibility with human tissues and biological functions. Factors such as degradation rate, mechanical strength, and toxicity are considered at this stage. With materials and models ready, the 3D printing process is executed. This involves the layer-by-layer deposition of material to fabricate the physical object, such as a tissue scaffold, bone implant, or organ prototype.

Post-printing, the construct undergoes material coating and stabilization, which may include surface modifications or the addition of functional coatings to enhance performance and longevity in biological environments. The printed product is then subjected to rigorous *in vitro* and *in vivo* testing, using lab-based systems and animal models to evaluate its functionality, safety, and biological response.

Following successful testing, the product moves through regulatory approval processes to ensure it meets all legal, ethical, and safety standards required for human use. Finally, once approved, the printed biomedical product reaches the commercial and surgical application stage, where it is implemented in clinical settings for patient treatment, such as implants, prosthetics, or tissue engineering constructs. Table 9.1 outlines studies investigating various biodegradable polymers, emphasizing their material types, degradation rates, mechanical performance, and associated clinical outcomes in regenerative and reconstructive applications.

9.2.1 Digital Model

The accuracy and performance of 3D-printed biomedical devices are influenced by various factors throughout the manufacturing process. CAD model parameters and STL file conversion resolution can affect the dimensional accuracy of printed parts [35, 36]. Slicing parameters, such as printing pattern, strand distance, and spatial fill, impact the porosity and mechanical strength of printed objects [37]. The choice of 3D printing technology, materials, and process parameters significantly affects surface roughness, printing accuracy, and mechanical properties [38]. Post-processing procedures also play a role in the final product quality [38]. For laser powder bed fusion, factors like part orientation and process parameters influence surface roughness [35]. In vat-polymerization, supporting structures and post-processing techniques affect the outcome [38]. Optimizing these parameters is crucial for producing high-quality,

Table 9.1 Comparative evaluation of biodegradable polymer-based implants: material properties, degradation, and biological performance.

Study	Material type	Degradation rate	Mechanical strength	Clinical outcomes	Reference
Seyednejad *et al.*, 2012	Poly(hydroxymethyl glycolide-co-ε-caprolactone) (PHMGCL, HMG:CL 8:92), Polycaprolactone (PCL)	PHMGCL: >60% weight loss in 3 months; PCL: no weight loss; PHMGCL molecular weight: 46.9–23.2 kDa	No mention found	Higher tissue integration and vascularization for PHMGCL	[26]
El-Habashy *et al.*, 2021	Gelatin/polyvinyl alcohol/ hyaluronic acid + hydroxyapatite/ polycaprolactone nanoparticles	Maintained integrity for 28 days	Efficient mechanical stiffness/strength (no quantitative data)	Osteoconduction, bone healing in rabbits	[27]
Shishatskaya *et al.*, 2023	Poly(3-hydroxybutyrate-co-3-hydroxy valerate)	40% decrease in molecular weight	Young's modulus, tensile strength, cell colonization, bone restoration	Full bone defect restoration in 150 days	[28]

(*Continued*)

Table 9.1 Comparative evaluation of biodegradable polymer-based implants: material properties, degradation, and biological performance. (*Continued*)

Study	Material type	Degradation rate	Mechanical strength	Clinical outcomes	Reference
Jeong *et al.*, 2024	Polycaprolactone (PCL), PCL-TIB	No mention found	Mechanical stability retained (no quantitative data)	>90% human umbilical vein endothelial cell (HUVEC) viability; radiopacity for 42 days	[29]
Barbeck *et al.*, 2017	Polylactic acid (PLA), PLA/bioglass G5	G5 reduces weight loss	PLA/G5 > PLA in compressive modulus; both decrease with degradation	Higher vascularization with G5	[30]
Oberdiek *et al.*, 2021	Polycaprolactone (PCL), PCL + biphasic calcium phosphate (BCP)	No mention found	No significant difference between PCL and PCL+BCP	Osteoconductivity, reduced swelling with BCP	[31]

(*Continued*)

Table 9.1 Comparative evaluation of biodegradable polymer-based implants: material properties, degradation, and biological performance. (*Continued*)

Study	Material type	Degradation rate	Mechanical strength	Clinical outcomes	Reference
Kirmanidou *et al.*, 2024	Polycaprolactone (PCL), PCL-based composites	Adjustable biodegradation rates (no quantitative data)	Mechanical properties suitable for load-bearing (no quantitative data)	Clinical use in craniofacial reconstruction; intraoral complications	[32]
Yeo *et al.*, 2008	Polycaprolactone (PCL)-20% tricalcium phosphate (TCP)	Porosity increased by 29.2% *in vivo* (24 weeks); number-average molecular weight decreased by 79.6%, weight-average molecular weight decreased by 88.7% *in vivo*	Similar to cancellous bone at 24 weeks (no quantitative data)	Suitable for dentoalveolar reconstruction	[33]
Thuaksuban *et al.*, 2018	Polycaprolactone (PCL)-20% biphasic calcium phosphate (BCP), PCL-20% tricalcium phosphate (TCP)	Degraded over 90 days; slight molecular weight change	No mention found	Collagen/vessel formation, low inflammation	[34]

patient-specific medical devices with desired properties and performance characteristics [39, 40].

9.2.2 High-Resolution Fabrication Using Inkjet-Based 3D Printing Techniques

Inkjet-based bioprinting is a promising technique for creating 3D tissue constructs but optimizing droplet formation and cell deposition remains challenging. Key printing parameters influencing droplet characteristics include polymer concentration, excitation voltage, dwell time, and rise time [41]. Ink properties such as viscosity and surface tension, as well as physical parameters like printhead diameter and actuation mechanisms, significantly affect the printing process [42]. Achieving cell homogeneity in bio-inks is crucial for consistent printing, while shear stress during printing impacts cell viability [43]. Other important factors include bio-ink composition, printing speed and pressure, nozzle characteristics, and crosslinking parameters. Researchers are exploring machine-learning techniques to optimize cell deposition and predict droplet formation [41, 43]. Future research should focus on enhancing the resolution, printability, and uniformity of printed structures for various biomedical applications [42, 44–47]. Figure 9.2 illustrates two main types of inkjet printers used

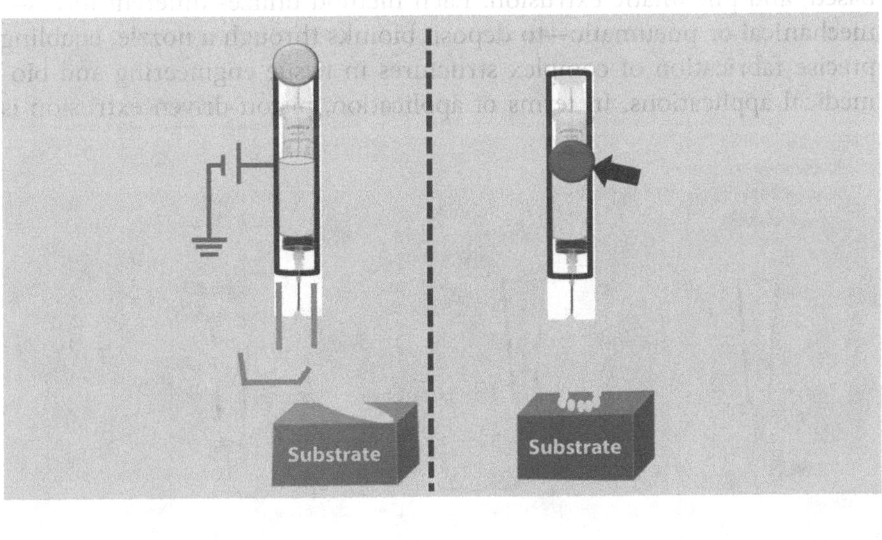

in 3D printing: Continuous Inkjet Printing (CIJ) and Drop-on-Demand (DOD) Inkjet Printing. In CIJ, a continuous stream of ink droplets is generated, and unwanted drops are deflected by an electric charge and collected, while selected drops reach the substrate. In contrast, DOD printers generate droplets only when needed using a resistive element or piezoelectric crystal, providing greater control and precision. DOD is widely used in bioprinting due to its reduced material waste and ability to deposit cells or biomaterials with high spatial resolution directly onto the substrate.

9.2.3 Extrusion Based 3D-Printing

When ink is extruded and dispensed *via* mechanical or pneumatic force, it is sometimes referred to as direct writing. The viscosity of the ink used in extrusion-based 3D printing is kept within the range of 30 to 107 mPas, depending on the material. Extrusion-based 3D printing involves the deposition of material directly onto a substrate that is kept beneath the syringe. Because of their compatibility with one another, a variety of materials can be employed as ink in this procedure. For the creation of the ink, hydrogel and biodegradable polymers such as PLA and PCL were used. In commercial extrusion-based 3D printing, screw and piston-type arrangements vary according to the mechanical force [6, 8]. Figure 9.3 illustrates three primary extrusion-based 3D printing mechanisms: piston-based, screw-based, and pneumatic extrusion. Each method utilizes different forces—mechanical or pneumatic—to deposit bioinks through a nozzle, enabling precise fabrication of complex structures in tissue engineering and biomedical applications. In terms of application, piston-driven extrusion is

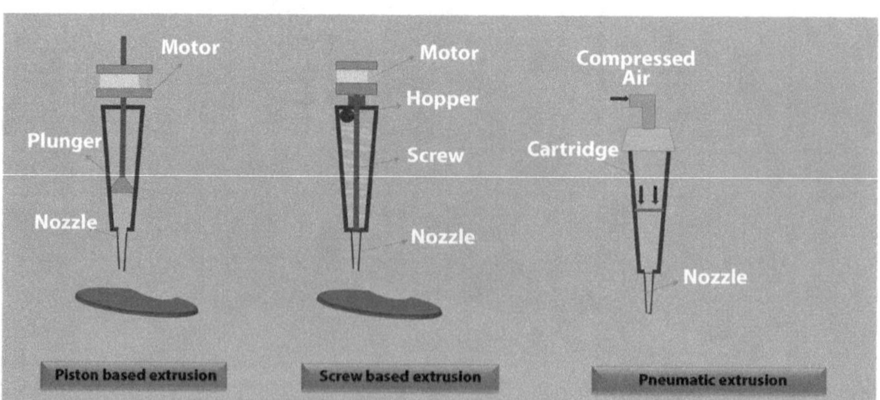

more versatile. A guiding motor is used to control the piston's motion, and it runs linearly. It is possible to get greater volumetric control in screw-driven extrusion. This is used when there are more viscous bioinks, and in this process, the screw serves as a direct conveyer. Using the appropriate adapter and pipes, the pneumatic mode uses air pressure to transform ink into filament. Air sterilization is required for bioprinting and works as expected for shear-thinning polymers [8]. Before the inks are extruded, it is crucial to understand their rheological characteristics. Gel's storage and loss characteristics dictate the solution's structural stability after printing, whereas viscosity establishes the necessary shear rate for extrusion [48]. Food materials (milk and starch) were printed using the extrusion technique, whereas other compositions were printed using pneumatic pressure.

Depending on the materials and mix, the product quality ranged from sticky to dry [49]. Five retractable extruder systems were used to create a complex vascular structure with gelatin as the main component [50]. It was found that the biodegradable PLA polymer loaded with magnesium and ceramics could be extruded for use in future product development [51, 52]. Extrusion-based 3D printers are increasingly being used for bioprinting. Alginate-gelatin hydrogel was used to successfully print mesenchymal stem cells with a 65% printing accuracy [53]. Although extrusion-based printing offers several benefits, the main challenges with this technology are structural failure and shape distortion [54].

9.2.4 Laser-Based 3D Printing

Implants are fabricated from metal powder and polymers using laser-assisted methods like selective laser melting (SLM) and selective laser sintering (SLS). In the case of the SLS technique, the ingredient materials are heated below the temperature of melting whereas in the case of SLM, the temperature is above the melting temperature. Pulse laser source and bulk powder materials are generally employed for the SLS technique. The SLS technique is applied for polyimides, polycarbonates, polystyrenes, etc. SLM process needs high laser power and prototypes of alloys can be made directly by melting metal powders. In the case of metallic implant making SLM technique is usually employed which gives the liberty of selecting different components with varying compositions for making alloys. Selective laser melting (SLM) is a promising technique for manufacturing titanium-based implants, particularly Ti6Al4V, for biomedical applications. SLM-produced Ti6Al4V components demonstrate good mechanical properties, biocompatibility, and corrosion resistance [55]. The surface roughness of SLM-fabricated implants can be significantly reduced through post-processing techniques like glass-blasting [56]. However, untreated SLM surfaces may be better suited for cell growth [56]. The surface chemistry of SLM-fabricated Ti6Al4V differs from traditionally manufactured components, but this does not negatively impact cell viability [57]. SLM allows for the creation of porous implants with customized geometry and mechanical properties, which can help address issues like stress shielding in dental implants by reducing the effective modulus [58]. Overall, SLM shows great potential for producing biomedical implants with tailored properties [57, 59–61].

9.2.5 Fused Deposition Modeling (FDM)

Thermoplastic filaments are heated and extruded through the nozzle for layer-by-layer deposition in this 3D-printing technique. In biomedical applications, bone, scaffolds, and stents fabrication FDM is emerging as a useful technique. The properties of the printed objects depend on printing speed, nozzle type, temperature, rate of extrusion, etc. Biodegradable polymers such as PCL, PLA, etc. are in use in making filaments and 3D printing. Vaquette *et al.*, fabricated a medical-grade polycaprolactone-based scaffold for application in jaw repair. FDM technique was utilized with 110°C and 0/90 lattice pattern for printing of PCL. The filament gap was kept at 1 mm with a layer thickness of 0.3 mm for the printing. Extraskeletal bone formation was observed for the scaffolds throughout the growth of bone [62]. Sahmani *et al.*, fabricated PLA- Hydroxyapatite (HA) composite filaments with a diameter of 1.75 mm using an extrusion technique before printing. Composite scaffolds were fabricated using the FDM technique with a temperature of 210°C and a nozzle diameter of 3 mm. The hexagonal structure was observed to have the highest compressive strength 7.55 MPa [63]. Wurm *et al.*, [64] used the FDM process for making PLA-based 3D-printed objects. The nozzle temperature was kept at 225°C for the fabrication of disc-shaped objects of a diameter of 14 mm. Cell viability was observed to be 95.3% for PLA scaffold which was comparable with titanium 91.7 %. Optimizing tools were observed to improve the performances of 3D-printing [64].

9.2.6 Bioplotting

This technique is mostly utilized for tissue engineering applications. Syringes are utilized to extrude polymeric or cellular substances. Sometimes photo or UV curing of polymers helps in printing the objects. Alginate and gelatine are mostly used in bioplotting. Mellor *et al.*, utilized a bio-plotter for the construction of a 3D-printed thin disk-shaped scaffold. Extrusion was conducted at 180°C with pressure of 0.5 N/mm^2. Printing was carried out through a 0.4 mm inner diameter nozzle with 0.4 mm/s printing speed. Significant improvement in cell proliferation was observed for the 3D-printed scaffold [65]. Warren *et al.* used an extrusion-based bio plotter for the fabrication of scaffolds. A nozzle of internal diameter 200 μm was employed for printing at 0.6 N/mm^2 pressure and a print head speed of 0.4 mm/s [66].

9.3 Materials for 3D-Printing

Implants and fixation elements are fabricated using different 3D printing strategies depending upon the need of the properties. Properties of the printed substances strongly rely on the materials, modifications of the ingredients, and composition. Polymers, metals, and ceramics with processing-oriented or application-oriented modifications are the main ingredients for the 3D printing of implants and fixation elements. Figure 9.4 categorizes the primary materials used in 3D printing for biomedical applications into four major classes: polymers, ceramics, metals, and composites. Polymers such as

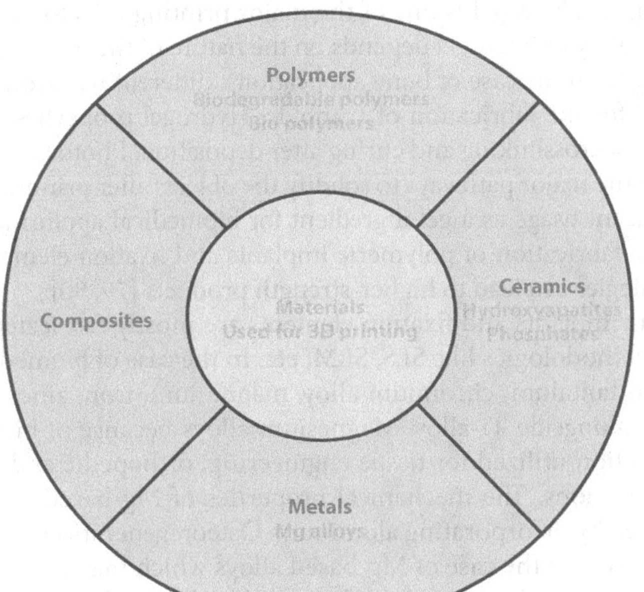

biodegradable and bio-polymers are widely used due to their flexibility and biocompatibility. Ceramics like hydroxyapatites and phosphates are essential for bone-related applications due to their bioactivity. Metals, particularly magnesium (Mg) alloys, offer mechanical strength and biodegradability, making them suitable for implants. Composites combine the properties of different materials to enhance performance and functionality in tissue engineering and drug delivery. Each category supports specific roles in fabricating complex and functional biomedical structures [67].

Bio-degradable polymers are one of the major ingredients for the preparation of raw materials for inkjet, extrusion, FDM, and bio-plotting-based 3D printing. Biodegradable polymers like PCL, PLA, and PHB are thermoplastic in nature and possess significant biocompatibility. The degradability of these polymers makes them lucrative for resorbable applications. These polymers have significant stability and mechanical strength under *in-vitro* and *in-vivo* exposure [68–73]. To improve the applicability of such polymers, composite fabrication incorporating fillers like HA, graphene, graphene oxide clays, bio-fibers, etc. observed to impart significant progress. Chitosan, silk, chondroitin sulphate, gelatine, and collagen are some of the resorbable biopolymers extensively used for 3D printing for tissue culture applications [74–78]. In the case of bone scaffold development also affectivity of these polymers is investigated with different crosslinking strategies. Hydrogel is one of the major printing inks for 3D printing. The printability of hydrogel depends on the nature of the polymer, viscosity, and strength. In the case of bone applications, different nanocomposite gels were used for the fabrication of scaffolds. Hydrogel properties depend on the nature of crosslinking and curing after deposition. Photo curing and UV curing are the major pathways to solidify the objects after printing. Chitosan has significant usage as a gel ingredient for biomedical applications. In the case of the fabrication of polymeric implants and fixation elements, FDM-based strategies can lead to higher-strength products [79, 80].

Metallic implants and fixation elements are mostly fabricated by laser-assisted methodologies like SLS, SLM, etc. In the case of biomedical object fabrication tantalum, chromium alloy, magnesium, iron, zinc, and nickel are in use alongside Ti-alloy. Magnesium alloys because of biosafety and biodegradation utilized for tissue engineering, orthopedic and cardiovascular applications. The mechanical properties of Mg-based implants can be improved by incorporating aluminum. Osteoregeneration was observed to be increased in the case of Mg-based alloys which makes them effective in bone-related applications. Mg-Ca and Mg-Mn-Zn alloys were observed to have low corrosion rates which makes them more useful for making orthopedic implants [80].

The incorporation of ceramic materials like hydroxyapatite (HAP) and β-tricalcium phosphate (TCP) into polymer matrices significantly enhances the mechanical and biological properties of 3D-printed scaffolds for bone tissue engineering [81]. HAP-reinforced polycaprolactone (PCL) scaffolds exhibit improved compressive strength, hydrophilicity, and osteogenic properties [82]. Similarly, infiltrating 3D-printed HAP scaffolds with biopolymers increases their compressive strength, making them suitable for load-bearing applications [83]. The addition of HAP to PCL enhances biological properties, while TCP improves mechanical properties [84]. 3D printing techniques allow for user-designed pore structures, ensuring bulk interconnectivity and creating inherent surface roughness that facilitates osteoconduction and osteointegration [85]. However, anisotropic behavior and interlayer bond failures in 3D-printed scaffolds remain challenges to be addressed [63, 86–89].

9.4 3D-Printing of Implants

Over time, the use of 3D printed materials in biomedicine has grown, becoming more efficient and tailored to individual patients. 3D printing technology is increasingly being used to make a variety of surgical instruments, implants, medications, and other products. Artificial kidneys, artificial hearts, hip and knee prostheses, pacemakers, and dental implants all need to be designed specifically for each patient. 3D printing has been incredibly helpful in creating and customizing these devices to meet individual demands. Additionally, biomolecular sensing, cancer treatment, and improved drug release products can all benefit from 3D printing [90, 91].

Stereolithography (SLA) 3D printing has emerged as a promising technique for fabricating microneedles for transdermal drug delivery. SLA-printed microneedles demonstrate excellent mechanical strength and skin penetration efficiency, with the ability to pierce up to 80% of skin depth [92]. These microneedles can be coated with drug formulations using inkjet dispensing, allowing for precise and uniform drug coating [92, 93]. Studies have shown that SLA-printed microneedles coated with cisplatin formulations exhibit rapid drug release rates, with 80-90% release within 1 hour [92]. The combination of SLA printing and inkjet coating enables the development of high-performance microneedle systems for transdermal drug delivery [94]. *In vitro* studies using Franz diffusion cells have demonstrated enhanced permeation of model drugs across human skin when using SLA-printed microneedles compared to untreated skin [92, 93, 95].

Bioprinting is also used for bone and dental purposes which are personalized and developed as per patient needs. Various metallic and polymeric ingredients are used to make implants and scaffolds. In the case of bone tissue printing, various organizations use vascularized 3D printing [96]. Zare *et al.* successfully cast hydroxyapatite/PLA-based 3D printed bone utilizing a hybrid 3D printing-milling machine. The HAP and PLA ratios in various formulations ranged from 95:5 to 50:50. It was discovered that 10:90 loading of HA and PLA resulted in the best mechanical properties [97].

Recent studies have explored the influence of pore geometry and material composition on 3D-printed bone scaffolds. Titanium-based alloys fabricated *via* selective laser melting (SLM) have shown promise for biomedical applications, with scaffold properties highly dependent on geometrical features and surface characteristics [98]. SLM-produced Ti6Al4V scaffolds demonstrated biocompatibility and the ability to support osteoblast growth, with pore width affecting cell occlusion, compressive strength, and porosity [99]. Further investigation of Ti6Al4V scaffolds revealed that aperture size and pore shape significantly impact porosity, compressive strength, and torsional strength [100]. In bioceramic scaffolds, pore geometry design can be used to tune mechanical properties and bio-dissolution rates. Cylindrical pore structures showed superior compressive and flexural resistance, while hexagonal cellular structures contributed to high specific compressive strength [101]. These findings highlight the importance of optimizing pore geometry for specific tissue engineering applications [63, 102–104].

The use of 3D printing in dentistry is also growing over time, especially in the fields of prosthodontics, maxillofacial prostheses, implantology, and oral surgery. The most often used techniques for printing dental components are SLM, stereolithography, FDM, and DLP [105, 106]. Branco *et al.*, used the robocasting technology to print Yttria (3% mol)-stabilized zirconia powder. It was noted that the specimens were non-toxic and had a denser structure and that their mechanical qualities met the requirements for dental replacements [107].

9.5 3D-Printing of Fixation Devices

Extrusion-printed bioresorbable scaffolds show favorable biocompatibility, degradation, and mechanical properties when configured with optimized polymer compositions, chemical modifications, and printing parameters [27]. Seyednejad *et al.* report rapid degradation in poly(hydroxymethylglycolide-co-ε-caprolactone) (>60% weight loss in 3 months) with enhanced tissue integration and vascularization, whereas polycaprolactone (PCL) scaffolds maintain stability [26]. Shishatskaya *et al.* document Young's moduli between 207.5 and 241.3 MPa and tensile strengths of 19.5–22.4 MPa in poly(3-hydroxybutyrate-co-3-hydroxyvalerate) grafts, correlating robust mechanical performance with bone defect restoration [28]. Studies using PLA or PCL composites enriched with ceramics—biphasic calcium phosphate, tricalcium phosphate, hydroxyapatite, or bioglass— demonstrate improved osteoconduction, mechanical tunability, and vascularization [30]. Chemical modifications such as hydroxyl-functionalization and radiopaque agent incorporation (PCL-TIB) yield enhanced cell viability (e.g., >90% HUVEC viability) and imaging properties. Fused deposition modeling and related techniques enable precise scaffold architectures influential in both degradation profiles and *in vivo* outcomes across bone, cardiovascular, craniofacial, and soft tissue applications [28].

9.5.1 Fixation Devices and Their Applications in Healthcare

It is essential to first understand what fixation devices are, along with their applications and advantages within the medical field. Fixation devices or Orthopedic fixation devices are used in the treatment of fractures, reconstructive surgeries, and soft-tissue injuries. These devices are divided into internal, external, or intramedullary fixation based on the purpose and mainly are used for fracture reduction, and to provide stability and support to maintain the alignment of bone fragments during the healing process. They must be strong and secure enough to allow early mobilization of the injured part, as well as the entire patient. By external means, compression is provided to the implanted device to reduce the stress, and increase stability, and the contact area for faster healing. This is done with the help of screws which provide inter-fragmental compression or attach plates, which can then provide compression, prevent displacement, and support the fragments during the healing process. Pins and wires can also be used to treat small bone fragments or fractures, as well as to attach external fixation devices and provide traction [108]. Table 9.2 summarizes common material types—titanium alloys, magnesium alloys, PLA, and PCL—used in biomedical fixation devices. It outlines examples, key properties like biodegradability and mechanical strength, and specific fixation applications such as plates, screws, and bone scaffolds, highlighting their suitability for various clinical and anatomical requirements.

Table 9.2 Materials used for fixation devices in biomedical applications and their key properties.

Material type	Examples	Key properties	Fixation use
Titanium Alloys	Ti6Al4V	Strong, corrosion-resistant, osseointegrative	Plates, screws, cages
Magnesium Alloys	Mg-Zn, Mg-Ca	Biodegradable, osteoconductive	Temporary implants (pins, screws)
PLA	PLA, PLA-CNC, PLA-HAP	Biodegradable, printable with FDM	Cranial plates, low-load screws
PCL	PCL composites	Flexible, slow degrading	Bone scaffolds, fixation in soft bone environments

9.5.2 Materials Used for Printing

Materials used in the 3D printing of fixation devices must combine biocompatibility, mechanical strength, and, in some cases, controlled biodegradability. Commonly used materials include titanium and its alloys (e.g., Ti6Al4V), known for their excellent strength, corrosion resistance, and osseointegration properties, making them ideal for permanent implants like bone plates and screws. Magnesium alloys are also gaining attention due to their lightweight and biodegradable nature, suitable for temporary fixation devices. On the polymer side, biodegradable thermoplastics such as polylactic acid (PLA) and polycaprolactone (PCL) are widely used for producing low-load bearing fixation elements, offering the advantage of gradual resorption in the body without the need for surgical removal. These materials can be further enhanced with bioactive fillers like hydroxyapatite to improve mechanical and biological performance. Table 9.3 discusses a range of studies on 3D-printed biomaterials, highlighting their study designs, material compositions, biomedical application areas, and key performance metrics. The table encompasses applications such as tissue engineering, bone regeneration, cardiovascular stents, and craniofacial reconstruction, emphasizing the versatility and effectiveness of various biomaterial formulations.

Table 9.3 Summary of studies on 3D-printed biomaterials and their biomedical applications.

Study	Study design	Material composition	Application focus	Key performance metrics	Reference
Seyednejad *et al.*, 2012	*In vivo* animal study, *ex vivo* analysis, biomaterial characterization	Poly(hydroxymethyl glycolide-co-ε-caprolactone) (PHMGCL, HMG:CL 8:92), Polycaprolactone (PCL)	Subcutaneous tissue engineering	Degradation rate, molecular weight loss, tissue integration, vascularization	[26]
El-Habashy *et al.*, 2021	Biomaterial characterization, *in vivo* animal study	Gelatin, polyvinyl alcohol, hyaluronic acid, hydroxyapatite/ polycaprolactone nanoparticles	Bone regeneration (rabbit tibia)	Structural integrity, mechanical stiffness, osteoconduction, bone healing	[27]
Shishatskaya *et al.*, 2023	Biomaterial characterization, *ex vivo, in vivo* animal study	Poly(3-hydroxybutyrate-co-3-hydroxyvalerate)	Bone grafting (pig femur defect)	Young's modulus, tensile strength, cell colonization, bone restoration	[28]
Jeong *et al.*, 2024	*In vivo* animal study, biomaterial characterization	Polycaprolactone (PCL), PCL-TIB (PCL-diol + 2,3,5-triiodobenzoic acid)	Cardiovascular stents	Radiopacity, mechanical stability, human umbilical vein endothelial cell (HUVEC) viability	[29]

(*Continued*)

Table 9.3 Summary of studies on 3D-printed biomaterials and their biomedical applications. (*Continued*)

Study	Study design	Material composition	Application focus	Key performance metrics	Reference
Barbeck *et al.*, 2017	Biomaterial characterization, *in vivo* animal study, preclinical experimental study	Polylactic acid (PLA), PLA/bioglass G5	Osteochondral regeneration (mice)	Compressive modulus, vascularization, tissue response	[30]
Oberdiek *et al.*, 2021	*Ex vivo* analysis, *in vivo* animal study, biomaterial characterization	Polycaprolactone (PCL), PCL + biphasic calcium phosphate (BCP)	Bone regeneration (subcutaneous, calvaria)	Mechanical properties, osteoconductivity, swelling, biodegradability	[31]
Kirmanidou *et al.*, 2024	Review	Polycaprolactone (PCL), PCL-based composites	Craniofacial reconstruction	Clinical outcomes, complications, mechanical/biodegradation properties	[32]
Yeo *et al.*, 2008	Biomaterial characterization, *in vivo* animal study	Polycaprolactone (PCL)-20% tricalcium phosphate (TCP)	Dentoalveolar reconstruction (rat abdomen)	Porosity, molecular weight loss, mechanical properties	[33]
Thuaksuban *et al.*, 2018	*In vivo* animal study, biomaterial characterization	Polycaprolactone (PCL)-20% biphasic calcium phosphate (BCP), PCL-20% tricalcium phosphate (TCP)	Bone substitute (rat, subcutaneous)	Collagen/vessel formation, inflammation, molecular weight	[34]

9.6 3D-Printing Methods and Advantages

Fixation devices such as cranial plates, orthopedic screws, maxillofacial implants, and spinal cages are key components that are used in surgical procedures needing structural stability3D printing technologies like Selective Laser Melting (SLM), Fused Deposition Modeling (FDM), and extrusion-based printing are frequently used due to their ability to build structurally strong and complicated structures. These methods of printing allow the fabrication of patient-specific implants by using CT or MRI scan data to develop structures and geometries that match closely with the patient's anatomy. Design requirements for fixation devices include high mechanical stress resistance to handle physiological demands and surface factors that encourage osseointegration, which helps and enable robust and long-term fixation. The precision and flexibility of 3D printing operations help increase the production of individualized implants, leading to better and more accurate surgical outcomes and decreased recovery time. Additionally, these methods help develop porous structures to promote bone formation and the use of composite materials or bioactive coatings to enhance biological response. The reliability, adaptability, and clinical efficacy of 3D-printed fixation devices are increasing because of continuous developments in material science and imaging technology [109]. Figure 9.5 illustrates various 3D-printed biomedical implants and fixation devices, including jaw, skull, spinal, hip, and knee implants, as well as heart stents and bone plates. These personalized, biocompatible devices enhance surgical outcomes, reduce recovery time, and improve patient-specific treatments in modern regenerative and orthopedic medicine.

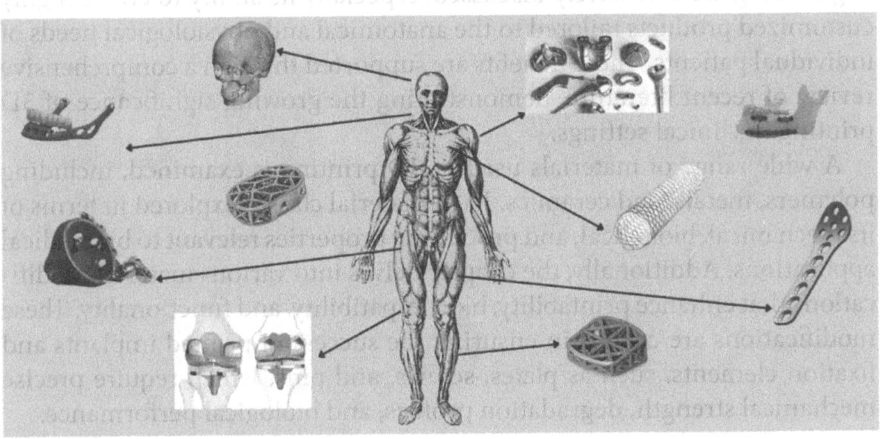

9.7 Challenges and Opportunities

Three-dimensional (3D) printing techniques have predominantly evolved through advancements in material development, printing methodologies, and digital modeling. Recently, a variety of composite systems, particularly those involving bioresorbable polymers, have emerged for biomedical applications. However, implant-related applications require further investigation through *in vivo* studies and human clinical trials to ensure efficacy and safety. Key challenges such as biocompatibility and degradation stability must be thoroughly addressed, particularly for polymeric scaffolds, implants, and fixation devices. Product-driven research and development are expected to yield more efficient and effective solutions, including bioresorbable screws, plates, pins, and other fixation elements. Additionally, extrusion-based printing techniques combined with post-processing methods such as curing offer promising potential for bioresorbable polymers that are not inherently thermoplastic. Nevertheless, challenges related to controlling viscosity, ensuring environmental stability, achieving adequate mechanical strength, and maintaining structural integrity necessitate further investigation [110, 111].

9.8 Outlook

This chapter provides an in-depth analysis of 3D printing strategies with a particular focus on their applications in the development of patient-specific implants and fixation devices. The advantages of 3D printing in biomedical engineering are extensively discussed, especially its ability to create highly customized products tailored to the anatomical and physiological needs of individual patients. These benefits are supported through a comprehensive review of recent literature, demonstrating the growing significance of 3D printing in clinical settings.

A wide range of materials used in 3D printing is examined, including polymers, metals, and ceramics. Each material class is explored in terms of its mechanical, biological, and processing properties relevant to biomedical applications. Additionally, the chapter delves into various material modifications that enhance printability, biocompatibility, and functionality. These modifications are critical in ensuring the success of printed implants and fixation elements, such as plates, screws, and pins, which require precise mechanical strength, degradation profiles, and biological performance.

The chapter also emphasizes the growing interest in bioresorbable materials for implantable devices. These materials, which degrade safely within the body over time, offer several advantages, including the elimination of secondary surgeries for implant removal and reduced long-term complications. Their inclusion in this chapter is supported by emerging research that highlights the effectiveness of bioresorbable polymers and composites in promoting tissue regeneration and healing.

Furthermore, the integration of extrusion-based printing techniques and post-processing methods, such as curing, is discussed as a promising approach for fabricating non-thermoplastic bioresorbable polymers. These techniques present new opportunities for expanding the material portfolio in biomedical 3D printing. The chapter concludes by identifying key challenges such as viscosity control, mechanical stability, and structural integrity, and emphasizes the need for continued research to overcome these barriers and translate innovative materials and methods into clinically viable products.

Acknowledgements

The authors gratefully acknowledge the National Institute of Technology Andhra Pradesh, India, for providing the necessary infrastructure and support for the preparation of this book chapter.

AI disclosure statement: The author confirms that no artificial intelligence (AI)-generated content or AI-assisted writingtools were used in the preparation of this chapter (including summaries and the abstract). All text,analysis, and conclusions are the result of the author's original work.

References

1. Szymczyk-Ziółkowska, P., Łabowska, M.B., Detyna, J., Michalak, I., Gruber, P., A review of fabrication polymer scaffolds for biomedical applications using additive manufacturing techniques. *Biocybern. Biomed. Eng.*, 40, 624–638, 2020, https://doi.org/10.1016/j.bbe.2020.01.015.
2. Deo, K.A., Singh, K.A., Peak, C.W., Alge, D.L., Gaharwar, A.K., Bioprinting 101: Design, Fabrication, and Evaluation of Cell-Laden 3D Bioprinted Scaffolds. *Tissue Eng. Part A*, 26, 318–338, 2020, https://doi.org/10.1089/ten.tea.2019.0298.

3. Dai, G. and Lee, V., Three-dimensional bioprinting and tissue fabrication: prospects for drug discovery and regenerative medicine. *Adv. Health Care Technol.*, 23, 2015, https://doi.org/10.2147/ahct.s69191.
4. Malekpour, A. and Chen, X., Printability and Cell Viability in Extrusion-Based Bioprinting from Experimental, Computational, and Machine Learning Views. *J. Funct. Biomater.*, 13, 2022, https://doi.org/10.3390/jfb13020040.
5. Ozbolat, I.T., Moncal, K.K., Gudapati, H., Evaluation of bioprinter technologies. *Addit. Manuf.*, 13, 179–200, 2017, https://doi.org/10.1016/j.addma.2016.10.003.
6. Mobaraki, M., Ghaffari, M., Yazdanpanah, A., Luo, Y., Mills, D.K., Bioinks and bioprinting: A focused review. *Bioprinting*, 18, 2020, https://doi.org/10.1016/j.bprint.2020.e00080.
7. Vanaei, S., Parizi, M.S., Vanaei, S., Salemizadehparizi, F., Vanaei, H.R., An Overview on Materials and Techniques in 3D Bioprinting Toward Biomedical Application. *Eng. Regen.*, 2, 1–18, 2021, https://doi.org/10.1016/j.engreg.2020.12.001.
8. Gu, Z., Fu, J., Lin, H., He, Y., Development of 3D bioprinting: From printing methods to biomedical applications. *Asian J. Pharm. Sci.*, 15, 529–557, 2020, https://doi.org/10.1016/j.ajps.2019.11.003.
9. Serra, T., Planell, J.A., Navarro, M., High-resolution PLA-based composite scaffolds *via* 3-D printing technology. *Acta Biomater.*, 9, 5521–5530, 2013, https://doi.org/10.1016/j.actbio.2012.10.041.
10. Xu, W., Wang, X., Sandler, N., Willför, S., Xu, C., Three-Dimensional Printing of Wood-Derived Biopolymers: A Review Focused on Biomedical Applications. *ACS Sustainable Chem. Eng.*, 6, 5663–5680, 2018, https://doi.org/10.1021/acssuschemeng.7b03924.
11. Mandrycky, C., Wang, Z., Kim, K., Kim, D.H., 3D bioprinting for engineering complex tissues. *Biotechnol. Adv.*, 34, 422–434, 2016, https://doi.org/10.1016/j.biotechadv.2015.12.011.
12. Ruchika, Bhardwaj, N., Yadav, S.K., Saneja, A., Recent advances in 3D bioprinting for cancer research: From precision models to personalized therapies. *Drug Discovery Today*, 29, 2024, https://doi.org/10.1016/j.drudis.2024.103924.
13. Agrawal, I. and Dua, P., Surface Modification of Advanced Biomaterials for Applications in the Pharmaceutical and Medical Fields. *Biotechnol. Kiosk*, 4, 1–16, 2022, https://doi.org/10.37756/bk.22.4.3.1.
14. Wu, G., Li, P., Feng, H., Zhang, X., Chu, P.K., Engineering and functionalization of biomaterials *via* surface modification. *J. Mater. Chem. B*, 3, 2024–2042, 2015, https://doi.org/10.1039/c4tb01934b.
15. Ahangar, P., Cooke, M.E., Weber, M.H., Rosenzweig, D.H., Current biomedical applications of 3D printing and additive manufacturing. *Appl. Sci.*, 9, 2019, https://doi.org/10.3390/app9081713.

16. Tetsuka, H. and Shin, S.R., Materials and technical innovations in 3D printing in biomedical applications. *J. Mater. Chem. B*, 8, 2930–2950, 2020, https://doi.org/10.1039/d0tb00034e.
17. Jin, Y., Li, X., Campbell, R.I., Ji, S., Visualizing the hotspots and emerging trends of 3D printing through scientometrics. *Rapid Prototyping J.*, 24, 801–812, 2018, https://doi.org/10.1108/RPJ-05-2017-0100.
18. Bănică, C.F., Sover, A., Anghel, D.C., Printing the Future Layer by Layer: A Comprehensive Exploration of Additive Manufacturing in the Era of Industry 4.0. *Appl. Sci.*, 14, 2024, https://doi.org/10.3390/app14219919.
19. Elhadad, A.A., Rosa-Sainz, A., Cañete, R., Peralta, E., Begines, B., Balbuena, M., Alcudia, A., Torres, Y., Applications and multidisciplinary perspective on 3D printing techniques: Recent developments and future trends. *Mater. Sci. Eng.: R: Rep.*, 156, 2023, https://doi.org/10.1016/j.mser.2023.100760.
20. Kharat, V.J., Singh, P., Sharath Raju, G., Kumar Yadav, D., Satyanarayana. Gupta, M., Arun, V., Hussein Majeed, A., Singh, N., Additive manufacturing (3D printing): A review of materials, methods, applications and challenges. *Mater. Today Proc.*, 2023, https://doi.org/10.1016/j.matpr.2023.11.033.
21. Chen, Z., Zhao, D., Liu, B., Nian, G., Li, X., Yin, J., Qu, S., Yang, W., 3D Printing of Multifunctional Hydrogels. *Adv. Funct. Mater.*, 29, 2019, https://doi.org/10.1002/adfm.201900971.
22. Cardoso, R.M., Silva, P.R.L., Lima, A.P., Rocha, D.P., Oliveira, T.C., do Prado, T.M., Fava, E.L., Fatibello-Filho, O., Richter, E.M., Muñoz, R.A.A., 3D-Printed graphene/polylactic acid electrode for bioanalysis: Biosensing of glucose and simultaneous determination of uric acid and nitrite in biological fluids. *Sens. Actuators B Chem.*, 307, 2020, https://doi.org/10.1016/j.snb.2019.127621.
23. Liu, Y., Li, T., Ma, H., Zhai, D., Deng, C., Wang, J., Zhuo, S., Chang, J., Wu, C., 3D-printed scaffolds with bioactive elements-induced photothermal effect for bone tumor therapy. *Acta Biomater.*, 73, 531–546, 2018, https://doi.org/10.1016/j.actbio.2018.04.014.
24. Vijayavenkataraman, S., Fuh, J.Y.H., Lu, W.F., 3D printing and 3D bioprinting in pediatrics. *Bioengineering*, 4, 2017, https://doi.org/10.3390/bioengineering4030063.
25. Katseli, V., Economou, A., Kokkinos, C., Single-step fabrication of an integrated 3D-printed device for electrochemical sensing applications. *Electrochem. Commun.*, 103, 100–103, 2019, https://doi.org/10.1016/j.elecom.2019.05.008.
26. Seyednejad, H., Gawlitta, D., Kuiper, R.V., De Bruin, A., Van Nostrum, C.F., Vermonden, T., Dhert, W.J.A., Hennink, W.E., *In vivo* biocompatibility and biodegradation of 3D-printed porous scaffolds based on a hydroxyl-functionalized poly(ε-caprolactone). *Biomaterials*, 33, 4309–4318, 2012, https://doi.org/10.1016/j.biomaterials.2012.03.002.
27. El-Habashy, S.E., El-Kamel, A.H., Essawy, M.M., Abdelfattah, E.Z.A., Eltaher, H.M., Engineering 3D-printed core-shell hydrogel scaffolds reinforced with

hybrid hydroxyapatite/polycaprolactone nanoparticles for *in vivo* bone regeneration. *Biomater. Sci.*, 9, 4019–4039, 2021, https://doi.org/10.1039/d1bm00062d.

28. Shishatskaya, E.I., Demidenko, A.V., Sukovatyi, A.G., Dudaev, A.E., Mylnikov, A.V., Kisterskij, K.A., Volova, T.G., Three-Dimensional Printing of Poly(3-hydroxybutyrate-co-3-hydroxyvalerate) [P(3HB-co-3HV)] Biodegradable Scaffolds: Properties, *In Vitro* and *In Vivo* Evaluation. *Int. J. Mol. Sci.*, 24, 2023, https://doi.org/10.3390/ijms241612969.
29. Jeong, Y.J., Choi, B., Kim, S., Jo, J., Shanmugasundaram, A., Kang, S., Kim, H., Choi, E., Lee, D.W., Fully bioresorbable, all-polymer-based, 3D-printed scaffolds with prolonged radiopacity and enhanced durability. *Chem. Eng. J.*, 494, 2024, https://doi.org/10.1016/j.cej.2024.152851.
30. Barbeck, M., Serra, T., Booms, P., Stojanovic, S., Najman, S., Engel, E., Sader, R., Kirkpatrick, C.J., Navarro, M., Ghanaati, S., Analysis of the *in vitro* degradation and the *in vivo* tissue response to bi-layered 3D-printed scaffolds combining PLA and biphasic PLA/bioglass components – Guidance of the inflammatory response as basis for osteochondral regeneration. *Bioact. Mater.*, 2, 208–223, 2017, https://doi.org/10.1016/j.bioactmat.2017.06.001.
31. Oberdiek, F., Vargas, C.I., Rider, P., Batinic, M., Görke, O., Radenković, M., Najman, S., Baena, J.M., Jung, O., Barbeck, M., *Ex vivo* and *in vivo* analyses of novel 3d-printed bone substitute scaffolds incorporating biphasic calcium phosphate granules for bone regeneration. *Int. J. Mol. Sci.*, 22, 2021, https://doi.org/10.3390/ijms22073588.
32. Kirmanidou, Y., Chatzinikolaidou, M., Michalakis, K., Tsouknidas, A., Clinical translation of polycaprolactone-based tissue engineering scaffolds, fabricated *via* additive manufacturing: A review of their craniofacial applications. *Biomater. Adv.*, 162, 2024, https://doi.org/10.1016/j.bioadv.2024.213902.
33. Yeo, A., Rai, B., Sju, E., Cheong, J.J., Teoh, S.H., The degradation profile of novel, bioresorbable PCL-TCP scaffolds: An *in vitro* and *in vivo* study. *J. Biomed. Mater. Res. Part A*, 84, 208–218, 2008, https://doi.org/10.1002/jbm.a.31454.
34. Thuaksuban, N., Pannak, R., Boonyaphiphat, P., Monmaturapoj, N., *In vivo* biocompatibility and degradation of novel Polycaprolactone-Biphasic Calcium phosphate scaffolds used as a bone substitute. *Bio-Med. Mater. Eng.*, 29, 253–267, 2018, https://doi.org/10.3233/BME-171727.
35. Calignano, F., Investigation of the accuracy and roughness in the laser powder bed fusion process. *Virtual Phys. Prototyp.*, 13, 97–104, 2018, https://doi.org/10.1080/17452759.2018.1426368.
36. Campioni, I., Cacciotti, I., Gupta, N., Additive manufacturing of reconstruction devices for maxillofacial surgery: design and accuracy assessment of a mandibular plate prototype. *Ann. Ist. Super. Sanita*, 56, 10–18, 2020, https://doi.org/10.4415/ANN_20_01_04.

37. Feuerbach, T., Kock, S., Thommes, M., Slicing parameter optimization for 3D printing of biodegradable drug-eluting tracheal stents. *Pharm. Dev. Technol.*, 25, 650–658, 2020, https://doi.org/10.1080/10837450.2020.1727921.
38. Piedra-Cascón, W., Krishnamurthy, V.R., Att, W., Revilla-León, M., 3D printing parameters, supporting structures, slicing, and post-processing procedures of vat-polymerization additive manufacturing technologies: A narrative review. *J. Dent.*, 109, 2021, https://doi.org/10.1016/j.jdent.2021.103630.
39. Awad, R.H., Habash, S.A., Hansen, C.J., 3D printing methods, 3D Print. *Appl. Cardiovasc. Med.*, 11–32, 2018, https://doi.org/10.1016/B978-0-12-803917-5.00002-X.
40. Jeong, Y.G., Lee, W.S., Lee, K.B., Accuracy evaluation of dental models manufactured by CAD/CAM milling method and 3D printing method. *J. Adv. Prosthodont.*, 10, 245–251, 2018, https://doi.org/10.4047/jap.2018.10.3.245.
41. Wu, D. and Xu, C., Predictive modeling of droplet formation processes in inkjet-based bioprinting. *J. Manuf. Sci. Eng. Trans. ASME*, 140, 2018, https://doi.org/10.1115/1.4040619.
42. Azizi Machekposhti, S., Movahed, S., Narayan, R.J., Physicochemical parameters that underlie inkjet printing for medical applications. *Biophys. Rev.*, 1, 2020, https://doi.org/10.1063/5.0011924.
43. Ng, W.L. and Shkolnikov, V., Optimizing cell deposition for inkjet-based bioprinting. *Int. J. Bioprinting*, 10, 182–206, 2024, https://doi.org/10.36922/ijb.2135.
44. Wang, J., Liu, Y., Fan, Z., Wang, W., Wang, B., Guo, Z., Ink-based 3D printing technologies for graphene-based materials: a review. *Adv. Compos. Hybrid Mater.*, 2, 2019, https://doi.org/10.1007/s42114-018-0067-9.
45. Brindha, J., Privita Edwina, R.A.G., Rajesh, P.K., Rani, P., Influence of rheological properties of protein bio-inks on printability: A simulation and validation study. *Mater. Today Proc.*, 3, 3285–3295, 2016, https://doi.org/10.1016/j.matpr.2016.10.010.
46. De Gans, B.J., Duineveld, P.C., Schubert, U.S., Inkjet printing of polymers: State of the art and future developments. *Adv. Mater.*, 16, 203–213, 2004, https://doi.org/10.1002/adma.200300385.
47. Habib, M.A. and Khoda, B., Development of clay based novel bio-ink for 3D bio-printing process. *Procedia Manuf.*, 26, 846–856, 2018, https://doi.org/10.1016/j.promfg.2018.07.105.
48. Duty, C., Ajinjeru, C., Kishore, V., Compton, B., Hmeidat, N., Chen, X., Liu, P., Hassen, A.A., Lindahl, J., Kunc, V., What makes a material printable? A viscoelastic model for extrusion-based 3D printing of polymers. *J. Manuf. Processes*, 35, 526–537, 2018, https://doi.org/10.1016/j.jmapro.2018.08.008.
49. Lille, M., Nurmela, A., Nordlund, E., Metsä-Kortelainen, S., Sozer, N., Applicability of protein and fiber-rich food materials in extrusion-based 3D printing. *J. Food Eng.*, 220, 20–27, 2018, https://doi.org/10.1016/j.jfoodeng.2017.04.034.

50. Fitzsimmons, R.E., Aquilino, M.S., Quigley, J., Chebotarev, O., Tarlan, F., Simmons, C.A., Generating vascular channels within hydrogel constructs using an economical open-source 3D bioprinter and thermoreversible gels. *Bioprinting*, 9, 7–18, 2018, https://doi.org/10.1016/j.bprint.2018.02.001.
51. Ferrández-Montero, A., Lieblich, M., Benavente, R., González-Carrasco, J.L., Ferrari, B., Study of the matrix-filler interface in PLA/Mg composites manufactured by Material Extrusion using a colloidal feedstock. *Addit. Manuf.*, 33, 2020, https://doi.org/10.1016/j.addma.2020.101142.
52. Faes, M., Valkenaers, H., Vogeler, F., Vleugels, J., Ferraris, E., Extrusion-based 3D printing of ceramic components. *Procedia CIRP*, 28, 76–81, 2015, https://doi.org/10.1016/j.procir.2015.04.028.
53. Koch, F., Tröndle, K., Finkenzeller, G., Zengerle, R., Zimmermann, S., Koltay, P., Generic method of printing window adjustment for extrusion-based 3D-bioprinting to maintain high viability of mesenchymal stem cells in an alginate-gelatin hydrogel. *Bioprinting*, 20, 2020, https://doi.org/10.1016/j.bprint.2020.e00094.
54. Wolfs, R.J.M. and Suiker, A.S.J., Structural failure during extrusion-based 3D printing processes. *Int. J. Adv. Manuf. Technol.*, 104, 565–584, 2019, https://doi.org/10.1007/s00170-019-03844-6.
55. Zhang, L.C., Attar, H., Calin, M., Eckert, J., Review on manufacture by selective laser melting and properties of titanium based materials for biomedical applications. *Mater. Technol.*, 31, 66–76, 2016, https://doi.org/10.1179/1753555715Y.0000000076.
56. Matouskova, L., Ackermann, M., Horakova, J., Capek, L., Henys, P., Safka, J., How does the surface treatment change the cytocompatibility of implants made by selective laser melting? *Expert Rev. Med. Devices*, 15, 313–321, 2018, https://doi.org/10.1080/17434440.2018.1456335.
57. Vaithilingam, J., Prina, E., Goodridge, R.D., Hague, R.J.M., Edmondson, S., Rose, F.R.A.J., Christie, S.D.R., Surface chemistry of Ti6Al4V components fabricated using selective laser melting for biomedical applications. *Mater. Sci. Eng., C*, 67, 294–303, 2016, https://doi.org/10.1016/j.msec.2016.05.054.
58. Cardaropoli, F., Alfieri, V., Caiazzo, F., Sergi, V., Manufacturing of porous biomaterials for dental implant applications through Selective Laser Melting. *Adv. Mater. Res.*, 535–537, 1222–1229, 2012, https://doi.org/10.4028/www.scientific.net/AMR.535-537.1222.
59. He, B., Wu, W., Zhang, L., Lu, L., Yang, Q., Long, Q., Chang, K., Microstructural characteristic and mechanical property of Ti6Al4V alloy fabricated by selective laser melting. *Vacuum*, 150, 79–83, 2018, https://doi.org/10.1016/j.vacuum.2018.01.026.
60. Mujian, D., Guanqun, D., Hongyu, C., Qimin, Selective laser melting 3D printing of Ni-based superalloy: understanding thermodynamic mechanisms. *Sci. Bull.*, 61, 1013–1022, 2016.
61. Allahham, N., Fina, F., Marcuta, C., Kraschew, L., Mohr, W., Gaisford, S., Basit, A.W., Goyanes, A., Selective laser sintering 3D printing of orally

disintegrating printlets containing ondansetron. *Pharmaceutics*, 12, 2020, https://doi.org/10.3390/pharmaceutics12020110.
62. Vaquette, C., Mitchell, J., Fernandez-Medina, T., Kumar, S., Ivanovski, S., Resorbable additively manufactured scaffold imparts dimensional stability to extraskeletally regenerated bone. *Biomaterials*, 269, 2021, https://doi.org/10.1016/j.biomaterials.2021.120671.
63. Sahmani, S., Khandan, A., Esmaeili, S., Saber-Samandari, S., Ghadiri Nejad, M., Aghdam, M.M., Calcium phosphate-PLA scaffolds fabricated by fused deposition modeling technique for bone tissue applications: Fabrication, characterization and simulation. *Ceram. Int.*, 46, 2447–2456, 2020, https://doi.org/10.1016/j.ceramint.2019.09.238.
64. Wurm, M.C., Möst, T., Bergauer, B., Rietzel, D., Neukam, F.W., Cifuentes, S.C., von Wilmowsky, C., *In-vitro* evaluation of Polylactic acid (PLA) manufactured by fused deposition modeling. *J. Biol. Eng.*, 11, 2017, https://doi.org/10.1186/s13036-017-0073-4.
65. Mellor, L.F., Huebner, P., Cai, S., Mohiti-Asli, M., Taylor, M.A., Spang, J., Shirwaiker, R.A., Loboa, E.G., Fabrication and Evaluation of Electrospun, 3D-Bioplotted, and Combination of Electrospun/3D-Bioplotted Scaffolds for Tissue Engineering Applications. *Biomed Res. Int.*, 2017, 2017, https://doi.org/10.1155/2017/6956794.
66. Warren, P.B., Huebner, P., Spang, J.T., Shirwaiker, R.A., Fisher, M.B., Engineering 3D-Bioplotted scaffolds to induce aligned extracellular matrix deposition for musculoskeletal soft tissue replacement. *Connect. Tissue Res.*, 58, 342–354, 2017, https://doi.org/10.1080/03008207.2016.1276177.
67. Paul, A.K., Migration Studies of Biodegradable Composites, in: *Biodegrad. Compos. Packag. Appl.*, pp. 193–212, CRC Press, Boca Raton, 2022, https://doi.org/10.1201/9781003227908-11.
68. Patel, S., Sanjay, M.S.S.S., Chakraborty, G., Paul, A.K., Nanocomposites/Nanomaterials and the Risk Management of Biofuels Production, in: *Nanomater. as a Catal. Biofuel Prod.*, V. Dave and A. Kuila (Eds.), pp. 151–184, Springer Nature Singapore, Singapore, 2025, https://doi.org/10.1007/978-981-96-1706-7_7.
69. Chakraborty, G., Valapa, R.B., Pugazhenthi, G., Katiyar, V., Investigating the properties of poly (lactic acid)/exfoliated graphene based nanocomposites fabricated by versatile coating approach. *Int. J. Biol. Macromol.*, 113, 1080–1091, 2018, https://doi.org/10.1016/j.ijbiomac.2018.03.037.
70. Chakraborty, G., Prasad, A., Kumar, A., Processing of Biodegradable Composites, in: *Biodegrad. Compos. Packag. Appl.*, pp. 33–48, CRC Press, Boca Raton, 2022, https://doi.org/10.1201/9781003227908-3.
71. Prasad, A., Chakraborty, G., Kumar, A., Kumar, K., Introduction to Biodegradable Polymers, in: *Biodegrad. Compos. Packag. Appl.*, pp. 1–12, CRC Press, Boca Raton, 2022, https://doi.org/10.1201/9781003227908-1.
72. Sringari, S.S., Chakraborty, G., Prasad, A., Biodegradable and Biocompatible Polymeric Nanocomposites for Tissue Engineering Applications, in:

Nanomanufacturing Nanomater. Des., pp. 205–225, CRC Press, Boca Raton, 2022, https://doi.org/10.1201/9781003220602-13.
73. Chakraborty, G., Bhattacharjee, S., Katiyar, V., Pugazhenthi, G., Magnetic nanomaterials reinforced PLA nanocomposite: impact of morphology and external field on magneto-rheological flow behaviour. *J. Polym. Res.*, 32, 2025, https://doi.org/10.1007/s10965-025-04383-3.
74. Chakraborty, G. and Prasad, A., Bio-based materials: origin, synthesis, and properties, in: *Sustain. Bio-Based Compos.*, pp. 1–24, De Gruyter, Walter de Gruyter GmbH, Berlin/Boston, 2024, https://doi.org/10.1515/9783111321530-001.
75. Das, P. and Chakraborty, G., 9 Biodegradable polymers-based proton exchange membrane for fuel cells, in: *Sustain. Bio-Based Compos.*, pp. 183–192, De Gruyter, Walter de Gruyter GmbH, Berlin/Boston, 2024, https://doi.org/10.1515/9783111321530-009.
76. Kumar, M.H., Paul, A.K., Chakraborty, G., Utilization of Gelatin, Collagen, and Chondroitin Sulfate in 3D Bioprinting, in: *Handb. 3D Print. Biomed. Appl.*, pp. 165–187, CRC Press, Boca Raton, 2025.
77. Paul, A.K., Chakraborty, G., Borkotoky, S.S., Prasad, A., Degradation Studies of Bioabsorbable Polymeric Composites, in: *Bioabsorbable Polym. Drug Carriers Fixat. Devices*, pp. 105–128, Apple Academic Press, New York, 2025.
78. Chakrabort, G. and Das, P., Bio-Absorbable Polymer-Based Scaffolds as Drug Carriers: Fabrication, Characterization, and Drug Release, in: *Bioabsorbable Polym. Drug Carriers Fixat. Devices*, pp. 31–59, Apple Academic Press, New York, 2025.
79. Prasad, A., Chakraborty, G., Kumar, A., Bio-based environmentally benign polymeric resorbable materials for orthopedic fixation applications, in: *Adv. Mater. Biomed. Appl.*, pp. 251–266, CRC Press, Boca Raton, 2022, https://doi.org/10.1201/9781003344810-15.
80. Chakraborty, G. and Paul, A.K., 3D Printing in Biomedical Applications: Techniques and Emerging Trends, in: *Adv. Mater. Manuf. Tech. Biomed. Appl.*, pp. 275–301, Scrivener Publishing LLC, Beverly, Massachusetts, 2023, https://doi.org/10.1002/9781394166985.ch11.
81. Paul, A.K., Borkotoky, S.S., Prasad, A., Nanohydroxyapatite-Based Composite Materials and Processing, in: *Adv. Mater. Manuf. Tech. Biomed. Appl.*, pp. 157–183, Scrivener Publishing LLC, Beverly, Massachusetts, 2023, https://doi.org/10.1002/9781394166985.ch7.
82. Xin, C., Qi, X., Zhu, M., Zhao, S.C., Zhu, Y.F., Hydroxyapatite Whisker-reinforced Composite Scaffolds Through 3D Printing for Bone Repair. *Wuji Cailiao Xuebao/J. Inorg. Mater.*, 32, 837–844, 2017, https://doi.org/10.15541/jim20160628.
83. Drescher, P., Spath, S., Seitz, H., Improvement of mechanical properties of bone tissue engineered scaffolds through sintering and infiltration with biopolymers. *Proc. IASTED Int. Conf. Biomed. Eng. BioMed 2013*, pp. 447–450, 2013, https://doi.org/10.2316/P.2013.791-169.

84. Huang, B., Caetano, G., Vyas, C., Blaker, J.J., Diver, C., Bártolo, P., Polymer-ceramic composite scaffolds: The effect of hydroxyapatite and β-tri-calcium phosphate. *Materials (Basel)*, 11, 2018, https://doi.org/10.3390/ma11010129.
85. Cox, S.C., Thornby, J.A., Gibbons, G.J., Williams, M.A., Mallick, K.K., 3D printing of porous hydroxyapatite scaffolds intended for use in bone tissue engineering applications. *Mater. Sci. Eng., C*, 47, 237–247, 2015, https://doi.org/10.1016/j.msec.2014.11.024.
86. Varma, M.V., Kandasubramanian, B., Ibrahim, S.M., 3D printed scaffolds for biomedical applications. *Mater. Chem. Phys.*, 255, 2020, https://doi.org/10.1016/j.matchemphys.2020.123642.
87. Han, Y., Wei, Q., Chang, P., Hu, K., Okoro, O.V., Shavandi, A., Nie, L., Three-dimensional printing of hydroxyapatite composites for biomedical application. *Crystals*, 11, 2021, https://doi.org/10.3390/cryst11040353.
88. Cao, Y., Shi, T., Jiao, C., Liang, H., Chen, R., Tian, Z., Zou, A., Yang, Y., Wei, Z., Wang, C., Shen, L., Fabrication and properties of zirconia/hydroxyapatite composite scaffold based on digital light processing. *Ceram. Int.*, 46, 2300–2308, 2020, https://doi.org/10.1016/j.ceramint.2019.09.219.
89. Esposito Corcione, C., Gervaso, F., Scalera, F., Padmanabhan, S.K., Madaghiele, M., Montagna, F., Sannino, A., Licciulli, A., Maffezzoli, A., Highly loaded hydroxyapatite microsphere/ PLA porous scaffolds obtained by fused deposition modelling. *Ceram. Int.*, 45, 2803–2810, 2019, https://doi.org/10.1016/j.ceramint.2018.07.297.
90. Rajendran, A. and Pattanayak, D.K., Bioactive and antimicrobial macro-/micro-nanoporous selective laser melted Ti–6Al–4V alloy for biomedical applications. *Heliyon*, 8, 2022, https://doi.org/10.1016/j.heliyon.2022.e09122.
91. Bozkurt, Y. and Karayel, E., 3D printing technology; methods, biomedical applications, future opportunities and trends. *J. Mater. Res. Technol.*, 14, 1430–1450, 2021, https://doi.org/10.1016/j.jmrt.2021.07.050.
92. Uddin, M.J., Scoutaris, N., Economidou, S.N., Giraud, C., Chowdhry, B.Z., Donnelly, R.F., Douroumis, D., 3D printed microneedles for anticancer therapy of skin tumours. *Mater. Sci. Eng., C*, 107, 2020, https://doi.org/10.1016/j.msec.2019.110248.
93. Gao, Y., Yu, G., Shu, T., Chen, Y., Yang, W., Liu, Y., Long, J., Xiong, W., Xuan, F., 3D-Printed Coaxial Fibers for Integrated Wearable Sensor Skin. *Adv. Mater. Technol.*, 4, 2019, https://doi.org/10.1002/admt.201900504.
94. Kaur, S. and Kaur, B., Stereolithography (SLA) 3D Printing Technology in Microneedles. *J. Pharm. Res. Int.*, 104–113, 2021, https://doi.org/10.9734/jpri/2021/v33i62b35176.
95. Xenikakis, I., Tzimtzimis, M., Tsongas, K., Andreadis, D., Demiri, E., Tzetzis, D., Fatouros, D.G., Fabrication and finite element analysis of stereolithographic 3D printed microneedles for transdermal delivery of model dyes across human skin *in vitro*. *Eur. J. Pharm. Sci.*, 137, 2019, https://doi.org/10.1016/j.ejps.2019.104976.

96. Adepu, S., Dhiman, N., Laha, A., Sharma, C.S., Ramakrishna, S., Khandelwal, M., Three-dimensional bioprinting for bone tissue regeneration. *Curr. Opin. Biomed. Eng.*, 2, 22–28, 2017, https://doi.org/10.1016/j.cobme.2017.03.005.
97. Zare, R.N., Doustkhah, E., Assadi, M.H.N., Three-dimensional bone printing using hydroxyapatite-PLA composite. *Mater. Today Proc.*, 42, 1531–1533, 2021. https://doi.org/10.1016/j.matpr.2019.12.046/.
98. Dziaduszewska, M. and Zieliński, A., Structural and material determinants influencing the behavior of porous Ti and its alloys made by additive manufacturing techniques for biomedical applications. *Materials (Basel)*, 14, 1–48, 2021, https://doi.org/10.3390/ma14040712.
99. Warnke, P.H., Douglas, T., Wollny, P., Sherry, E., Steiner, M., Galonska, S., Becker, S.T., Springer, I.N., Wiltfang, J., Sivananthan, S., Rapid prototyping: Porous titanium alloy scaffolds produced by selective laser melting for bone tissue engineering. *Tissue Eng. - Part C Methods*, 15, 115–124, 2009, https://doi.org/10.1089/ten.tec.2008.0288.
100. Yang, J., Qin, C., Lu, J., Shi, X., Shi, K., Cui, Y., Xiong, X., Wan, K., Shen, M., Investigating mechanical properties of 3D printed porous titanium scaffolds for bone tissue engineering. *Mater. Res. Express*, 11, 2024, https://doi.org/10.1088/2053-1591/ad64af.
101. Lu, F., Wu, R., Shen, M., Xie, L., Liu, M., Li, Y., Xu, S., Wan, L., Yang, X., Gao, C., Gou, Z., Rational design of bioceramic scaffolds with tuning pore geometry by stereolithography: Microstructure evaluation and mechanical evolution. *J. Eur. Ceram. Soc.*, 41, 1672–1682, 2021, https://doi.org/10.1016/j.jeurceramsoc.2020.10.002.
102. Zhao, L., Pei, X., Jiang, L., Hu, C., Sun, J., Xing, F., Zhou, C., Fan, Y., Zhang, X., Bionic design and 3D printing of porous titanium alloy scaffolds for bone tissue repair. *Compos. Part B Eng.*, 162, 154–161, 2019, https://doi.org/10.1016/j.compositesb.2018.10.094.
103. Mondal, S., Nguyen, T.P., Pham, V.H., Hoang, G., Manivasagan, P., Kim, M.H., Nam, S.Y., Oh, J., Hydroxyapatite nano bioceramics optimized 3D printed poly lactic acid scaffold for bone tissue engineering application. *Ceram. Int.*, 46, 3443–3455, 2020, https://doi.org/10.1016/j.ceramint.2019.10.057.
104. Wang, G., Influence of polydopamine/polylactic acid coating on mechanical properties and cell behavior of 3D-printed calcium silicate scaffolds. *Mater. Lett.*, 275, 2020, https://doi.org/10.1016/j.matlet.2020.128131.
105. Rahman, Z.A.B.A., 3D printing in dentistry. *Oral Health Dent. Manag.*, 15, 2016, https://doi.org/10.4172/2247-2452.c1.032.
106. Zaharia, C., Gabor, A.G., Gavrilovici, A., Stan, A.T., Idorasi, L., Sinescu, C., Negruțiu, M.L., Digital Dentistry—3D Printing Applications. *J. Interdiscip. Med.*, 2, 50–3, 2017.
107. Branco, A.C., Silva, R., Santos, T., Jorge, H., Rodrigues, A.R., Fernandes, R., Bandarra, S., Barahona, I., Matos, A.P.A., Lorenz, K., Polido, M., Colaço, R., Serro, A.P., Figueiredo-Pina, C.G., Suitability of 3D printed pieces of

nanocrystalline zirconia for dental applications. *Dent. Mater.*, 36, 442–455, 2020, https://doi.org/10.1016/j.dental.2020.01.006.
108. Slone, R.M., Heare, M.M., Vander Griend, R.A., Montgomery, W.J., Orthopedic fixation devices. *Radiographics*, 11, 823–847, 1991, https://doi.org/10.1148/radiographics.11.5.1947319.
109. Ni, J., Ling, H., Zhang, S., Wang, Z., Peng, Z., Benyshek, C., Zan, R., Miri, A.K., Li, Z., Zhang, X., Lee, J., Lee, K.J., Kim, H.J., Tebon, P., Hoffman, T., Dokmeci, M.R., Ashammakhi, N., Li, X., Khademhosseini, A., Three-dimensional printing of metals for biomedical applications. *Mater. Today Bio*, 3, 2019, https://doi.org/10.1016/j.mtbio.2019.100024.
110. Naghieh, S. and Chen, X., Printability–A key issue in extrusion-based bio-printing. *J. Pharm. Anal.*, 11, 564–579, 2021, https://doi.org/10.1016/j.jpha.2021.02.001.
111. Ghilan, A., Chiriac, A.P., Nita, L.E., Rusu, A.G., Neamtu, I., Chiriac, V.M., Trends in 3D Printing Processes for Biomedical Field: Opportunities and Challenges. *J. Polym. Environ.*, 28, 1345–1367, 2020, https://doi.org/10.1007/s10924-020-01722-x.

10

Advanced Scaffolding for Bone Regeneration: Role of Hybridization

Atanu Kumar Paul[1]*, Gourhari Chakraborty[2] and Arbind Prasad[3]

[1]Department of Chemical Engineering, National Institute of Technology Andhra Pradesh, Tadepalligudem, Andhra Pradesh, India
[2]School of Chemical and Biotechnology, SASTRA Deemed University, Thirumalaisamudram, Tamil Nadu, India
[3]Mechanical Engineering Department, Katihar Engineering College (Under Department of Science, Technology and Technical Education, Government of Bihar), Katihar, Bihar, India

Abstract

Bone regeneration of critical-sized defects remains a major clinical challenge due to the complex hierarchical structure, load-bearing function, and dynamic remodeling behavior of native bone. This chapter presents a comprehensive overview of advanced hybrid scaffolding strategies for bone regeneration, with a particular emphasis on the role of material hybridization in addressing the inherent limitations of conventional single-component scaffolds. Hybrid scaffolds, integrating biodegradable polymers with bioactive ceramics, nanomaterials, and bioinspired fillers, enable synergistic enhancement of mechanical integrity, osteoconductivity, degradation control, and biological performance. The chapter systematically discusses structure–property–function relationships, multiscale design principles, and material platforms underpinning hybrid scaffold development. Contemporary fabrication techniques, including additive manufacturing, electrospinning, and freeze-drying, are examined with respect to their ability to generate hierarchical, anisotropic, and gradient architectures that more closely mimic native bone tissue. Furthermore, biological evaluation methodologies, degradation behavior, and load-transfer mechanisms during healing are critically analyzed to highlight translational considerations. The chapter also addresses regulatory, sterilization, and scale-up challenges that constrain clinical implementation. Finally, emerging trends such as smart stimuli-responsive scaffolds, patient-specific data-driven

**Corresponding author*: atanukpaul@outlook.com

Arbind Prasad (ed.) Resorbable Composites for Bioimplants and Fixation Devices, (287–328) © 2026 Scrivener Publishing LLC

design, and intelligent therapeutic platforms are outlined as future directions for next-generation bone implants. Collectively, this chapter provides a cohesive framework linking materials science, bioengineering, and clinical translation to guide the rational design of hybrid scaffolds for effective and reliable bone regeneration.

***Keywords*:** Hybrid bone scaffolds, bone tissue engineering, resorbable composites, polymer–ceramic hybrids, multiscale scaffold design, additive manufacturing, osteogenic differentiation, clinical translation

10.1 Introduction to Hybrid Scaffolding in Bone Regeneration

10.1.1 Clinical Need for Advanced Bone Regeneration Strategies

Critical-sized bone defects represent a significant clinical challenge as they fail to heal spontaneously within a patient's lifetime [1]. These defects, resulting from trauma, disease, or surgical resection, require advanced treatment strategies due to the complex anatomy and physiology of bone tissue [2]. Current clinical approaches include vascularized bone grafts, distraction osteogenesis, and the induced membrane technique, with the latter showing favorable union rates [1]. Bone tissue engineering has emerged as a promising alternative, utilizing osteogenic cells, growth factors, and biomaterial scaffolds to promote regeneration [2]. The healing of large defects requires the formation of substantial bone tissue in environments often compromised by soft tissue and vascular damage [3]. Three-dimensional bioprinting technology represents a cutting-edge approach, allowing incorporation of living cells and growth factors into scaffolds that mimic native bone structure and properties [4].

10.1.2 Limitations of Conventional Scaffold Materials

Conventional single-component scaffolds face significant limitations in bone tissue engineering, exhibiting trade-offs between mechanical strength, bioactivity, and degradation kinetics that compromise their effectiveness. Single hydrogel scaffolds cannot achieve the superior mechanical strength required for large load-bearing bone defects [5]. Load-bearing scaffolds can actually reduce bone formation compared to non-structural alternatives due to stress shielding and inhibited cellular infiltration [6]. To address these challenges, researchers have developed multi-layered composite approaches that combine the advantages of different materials while minimizing their individual disadvantages [7]. For example, triple-layered gel/PCL/gel scaffolds demonstrate compression strengths exceeding 13 MPa with significantly improved bioactivity compared to single-component designs, achieving over 3000% higher mechanical strength than gel scaffolds alone while maintaining enhanced cellular attachment and mineralization [8]. These composite strategies enable biomimetic scaffolds that better replicate the natural extracellular matrix environment.

10.1.3 Rationale for Hybridization in Resorbable Scaffolds

Hybrid organic-inorganic biomaterials represent a promising approach for bone tissue engineering by combining the advantages of polymers and bioactive ceramics to overcome individual material limitations. These hybrid systems create synergistic effects that enhance bioactivity in polymers while improving the stiffness-toughness balance in ceramics [9]. The integration of calcium phosphates, particularly hydroxyapatite and bioactive glass, with biodegradable polymers mimics natural bone composition and provides suitable mechanical properties [10]. This biomimetic strategy is

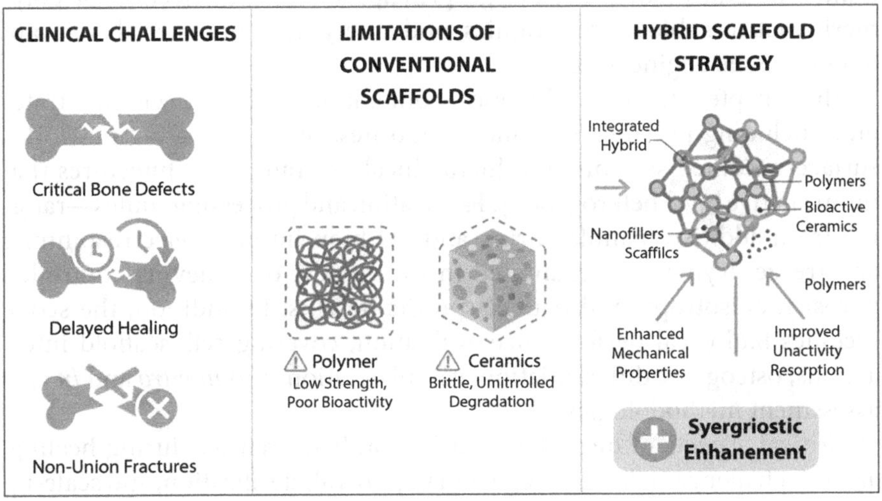

Figure 10.1 Hybrid scaffolding concept and clinical rationale.

inspired by bone's hierarchical structure, where nanometer-sized hydroxyapatite crystals are precisely organized within collagen fibrils, creating nanomechanical heterogeneities that enable high energy dissipation and fracture resistance [11]. Recent developments focus on osteoconductive, elastomeric, photoluminescent, and electroactive hybrid polymers that can be tailored through controlled synthesis to regulate cellular behavior and incorporate biomolecule release capabilities [12]. Figure 10.1 illustrates the clinical challenges associated with bone defects and the limitations of conventional single-phase scaffolds, underscoring the need for advanced regenerative strategies. It depicts how hybrid scaffolds integrating polymers, ceramics, and nanofillers synergistically enhance mechanical performance, bioactivity, and controlled resorption.

10.1.4 Scope of this Chapter

This chapter focuses on the design, development, and translational potential of advanced hybrid scaffolds for bone regeneration, with particular emphasis on resorbable composite systems. It systematically covers the scientific rationale for scaffold hybridization and its role in overcoming the mechanical, biological, and degradation-related limitations of conventional single-phase biomaterials. The scope encompasses fundamental structure–property–function relationships governing hybrid scaffolds, multiscale and biomimetic design principles, and the selection of polymeric matrices, bioactive ceramics, and emerging nanomaterials relevant to bone tissue engineering.

The chapter further addresses contemporary hybridization strategies, including polymer–ceramic composites, interpenetrating networks, surface-modified systems, and hierarchical or gradient architectures that mimic native bone heterogeneity. Fabrication and processing routes—ranging from additive manufacturing and 3D bioprinting to electrospinning and freeze-drying—are examined in the context of achieving controlled porosity, anisotropy, and patient-specific designs. In addition, the scope includes biological performance evaluation, covering cell–scaffold interactions, osteogenic differentiation, vascularization, and *in vitro* and *in vivo* assessment methodologies.

Critical aspects of degradation behavior, load transfer during healing, and key challenges related to regulatory approval, sterilization, and scale-up are also discussed to bridge the gap between laboratory research and clinical application. Finally, the chapter outlines future perspectives, including smart and stimuli-responsive scaffolds, data-driven and personalized design approaches, and emerging translational roadmaps, providing a comprehensive reference for researchers, clinicians, and engineers engaged in next-generation bone regenerative therapies.

10.2 Design Principles of Hybrid Bone Scaffolds

10.2.1 Structure–Property Relationships in Hybrid Systems

Hybrid bone scaffolds demonstrate complex structure-property relationships that enable precise control over mechanical and biological performance. Peng *et al.* showed that hollow-strut architectures decouple mechanical and mass-transport properties, achieving elastic moduli ranging from 1428.6-3924.4 MPa while maintaining consistent permeability ($11.08\text{-}12.35 \times 10^{-9}$ m^2), providing design freedom for optimizing multi-physical properties [13]. Fan *et al.* resolved the strength-toughness conflict in silica-polymer hybrids by controlling noncovalent intermolecular interactions, achieving optimized mechanical properties including 1.75 MPa strength, 31% elongation, and excellent biocompatibility [14]. Martin *et al.* emphasized that sol-gel hybrid scaffolds incorporate polymers at the molecular level, enabling control over mechanical properties and degradation rates through hierarchical structures spanning atomic to macroporous scales [15]. Barthelat demonstrated that architectural materials achieve unusual property combinations through controlled building block arrangements and interfacial behaviors, with natural materials like bone exemplifying how structure-property relationships generate exceptional stiffness, strength, and toughness combinations [16].

10.2.2 Mechanical, Biological, and Degradation Requirements

Hybrid scaffolds for bone tissue engineering must balance mechanical strength with biological functionality to achieve successful regeneration [17]. Shalumon *et al.* demonstrated that combining PLGA/nanohydroxyapatite microspheres with gelatin/nanohydroxyapatite cryogels creates scaffolds with 25-fold increased ultimate stress while promoting osteogenic differentiation and successfully repairing rabbit tibia defects [18]. The dual responsibility of providing load-bearing capability while interacting with the extracellular matrix presents significant challenges for synthetic scaffolds [19]. Aldemir Dikici *et al.* showed that decorating multiscale porous polycaprolactone scaffolds with *in vitro* generated extracellular matrix significantly enhanced both osteogenic and angiogenic performance

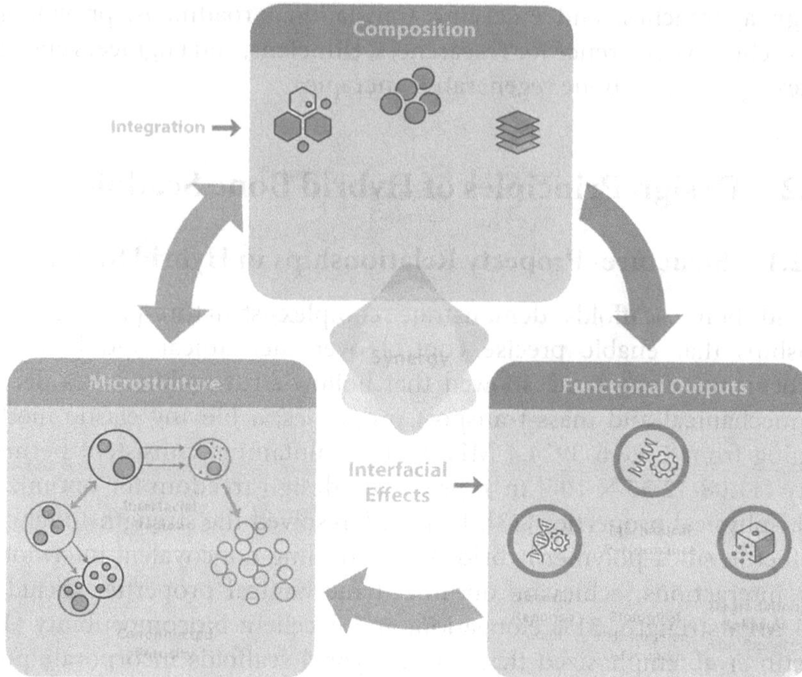

compared to polymer-only scaffolds [20]. Hybrid constructs incorporating multiple components from native bone matrix enhance the osteogenicity of otherwise non-osteogenic biocompatible materials [21]. These approaches demonstrate that multi-material hierarchical structures can effectively mimic natural bone tissue properties, providing both structural integrity and biological functionality essential for bone healing. Figure 10.2 depicts the structure–property–function relationships governing hybrid bone scaffolds, linking composition and microstructural features to functional outcomes. It highlights the role of interfacial interactions and synergistic effects in optimizing mechanical, biological, and degradation behavior.

10.2.3 Multiscale Design Considerations for Bone-Mimetic Scaffolds

Bone-mimetic scaffold design requires hierarchical optimization across multiple scales to achieve effective tissue engineering outcomes. Recent advances demonstrate that scaffolds must incorporate biomimetic features ranging from molecular composition to macro-scale architecture [22].

Multiscale porosity is critical, with macropores (>100 μm) enabling cellular infiltration, micropores (1-10 μm) promoting cell adhesion and nutrient transport, and nanopores (1-50 nm) enhancing cell-material interactions through increased surface area [23]. Computational modeling approaches enable optimization of these hierarchical structures, with first-principles calculations informing nanoscale properties that are then integrated into microscale and macroscale designs [24]. Image-based homogenization optimization can resolve conflicting design requirements, such as balancing mechanical support with porosity for cell delivery, while ensuring scaffolds match native bone properties [25]. This multiscale approach successfully creates scaffolds that mimic natural bone's hierarchical organization and functional requirements. Table 10.1 summarizes the key biological, structural, and mechanical requirements that govern the rational design of hybrid bone scaffolds for effective regeneration.

Table 10.1 Critical biological, structural, and mechanical attributes of bone tissue engineering scaffolds.

Category	Scaffold attribute	Reformulated description	Functional outcome	Ref.
Biological	Biocompatibility	Materials must exhibit cytocompatibility, enabling cellular adhesion, proliferation, and normal metabolic activity without eliciting toxic or inflammatory responses.	Supports cell survival and harmonious integration with host tissue.	[26]
Biological	Controlled Biodegradation	The scaffold should progressively resorb in synchrony with new bone formation, avoiding long-term foreign body persistence.	Facilitates gradual replacement by native bone tissue.	[27]

(*Continued*)

Table 10.1 Critical biological, structural, and mechanical attributes of bone tissue engineering scaffolds. (*Continued*)

Category	Scaffold attribute	Reformulated description	Functional outcome	Ref.
Biological	Osteoconductive Nature	Surface chemistry and composition should favor bone cell attachment and guided migration along the scaffold framework.	Promotes extracellular matrix deposition and mineralized tissue formation.	[28]
Biological	Osteoinductive Capability	Scaffold architecture and chemistry should stimulate progenitor cells toward osteogenic differentiation.	Enhances new bone generation and maturation.	[29]
Biological	Angiogenic Support	Design must enable vascular infiltration to ensure oxygen and nutrient transport within the regenerating tissue.	Improves tissue viability and regenerative efficiency.	[30]
Structural	Porosity and Pore Architecture	An interconnected porous network with optimized pore dimensions is required for cellular infiltration and mass transport.	Enables efficient nutrient diffusion and tissue ingrowth.	[31]
Structural	Interconnectivity	Continuous pore connectivity should be ensured to permit cell migration and waste removal throughout the scaffold.	Prevents hypoxic regions and supports homogeneous regeneration.	[32]

(*Continued*)

Table 10.1 Critical biological, structural, and mechanical attributes of bone tissue engineering scaffolds. (*Continued*)

Category	Scaffold attribute	Reformulated description	Functional outcome	Ref.
Structural	Hierarchical Architecture	Multiscale structural organization should emulate native bone morphology from micro- to macro-levels.	Guides organized tissue development and load distribution.	[33]
Structural	Surface Topography	Micro- and nanoscale surface features should be engineered to modulate cell–material interactions.	Enhances adhesion, proliferation, and osteogenic signaling.	[34]
Mechanical	Elastic Modulus Matching	Scaffold stiffness should approximate that of host bone to ensure physiological load sharing.	Prevents stress shielding and supports remodeling.	[26]
Mechanical	Mechanical Strength	Adequate compressive and tensile strength must be maintained during early healing stages.	Ensures structural integrity under physiological loads.	[35]
Mechanical	Fatigue Resistance	The material should withstand cyclic mechanical loading without premature failure.	Maintains long-term functionality during regeneration.	[36]
Mechanical	Fracture Toughness	Resistance to crack initiation and propagation is necessary to avoid catastrophic failure.	Improves durability and clinical reliability.	[37]

10.3 Material Platforms for Hybrid Scaffold Fabrication

10.3.1 Polymeric Matrices for Resorbable Hybrid Scaffolds

Biodegradable polymeric matrices serve as crucial structural frameworks in tissue engineering applications, offering controllable physicochemical properties that directly influence scaffold performance. Biopolymers provide essential control over structure, morphology, and chemistry while serving as extracellular matrix mimics, with their inherent biodegradability helping regulate cell and tissue remodeling rates [38, 39]. The selection of natural versus synthetic polymers requires careful consideration of cellular environments and interactions, as scaffold materials must balance biocompatibility, biodegradability, mechanical strength, and porosity [40]. Polymer blend miscibility significantly impacts physicochemical properties and biological responses, with immiscible blends generally promoting better cell adhesion and proliferation, while compatible blends enhance cellular infiltration [41, 42]. These biodegradable scaffolds function as temporary templates providing physical and biochemical signals to cells, with their time-dependent property changes introducing a fourth dimension to tissue engineering that becomes increasingly complex in composite scaffold systems [43, 44].

10.3.2 Bioactive Ceramics and Inorganic Reinforcements

Bioactive ceramics, particularly calcium phosphate-based materials, serve as crucial inorganic reinforcements in polymer composites for bone tissue engineering applications. These materials enhance osteoconductivity, stiffness, and mineralization potential while promoting bone-implant interfacial bonding [45, 46]. Polycaprolactone (PCL) composites incorporating calcium phosphate ceramics and bioactive glasses demonstrate remarkably improved mechanical properties, controllable degradation rates, and enhanced bioactivity compared to pure polymers [47]. The combination addresses inherent limitations of individual materials: while bioceramics provide bioactivity and bone-bonding capability, they suffer from brittleness and low fracture toughness, whereas polymers offer better mechanical properties but lack appropriate surface characteristics for bone integration [48]. These polymer-ceramic composites can be fabricated into three-dimensional porous structures that closely match the desirable properties of autograft tissue, including osteoconductivity and osteoinductivity, making them attractive candidates for synthetic bone graft substitutes [46].

10.3.3 Emerging Nanomaterials and Bioinspired Fillers

Emerging nanomaterials enable precise control over scaffold functionality by mimicking native bone's hierarchical nanostructure, which comprises organic collagen and inorganic nano-hydroxyapatite components [49–51]. These bionic nanostructures can be engineered with tailored properties, including stiffness, pore size, nanomorphology, and biochemical functionalities to control stem cell behavior with spatial and temporal precision [52]. Nanostructured scaffolds provide enhanced surface area and wettability compared to conventional materials, leading to increased protein adsorption and improved cell-scaffold interactions mediated by

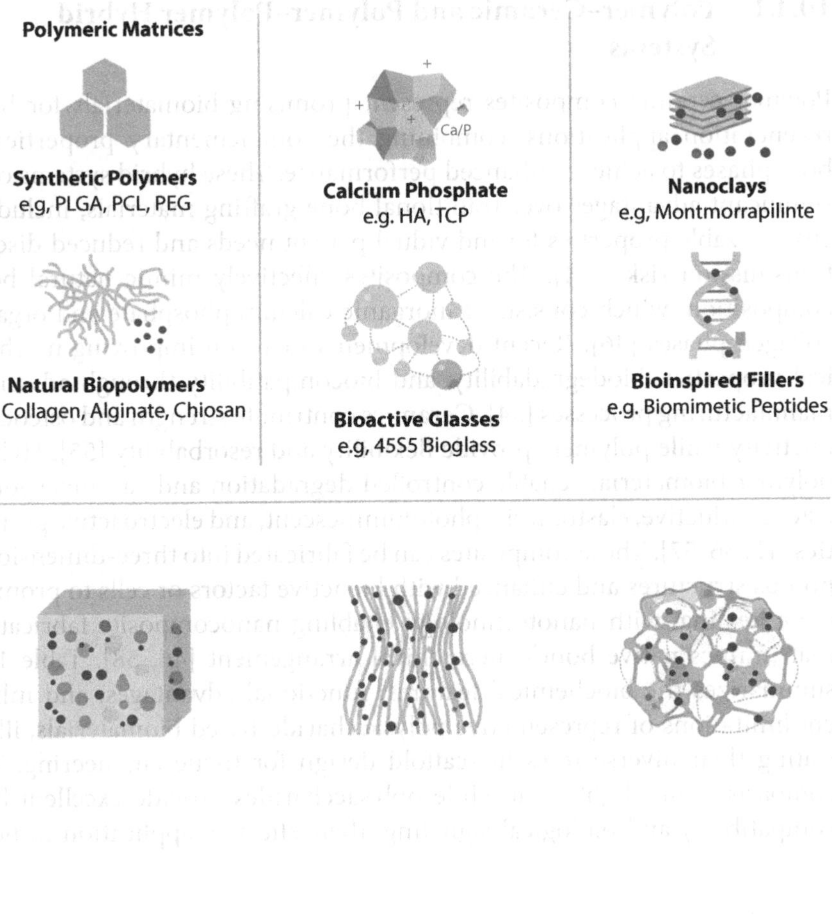

integrin-triggered signaling pathways [50]. Biomimetic nanocomposites, particularly those combining type I collagen with nano-hydroxyapatite, demonstrate remarkable osteoinductive properties, stimulating human mesenchymal stem cells to rapidly produce bone minerals even without osteogenic supplements [53]. These materials regulate cell proliferation, differentiation, and migration, resulting in functional tissue formation and accelerated bone matrix development within one to two weeks *in vivo* [51, 53]. Figure 10.3 presents the principal material platforms employed in hybrid scaffold fabrication, including polymeric matrices, bioactive ceramics, and emerging nanofillers. It further outlines how these components are combined into distinct hybrid architectures to achieve tailored scaffold performance.

10.4 Hybridization Strategies and Architectures

10.4.1 Polymer–Ceramic and Polymer–Polymer Hybrid Systems

Polymer-ceramic composites represent promising biomaterials for bone regeneration applications, combining the complementary properties of both phases to achieve enhanced performance. These hybrid systems offer significant advantages over traditional bone grafting materials, including customizable properties for individual patient needs and reduced disease transmission risks [54]. The composites effectively mimic natural bone composition, which consists of inorganic calcium phosphate and organic collagen phases [46]. Recent developments focus on improving mechanical properties, biodegradability, and biocompatibility through advanced manufacturing processes [54]. Ceramics contribute strength and osteoconductivity while polymers provide flexibility and resorbability [55]. Hybrid polymer biomaterials enable controlled degradation and can incorporate osteoconductive, elastomeric, photoluminescent, and electroactive properties [12, 56, 57]. These composites can be fabricated into three-dimensional porous structures and enhanced with bioactive factors or cells to promote bone healing, with nanotechnology enabling nanocomposite fabrication that mimics native bone's hierarchical arrangement [46, 58]. Table 10.2 summarizes the biochemical features, functional advantages, and inherent limitations of representative polysaccharide-based biomaterials, illustrating their diverse roles in scaffold design for tissue engineering. The comparison highlights that while polysaccharides provide excellent biocompatibility and biological signaling, their effective application in bone

Table 10.2 Biochemical characteristics, advantages, and limitations of polysaccharide-based biomaterials for tissue engineering.

Polysaccharide material	Key biochemical and functional attributes	Primary limitations and challenges	Ref.
Sodium alginate	Readily forms ionically cross-linked hydrogels (e.g., *via* divalent cations); exhibits good biocompatibility, biodegradability, and low cytotoxicity	Requires chemical or biological modification to tailor functionality; inherently weak cell-adhesive properties	[59, 60]
Hyaluronic acid	Endogenous component of the extracellular matrix; highly hydrophilic; capable of hydrogel formation through chemical crosslinking	Rapid degradation under physiological conditions; necessitates crosslinking or blending for structural stability	[61, 62]
Chitosan	Forms hydrogels in mildly acidic environments; biocompatible, biodegradable, and antimicrobial; supports cellular attachment	Limited solubility at neutral pH; often requires functionalization to improve bioactivity and processability	[63]
Dextran	Hydrophilic and biocompatible polysaccharide; can be converted into hydrogel networks following chemical modification	Rapid clearance *in vivo*; structural modification needed to prolong residence time and functionality	[64, 65]
Heparin	Frequently combined with other polymers; stabilizes and binds growth factors; biocompatible and chemically versatile	Risk of immunogenic responses; relatively fast degradation in physiological environments	[66, 67]

(Continued)

Table 10.2 Biochemical characteristics, advantages, and limitations of polysaccharide-based biomaterials for tissue engineering. (*Continued*)

Polysaccharide material	Key biochemical and functional attributes	Primary limitations and challenges	Ref.
Cellulose derivatives	Biocompatible and readily modifiable polysaccharide derivatives with good physicochemical stability	Low intrinsic bioactivity; requires blending with bioactive materials to enhance cell adhesion	[68, 69]
kappa-Carrageenan	Gel-forming polysaccharide with high biocompatibility; reported antioxidant, anticoagulant, antitumor, and immunomodulatory effects	Variable degradation behavior; potential to induce immune reactions depending on formulation	[70, 71]
Pullulan	Water-soluble, non-toxic, and biodegradable polysaccharide suitable for hydrogel and film formation	Insufficient mechanical robustness for load-bearing applications without reinforcement	[72]
Agarose	Exhibits thermoreversible gelation; biocompatible and structurally stable under mild conditions	Limited biological signaling capability; typically combined with other biomaterials for enhanced performance	[73]
Pectin	Biodegradable and non-toxic polysaccharide with gel-forming ability under specific conditions	Poor mechanical integrity; chemical modification often required to improve functionality	[74, 75]
Starch	Abundant, low-cost, biocompatible, and biodegradable natural polymer	Weak mechanical properties and limited bioactivity; requires modification or blending for biomedical use	[76, 77]

regeneration typically requires hybridization or chemical modification to achieve adequate mechanical stability and controlled degradation.

10.4.2 Surface-Modified and Interpenetrating Network Scaffolds

Surface modifications of biomaterial scaffolds significantly enhance their biological functionality and interfacial properties for tissue engineering applications. Aliphatic polyester scaffolds, while offering desirable mechanical properties and biocompatibility, require surface modifications to overcome limitations such as hydrophobicity and lack of biochemical binding sites [78]. Multiple modification approaches, including topographical alterations, protein adsorption, mineral coating, and functional group incorporation, improve cell-surface interactions and protein attachment [79]. For multicomponent biopolymer scaffolds, interfacial bonding mechanisms such as interface phase introduction, surface modification, and *in situ* growth strategies are crucial for enhancing mechanical properties and preventing compatibility issues between different components [80]. These surface modifications create optimized microenvironments that promote cellular adhesion, growth, proliferation, and differentiation while maintaining scaffold biocompatibility and structural integrity [81]. Understanding scaffold interfacial properties is essential for advancing tissue engineering toward successful clinical applications.

10.4.3 Hierarchical and Gradient Hybrid Scaffold Architectures

Hierarchical and gradient scaffold architectures represent a promising approach for bone tissue engineering by mimicking native bone's structural heterogeneity. Di Luca *et al.* developed scaffolds with radial porosity gradients, featuring outer regions with 29.6% porosity (500 μm pores) transitioning to inner regions with 77.6% porosity (1000 μm pores), successfully replicating bone's cortical-to-cancellous transition and promoting osteogenic differentiation [82]. Egan *et al.* demonstrated that hierarchical scaffolds with large central pores increase porosity while affecting elastic moduli and surface area, emphasizing the importance of balancing mechanical and biological factors in scaffold design [83]. Fang *et al.* reviewed current strategies for creating compositional and structural gradients in scaffolds for musculoskeletal tissue regeneration, highlighting their potential for improved therapeutic outcomes [84]. Daskalakis *et al.* introduced "bone bricks" with nonuniform pore dimensions and gradient designs, achieving superior mechanical properties and biological characteristics compared to conventional uniform structures, demonstrating the effectiveness of anatomically-inspired architectures [85]. Table 10.3 highlights the breadth of natural polymer–based scaffold systems

Table 10.3 Natural polymer–based scaffold systems and their applications in tissue engineering.

Material class	Scaffold composition and form	Target tissue/ application	Incorporated bioactive factors	Cell type utilized	Ref.
Collagen-derived scaffolds	Collagen–hydroxyapatite composite matrices	Bone regeneration	Nerve growth factor	—	[86]
Collagen-derived scaffolds	Porous collagen sponge structures	Cartilage repair	Basic fibroblast growth factor	Chondrocytes	[87]
Collagen-derived scaffolds	Electrospun collagen nanofiber mats	Bone tissue engineering	—	Adult bone marrow–derived mesenchymal stem cells	[88]
Collagen-derived scaffolds	Collagen-based barrier membranes	Cartilage regeneration	—	Chondrocytes	[88]
Collagen-derived scaffolds	Injectable collagen gel systems	Bone and cartilage repair	Bone morphogenetic protein-2 (BMP-2)	Bone marrow stromal cells	[89]
Chitosan-based scaffolds	Chitosan–collagen composite constructs	Periodontal bone regeneration	Transforming growth factor-β1 plasmid	Human periodontal ligament cells	[90]
Chitosan-based scaffolds	Chitosan–coral hybrid scaffolds	Periodontal bone regeneration	PDGFB plasmid	Human periodontal ligament cells	[91]
Chitosan-based scaffolds	Particle-aggregated chitosan frameworks	Bone and cartilage regeneration	—	Human adipose-derived stem cells	[92]

employed in tissue engineering, demonstrating how variations in scaffold composition, architecture, and incorporated bioactive cues enable targeted regeneration of bone, cartilage, vascular, and ligament tissues. The summarized studies underscore the versatility of collagen, chitosan, fibrin, and silk fibroin platforms in supporting cell-specific responses and enhancing tissue-specific healing outcomes.

10.5 Fabrication and Processing Techniques

10.5.1 Additive Manufacturing and 3D Bioprinting Approaches

Additive manufacturing (AM) techniques have revolutionized tissue engineering by enabling precise control over scaffold geometry, composition, and internal architecture. These layer-by-layer manufacturing processes, including stereolithography, selective laser sintering, and 3D printing, allow for complex shapes and microstructures with high accuracy and reproducibility [93, 94]. AM addresses major challenges in developing bioactive, patient-specific scaffolds with controlled pore architecture, optimized porosity, and microstructure that support cell proliferation, adhesion, and differentiation [95]. Computer-aided design systems enable customized, image-based scaffolds with spatial patterning of internal architecture, while 3D bioprinting enhances functionalization through incorporation of viable cells and bioactive molecules [96]. The technology provides precise control over the internal structure of porous materials, enabling specialized scaffold development and personalized implants that improve patient comfort and outcomes in orthopedic and dental applications [97, 98].

10.5.2 Electrospinning, Freeze-Drying, and Solvent-Based Methods

Electrospinning, freeze-drying, and solvent-based techniques are fundamental methods for fabricating porous scaffolds in tissue engineering applications. These techniques enable the creation of three-dimensional structures with high porosity (93-98%) and interconnected pore networks essential for cell proliferation, migration, and nutrient transport [99, 100]. Freeze-drying processes can produce highly porous structures with controllable morphology, where freezing temperature affects final scaffold architecture [100]. Solution-based processing methods, including freeze-drying, phase separation techniques, and electrospinning, offer controlled fabrication parameters that allow tuning of pore size distribution, porosity, and mechanical properties [101]. Hybrid approaches combining multiple techniques demonstrate particular promise, as shown by electrospinning-freeze-drying combinations that create scaffolds with both microfibers and macroporosity, mimicking natural extracellular matrix architecture while achieving favorable mechanical properties and enhanced cell proliferation [102].

10.5.3 Advanced Processing for Controlled Porosity and Anisotropy

Advanced processing strategies enable precise control of scaffold architecture to mimic native bone structure and enhance regenerative outcomes. Three-dimensional plotting techniques can produce scaffolds with controlled pore size gradients, improving cell seeding efficiency from ~35% to ~70% and promoting more homogeneous cell distribution compared to uniform pore architectures [103]. Anatomically designed scaffolds with nonuniform pore dimensions and gradient designs achieve superior mechanical properties and biological characteristics when incorporating ceramic materials [85]. Optimal pore sizes for bone formation range from 390-590 μm, with bimodal pore topologies—alternating large and small pores—substantially enhancing bone volume and functionality [104]. Table 10.4 presents a comparative overview of widely adopted fabrication techniques, associated material systems, and their reported biological and clinical performance in bone regeneration applications. Pore size influences osteogenic pathways, where small pores favor hypoxic conditions leading to osteochondral formation, while larger, well-vascularized pores promote direct osteogenesis. Gradient pore architectures are recommended for multi-tissue formation and interface development [105]. Figure 10.4 summarizes the fabrication and processing strategies used to generate hierarchical and anisotropic hybrid scaffolds. It demonstrates how different manufacturing techniques influence pore architecture, orientation, and structural organization to mimic native bone.

Table 10.4 Fabrication strategies, material platforms, and reported clinical outcomes in bone regeneration.

Fabrication approach	Representative material systems	Distinctive technical attributes	Reported biological/ clinical performance	Ref.
Additive Manufacturing (3D Printing)	Biodegradable polymers (PCL, PLA, PLGA), calcium phosphate ceramics (HA, β-TCP, BCP), and titanium-based metallic systems	Allows accurate regulation of scaffold architecture, pore geometry, and patient-specific defect matching while enabling multimaterial combinations	Customized polymeric and ceramic constructs improved alveolar bone regeneration, enhanced implant stability, and promoted reliable osseointegration in translational studies	[106–108]
Electrospinning	Natural polymers (collagen, gelatin), synthetic polymers (PCL), polymer–ceramic nanocomposites	Generates fibrous matrices that closely emulate the extracellular matrix with high surface area and controllable fiber diameters	Nanofibrous hybrid scaffolds supported osteoblast attachment and proliferation and demonstrated suitability for localized drug or growth factor delivery	[109–111]
Freeze-Drying (Lyophilization)	Natural and synthetic polymer matrices, polymer–ceramic composites	Produces highly porous, interconnected structures with pore morphology dictated by freezing kinetics and sublimation conditions	Cell-integrated composite scaffolds facilitated vascular infiltration and substantial new bone formation while preserving the biological activity of incorporated agents	[112–114]
Solvent Casting and Particulate Leaching	Biodegradable polymers such as PCL and PLGA	A reproducible and scalable approach where pore size and overall porosity are governed by porogen characteristics and polymer concentration	Resulting scaffolds exhibited good cytocompatibility and notable bone ingrowth in animal models, making the method suitable for preliminary scaffold development	[112, 115]

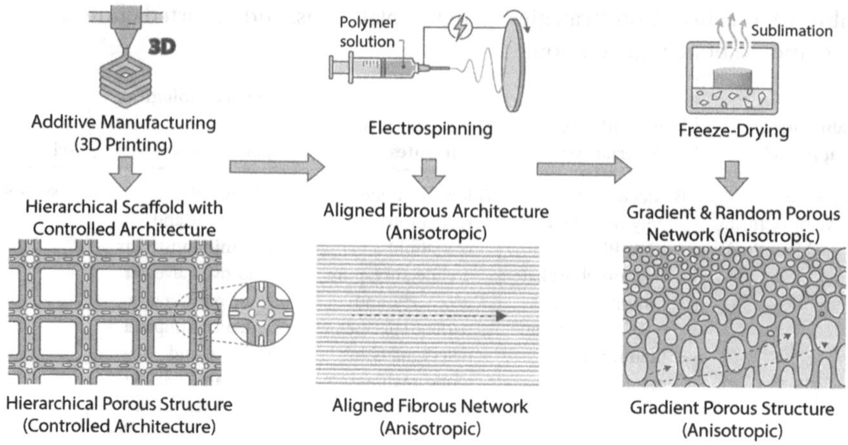

10.6 Biological Performance and Functional Evaluation

10.6.1 Cell–Scaffold Interactions and Osteogenic Differentiation

Cell-scaffold interactions are fundamental to bone tissue engineering success, with scaffold properties directly influencing cellular behavior and osteogenic outcomes. Scaffold architecture, particularly pore size, affects cell attachment and invasion, while different materials exert varying biomechanical forces that influence cellular differentiation and migration in a cell type-specific manner [116]. Surface chemistry plays a critical role in modulating osteoblastic differentiation and matrix mineralization independently of cell proliferation effects [117]. Integrin binding specificity serves as a key mechanism regulating these surface chemistry-dependent cellular responses, with specific integrin-fibronectin interactions controlling differentiation outcomes [117]. Biomimetic surface engineering approaches that target integrin receptors can activate signaling pathways directing osteoblast differentiation programs [118]. The synergistic combination of biochemical and physical cues proves most effective for regulating cell adhesion, proliferation, osteogenic differentiation, and mineralization compared to single-cue approaches [119].

10.6.2 Vascularization and Integration with Host Tissue

Effective vascularization is critical for successful tissue engineering and scaffold integration. Research demonstrates that pre-seeded osteoblasts significantly enhance scaffold vascularization by producing matrix and soluble factors that instruct host endothelial cells to migrate and proliferate, with this effect occurring in concert with pro-angiogenic stimuli from host immune cells [120]. Poor scaffold perfusion leads to lower success rates, necessitating improved vascular components for nutrient delivery, oxygen transport, and waste elimination [121]. Structural modifications play a crucial role, as channels in porous scaffolds promote cell growth and rapid vascularization, preventing necrotic core formation in large scaffolds that lack perfusable networks [122]. Electrospun polymer and hybrid scaffolds can be designed with specific architectural features and porosity to enhance cellular penetration, proliferation, and vascularization by mimicking the extracellular matrix, though insufficient vascularization remains a primary limitation in bone organoids [123].

10.6.3 *In Vitro* and *In Vivo* Assessment Methodologies

Research demonstrates that comprehensive *in vitro* and *in vivo* evaluation methodologies are essential for assessing bone regenerative biomaterials. *In vitro* assays evaluate cytocompatibility through cell proliferation, viability, and differentiation studies, with scanning electron microscopy widely used to assess cell morphology and adhesion [124, 125]. Standardized tests like MTT assays and alkaline phosphatase activity measurements provide critical pre-screening before *in vivo* studies [126, 127]. Animal models validate translational performance by assessing osteoinductivity and degradation behavior in physiological conditions. Studies show that materials like PLA/hydroxyapatite composites demonstrate enhanced osteogenic differentiation *in vitro* and successful heterotopic bone formation in canine models [126]. Systematic evaluation protocols following international standards ensure material safety and efficacy before clinical translation [127]. However, standardization of biocompatibility assessment methods remains crucial to avoid inconsistencies and bias between studies [125]. Table 10.5 summarizes the key scaffold-related, *in vitro*, and *in vivo* parameters that collectively regulate the rate and mode of bioceramic biodegradation. Scaffold-intrinsic factors such as fabrication route, pore architecture, surface roughness, and surface-area-to-volume ratio directly influence fluid penetration, ion exchange, and dissolution kinetics, thereby dictating early-stage degradation behavior. Figure 10.5 outlines the

Table 10.5 Parameters governing the biodegradation behavior of bioceramic scaffolds.

Parameter category	Key influencing factors
Scaffold-intrinsic characteristics	Fabrication method and processing conditions; macroscopic scaffold dimensions; pore size distribution and architecture beyond overall porosity; surface roughness and topographical features; surface-area-to-volume ratio; incorporation of additives, dopants, or unintended impurities
In vitro environmental conditions	Physicochemical properties of the surrounding medium, including pH and ionic strength; type of cells present and their population density; applied or cell-induced mechanical loading; composition of the biological or simulated culture medium
In vivo physiological conditions	Degree of tissue formation and remodeling at the implant site; physiological mechanical stresses; local enzymatic activity and concentration; anatomical location and orientation of implantation

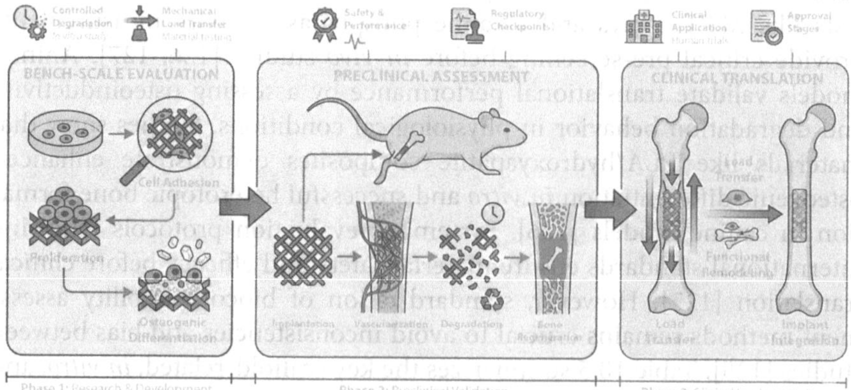

biological evaluation, degradation behavior, and translational pathway of hybrid bone scaffolds from bench-scale studies to clinical application. It integrates *in vitro* assessment, preclinical validation, load transfer during healing, and regulatory considerations toward clinical implementation.

10.7 Degradation Behavior and Clinical Translation Considerations

10.7.1 Degradation Kinetics and Byproduct Biocompatibility

Research demonstrates that matching scaffold degradation kinetics with bone formation rates is critical for successful bone regeneration. Chen *et al.* developed elastomer-toughened composite scaffolds from poly(glycerol sebacate) and Bioglass® that exhibit mechanically steady states with three-stage degradation profiles (lag, log, and plateau phases), providing essential structural support during the initial healing period [128]. Bossard *et al.* created polycaprolactone/bioactive glass hybrid scaffolds showing suitable degradation rates (13.2% weight loss after 8 weeks) that matched bone regeneration timelines while maintaining biocompatibility and avoiding cytotoxicity [129]. Park *et al.* demonstrated that scaffold degradation rates directly impact cellular metabolism and osteogenesis, with rapidly degrading silk scaffolds supporting higher glucose consumption and enhanced mineralized extracellular matrix formation [130]. Ozbolat *et al.* emphasized the importance of controlling degradation profiles and permeability to synchronize release kinetics with tissue regeneration requirements, highlighting the need for spatiotemporal control in hybrid scaffold design [131].

10.7.2 Load Transfer and Mechanical Integrity During Healing

Effective load transfer between scaffolds and regenerating bone tissue is essential for successful bone tissue engineering, as mechanical forces play a crucial role in bone regeneration through mechanotransduction [132]. However, scaffold design faces conflicting requirements: providing temporary mechanical support while maintaining porosity for cell delivery and tissue ingrowth [25]. To address these challenges, researchers have developed hybrid scaffold approaches that combine mechanical integrity with biological functionality. Entezari *et al.* demonstrated a modular design strategy using fenestrated polymeric shells with ceramic pillars that increased alginate hydrogel failure loads by 3200-fold in compression while generating osteoinductive strains of 0.2-0.4% [133]. Similarly, Igwe *et al.* developed polymer-hydrogel hybrid scaffolds that synergistically combined load-bearing polymer matrices with osteoinductive peptide hydrogels, showing enhanced osteoblastic gene expression [134]. These hybrid approaches enable gradual stress transfer to newly formed tissue while maintaining scaffold mechanical integrity throughout the regeneration process.

10.7.3 Regulatory, Sterilization, and Scale-Up Challenges

Clinical translation of scaffold-based tissue engineering faces significant barriers despite decades of research and substantial funding [135]. The primary challenges include regulatory approval requirements, particularly design control processes that demand precise clinical indication definition and systematic verification through sequential testing phases [136]. Current scaffolds exhibit technical limitations, including insufficient vascularization, inadequate mechanical strength, and osseointegration issues [137]. Manufacturing scalability and reproducibility present additional hurdles, as regulatory frameworks require simple, reproducible solutions that can be consistently produced [138]. The transition from academic research paradigms, which focus on discovery and generality, to targeted clinical applications requires addressing specific market needs and obtaining private investment [135]. Successful translation necessitates integrated technology platforms rather than single-technology approaches, with modular scaffold systems offering promising pathways to overcome these multifaceted technical, business, and regulatory challenges [136].

10.8 Future Perspectives and Emerging Trends

10.8.1 Smart and Stimuli-Responsive Hybrid Scaffolds

Smart and stimuli-responsive materials represent an emerging class of biomaterials designed to dynamically respond to various triggers for tissue engineering and drug delivery applications [139]. These materials can respond to diverse stimuli, including temperature, pH, redox-state, light, magnetic fields, enzymes, and electromagnetic energy, enabling controlled drug release with improved spatiotemporal specificity [139, 140]. Stimuli-responsive scaffolds offer significant advantages over static systems by recapitulating the dynamic nature of native extracellular matrices through on-demand manipulation of cell microenvironments [141].

These biomaterials can deliver temporal and spatial cues to stem cells through multiple pathways, including biochemical signaling, scaffold property changes, drug release, mechanical stress, and electrical stimulation [142]. Integration with biofabrication technologies enables the development of hierarchical scaffolds with organizational features similar to native tissues, allowing multiple spatiotemporal modulation of cell-directing cues for controlled tissue repair and remodeling [139, 141].

10.8.2 Patient-Specific and Data-Driven Scaffold Design

Patient-specific scaffold design represents a transformative approach in regenerative medicine, integrating medical imaging, computational modeling, and additive manufacturing to create tailored therapeutic solutions. This methodology enables precise replication of complex anatomical geometries by utilizing CT or MRI data to define scaffold topology through voxel density distribution [143]. The approach has been successfully applied across diverse applications, from craniofacial reconstructive surgery to meniscal repair, where contralateral anatomy serves as a reference for creating anatomically precise scaffolds [144]. Additive manufacturing technologies provide unprecedented freedom in scaffold design and composition, allowing customization of geometries, mechanical properties, and biological responses to match individual patient anatomy and biology [145]. The integration of *in silico* modeling, omics data, and bioinformatics with additive manufacturing creates synergistic possibilities for personalized tissue regeneration strategies [146]. This interdisciplinary approach, requiring collaboration between patients, clinicians, and engineers, promises enhanced anatomical fit, mechanical compatibility, and improved regenerative outcomes in precision medicine applications.

10.8.3 Translational Roadmap toward Next-Generation Bone Implants

The translational roadmap for next-generation bone implants requires comprehensive multidisciplinary integration spanning materials science, biology, and clinical validation. Current bone tissue engineering faces significant translation barriers despite decades of research and substantial funding, with limited clinical success [135, 147]. Key challenges include regulatory approval complexities, lack of standardized guidelines, and gaps in translational research that hinder clinical implementation of advanced technologies like 3D-printed patient-specific implants [148].

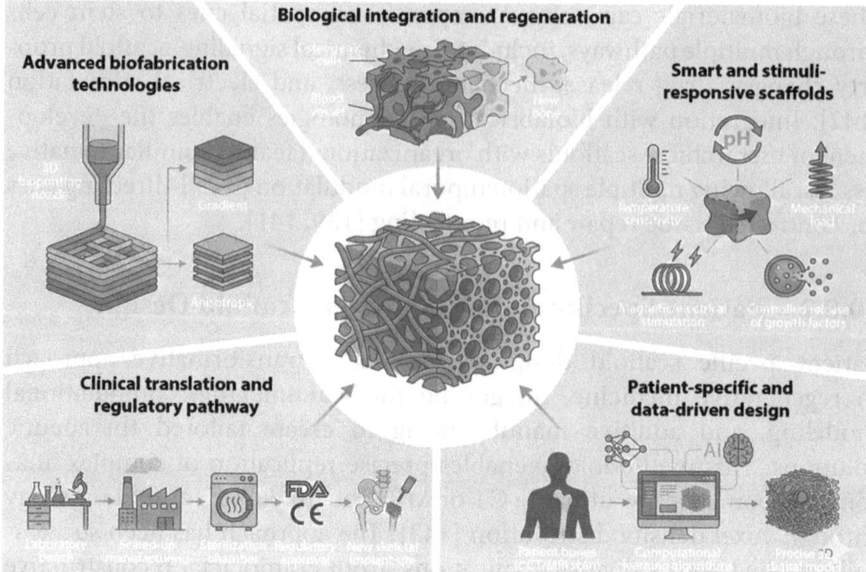

Successful translation demands addressing technical challenges in scaffold design, manufacturing, and functionalization while simultaneously tackling business considerations including regulatory pathways, market needs, and private investment requirements [135]. The field requires strengthened collaboration among policymakers, regulatory agencies, and clinicians to overcome implementation barriers [148]. Emerging intelligent therapeutic platforms integrating artificial intelligence and machine learning show promise for accelerating material discovery and clinical translation, though scalability, reproducibility, and regulatory approval remain persistent challenges [149]. Figure 10.6 illustrates the convergence of smart stimuli-responsive materials, patient-specific data-driven design, advanced biofabrication, and translational frameworks that collectively define next-generation hybrid scaffolds for personalized and clinically relevant bone regeneration.

10.9 Conclusions

This chapter has comprehensively examined advanced hybrid scaffolding strategies for bone regeneration, emphasizing the pivotal role of material hybridization in addressing the multifaceted challenges associated with critical-sized bone defects. Conventional single-component scaffolds are inherently limited by trade-offs between mechanical strength, bioactivity, and degradation behavior. Hybrid scaffold systems, integrating biodegradable polymers with bioactive ceramics, nanomaterials, and bioinspired fillers, provide synergistic solutions that more closely replicate the hierarchical structure and functional complexity of native bone tissue.

The discussion highlights that rational control of structure–property–function relationships across multiple length scales is central to effective scaffold design. Hierarchical and gradient architectures, coupled with tailored interfacial interactions, enable simultaneous optimization of load-bearing capacity, cellular infiltration, vascularization, and controlled resorption. Advances in fabrication technologies, particularly additive manufacturing, electrospinning, and hybrid processing approaches, have significantly expanded the design space, allowing precise control over porosity, anisotropy, and patient-specific geometries. These capabilities are critical for aligning scaffold performance with anatomical and biomechanical requirements.

Biological performance evaluation underscores the importance of integrating biochemical, topographical, and mechanical cues to regulate cell–scaffold interactions, osteogenic differentiation, and host tissue integration. Equally important is the synchronization of scaffold degradation kinetics with new bone formation to ensure effective load transfer during healing and to avoid adverse biological responses. Despite notable progress, the chapter identifies persistent barriers to clinical translation, including regulatory complexity, sterilization constraints, manufacturing scalability, and the lack of standardized evaluation protocols.

Looking forward, the convergence of smart and stimuli-responsive materials, data-driven design, and personalized fabrication is poised to redefine next-generation bone implants. Hybrid scaffolds, when developed within an integrated translational framework, hold substantial promise for advancing reliable, clinically relevant solutions for bone regeneration and repair.

AI disclosure statement: The author confirms that no artificial intelligence (AI)-generated content or AI-assisted writing tools were used in the preparation of this chapter (including summaries and the abstract). All text, analysis, and conclusions are the result of the author's original work.

References

1. Roddy, E., DeBaun, M.R., Daoud-Gray, A., Yang, Y.P., Gardner, M.J., Treatment of critical-sized bone defects: clinical and tissue engineering perspectives. *Eur. J. Orthop. Surg. Traumatol.*, 28, 351–362, 2018, https://doi.org/10.1007/s00590-017-2063-0.
2. Mistry, A.S. and Mikos, A.G., Tissue engineering strategies for bone regeneration, in: *Adv. Biochem. Eng. Biotechnol.*, pp. 1–22, Springer, Berlin/Heidelberg (Springer-Verlag), Germany, 2005, https://doi.org/10.1007/b99997.
3. Verrier, S., Alini, M., Alsberg, E., Buchman, S.R., Kelly, D., Laschke, M.W., Menger, M.D., Murphy, W.L., Stegemann, J.P., Schütz, M., Miclau, T., Stoddart, M.J., Evans, C., Tissue engineering and regenerative approaches to improving the healing of large bone defects. *Eur. Cells Mater.*, 32, 87–110, 2016, https://doi.org/10.22203/eCM.v032a06.
4. Yazdanpanah, Z., Johnston, J.D., Cooper, D.M.L., Chen, X., 3D Bioprinted Scaffolds for Bone Tissue Engineering: State-Of-The-Art and Emerging Technologies. *Front. Bioeng. Biotechnol.*, 10, 2022, https://doi.org/10.3389/fbioe.2022.824156.
5. Yang, J., Fatima, K., Zhou, X., He, C., Meticulously engineered three-dimensional-printed scaffold with microarchitecture and controlled peptide release for enhanced bone regeneration. *Biomater. Transl.*, 5, 69–83, 2024, https://doi.org/10.12336/biomatertransl.2024.01.007.
6. McDermott, A.M., Mason, D.E., Lin, A.S.P., Guldberg, R.E., Boerckel, J.D., Influence of structural load-bearing scaffolds on mechanical load- and BMP-2-mediated bone regeneration. *J. Mech. Behav. Biomed. Mater.*, 62, 169–181, 2016, https://doi.org/10.1016/j.jmbbm.2016.05.010.
7. John, P., Gleeson, Fergal, J., O'Brien as the authors of Composite Scaffolds for Orthopaedic Regenerative Medicine, in: *Advances in Composite Materials for Medicine and Nanotechnology*, InTech, London, Greater London, United Kingdom, 2011, https://doi.org/10.5772/15033.
8. Gupta, D., Singh, A.K., Bellare, J., Natural bone inspired core-shell triple-layered gel/PCL/gel 3D printed scaffolds for bone tissue engineering. *Biomed. Mater.*, 18, 065027, 2023, https://doi.org/10.1088/1748-605X/ad06c2.
9. Ferreira, F.V., Otoni, C.G., Lopes, J.H., de Souza, L.P., Mei, L.H.I., Lona, L.M.F., Lozano, K., Lobo, A.O., Mattoso, L.H.C., Ultrathin polymer fibers hybridized with bioactive ceramics: A review on fundamental pathways of electrospinning towards bone regeneration. *Mater. Sci. Eng., C*, 123, 111853, 2021, https://doi.org/10.1016/j.msec.2020.111853.
10. Sachot, N., Engel, E., Castano, O., Hybrid Organic-Inorganic Scaffolding Biomaterials for Regenerative Therapies. *Curr. Org. Chem.*, 18, 2299–2314, 2014, https://doi.org/10.2174/13852728196661408062003 55.

11. Liu, Y., Luo, D., Wang, T., Hierarchical Structures of Bone and Bioinspired Bone Tissue Engineering. *Small*, 12, 4611–4632, 2016, https://doi.org/10.1002/smll.201600626.
12. Lei, B., Guo, B., Rambhia, K.J., Ma, P.X., Hybrid polymer biomaterials for bone tissue regeneration. *Front. Med.*, 13, 189–201, 2019, https://doi.org/10.1007/s11684-018-0664-6.
13. Peng, X., Huo, Y., Zhang, G., Cheng, L., Lu, Y., Li, J., Jin, Z., Controlled mechanical and mass-transport properties of porous scaffolds through hollow strut. *Int. J. Mech. Sci.*, 248, 108202, 2023, https://doi.org/10.1016/j.ijmecsci.2023.108202.
14. Fan, W., Du, T., Droce, A., Jensen, L.R., Youngman, R.E., Ren, X., Gurevich, L., Bauchy, M., Kristensen, P., Xing, B., Yu, D., Smedskjaer, M.M., Resolving the Conflict between Strength and Toughness in Bioactive Silica-Polymer Hybrid Materials. *ACS Nano*, 16, 9748–9761, 2022, https://doi.org/10.1021/acsnano.2c03440.
15. Martin, R.A., Yue, S., Hanna, J.V., Lee, P.D., Newport, R.J., Smith, M.E., Jones, J.R., Characterizing the hierarchical structures of bioactive sol-gel silicate glass and hybrid scaffolds for bone regeneration. *Philos. Trans. R. Soc. A Math. Phys. Eng. Sci.*, 370, 1422–1443, 2012, https://doi.org/10.1098/rsta.2011.0308.
16. Barthelat, F., Architectured materials in engineering and biology: Fabrication, structure, mechanics and performance. *Int. Mater. Rev.*, 60, 413–430, 2015, https://doi.org/10.1179/1743280415Y.0000000008.
17. Paul, A.K., Borkotoky, S.S., Prasad, A., Nanohydroxyapatite-Based Composite Materials and Processing, in: *Adv. Mater. Manuf. Tech. Biomed. Appl.*, pp. 157–183, Wiley, Beverly, Massachusetts, USA, 2023, https://doi.org/10.1002/9781394166985.ch7.
18. Shalumon, K.T., Kuo, C.Y., Wong, C.B., Chien, Y.M., Chen, H.A., Chen, J.P., Gelatin/nanohyroxyapatite cryogel embedded poly(lactic-co-glycolic acid)/nanohydroxyapatite microsphere hybrid scaffolds for simultaneous bone regeneration and load-bearing. *Polym. (Basel).*, 8, 620, 2018, https://doi.org/10.3390/polym10060620.
19. Koushik, T.M., Miller, C.M., Antunes, E., Bone Tissue Engineering Scaffolds: Function of Multi-Material Hierarchically Structured Scaffolds. *Adv. Healthc. Mater.*, 12, 2023, https://doi.org/10.1002/adhm.202202766.
20. Aldemir Dikici, B., Reilly, G.C., Claeyssens, F., Boosting the Osteogenic and Angiogenic Performance of Multiscale Porous Polycaprolactone Scaffolds by *in Vitro* Generated Extracellular Matrix Decoration. *ACS Appl. Mater. Interfaces*, 12, 12510–12524, 2020, https://doi.org/10.1021/acsami.9b23100.
21. Thibault, R.A., Mikos, A.G., Kasper, F.K., Scaffold/Extracellular Matrix Hybrid Constructs for Bone-Tissue Engineering. *Adv. Healthc. Mater.*, 2, 13–24, 2013, https://doi.org/10.1002/adhm.201200209.

22. Du, Y., Guo, J.L., Wang, J., Mikos, A.G., Zhang, S., Hierarchically designed bone scaffolds: From internal cues to external stimuli. *Biomaterials*, 218, 119334, 2019, https://doi.org/10.1016/j.biomaterials.2019.119334.
23. Jeyachandran, D. and Cerruti, M., Glass, Ceramic, Polymeric, and Composite Scaffolds with Multiscale Porosity for Bone Tissue Engineering. *Adv. Eng. Mater.*, 25, 2023, https://doi.org/10.1002/adem.202201743.
24. Chan, K.S., Liang, W., Francis, W.L., Nicolella, D.P., A multiscale modeling approach to scaffold design and property prediction. *J. Mech. Behav. Biomed. Mater.*, 3, 584–593, 2010, https://doi.org/10.1016/j.jmbbm.2010.07.006.
25. Hollister, S.J., Maddox, R.D., Taboas, J.M., Optimal design and fabrication of scaffolds to mimic tissue properties and satisfy biological constraints. *Biomaterials*, 23, 4095–4103, 2002, https://doi.org/10.1016/S0142-9612(02)00148-5.
26. Hollister, S.J., Porous scaffold design for tissue engineering. *Nat. Mater.*, 4, 518–524, 2005, https://doi.org/10.1038/nmat1421.
27. Taniguchi, N., Fujibayashi, S., Takemoto, M., Sasaki, K., Otsuki, B., Nakamura, T., Matsushita, T., Kokubo, T., Matsuda, S., Effect of pore size on bone ingrowth into porous titanium implants fabricated by additive manufacturing: An *in vivo* experiment. *Mater. Sci. Eng., C*, 59, 690–701, 2016, https://doi.org/10.1016/j.msec.2015.10.069.
28. Vijayavenkataraman, S., Zhang, L., Zhang, S., Fuh, J.Y.H., Lu, W.F., Triply periodic minimal surfaces sheet scaffolds for tissue engineering applications: An optimization approach toward biomimetic scaffold design. *ACS Appl. Bio Mater.*, 1, 259–269, 2018, https://doi.org/10.1021/acsabm.8b00052.
29. Bose, S., Vahabzadeh, S., Bandyopadhyay, A., Bone tissue engineering using 3D printing. *Mater. Today*, 16, 496–504, 2013, https://doi.org/10.1016/j.mattod.2013.11.017.
30. Warnke, P.H., Douglas, T., Wollny, P., Sherry, E., Steiner, M., Galonska, S., Becker, S.T., Springer, I.N., Wiltfang, J., Sivananthan, S., Rapid prototyping: Porous titanium alloy scaffolds produced by selective laser melting for bone tissue engineering. *Tissue Eng. Part C Methods*, 15, 115–124, 2009, https://doi.org/10.1089/ten.tec.2008.0288.
31. Van Bael, S., Chai, Y.C., Truscello, S., Moesen, M., Kerckhofs, G., Van Oosterwyck, H., Kruth, J.P., Schrooten, J., The effect of pore geometry on the *in vitro* biological behavior of human periosteum-derived cells seeded on selective laser-melted Ti6Al4V bone scaffolds. *Acta Biomater.*, 8, 2824–2834, 2012, https://doi.org/10.1016/j.actbio.2012.04.001.
32. Xue, X., Hu, Y., Deng, Y., Su, J., Recent Advances in Design of Functional Biocompatible Hydrogels for Bone Tissue Engineering. *Adv. Funct. Mater.*, 31, 2021, https://doi.org/10.1002/adfm.202009432.
33. Zhi, W., Wang, X., Sun, D., Chen, T., Yuan, B., Li, X., Chen, X., Wang, J., Xie, Z., Zhu, X., Zhang, K., Zhang, X., Optimal regenerative repair of large segmental bone defect in a goat model with osteoinductive calcium

phosphate bioceramic implants. *Bioact. Mater.*, 11, 240–253, 2022, https://doi.org/10.1016/j.bioactmat.2021.09.024.

34. Yin, S., Zhang, W., Zhang, Z., Jiang, X., Recent Advances in Scaffold Design and Material for Vascularized Tissue-Engineered Bone Regeneration. *Adv. Healthc. Mater.*, 8, 2019, https://doi.org/10.1002/adhm.201801433.
35. Chao, L., Jiao, C., Liang, H., Xie, D., Shen, L., Liu, Z., Analysis of Mechanical Properties and Permeability of Trabecular-Like Porous Scaffold by Additive Manufacturing. *Front. Bioeng. Biotechnol.*, 9, 2021, https://doi.org/10.3389/fbioe.2021.779854.
36. Hallman, M., Driscoll, J.A., Lubbe, R., Jeong, S., Chang, K., Haleem, M., Jakus, A., Pahapill, R., Yun, C., Shah, R., Hsu, W.K., Stock, S.R., Hsu, E.L., Influence of Geometry and Architecture on the *in Vivo* Success of 3D-Printed Scaffolds for Spinal Fusion. *Tissue Eng. Part A*, 27, 26–36, 2021, https://doi.org/10.1089/ten.tea.2020.0004.
37. Feng, C., Xue, J., Yu, X., Zhai, D., Lin, R., Zhang, M., Xia, L., Wang, X., Yao, Q., Chang, J., Wu, C., Co-inspired hydroxyapatite-based scaffolds for vascularized bone regeneration. *Acta Biomater.*, 119, 419–431, 2021, https://doi.org/10.1016/j.actbio.2020.11.010.
38. Prasad, A. and Paul, A.K., *Biodegradable waste processing for sustainable developments*, CRC Press, Boca Raton, 2024, https://doi.org/10.1201/9781003502012.
39. Velema, J. and Kaplan, D., Biopolymer-based biomaterials as scaffolds for tissue engineering, in: *Adv. Biochem. Eng. Biotechnol.*, pp. 187–238, Springer Berlin Heidelberg, Berlin and Heidelberg, Germany, 2006, https://doi.org/10.1007/10_013.
40. Reddy, M.S.B., Ponnamma, D., Choudhary, R., Sadasivuni, K.K., A comparative review of natural and synthetic biopolymer composite scaffolds. *Polym. (Basel).*, 13, 1105, 2021, https://doi.org/10.3390/polym13071105.
41. Kumar, M.H., Paul, A.K., Prasad, A., Chakraborty, G., Rheological Studies of Resorbable Polymeric Nanocomposites, in: *Resorbable Polym. Bioimplants Fixat. Devices*, pp. 163–183, Springer Nature, Singapore, Singapore, 2025, https://doi.org/10.1007/978-981-96-9306-1_7.
42. Goonoo, N., Bhaw-Luximon, A., Jhurry, D., Biodegradable polymer blends: Miscibility, physicochemical properties and biological response of scaffolds. *Polym. Int.*, 64, 1289–1302, 2015, https://doi.org/10.1002/pi.4937.
43. Paul, A.K., Chakraborty, G., Borkotoky, S.S., Prasad, A., Degradation Studies of Bioabsorbable Polymeric Composites, in: *Bioabsorbable Polym. Drug Carriers Fixat. Devices*, pp. 105–128, Apple Academic Press, New York, 2025, https://doi.org/10.1201/9781003569817-5.
44. Zhang, F. and King, M.W., Biodegradable Polymers as the Pivotal Player in the Design of Tissue Engineering Scaffolds. *Adv. Healthc. Mater.*, 9, 2020, https://doi.org/10.1002/adhm.201901358.

45. Tavoni, M., Dapporto, M., Tampieri, A., Sprio, S., Bioactive calcium phosphate-based composites for bone regeneration. *J. Compos. Sci.*, 5, 227, 2021, https://doi.org/10.3390/jcs5090227.
46. Deng, M., Kumbar, S.G., Lo, K.W.H., Ulery, B.D., Laurencin, C.T., Novel polymer-ceramics for bone repair and regeneration. *Recent Pat. Biomed. Eng.*, 4, 168–184, 2011, https://doi.org/10.2174/1874764711104030168.
47. Hajiali, F., Tajbakhsh, S., Shojaei, A., Fabrication and Properties of Polycaprolactone Composites Containing Calcium Phosphate-Based Ceramics and Bioactive Glasses in Bone Tissue Engineering: A Review. *Polym. Rev.*, 58, 164–207, 2018, https://doi.org/10.1080/15583724.2017.1332640.
48. Katsanevakis, E., Wen, X., Shi, D., Zhang, N., Biomineralization of polymer scaffolds. *Key Eng. Mater.*, 441, 269–295, 2010, https://doi.org/10.4028/www.scientific.net/KEM.441.269.
49. Patel, S., Sanjay, M.S.S.S., Chakraborty, G., Paul, A.K., Nanocomposites/Nanomaterials and the Risk Management of Biofuels Production, in: *Nanomater. as a Catal. Biofuels Prod.*, pp. 151–184, Springer Nature Singapore, Singapore, 2025, https://doi.org/10.1007/978-981-96-1706-7_7.
50. McMahon, R.E., Wang, L., Skoracki, R., Mathur, A.B., Development of nanomaterials for bone repair and regeneration. *J. Biomed. Mater. Res. Part B Appl. Biomater.*, 101, B, 387–397, 2013, https://doi.org/10.1002/jbm.b.32823.
51. Gong, T., Xie, J., Liao, J., Zhang, T., Lin, S., Lin, Y., Nanomaterials and bone regeneration. *Bone Res.*, 3, 15029, 2015, https://doi.org/10.1038/boneres.2015.29.
52. Pei, Y., Wang, Y., Chen, J., Zhou, J., Han, Y., Liu, X., Chen, S., Chen, S., He, D., Wu, Y., Lv, H., Zhou, Y., Bionic Nanostructures Create Mechanical Signals to Mediate the Composite Structural Bone Regeneration Through Multi-System Regulation. *Adv. Sci.*, 12, 2025, https://doi.org/10.1002/advs.202502299.
53. Liao, S., Nguyen, L.T.H., Ngiam, M., Wang, C., Cheng, Z., Chan, C.K., Ramakrishna, S., Biomimetic nanocomposites to control osteogenic differentiation of human mesenchymal stem cells. *Adv. Healthc. Mater.*, 3, 737–751, 2014, https://doi.org/10.1002/adhm.201300207.
54. Monia, T. and Ridha, B.C., Polymer-ceramic composites for bone challenging applications: Materials and manufacturing processes. *J. Thermoplast. Compos. Mater.*, 37, 1540–1557, 2024, https://doi.org/10.1177/08927057231190066.
55. Basha, R.Y., Sampath, S.K., Doble, M., Design of biocomposite materials for bone tissue regeneration. *Mater. Sci. Eng., C*, 57, 452–463, 2015, https://doi.org/10.1016/j.msec.2015.07.016.
56. Bhattacharjee, M., Sai, R.M., Chakraborty, G., Paul, A.K., Degradation and Toxicological Studies of Polymeric Nanocomposites, in: *Resorbable Polym. Bioimplants Fixat. Devices*, pp. 283–314, Springer Nature, Singapore, Singapore, 2025, https://doi.org/10.1007/978-981-96-9306-1_12.
57. Girish, B., Rakshith, G.R., Paul, A.K., Raja, V.K., Chakraborty, G., Hydrolytic and soil degradation of cellulosic material (paper): optimization

of parameters using ANN and RSM. *Polym. Bull.*, 81, 12893–12920, 2024, https://doi.org/10.1007/s00289-024-05319-5.
58. Paul, A.K., Migration Studies of Biodegradable Composites, in: *Biodegrad. Compos. Packag. Appl.*, pp. 193–212, CRC Press, Boca Raton, 2022, https://doi.org/10.1201/9781003227908-11.
59. Sharma, R., Malviya, R., Singh, S., Prajapati, B., A Critical Review on Classified Excipient Sodium-Alginate-Based Hydrogels: Modification, Characterization, and Application in Soft Tissue Engineering. *Gels*, 9, 430, 2023, https://doi.org/10.3390/gels9050430.
60. Sarker, B., Rompf, J., Silva, R., Lang, N., Detsch, R., Kaschta, J., Fabry, B., Boccaccini, A.R., Alginate-based hydrogels with improved adhesive properties for cell encapsulation. *Int. J. Biol. Macromol.*, 78, 72–78, 2015, https://doi.org/10.1016/j.ijbiomac.2015.03.061.
61. Highley, C.B., Prestwich, G.D., Burdick, J.A., Recent advances in hyaluronic acid hydrogels for biomedical applications. *Curr. Opin. Biotechnol.*, 40, 35–40, 2016, https://doi.org/10.1016/j.copbio.2016.02.008.
62. Baek, J., Fan, Y., Jeong, S.H., Lee, H.Y., Do Jung, H., Kim, H.E., Kim, S., Jang, T.S., Facile strategy involving low-temperature chemical cross-linking to enhance the physical and biological properties of hyaluronic acid hydrogel. *Carbohydr. Polym.*, 202, 545–553, 2018, https://doi.org/10.1016/j.carbpol.2018.09.014.
63. Ardean, C., Davidescu, C.M., Nemeş, N.S., Negrea, A., Ciopec, M., Duteanu, N., Negrea, P., Duda-seiman, D., Musta, V., Factors influencing the antibacterial activity of chitosan and chitosan modified by functionalization. *Int. J. Mol. Sci.*, 22, 7449, 2021, https://doi.org/10.3390/ijms22147449.
64. Luanda, A. and Badalamoole, V., Past, present and future of biomedical applications of dextran-based hydrogels: A review. *Int. J. Biol. Macromol.*, 228, 794–807, 2023, https://doi.org/10.1016/j.ijbiomac.2022.12.129.
65. Sun, G. and Mao, J.J., Engineering dextran-based scaffolds for drug delivery and tissue repair. *Nanomedicine*, 7, 1771–1784, 2012, https://doi.org/10.2217/nnm.12.149.
66. Roberts, J.J., Farrugia, B.L., Green, R.A., Rnjak-Kovacina, J., Martens, P.J., *In situ* formation of poly(vinyl alcohol)–heparin hydrogels for mild encapsulation and prolonged release of basic fibroblast growth factor and vascular endothelial growth factor. *J. Tissue Eng.*, 7, 2016, https://doi.org/10.1177/2041731416677132.
67. He, C., Ji, H., Qian, Y., Wang, Q., Liu, X., Zhao, W., Zhao, C., Heparin-based and heparin-inspired hydrogels: size-effect, gelation and biomedical applications. *J. Mater. Chem. B*, 7, 1186–1208, 2019, https://doi.org/10.1039/c8tb02671h.
68. Chen, C., Xi, Y., Weng, Y., Recent Advances in Cellulose-Based Hydrogels for Tissue Engineering Applications. *Polym. (Basel).*, 14, 3335, 2022, https://doi.org/10.3390/polym14163335.

69. Singh, P., Baniasadi, H., Gupta, S., Ghosh, R., Shaikh, S., Seppälä, J., Kumar, A., 3D-printed cellulose nanocrystals and gelatin scaffolds with bioactive cues for regenerative medicine: Advancing biomedical applications. *Int. J. Biol. Macromol.*, 278, 134402, 2024, https://doi.org/10.1016/j.ijbiomac.2024.134402.
70. Mokhtari, H., Tavakoli, S., Safarpour, F., Kharaziha, M., Bakhsheshi-Rad, H.R., Ramakrishna, S., Berto, F., Recent advances in chemically-modified and hybrid carrageenan-based platforms for drug delivery, wound healing, and tissue engineering. *Polym. (Basel).*, 13, 1744, 2021, https://doi.org/10.3390/polym13111744.
71. Liu, J., Zhan, X., Wan, J., Wang, Y., Wang, C., Review for carrageenan-based pharmaceutical biomaterials: Favourable physical features versus adverse biological effects. *Carbohydr. Polym.*, 121, 27–36, 2015, https://doi.org/10.1016/j.carbpol.2014.11.063.
72. Elangwe, C.N., Morozkina, S.N., Podshivalov, A.V., Uspenskaya, M.V., Evaluation of composition effects on the tissue-adhesive, mechanical and physical properties of physically crosslinked hydrogels based on chitosan and pullulan for wound healing applications. *Int. J. Biol. Macromol.*, 276, 133857, 2024, https://doi.org/10.1016/j.ijbiomac.2024.133857.
73. Su, T., Zhang, M., Zeng, Q., Pan, W., Huang, Y., Qian, Y., Dong, W., Qi, X., Shen, J., Mussel-inspired agarose hydrogel scaffolds for skin tissue engineering. *Bioact. Mater.*, 6, 579–588, 2021, https://doi.org/10.1016/j.bioactmat.2020.09.004.
74. Wu, X., Sun, H., Qin, Z., Che, P., Yi, X., Yu, Q., Zhang, H., Sun, X., Yao, F., Li, J., Fully physically crosslinked pectin-based hydrogel with high stretchability and toughness for biomedical application. *Int. J. Biol. Macromol.*, 149, 707–716, 2020, https://doi.org/10.1016/j.ijbiomac.2020.01.297.
75. Li, D., Li, J., lin Dong, H., Li, X., Zhang, J., Ramaswamy, S., Xu, F., Pectin in biomedical and drug delivery applications: A review. *Int. J. Biol. Macromol.*, 185, 49–65, 2021, https://doi.org/10.1016/j.ijbiomac.2021.06.088.
76. Qamruzzaman, M., Ahmed, F., Mondal, M.I.H., An Overview on Starch-Based Sustainable Hydrogels: Potential Applications and Aspects. *J. Polym. Environ.*, 30, 19–50, 2022, https://doi.org/10.1007/s10924-021-02180-9.
77. Lee, C.S. and Hwang, H.S., Starch-Based Hydrogels as a Drug Delivery System in Biomedical Applications. *Gels*, 9, 951, 2023, https://doi.org/10.3390/gels9120951.
78. Richbourg, N.R., Peppas, N.A., Sikavitsas, V.I., Tuning the biomimetic behavior of scaffolds for regenerative medicine through surface modifications. *J. Tissue Eng. Regen. Med.*, 13, 1275–1293, 2019, https://doi.org/10.1002/term.2859.
79. Katti, D., Vasita, R., Shanmugam, K., Improved Biomaterials for Tissue Engineering Applications: Surface Modification of Polymers. *Curr. Top. Med. Chem.*, 8, 341–353, 2008, https://doi.org/10.2174/156802608783790893.

80. Feng, P., He, J., Peng, S., Gao, C., Zhao, Z., Xiong, S., Shuai, C., Characterizations and interfacial reinforcement mechanisms of multicomponent biopolymer based scaffold. *Mater. Sci. Eng., C*, 100, 809–825, 2019, https://doi.org/10.1016/j.msec.2019.03.030.
81. Dave, K. and Gomes, V.G., Interactions at scaffold interfaces: Effect of surface chemistry, structural attributes and bioaffinity. *Mater. Sci. Eng., C*, 105, 110078, 2019, https://doi.org/10.1016/j.msec.2019.110078.
82. Di Luca, A., Longoni, A., Criscenti, G., Mota, C., Van Blitterswijk, C., Moroni, L., Toward mimicking the bone structure: Design of novel hierarchical scaffolds with a tailored radial porosity gradient. *Biofabrication*, 8, 045007, 2016, https://doi.org/10.1088/1758-5090/8/4/045007.
83. Egan, P.F., Ferguson, S.J., Shea, K., Design of Hierarchical Three-Dimensional Printed Scaffolds Considering Mechanical and Biological Factors for Bone Tissue Engineering. *J. Mech. Des.*, 139, 2017, https://doi.org/10.1115/1.4036396.
84. Fang, L., Lin, X., Xu, R., Liu, L., Zhang, Y., Tian, F., Li, J.J., Xue, J., Advances in the Development of Gradient Scaffolds Made of Nano-Micromaterials for Musculoskeletal Tissue Regeneration. *Nano Micro Lett.*, 17, 75, 2025, https://doi.org/10.1007/s40820-024-01581-4.
85. Daskalakis, E., Huang, B., Vyas, C., Acar, A.A., Liu, F., Fallah, A., Cooper, G., Weightman, A., Blunn, G., Koç, B., Bartolo, P., Bone Bricks: The Effect of Architecture and Material Composition on the Mechanical and Biological Performance of Bone Scaffolds. *ACS Omega*, 7, 7515–7530, 2022, https://doi.org/10.1021/acsomega.1c05437.
86. Letic-Gavrilovic, A., Piattelli, A., Abe, K., Nerve growth factor β(NGF β) delivery *via* a collagen/hydroxyapatite (Col/HAp) composite and its effects on new bone ingrowth. *J. Mater. Sci. Mater. Med.*, 14, 95–102, 2003, https://doi.org/10.1023/A:1022099208535.
87. Fujisato, T., Sajiki, T., Qiang, L., Ikada, Y., Effect of basic fibroblast growth factor on cartilage regeneration in chondrocyte-seeded collagen sponge scaffold. *Biomaterials*, 17, 155–162, 1996, https://doi.org/10.1016/0142-9612(96)85760-7.
88. Shih, Y.V., Chen, C., Tsai, S., Wang, Y.J., Lee, O.K., Growth of Mesenchymal Stem Cells on Electrospun Type I Collagen Nanofibers. *Stem Cells*, 24, 2391–2397, 2006, https://doi.org/10.1634/stemcells.2006-0253.
89. Xu, X.L., Lou, J., Tang, T., Ng, K.W., Zhang, J., Yu, C., Dai, K., Evaluation of different scaffolds for BMP-2 genetic orthopedic tissue engineering. *J. Biomed. Mater. Res. Part B Appl. Biomater.*, 75, 289–303, 2005, https://doi.org/10.1002/jbm.b.30299.
90. Ahmadipour, M., Mohammadi, H., Pang, A.L., Arjmand, M., Ayode Otitoju, T., Okoye, P.U., Rajitha, B., A review: silicate ceramic-polymer composite scaffold for bone tissue engineering. *Int. J. Polym. Mater. Polym. Biomater.*, 71, 180–195, 2022, https://doi.org/10.1080/00914037.2020.1817018.

91. Robey, P.G. and Boskey, A.L., The Composition of Bone, in: *Prim. Metab. Bone Dis. Disord. Miner. Metab.*, Seventh Ed., pp. 32–38, Wiley, Wiley-Blackwell (John Wiley & Sons), based in Hoboken, New Jersey, United States, 2009, https://doi.org/10.1002/9780470623992.ch6.
92. Shi, C., Yuan, Z., Han, F., Zhu, C., Li, B., Polymeric biomaterials for bone regeneration. *Ann. Jt.*, 1, 27–27, 2016, https://doi.org/10.21037/aoj.2016.11.02.
93. Mota, C., Puppi, D., Chiellini, F., Chiellini, E., Additive manufacturing techniques for the production of tissue engineering constructs. *J. Tissue Eng. Regen. Med*, 9, 174–190, 2015, https://doi.org/10.1002/term.1635.
94. Kumar, M.H., Paul, A.K., Chakraborty, G., Utilization of gelatin, collagen, and chondroitin sulfate in 3D bioprinting, in: *Handb. 3D Print. Biomed. Appl.*, pp. 165–187, CRC Press, Boca Raton, 2025, https://doi.org/10.1201/9781003521792-10.
95. Kumar, M. and Sharma, V., Additive manufacturing techniques for the fabrication of tissue engineering scaffolds: a review. *Rapid Prototyp. J.*, 27, 1230–1272, 2021, https://doi.org/10.1108/RPJ-01-2021-0011.
96. Latimer, J.M., Maekawa, S., Yao, Y., Wu, D.T., Chen, M., Giannobile, W.V., Regenerative Medicine Technologies to Treat Dental, Oral, and Craniofacial Defects. *Front. Bioeng. Biotechnol.*, 9, 2021, https://doi.org/10.3389/fbioe.2021.704048.
97. Paul, A.K., Chakraborty, G., Prasad, A., Toxicology Assessment of Polymeric Material for Implants, in: *Resorbable Polym. Bioimplants Fixat. Devices*, pp. 349–369, Springer Nature, Singapore, Singapore, 2025, https://doi.org/10.1007/978-981-96-9306-1_14.
98. Mobarak, M.H., Islam, M.A., Hossain, N., Al Mahmud, M.Z., Rayhan, M.T., Nishi, N.J., Chowdhury, M.A., Recent advances of additive manufacturing in implant fabrication – A review. *Appl. Surf. Sci. Adv.*, 18, 100462, 2023, https://doi.org/10.1016/j.apsadv.2023.100462.
99. Dehghani, F. and Annabi, N., Engineering porous scaffolds using gas-based techniques. *Curr. Opin. Biotechnol.*, 22, 661–666, 2011, https://doi.org/10.1016/j.copbio.2011.04.005.
100. Hou, Q., Grijpma, D.W., Feijen, J., Preparation of Interconnected Highly Porous Polymeric Structures by a Replication and Freeze-Drying Process. *J. Biomed. Mater. Res. Part B Appl. Biomater.*, 67, 732–740, 2003, https://doi.org/10.1002/jbm.b.10066.
101. Capuana, E., Lopresti, F., Carfì Pavia, F., Brucato, V., La Carrubba, V., Solution-based processing for scaffold fabrication in tissue engineering applications: A brief review. *Polym. (Basel).*, 13, 2041, 2021, https://doi.org/10.3390/polym13132041.
102. Ozkan, O. and Turkoglu Sasmazel, H., Hybrid polymeric scaffolds prepared by micro and macro approaches. *Int. J. Polym. Mater. Polym. Biomater.*, 66, 853–860, 2017, https://doi.org/10.1080/00914037.2016.1278218.
103. Sobral, J.M., Caridade, S.G., Sousa, R.A., Mano, J.F., Reis, R.L., Three-dimensional plotted scaffolds with controlled pore size gradients: Effect

of scaffold geometry on mechanical performance and cell seeding efficiency. *Acta Biomater.*, 7, 1009–1018, 2011, https://doi.org/10.1016/j.actbio.2010.11.003.
104. Entezari, A., Roohani, I., Li, G., Dunstan, C.R., Rognon, P., Li, Q., Jiang, X., Zreiqat, H., Architectural Design of 3D Printed Scaffolds Controls the Volume and Functionality of Newly Formed Bone. *Adv. Healthc. Mater.*, 8, 2019, https://doi.org/10.1002/adhm.201801353.
105. Karageorgiou, V. and Kaplan, D., Porosity of 3D biomaterial scaffolds and osteogenesis. *Biomaterials*, 26, 5474–5491, 2005, https://doi.org/10.1016/j.biomaterials.2005.02.002.
106. Chakraborty, G. and Paul, A.K., 3D Printing in Biomedical Applications: Techniques and Emerging Trends, in: *Adv. Mater. Manuf. Tech. Biomed. Appl.*, pp. 275–301, Wiley, Beverly, Massachusetts, USA, 2023, https://doi.org/10.1002/9781394166985.ch11.
107. Rstakyan, V., Mkhitaryan, L., Baghdasaryan, L., Ghaltaghchyan, T., Karabekian, Z., Sevoyan, G., Aghayan, M., Rodríguez, M.A., Stereolithography of ceramic scaffolds for bone tissue regeneration: Influence of hydroxyapatite/silica ratio on mechanical properties. *J. Mech. Behav. Biomed. Mater.*, 152, 106421, 2024, https://doi.org/10.1016/j.jmbbm.2024.106421.
108. Verisqa, F., Cha, J.R., Nguyen, L., Kim, H.W., Knowles, J.C., Digital Light Processing 3D Printing of Gyroid Scaffold with Isosorbide-Based Photopolymer for Bone Tissue Engineering. *Biomolecules*, 12, 1692, 2022, https://doi.org/10.3390/biom12111692.
109. Suamte, L. and Babu, P.J., Electrospun based functional scaffolds for biomedical engineering: A review. *Nano TransMed*, 4, 100055, 2024, https://doi.org/10.1016/j.ntm.2024.100055.
110. Gedik, B. and Erdem, M.A., Electrospun PCL membranes for localized drug delivery and bone regeneration. *BMC Biotechnol.*, 25, 31, 2025, https://doi.org/10.1186/s12896-025-00965-7.
111. Chen, S., Wu, Z., Huang, Z., Liang, C., Lee, S.J., Implantable Dental Barrier Membranes as Regenerative Medicine in Dentistry: A Comprehensive Review. *Tissue Eng. Regen. Med.*, 22, 527–549, 2025, https://doi.org/10.1007/s13770-025-00704-1.
112. Puppi, D., Pecorini, G., Chiellini, F., Biomedical processing of polyhydroxyalkanoates. *Bioengineering*, 6, 108, 2019, https://doi.org/10.3390/bioengineering6040108.
113. Katti, D.R., Sharma, A., Katti, K.S., Predictive Methodologies for Design of Bone Tissue Engineering Scaffolds, in: *Mater. Devices Bone Disord.*, pp. 453–492, Elsevier, Amsterdam, The Netherlands, 2017, https://doi.org/10.1016/B978-0-12-802792-9.00010-0.
114. Eltom, A., Zhong, G., Muhammad, A., Scaffold Techniques and Designs in Tissue Engineering Functions and Purposes: A Review. *Adv. Mater. Sci. Eng.*, 2019, 1–13, 2019, https://doi.org/10.1155/2019/3429527.

115. Sola, A., Bertacchini, J., D'Avella, D., Anselmi, L., Maraldi, T., Marmiroli, S., Messori, M., Development of solvent-casting particulate leaching (SCPL) polymer scaffolds as improved three-dimensional supports to mimic the bone marrow niche. *Mater. Sci. Eng., C*, 96, 153–165, 2019, https://doi.org/10.1016/j.msec.2018.10.086.
116. Murphy, C.M., O'Brien, F.J., Little, D.G., Schindeler, A., Cell-scaffold interactions in the bone tissue engineering triad. *Eur. Cells Mater.*, 26, 120–132, 2013, https://doi.org/10.22203/eCM.v026a09.
117. Keselowsky, B.G., Collard, D.M., García, A.J., Integrin binding specificity regulates biomaterial surface chemistry effects on cell differentiation. *Proc. Natl. Acad. Sci. U.S.A.*, 102, 5953–5957, 2005, https://doi.org/10.1073/pnas.0407356102.
118. García, A.J. and Keselowsky, B.G., Biomimetic surfaces for control of cell adhesion to facilitate bone formation. *Crit. Rev. Eukaryot. Gene Expr.*, 12, 151–162, 2002, https://doi.org/10.1615/CritRevEukaryotGeneExpr.v12.i2.50.
119. Jesus, D., Pinho, A.R., Gomes, M.C., Oliveira, C.S., Mano, J.F., Emerging modulators for osteogenic differentiation: a combination of chemical and topographical cues for bone microenvironment engineering. *Soft Matter*, 18, 3107–3119, 2022, https://doi.org/10.1039/d2sm00009a.
120. Ghanaati, S., Unger, R.E., Webber, M.J., Barbeck, M., Orth, C., Kirkpatrick, J.A., Booms, P., Motta, A., Migliaresi, C., Sader, R.A., Kirkpatrick, C.J., Scaffold vascularization *in vivo* driven by primary human osteoblasts in concert with host inflammatory cells. *Biomaterials*, 32, 8150–8160, 2011, https://doi.org/10.1016/j.biomaterials.2011.07.041.
121. Nambiar, J., Jana, S., Nandi, S.K., Strategies for Enhancing Vascularization of Biomaterial-Based Scaffold in Bone Regeneration. *Chem. Rec.*, 22, 2022, https://doi.org/10.1002/tcr.202200008.
122. Kang, Y. and Chang, J., Channels in a porous scaffold: A new player for vascularization. *Regen. Med.*, 13, 705–715, 2018, https://doi.org/10.2217/rme-2018-0022.
123. Polak, M., Karbowniczek, J.E., Stachewicz, U., Strategies in Electrospun Polymer and Hybrid Scaffolds for Enhanced Cell Integration and Vascularization for Bone Tissue Engineering and Organoids. *Wiley Interdiscip. Rev. Nanomed. Nanobiotechnol.*, 16, 2024, https://doi.org/10.1002/wnan.2022.
124. Fernandes, M.H. and de Sousa Gomes, P., Biological Assessment of Bioceramics: *In Vitro* and *In Vivo* Tests. *Encycl. Mater. Tech. Ceram. Glas.*, 1-3, 3, V3–798-V3-816, 2021, https://doi.org/10.1016/B978-0-12-803581-8.12110-1.
125. Salètes, M., Vartin, M., Mocquot, C., Chevalier, C., Grosgogeat, B., Colon, P., Attik, N., Mesoporous bioactive glasses cytocompatibility assessment: A review of *in vitro* studies. *Biomimetics*, 6, 1–32, 2021, https://doi.org/10.3390/biomimetics6010009.

126. Habibovic, P., *In vitro* and *in vivo* bioactivity assessment of a polylactic acid/hydroxyapatite composite for bone regeneration. *Biomatter*, 4, e27664, 2014, https://doi.org/10.4161/biom.27664.
127. Gut, G., Ambroziak, M., Bojar, W., Szaraniec, B., Chłopek, J., Flis, S., Koronkiewicz, M., Jastrzębski, Z., *In Vitro* and *In Vivo* (Rabbit, Guinea Pig, Mouse) Properties of a Novel Resorbable Polymer and Allogenic Bone Composite for Guided Bone Regeneration and Orthopedic Implants. *Transplant. Proc.*, 50, 2223–2228, 2018, https://doi.org/10.1016/j.transproceed.2018.02.121.
128. Chen, Q.Z., Quinn, J.M.W., Thouas, G.A., Zhou, X., Komesaroff, P.A., Bone-like elastomer-toughened scaffolds with degradability kinetics matching healing rates of injured bone. *Adv. Eng. Mater.*, 12, 2010, https://doi.org/10.1002/adem.201080002.
129. Bossard, C., Granel, H., Wittrant, Y., Jallot, E., Lao, J., Vial, C., Tiainen, H., Polycaprolactone/bioactive glass hybrid scaffolds for bone regeneration. *Biomed. Glas.*, 4, 108–122, 2018, https://doi.org/10.1515/bglass-2018-0010.
130. Park, S.H., Gil, E.S., Shi, H., Kim, H.J., Lee, K., Kaplan, D.L., Relationships between degradability of silk scaffolds and osteogenesis. *Biomaterials*, 31, 6162–6172, 2010, https://doi.org/10.1016/j.biomaterials.2010.04.028.
131. Ozbolat, I.T., Khoda, A.K.M.B., Marchany, M., Gardella, J.A., Koc, B., Hybrid tissue scaffolds for controlled release applications: A study on design and fabrication of hybrid and heterogeneous tissue scaffolds for controlled release applications is presented in this paper. *Virtual Phys. Prototyp.*, 7, 37–47, 2012, https://doi.org/10.1080/17452759.2012.668700.
132. Sikavitsas, V.I., Temenoff, J.S., Mikos, A.G., Biomaterials and bone mechanotransduction. *Biomaterials*, 22, 2581–2593, 2001, https://doi.org/10.1016/S0142-9612(01)00002-3.
133. Entezari, A., Swain, M.V., Gooding, J.J., Roohani, I., Li, Q., A modular design strategy to integrate mechanotransduction concepts in scaffold-based bone tissue engineering. *Acta Biomater.*, 118, 100–112, 2020, https://doi.org/10.1016/j.actbio.2020.10.012.
134. Igwe, J.C., Mikael, P.E., Nukavarapu, S.P., Design, fabrication and *in vitro* evaluation of a novel polymer-hydrogel hybrid scaffold for bone tissue engineering. *J. Tissue Eng. Regen. Med.*, 8, 131–142, 2014, https://doi.org/10.1002/term.1506.
135. Hollister, S.J. and Murphy, W.L., Scaffold translation: Barriers between concept and clinic. *Tissue Eng. Part B Rev.*, 17, 459–474, 2011, https://doi.org/10.1089/ten.teb.2011.0251.
136. Hollister, S.J., Flanagan, C.L., Zopf, D.A., Morrison, R.J., Nasser, H., Patel, J.J., Ebramzadeh, E., Sangiorgio, S.N., Wheeler, M.B., Green, G.E., Design Control for Clinical Translation of 3D Printed Modular Scaffolds. *Ann. Biomed. Eng.*, 43, 774–786, 2015, https://doi.org/10.1007/s10439-015-1270-2.

137. Liu, Y., Lim, J., Teoh, S.H., Review: Development of clinically relevant scaffolds for vascularised bone tissue engineering. *Biotechnol. Adv.*, 31, 688–705, 2013, https://doi.org/10.1016/j.biotechadv.2012.10.003.
138. Li, J.J., Kaplan, D.L., Zreiqat, H., Scaffold-based regeneration of skeletal tissues to meet clinical challenges. *J. Mater. Chem. B*, 2, 7272–7306, 2014, https://doi.org/10.1039/c4tb01073f.
139. Municoy, S., Álvarez Echazú, M.I., Antezana, P.E., Galdopórpora, J.M., Olivetti, C., Mebert, A.M., Foglia, M.L., Tuttolomondo, M.V., Alvarez, G.S., Hardy, J.G., Desimone, M.F., Stimuli-responsive materials for tissue engineering and drug delivery. *Int. J. Mol. Sci.*, 21, 1–39, 2020, https://doi.org/10.3390/ijms21134724.
140. Wells, C.M., Harris, M., Choi, L., Murali, V.P., Guerra, F.D., Jennings, J.A., Stimuli-responsive drug release from smart polymers. *J. Funct. Biomater.*, 10, 34, 2019, https://doi.org/10.3390/jfb10030034.
141. Mohamed, M.A., Fallahi, A., El-Sokkary, A.M.A., Salehi, S., Akl, M.A., Jafari, A., Tamayol, A., Fenniri, H., Khademhosseini, A., Andreadis, S.T., Cheng, C., Stimuli-responsive hydrogels for manipulation of cell microenvironment: From chemistry to biofabrication technology. *Prog. Polym. Sci.*, 98, 101147, 2019, https://doi.org/10.1016/j.progpolymsci.2019.101147.
142. Gelmi, A. and Schutt, C.E., Stimuli-Responsive Biomaterials: Scaffolds for Stem Cell Control. *Adv. Healthc. Mater.*, 10, 2021, https://doi.org/10.1002/adhm.202001125.
143. Hollister, S.J., Levy, R.A., Chu, T.-M., Halloran, J.W., Feinberg, S.E., An image-based approach for designing and manufacturing craniofacial scaffolds. *Int. J. Oral Maxillofac. Surg.*, 29, 67–71, 2000, https://doi.org/10.1034/j.1399-0020.2000.290115.x.
144. da Silva, F.T., Tsumura, W.G., de C.S. e. Silva, M., Lobo, A.O., Stocco, T.D., Establishing a framework for the design and fabrication of patient-specific scaffolds targeting partial meniscal defects. *Int. J. Bioprint.*, 11, 556–572, 2025, https://doi.org/10.36922/ijb.8109.
145. Zhou, J., See, C.W., Sreenivasamurthy, S., Zhu, D., Customized Additive Manufacturing in Bone Scaffolds—The Gateway to Precise Bone Defect Treatment. *Research*, 6, 2023, https://doi.org/10.34133/research.0239.

146. Poh, P.S., Lingner, T., Kalkhof, S., Märdian, S., Baumbach, J., Dondl, P., Duda, G.N., Checa, S., Enabling technologies towards personalization of scaffolds for large bone defect regeneration. *Curr. Opin. Biotechnol.*, 74, 263–270, 2022, https://doi.org/10.1016/j.copbio.2021.12.002.
147. O'Keefe, R.J. and Mao, J., Bone tissue engineering and regeneration: From discovery to the clinic-an overview. *Tissue Eng. Part B Rev.*, 17, 389–392, 2011, https://doi.org/10.1089/ten.teb.2011.0475.
148. Laubach, M., Cheers, G.M., Frankenbach-Désor, T., Weimer, L.P., Baumgartner, H., Böcker, W., Burgkart, R., Cidonio, G., D'Este, M., Dirnagl, U., Drude, N., Eschweiler, J., Friebe, M., Ganse, B., Hartmann, H., Hildebrand, F., Hoog Antink, C., Kim, M., Kneser, U., Łojkowski, W., Marmitt, G., Mayer-Wagner, S., Praster, M., Reimers, N., Schenke-Layland, K., Schulz, A.P., Spicher, N., Stoppe, C., Toelch, U., van Griensven, M., Wehrle, E., Weschke, S., Holzapfel, B.M., Hutmacher, D.W., Clinical translation of 3D-printed patient-specific bone implants: a consensus statement. *Int. J. Surg.*, 111, 7497–7506, 2025, https://doi.org/10.1097/js9.0000000000002944.
149. Chakrapani, G., Vijay Kumar, V., Zare, M., Next-Generation Biomaterials: Design Strategies, Clinical Translation, and the Rise of Intelligent Therapeutic Platforms. *ACS Omega*, 10, 46272–46293, 2025, https://doi.org/10.1021/acsomega.5c06034.

Part III

STUDIES AND CHALLENGES

11

Degradation Studies of Resorbable Composites

Kandala Chandra Sekhar[1], Atanu Kumar Paul[1] and Gourhari Chakraborty[2*]

[1]Department of Chemical Engineering, National Institute of Technology Andhra Pradesh, Tadepalligudem, Andhra Pradesh, India
[2]School of Chemical and Biotechnology, SASTRA Deemed University, Thirumalaisamudram, Tamil Nadu, India

Abstract
Resorbable composites are used for numerous biomedical applications including scaffolds, implants, drug carriers, sutures, etc. Polylactic acid (PLA), Polycaprolactone (PCL), PLA-based copolymers, Gelatin, Collagen, and Cellulose are some of the resorbable polymers primarily utilized to develop composites for *in-vivo* applications. This present chapter tells about the degradation properties of different bio-resorbable composites. This chapter, emphasizes three major composite materials fabricated based on the reinforcement Tri-calcium phosphate (TCP), Hydroxyapatite (HA), and Calcium Carbonate (CC) for various applications. Composite degradation mechanisms, properties, degradation behaviors, factors affecting degradation, degradation models, their applications, and prospects are suitably incorporated in corroboration with recent literature.

***Keywords*:** Resorbable polymer, PLA, degradation, *in vivo*, *in vitro*, composites

11.1 Introduction

Fabrication of composite materials is carried out by combining two or more constituent materials with significantly different qualities to form a material with unique properties [1].

Resorbable composites gained much importance in the modern-day world because of their ability to degrade and get replaced by natural tissues.

**Corresponding author*: gourhari115@gmail.com

Arbind Prasad (ed.) Resorbable Composites for Bioimplants and Fixation Devices, (331–354) © 2026 Scrivener Publishing LLC

There is a very slight difference between the terms bioabsorbable composites and resorbable composites, both seem to be similar but the difference is that bioabsorbable composites get absorbed in the body over time whereas resorbable composites degrade and are replaced by natural tissues over time. Because of their ability to be replaced by natural tissues, resorbable composites find their application in bone tissues, sutures, etc. The use of polymeric composites has been expanded starting from small screws to large tissues in the biomedical domain [2, 3]. The main advantage of using these resorbable composites as implants is the elimination of surgical removal and long-term consequences. Researchers from around the globe focus on various resorbable composites to learn their degradation behavior, and to expand their applications in the relevant field. Some resorbable composites include Tri-calcium phosphate (TCP), Hydroxyapatite (HA), $CaCO_3$-reinforced nanocomposites, metal-reinforced composites, etc. The ability of resorbable composites to degrade and form natural tissues depends on the mechanisms adopted by the composites and the polymer matrices formed by them. The parameters which determine the nature and rate of degradation are biocompatibility, rate of degradation, mechanical properties, material composition, etc. [4, 5]. Resorbable composites must be biocompatible, which means they should not cause an immune response, toxicity, or inflammation in the body. The rate at which the composite breaks down and is resorbed by the body varies depending on the material and application [5]. Composites must have sufficient mechanical strength to impart structural support at the time of the healing process. The strength should match the requirements of the tissue being repaired, such as bone or cartilage [6].

Resorbable composites can be prepared from a range of materials, comprised of bio-resorbable polymers and bioactive ceramics, etc. The material composition affects the mechanical properties, degradation rate, and biological activity of the composite. For example, calcium-phosphate-based composites promote bone regeneration, while polymers may be used in soft tissue repair [6].

The degrading pattern of resorbable composites indicates their continuous breakdown and absorption by the body over time. This process usually involves specific reactions that fragment the substance into smaller molecules that the body may expel or replace with naturally occurring tissues. Polymers like polylactic acid (PLA) degrade slower than polyglycolic acid (PGA) while incorporating calcium phosphates showed different degradation behaviors. The body's pH, temperature, and enzyme activity can accelerate or slow the degradation process. More porous materials degrade faster due to increased surface area exposed to body fluids [7, 8]. It is a complicated phenomenon of how resorbable composites degrade and which is comprised of successive steps. This chapter gives an overview of the degradation studies of resorbable composites including the approaches used and the characterization techniques used to evaluate these materials in corroboration with recent works.

11.2 Resorbable Polymers

Biodegradable polymers because of their superior properties like biocompatibility, degradability, processability grabbed the attention for product developments in all the fields [9–13]. The first clinical application of biodegradable polymers is marked sutures [1]. Polymers of α-hydroxy-acids, such as lactic acid, glycolic acids, and ε-caprolactone were approved by FDA for use in the biomedical field [2–4]. Biodegradable polyesters are thermoplastic which makes them suitable for melt processing like extrusion, molding, fiber spinning, etc. Polylactic acid is a polymer of lactic acid made by melt polycondensation or ring-opening polymerization (ROP). PLA-based degradable composites are in use in almost all fields including biomedical. Collagen, gelatin, and starch-based resorbable polymers are also turning out to be effective materials for biomedical products like scaffolds, artificial skin, organs, blood vessels, nerves, etc. (Figure 11.1). These polymers are extracted from various bio-resources like chicken skin, bone, fish, plants resources, etc. based on the nature of the polymer [14]. Biodegradable polymers also can be processed through advanced

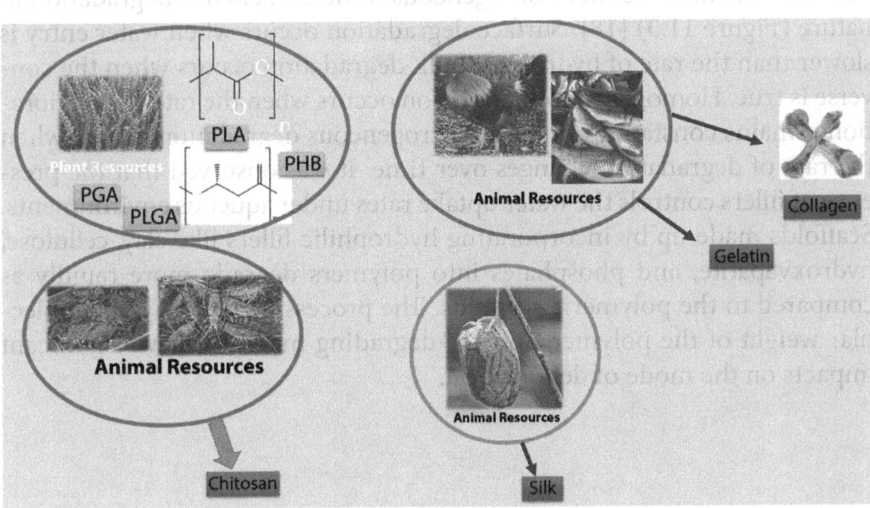

techniques 3D-printing, electrospinning, etc. which promotes their applicability as a drug carrier for *in vivo* release or external release [15, 16]. Resorbable polymers lack stability and product-oriented processability which are addressed by reinforcing different bioactive fillers, additives.

11.3 Degradation Mechanisms

11.3.1 Degradation Mechanisms: Molecular Level

Poly(α-hydroxy-acids) and poly(ε-caprolactone) degrade by esterification in aqueous medium producing carboxylic acid (R-COOH) and alcohol (R'-OH) [5]. An acid or a base can catalyze the fragmentation process (Figure 11.2). The hydrolysis reaction is catalyzed by the dissociation of the carboxylic acid end groups, which lowers the pH level. Autocatalysis is a process that has been commonly reported in poly(α-hydroxy-acids) [6, 7].

It was observed that the hydrolysis of ester bonds can either be random or the final ester bond in the chain can undergo hydrolysis. PCL experiences random hydrolysis of ester bond (random-scission) while in case of poly(D,L-lactide) a significantly higher amount of end-scission (hydrolysis of final ester bond in the chain), also called unzipping can be seen [8, 17].

11.3.2 Degradation Mechanisms: Macroscopic Level

Macroscopic degradation is of two types: surface and bulk degradation. Bulk degradation is either homogeneous or heterogeneous degradation in nature (Figure 11.3) [18]. Surface degradation occurs when water entry is slower than the rate of hydrolysis; bulk degradation occurs when the converse is true. Homogeneous degradation occurs when the rate of deterioration remains constant over time; heterogeneous degradation occurs when the rate of degradation changes over time. It was observed that the presence of fillers controls the water uptake rates under aqueous environments. Scaffolds made up by incorporating hydrophilic fillers like clay, cellulose, hydroxyapatite, and phosphates into polymers degrade more rapidly as compared to the polymeric scaffolds. The processing pathway, the molecular weight of the polymer, and the degrading media all have significant impacts on the mode of degradation.

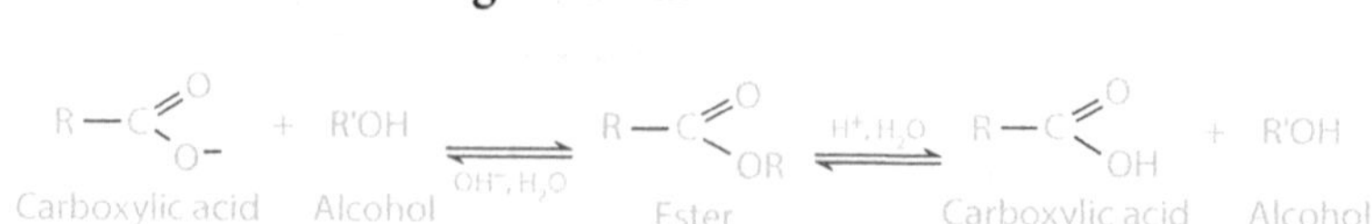

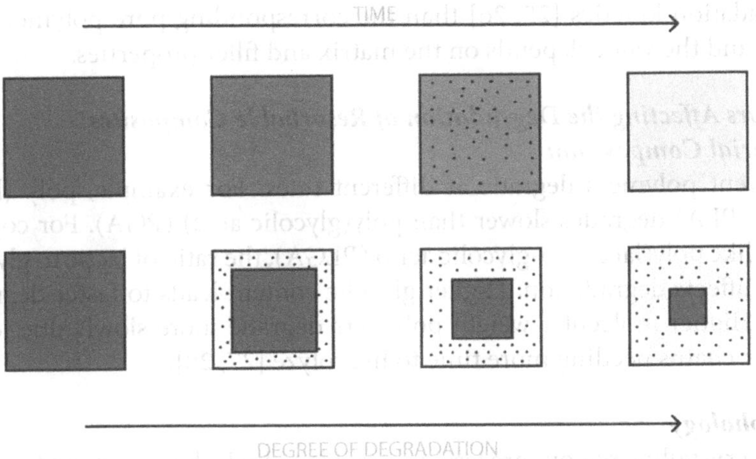

Figure 11.3 Schematic models of surface, homogeneous, and heterogeneous degradation, respectively [18].

11.3.3 Degradation Mechanism: Dissolution

The dissolution of both calcium phosphates and calcium carbonate takes place by dissolution which is influenced by various factors such as Ca^{2+}: anion ratio, crystal structure, crystallinity, lattice defects, phase purity, presence of foreign ions, chemical composition and porosity [19–22]. It was observed that environmental factors like pH, activity of biological macromolecules, temperature and media ionic strength has significant impact in dissolution of resorbable composites [20].

The dissolution of calcium-based ceramic composites is extensively researched by correcting the above-mentioned factors for various composite materials like HA, TCP, and calcium carbonate composite materials. The degradation of resorbable composites depends upon the nature of the filer as well as the loading of the filler. Dissolution is one of the prime factors that control the stability of the composites in the *in-vivo* and *in-vitro* conditions.

11.4 Degradation Behavior of Resorbable Composites

The degradation behavior of resorbable composites is a complex phenomenon with numerous factors pertaining to the polymer matrix, the ceramic filler, and the interactions between the both. Biocomposites have different degradation behavior than their pure polymer counterparts. Experimental studies reveal that biocomposites with both lower [23, 24] and higher degradation kinetics [25, 26] than the corresponding pure polymer scaffolds and the value depends on the matrix and filler properties.

Factors Affecting the Degradation of Resorbable Composites
Material Composition

Different polymers degrade at different rates. For example, poly (lactic acid) (PLA) degrades slower than poly(glycolic acid) (PGA). For copolymers like poly(lactic-co-glycolic acid) (PLGA), the ratio of lactic to glycolic units affects degradation. Higher glycolic content leads to faster degradation. Higher molecular weight polymers degrade more slowly due to the longer chains needing more time to hydrolyze [27, 28].

Morphology

More crystalline regions are less susceptible to hydrolysis contrast to amorphous regions, as water molecules cannot penetrate crystalline regions easily. Highly porous structures allow more water infiltration, increasing the rate of hydrolytic degradation. Composites with larger surface areas degrade faster due to increased exposure to water or other fluids. Resorbable polymeric scaffolds degrade faster compared to films because of their porous nature.

Environmental Factors

Acidic or basic conditions can accelerate the degradation rate. For example, acidic by-products from degradation can further catalyze degradation in the vicinity. Higher temperatures generally increase degradation rates through faster hydrolysis or enzymatic activity. Water is crucial for hydrolytic degradation. Increased water content leads to faster degradation. In biological environments, enzymes (e.g., esterases) can catalyze degradation. This is particularly important in natural resorbable materials like collagen [29]. *In-vivo* drug release depends on the pH of the body fluid. Selective release of drugs is primarily controlled by the pH of the environment.

Mechanical Loading

Mechanical stresses can create microcracks or increase porosity, accelerating degradation by allowing more fluid penetration. Cyclic loading can weaken the material over time and increase the rate of breakdown.

Filler Properties

The type, size, and distribution of fillers (e.g., bioactive ceramics, fibers) influence the overall degradation. For instance, bioactive glass may degrade by releasing ions that promote faster polymer degradation. Poor interfacial bonding between fillers and the polymer matrix can cause microstructural weaknesses, accelerating degradation.

Drug Loading
If the composite is used as a drug delivery system, the presence of certain drugs can alter the polymer's degradation profile. Some drugs may act as plasticizers, softening the matrix and accelerating degradation.

Implantation Site
The specific tissue or organ where the composite is implanted plays a role in degradation. Different areas of the body have varying pH levels, fluid dynamics, and enzyme concentrations.

11.5 Models for Bioresorbable Composites Degradation

The number of computational studies in practically every field of composites is growing. The application of mechanical property modeling in resorbable composites has received significant interest lately. Researchers employed two simple semi-empirical models to demonstrate elastic modulus of poly(lactide-co-glycolide)–nano-biphasic calcium phosphate composites [30] and a finite element model to characterize the elastic properties of poly(ε-caprolactone)–β-TCP scaffolds [31].

Only two significant attempts have been made to model the degradation of bioresorbable composites: a model for the degradation of composites made of polyesters and TCP with equations for both polymer hydrolysis and ceramic dissolution [32] and a simple mechanistic model for poly(L-lactide)–β-TCP composites in which polymer hydrolysis is water-controlled, and the special function that the ceramic filler is thought to have is the rise in water content in the areas of the matrix that surround the particles [26].

11.6 Recent Studies on Degradation of Resorbable Composites

11.6.1 Tri-Calcium Phosphate Incorporated Composites

The study of composites mainly focused on three major composite materials. They are poly(α-hydroxy) acids, and three different ceramic fillers: two calcium orthophosphates, namely tricalcium phosphate (TCP) and hydroxyapatite (HA), and a carbonate, namely calcium carbonate (CC). Tri-Calcium Phosphate composites are studied extensively in Bio-medical applications because of their biocompatibility, osteoconductivity and bioresorbability [33]. It plays a crucial role in bone regeneration and tissue engineering [34]. The *in-vivo* applications include bone regeneration, dental applications, orthopedic implants, drug delivery in animal models, and cartilage tissue engineering [4, 35–51].

The *in-vitro* applications include bone-cell proliferation and differentiation, 3D tissue models, drug delivery studies, cytotoxicity and biocompatibility testing [45–49].

11.6.1.1 Properties of TCP

Chemical composition and phases: The chemical formula of TCP is $Ca_3(PO_4)_2$. TCP exists in 2 crystalline forms: α-TCP and β-TCP. Each has distinct resorption rates and mechanical properties [35].
Biodegradability: TCP breaks down into calcium and phosphate ions which are absorbed by the body, aiding bone mineralization [36].
Mechanical Properties: Pure TCP lacks sufficient mechanical strength for load-bearing applications, necessitating its use in composites [37]. Adding polymers improves TCP's flexibility and toughness [38].
Biological Response: The material promotes cellular activity and integrates seamlessly with surrounding tissues [44].

11.6.1.2 Synthesis Techniques for TCP-Based Composites

Sol-Gel process: This technique involves the chemical reaction of calcium and phosphorus precursors in an aqueous solution to form TCP. The sol-gel technique provides exact control over particle shape and size, which is essential for maximizing mechanical qualities and bioactivity [39].
Powder Processing: The powder route involves mixing TCP with other materials, such as biopolymers or ceramics, followed by sintering to form solid composite structures. This method is used to tailor composite properties for various biomedical applications [40].
3D-Printing: Recent advancements in additive manufacturing allow for the creation of highly customizable TCP scaffolds. This method uses computer-aided design (CAD) to print structures that replicate the porosity and architecture of natural bone, making it an ideal solution for tissue engineering [37].

11.6.1.3 Polymers Combined with TCP

Natural Polymers

- Collagen: Enhances biocompatibility and mimics natural bone structure [41].
- Chitosan: Improves antimicrobial properties and promotes cell adhesion [42].

Synthetic Polymers

- Polylactic Acid (PLA) and Polyglycolic Acid (PGA): Provide structural integrity and controlled degradation [43].
- Polycaprolactone (PCL): Extends degradation time and enhances mechanical stability [4].

11.6.1.4 In Vivo Degradation Behavior

In vivo degradation of TCP composites is influenced by several factors, including the biological environment, load-bearing conditions, and interaction with bone cells. In bone regeneration applications, the degradation of TCP composites should match the rate of bone healing. If the composite degrades too rapidly, it may not give sufficient support, while if it degrades too slowly, it may interfere with the new bone formation. Studies have shown that TCP composites with a controlled degradation rate can provide effective bone substitutes by supporting cell growth, promoting angiogenesis, and gradually transferring the load to the newly formed bone [52, 53].

Strategies to Control Degradation

Several strategies have been developed to control the degradation behavior of TCP composites:

- ***Surface Modification***: Surface treatments of TCP particles, such as coating with biodegradable polymers, can help slow down the degradation rate by reducing the surface area exposed to body fluids. This approach can improve the material's mechanical stability and enhance its biocompatibility [54].
- ***Composite Design***: The design of TCP composites, such as the incorporation of different types of ceramics (e.g., hydroxyapatite, bioactive glass) or the use of composite fibers, can further regulate the degradation rate. These materials can either slow down or accelerate the degradation of TCP composites depending on the desired application [55].

- ***Incorporation of Bioactive Agents:*** The addition of bioactive agents like growth factors or other osteoconductive substances can increase the biological response to the degradation of TCP composites. These agents can help stimulate bone regeneration while the composite degrades, improving the overall healing process [56].

11.6.1.5 Future Prospects of TCP Composites

Tri-calcium phosphate (TCP) composites hold immense potential in regenerative medicine and other biomedical fields, driven by advancements in material science and nanotechnology.

1. ***Customization Through Additive Manufacturing***
   3D printing technology is set to revolutionize TCP composites, allowing precise customization of scaffold structures. These advancements enhance mechanical properties and mimic native bone porosity and architecture [37, 43].
2. ***Functionalized TCP Composites***
   Introduction of bioactive molecules like growth factors and peptides, can transform TCP into multifunctional systems. These composites will provide not only structural support but also enhanced cellular responses for targeted applications [35, 46].
3. ***Nanotechnology Integration***
   Nano-engineering of TCP surfaces is expected to improve osteoconductivity, reduce brittleness, and enhance drug delivery capabilities. Studies indicate that nanoscale modifications allow for faster tissue integration [38, 40].
4. ***Hybrid and Smart Materials***
   Future composites combining TCP with polymers, ceramics, and even metallic phases are expected to offer improved load-bearing capacities and controlled degradation rates. Smart TCP composites that respond to physiological stimuli could be game-changers in personalized medicine [42, 44].
5. ***Advanced Clinical Applications***
   Beyond orthopedic and dental implants, TCP composites are being explored for cartilage repair, vascular grafts, and cancer therapy, significantly broadening their biomedical scope [47].

11.6.2 HA Incorporated Composites

Hydroxyapatite (HA) is a naturally occurring mineral $Ca_{10}(PO_4)_6(OH)_2$ with bone like structure has profound applicability in biomedical field. HA-incorporated composites have gained significant attention in tissue engineering, bone repair, and drug delivery systems. The inclusion of HA into synthetic composites enhances their mechanical properties, bioactivity, and biocompatibility, making them ideal candidates for bone-related applications [58–62]. The *in-vivo* applications include bone tissue engineering, dental applications, drug delivery systems, spinal fusion, and repair [20, 63–65]. The *in-vivo* applications include biocompatibility testing, cell culture models, drug development, and gene therapy vectors [66–69].

11.6.2.1 Properties of HA Incorporated Composites

- ***Mechanical Properties:*** HA-incorporated composites are designed to mimic the mechanical properties of bone. The stiffness, hardness, and strength of HA-based composites are influenced by the type of matrix material, the HA content, and the manufacturing method. Polymer-based composites (e.g., poly(lactic acid), polycaprolactone) reinforced with HA exhibit improved tensile and compressive strength compared to the pure polymer matrix. The HA particles enhance load-bearing capabilities and promote better interaction with biological tissues. However, the mechanical properties of HA composites often depend on the particle size, distribution, and the interface between HA and the matrix material [57, 58].
- ***Biodegradability and Biocompatibility***: HA composites are typically designed to be biodegradable, allowing for gradual integration into the body as the material decomposes and is replaced by natural bone tissue. The rate of biodegradation relies on the type of polymer employed as the matrix and the properties of the HA phase. Biocompatibility is a critical aspect of HA composites, as they must not induce harmful immune responses. Studies have shown that HA-incorporated composites generally exhibit favorable biocompatibility, promoting osteoblast proliferation and osteogenesis. The release of ions from HA, such as calcium and phosphate, further aids in bone regeneration by boosting the formation of new bone tissue [59, 60].

- ***Osteoconductivity:*** One of the key properties of HA composites is osteoconductivity, which refers to the capability to support bone cell attachment, growth, and differentiation. The HA component plays a crucial role in promoting osteoconductivity by imparting a scaffold for cell adhesion and stimulating the formation of new bone tissue. The ionic exchange between HA and the surrounding environment also contributes to bone mineralization, further enhancing its osteoconductive properties [58, 61].
- ***Surface Properties and Functionalization:*** The surface characteristics of HA composites significantly impact their biological behavior. Surface roughness and porosity are essential for cell attachment, proliferation, and differentiation. The incorporation of HA enhances the surface properties of composites, promoting better cell-material interactions. Additionally, HA surfaces can be functionalized with bioactive molecules, such as growth factors, to improve their biological activity. Functionalization of HA composites enhances the osteoinductive potential, accelerating bone regeneration [61, 62].
- ***Antibacterial Properties:*** Another important feature of HA-based composites is their antibacterial properties. HA is known to exhibit antimicrobial activity, which can be enhanced by incorporating silver, zinc oxide, or other antibacterial agents into the composite. The antimicrobial activity of HA composites helps reduce the risk of infection in bone implants and supports the healing process. These composites are being explored for their potential in preventing postoperative infections and improving the overall success of orthopedic implants [63].

11.6.2.2 Degradation Studies of HA Composites

Case 1: PLGA/HA Degradation in Phosphate Buffer Saline (PBS)
In order to assess how HA effects, the degradation of composites hydrolytic degradation in PBS was examined by Naik *et al.*, 2017. The PBS was unchanged during the period of degradation. Pure polymer was used to control degradation. The degradation was studied in time points examining 3 samples per time point. The pH changes were observed till 45 days. The degradation was studied for 50 days and was characterized by measuring water absorption, mass loss, calcium concentration in solution and pH changes. A significant mass loss was observed after 21 days for the composites [70].

Case 2: Degradability of Poly(ε-Caprolatone)/Hydroxyapatite Composites

Hydroxyapatite was used as a filler in the Polycaprolactone (PCL) matrix to improve to Bio-activity as well as compressive properties of composites that could be used in scaffolds. Different properties of PCL/HA composites of various compositions were examined in conjunction with the rate of degradation. PCL has a low rate of degradation hence concentrated sodium hydroxide medium was used to promote the degradation. The compressive strength and other properties were observed to decrease with the increase of degradation time. Pure PCL samples retained their properties well in comparison with those of mixed composites. Although HA was used as a filler to improve the properties of composites, a compromise during hydrolytic degradation was observed [71].

Future Prospects of HA Incorporated Resorbable Composites

Hydroxyapatite (HA) composites are rapidly emerging as a cornerstone material in various fields due to their remarkable biocompatibility, bioactivity, and versatility. These composites hold immense promise for applications in biomedical engineering, dentistry, orthopedics, and even environmental science. With growing advancements in material science and manufacturing techniques, the future of HA composites looks exceptionally promising.

- ***Biomedical Applications***: The most significant potential lies in the biomedical sector, where HA composites are extensively utilized for bone and dental implants due to their similarity to human bone minerals. Advances in 3D printing and nanotechnology are expected to revolutionize the development of HA-based scaffolds and coatings, enabling more precise and patient-specific solutions. These materials can be functionalized with growth factors, antibiotics, or other bioactive agents, making them ideal for promoting bone regeneration and preventing post-surgical infections [72, 73].
- ***Orthopedic Innovations:*** In orthopedics, HA composites are becoming integral to developing load-bearing implants. Future research is focused on improving the mechanical properties of HA by combining it with polymers, metals, and ceramics to create composites that mimic the natural strength and flexibility of bone. The ability to engineer porous structures in these composites is expected to enhance their osseointegration and reduce rejection rates [74, 75].

- ***Dental Applications:*** HA composites have significant potential in dental applications, including fillings, crowns, and adhesives. Innovations in nanotechnology will likely lead to HA nanoparticles with superior mechanical and aesthetic properties, providing more durable and visually appealing dental solutions. Furthermore, the antimicrobial properties of functionalized HA can help reduce the incidence of dental infections [76,77].
- ***Environmental and Industrial Applications***: Beyond healthcare, HA composites have potential applications in environmental remediation, particularly in water purification. HA's ability to adsorb heavy metals and other contaminants is being explored for large-scale filtration systems. Additionally, its bioactivity and compatibility with other materials make it a candidate for various industrial uses, including catalysis and sensors [78, 79].
- ***Challenges and Opportunities:*** Despite their promising future, HA composites face challenges, such as high production costs, variability in properties due to synthesis methods, and limited mechanical strength in pure forms. Addressing these issues through innovative fabrication methods, such as sol-gel techniques, plasma spraying, and additive manufacturing, will open up new avenues. The integration of artificial intelligence and machine learning in material design could accelerate the discovery of optimized HA composites for specific applications [80, 81].

11.6.3 Calcium Carbonate Incorporated Resorbable Composites

Calcium carbonate ($CaCO_3$) is a naturally abundant mineral widely used in composite materials due to its versatility, affordability, and functional properties. It is found in several forms, such as calcite, aragonite, and vaterite, each with distinct structural characteristics that influence its performance in composites. These composites are employed in numerous industries, including construction, packaging, automotive, and biomedical applications [82–94]. The *in-vivo* applications of these composites include bone regeneration, orthopedic Implants, controlled drug delivery, and dental and Periodontal Applications [95–102]. The *in-vitro* applications include bone tissue engineering scaffolds, drug delivery systems, and dental regeneration studies [91–95].

Sources of Calcium Carbonate

- Natural sources: Limestone, marble, chalk, and shells [83].
- Synthetic sources: Precipitated calcium carbonate (PCC), obtained by chemical precipitation processes, which offers controlled particle sizes and shapes [83, 84].

Advantages of Calcium Carbonate in Composites

- **Economic:** Cost-effective filler material that reduces the overall cost of composites [85].
- **Availability:** Readily available in large quantities [85, 86].
- **Eco-friendly:** Often derived from natural sources and recyclable [86].

11.6.3.1 Properties of Calcium Carbonate-Based Composites

Physical Properties

1. **Particle Size and Morphology**
   - Calcium carbonate particles range from nanometer to micrometer sizes.
   - Shapes include spherical, needle-like, and rhombohedral, depending on the source and treatment [87].
2. **Density**
   - Typically around 2.7 g/cm^3, contributing to its use as a filler for lightweight yet strong composites [87].
3. **Surface Area**
   - Higher surface areas are achieved in PCC compared to natural $CaCO_3$, which enhances interaction with polymer matrices [87].

Mechanical Properties

1. **Strength and Rigidity**
   - Improves the stiffness of composites while maintaining adequate flexibility [88].
2. **Toughness**
   - Acts as a crack arrester in composites, improving fracture resistance [89].
3. **Reinforcement**
   - Enhances load-bearing capabilities when well-dispersed in the matrix [89].

Thermal Properties

1. **Thermal Stability**
   - Provides heat resistance, making it suitable for high-temperature applications [90].
2. **Fire Retardancy**
   - Calcium carbonate decomposes upon heating, releasing CO_2, which helps in fire suppression [90].

Chemical Properties

1. **Surface Chemistry**
   - High reactivity, allowing surface modifications to improve compatibility with polymer matrices [89, 90].
2. **Neutrality**
   - Chemically stable and non-toxic, suitable for sensitive applications like food packaging and medical devices [89, 90].

Optical Properties

- Enhances the whiteness and opacity of composites, making them ideal for aesthetic applications such as paints and coatings [91].

11.6.3.2 Degradation Studies of Calcium Carbonate Incorporated Composites

Controlled Degradation of HA/Calcium Carbonate Composites
Spherical calcium carbonate was prepared in presence of sericin which was transformed into HAP microsphere in phosphate solution in presence of microwave irradiation. The changes in $CaCO_3$ was investigated including reaction time, chemical composition of phosphate solution and microwave power to get a series of HAP/$CaCO_3$ composites. The degradation properties were recorded in *in-vitro* conditions. Results showed that degradation speed of composites with higher HAP content was lower. The degradation rate could be altered by altering the composition of HAP and $CaCO_3$. The composites were turned out to be useful for *in-vivo* and *in-vitro* applications [102].

11.6.3.3 *Future Prospects*

Advancements in nanotechnology and composite engineering are expected to enhance the application spectrum of calcium carbonate resorbable composites [100]. Ongoing research focuses on improving their mechanical strength, tuning degradation rates, and integrating additional functionalities such as antibacterial properties and enhanced bioactivity to broaden their utility in both *in vitro* and *in vivo* applications [101].

11.7 Outlook

The degradation behavior of bioresorbable composites comprising tricalcium phosphate (TCP), hydroxyapatite (HA), and calcium carbonate (CC) is critical in determining their suitability for biomedical applications, particularly in bone tissue engineering. These materials are designed to degrade *in vivo*, enabling gradual replacement by natural bone tissue while minimizing the risk of adverse reactions. An in-depth understanding of their degradation mechanisms, rates, and byproducts is crucial for optimizing their performance and ensuring compatibility with physiological conditions.

The degradation process of these composites is influenced by several factors, including the composition, microstructure, porosity, and the ratio of the constituents. TCP exhibits faster degradation due to its higher solubility compared to HA, making it suitable for applications requiring rapid resorption. HA, with its chemical similarity to natural bone mineral, provides long-term structural stability but degrades more slowly. Calcium carbonate, being highly soluble, significantly accelerates the degradation process when incorporated, enhancing ion release and promoting osteogenesis.

Environmental conditions such as pH, temperature, and the presence of biological fluids also play a significant role. Acidic environments, often associated with inflammation, can accelerate the degradation of these composites, particularly for TCP and CC. In contrast, HA maintains its structural integrity better under such conditions, acting as a stabilizing agent. The interplay between these materials ensures a balance between resorption and structural support, a crucial requirement for successful bone regeneration.

Experimental studies demonstrate that the degradation rates can be tailored by adjusting the composite formulation. For instance, increasing the proportion of HA can slow down the overall degradation, providing prolonged mechanical support. Conversely, a higher TCP or CC content can expedite resorption, facilitating quicker bone in-growth. The release of calcium and phosphate ions during degradation further stimulates cellular activities, enhancing osteoblast proliferation and matrix mineralization.

Despite these advantages, challenges such as uneven degradation, potential toxicity of byproducts, and brittleness need to be addressed. Future research should focus on optimizing composite design through advanced fabrication techniques like 3D printing and incorporating bioactive agents to control degradation kinetics and enhance biological responses.

In conclusion, TCP, HA, and CC-based composites exhibit promising potential as bioresorbable materials for bone repair. Their tunable degradation properties, biocompatibility, and osteoconductivity make them applicable for an extensive range of orthopedic and dental applications. However, a comprehensive understanding of their long-term behavior *in vivo* is essential for their effective clinical translation.

AI disclosure statement: The author confirms that no artificial intelligence (AI)-generated content or AI-assisted writing tools were used in the preparation of this chapter (including summaries and the abstract). All text, analysis, and conclusions are the result of the author's original work.

References

1. Horton, C.E., Adamson, J., Mladick, R., Carraway, J., Vicryl synthetic absorbable sutures. *Am. Surg.*, 40, 12, 729–731, 1974.
2. Middleton, J.C. and Tipton, A.J., Synthetic biodegradable polymers as orthopedic devices. *Biomaterials*, 21, 23, 2335–2346, 2000.
3. Van de Velde, K. and Kiekens, P., Biopolymers: Overview of several properties and consequences on their applications. *Polym. Test.*, 21, 4, 433–442, 2002.
4. Woodruff, M.A. and Hutmacher, D.W., The return of a forgotten polymer polycaprolactone in the 21st century. *Prog. Polym. Sci.*, 35, 10, 1217–1256, 2010.
5. Cameron, R. and Kamvari-Moghaddam, A., Synthetic bioresorbable polymers, in: *Durability and Reliability of Medical Polymers*, pp. 96–118, Elsevier, Woodhead Publishing (CRC Press/Maney), Sawston, United Kingdom, 2012.
6. Li, S., *et al.*, Hydrolytic degradation characteristics of aliphatic polyesters derived from lactic and glycolic acids. *J. Biomed. Mater. Res.*, 48, 3, 342–353, 1999.
7. Pitt, G., Gratzl, M., Kimmel, G., Surles, J., Sohindler, A., Aliphatic polyesters II.The degradation of poly(DL-lactide), poly(ε-caprolactone), and their copolymers *in vivo*. *Biomaterials*, 2, 4, 215–220, 1981.

8. Shih, C., Chain-end scission in acid catalyzed hydrolysis of poly(D, L-lactide) in solution. *J. Controlled Release*, 34, 1, 9–15, 1995.
9. Chakraborty, G., Valapa, R.B., Pugazhenthi, G., Katiyar, V., Investigating the properties of poly (lactic acid)/exfoliated graphene based nanocomposites fabricated by versatile coating approach. *Int. J. Biol. Macromol.*, 113, 1080–1091, 2018.
10. Chakraborty, G., Gupta, A., Pugazhenthi, G., Katiyar, V., Facile dispersion of exfoliated graphene/PLA nanocomposites *via in situ* polycondensation with a melt extrusion process and its rheological studies. *J. Appl. Polym. Sci.*, 135, 33, 46476, 2018.
11. Chakraborty, G., Pugazhenthi, G., Katiyar, V., Exfoliated graphene-dispersed poly (lactic acid)-based nanocomposite sensors for ethanol detection. *Polym. Bull.*, 76, 2367–2386, 2019.
12. Chakraborty, G., Katiyar, V., Pugazhenthi, G., Improvisation of polylactic acid (PLA)/exfoliated graphene (GR) nanocomposite for detection of metal ions (Cu2+). *Compos. Sci. Technol.*, 213, 108877, 2021.
13. Hazarika, D., Chakraborty, G., Kumar, A., Katiyar, V., Role of silk nanocrystal (SNC)-ZnO as an antibacterial nucleating nanohybrid for a patterned mimic poly (lactic acid) based nanofabric. *Int. J. Biol. Macromol.*, 242, 125126, 2023.
14. Chakraborty, G. and Prasad, A., 1 Bio-based materials: origin, synthesis, and properties, in: *Sustainable Bio-Based Composites: Biomedical and Engineering applications*, A. Prasad, G. Chakraborty, J. Davim (Eds.), pp. 1–24, De Gruyter, Berlin, Boston, 2024.
15. Chakraborty, G., Prasad, A., Kumar, A., Processing of biodegradable composites, in: *Biodegradable Composites for Packaging Applications*, pp. 33–48, CRC Press, Boca Raton, 2022.
16. Chakraborty, G. and Paul, A.K., 3D Printing in Biomedical Applications: Techniques and Emerging Trends, in: *Advanced Materials and Manufacturing Techniques for Biomedical Applications*, pp. 275–301, Scrivener Publishing LLC, Beverly, MA, 2023.
17. Shih, C., A graphical method for the determination of the mode of hydrolysis of biodegradable polymers. *Pharm. Res.*, 12, 12, 2036–2040, 1995b.
18. Farrar, D., Modeling of the degradation process for bioresorbable polymers, in: *Degradation rate of bioresorbable materials*, pp. 183–206, Elsevier, Woodhead Publishing/CRC Press, Sawston, United Kingdom, 2008.
19. Dorozhkin, S.V., Calcium orthophosphates in nature, biology and medicine. *Materials*, 2, 2, 399–498, 2009.
20. Dorozhkin, S.V., Bioceramics of calcium orthophosphates. *Biomaterials*, 31, 7, 1465–1485, 2010.
21. Morse, J.W., Arvidson, R.S., Lüttge, A., Calcium carbonate formation and dissolution. *Chem. Rev.*, 107, 2, 342–381, 2007.
22. Bohner, M., Calcium orthophosphates in medicine: From ceramics to calcium phosphate cements. *Injury*, 31, D37–D47, 2000.

23. Agrawal, C.M. and Athanasiou, K.A., Technique to control pH in vicinity of biodegrading PLA-PGA implants. *J. Biomed. Mater. Res.*, 38, 2, 105–114, 1997.
24. Yang, Z., Best, S.M., Cameron, R.E., The influence of α-tricalcium phosphate nanoparticles and microparticles on the degradation of poly(D, L-lactide-co-glycolide). *Adv. Mater.*, 21, 38–39, 3900–3904, 2009.
25. Furukawa, T., Matsusue, Y., Yasunaga, T., Shikinami, Y., Okuno, M., Nakamura, T., Biodegradation behavior of ultra-high-strength hydroxyapatite/poly(L-lactide) composite rods for internal fixation of bone fractures. *Biomaterials*, 21, 9, 889–898, 2000.
26. Kobayashi, S. and Yamaji, S., Analytical prediction of hydrolysis behavior of tricalcium phosphate/poly-L-lactic acid composites in simulated body environment. *Adv. Compos. Mater.*, 23, 3, 211–223, 20142014.
27. Marras, S.I., Zuburtikudis, I., Panayiotou, C., Nanostructure vs. Microstructure: Morphological and Thermomechanical Characterization of Poly(l-Lactic Acid)/Layered Silicate Hybrids. *Eur. Polym. J.*, *43*, 2191–2206, 2007, doi: 10.1016/j.eurpolymj.2007.03.013.
28. Eslami, H., Azimi Lisar, H., Jafarzadeh Kashi, T.S., Tahriri, M., Ansari, M., Rafiei, T., Bastami, F., Shahin-Shamsabadi, A., Mashhadi Abbas, F., Tayebi, L., Poly(Lactic-Co-Glycolic Acid)(PLGA)/TiO2 Nanotube Bioactive Composite as a Novel Scaffold for Bone Tissue Engineering: *In Vitro* and *in Vivo* Studies. *Biologicals*, *53*, 51–62, 2018, doi: 10.1016/j.biologicals.2018.02.004.
29. Chen, C.X. and Yoon, J.S., Morphology and Thermal Properties of Poly(L-Lactide)/Poly (Butylene Succinate-Co-Butylene Adipate) Compounded with Twice Functionalized Clay. *J. Polym. Sci. Part B Polym. Phys.*, *43*, 478–487, 2005, doi: 10.1002/polb.20345.
30. Ebrahimian-Hosseinabadi, M., Ashrafizadeh, F., Etemadifar, M., Venkatraman, S.S., Evaluating and modeling the mechanical properties of the prepared PLGA/nano-BCP composite scaffolds for bone tissue engineering. *J. Mater. Res. Technol.*, 27, 12, 1105–1112, 2011.
31. Doyle, H., Lohfeld, S., McHugh, P., Predicting the elastic properties of selective laser sintered PCL/β-TCP bone scaffold materials using computational modeling. *Ann. Biomed. Eng.*, 42, 3, 661–677, 2014.
32. Pan, J., *Modeling degradation of bioresorbable polymeric medical devices*, Elsevier, Woodhead Publishing, Sawston, United Kingdom, 2014.
33. Landi, E., Tampieri, A., Celotti, G., Sprio, S., Densification behaviour and mechanisms of synthetic hydroxyapatites. *J. Eur. Ceram. Soc.*, *20*, 14-15, 2377–2387, 2000.
34. Zhao, J., *et al.*, Preparation and properties of nano-β-TCP/collagen scaffold for bone tissue engineering. *J. Biomed. Mater. Res. A*, *72*, 1, 156–165, 2005.
35. Rouwkema, J., Rivron, N.C., van Blitterswijk, C.A., Vascularization in tissue engineering. *Trends Biotechnol.*, *29*, 10, 452–460, 2011.
36. Legeros, R.Z., Properties of osteoconductive biomaterials: calcium phosphates. *Clin. Orthop. Relat. Res.*, *395*, 81–98, 2008.

37. Huang, Y., *et al.*, 3D printing of biomaterials and scaffolds in regenerative medicine. *Adv. Funct. Mater.*, *25*, 30, 4639–4663, 2015.
38. Tampieri, A., *et al.*, A biomimetic approach to hydroxyapatite composites. *Biomaterials*, *22*, 11, 1365–1370, 2001.
39. Kukovecz, A., *et al.*, Sol-gel synthesis of calcium phosphate bioceramics. *J. Mater. Chem.*, *15*, 3, 221–225, 2005.
40. Dorozhkin, S.V., Biphasic, triphasic, and multiphasic calcium orthophosphates. *Ceram. Int.*, *38*, 8, 6075–6100, 2012.
41. Liao, S., *et al.*, Biomimetic Ca-P deposition on surface modified PCL. *Biomaterials*, *28*, 3, 439–449, 2007.
42. Muzzarelli, R.A., *et al.*, Chitosan, a versatile material in biomedical applications. *Carbohydr. Polym.*, *84*, 1, 54–63, 2012.
43. Vert, M., *et al.*, Degradable polyesters for use in tissue engineering. *J. Controlled Release*, *30*, 1, 3–12, 1994.
44. Xue, W., *et al.*, Development of hydroxyapatite composites for biomedical applications. *J. Biomed. Mater. Res. Part B*, *74*, 1, 1–8, 2005.
45. Hench, L.L. and Wilson, J., Analyzes the biocompatibility and bone-bonding ability of silica-based biomaterials, applicable to TCP composite research. *Science*, *226*, 630–636, 1993.
46. Chai, C. and Ben-Nissan, B., Focuses on bioactive materials for drug delivery, analyzing TCP as a matrix for controlled therapeutic release. *Biomaterials*, 1999.
47. Ripamonti, U., *et al.*, Reviews osteoinduction mechanisms, emphasizing TCP's role in promoting bone morphogenesis. *J. Bone Jt. Surg.*, 1997.
48. Denry, I. and Kuhn, L.T., Design and characterization of calcium phosphate ceramic scaffolds for bone tissue engineering. *Dent. Mater.*, 32, 1, 43–53, 2016.
49. Eggli, P.S., *et al.*, Examines TCP's osteoconductive properties through *in vivo* models for bone grafting. *J. Biomed. Mater. Res.*, 1988.
50. Mayakrishnan, V. and Murugan, P.A., Degradation Studies of Resorbable Materials for Biomedical Applications, in: *Nanomanufacturing Techniques in Sustainable Healthcare Applications*, pp. 258–277, CRC Press, Boca Raton, 2024.
51. Wang, G., Moya, S., Lu, Z., Gregurec, D., Zreiqat, H., Enhancing orthopedic implant bioactivity: refining the nanotopography. *Nanomedicine*, 10, 8, 1327–1341, CRC Press,Boca Raton, 2015.
52. Huang, Y., *et al.*, Degradation behavior of novel Fe/β-TCP composites produced by powder metallurgy for orthopedic applications. *J. Mater. Sci.*, 52, 9785–9797, 2017.
53. Lee, S.H., *et al.*, Characterization of the degradation behavior of polycaprolactone/β-tricalcium phosphate composites *in vitro*. *Biomaterials*, 30, 1007–1016, 2009.
54. Hu, Z., *et al.*, Development of PCL/β-TCP composites and their degradation behavior in simulated body fluid. *J. Biomater. Sci., Polym. Ed.*, 29, 1245–1256, 2018.

55. Jang, J.H., *et al.*, Degradation of β-tricalcium phosphate/polymer composites for bone regeneration. *J. Mater. Sci. - Mater. Med.*, 25, 1687–1695, 2014.
56. Xie, J., *et al.*, Degradation of β-TCP composites with various additives for enhanced osteogenesis. *Biomed. Mater. Res. B Appl. Biomater.*, 104, 1764–1773, 2016.
57. Zhang, L., Feng, X., Liu, H., *et al.*, Hydroxyapatite/collagen composite materials formation in simulated body fluid environment. *Mater. Lett.*, 58, 5, 719–722, 2004.
58. Manivasagam, G., Dhinasekaran, D., Rajamanickam, A., Biomedical Implants: Corrosion and its Prevention—A Review. *Recent Pat. Corros. Sci.*, 2, 40–54, 2010.
59. De Santis, R., Ambrosio, L., Nicolais, L., Polymer-based composites for hip prostheses. *J. Inorg. Biochem.*, 79, 2, 97–102, 2000.
60. Bui, H.T., Nguyen, D.T., Tran, T.T., Mechanical properties and degradation behavior of hydroxyapatite/polymer composites. *J. Biomed. Mater. Res. B Appl. Biomater.*, 110, 5, 1575–1588, 2022.
61. Aghayev, A. and Jabbari, E., Hydroxyapatite-based composites for bone tissue engineering. *J. Biomater. Sci., Polym. Ed.*, 29, 12, 1609–1630, 2018.
62. Chung, W.H., Leong, F.M., *et al.*, Hydroxyapatite/Polymer Composite Nanomaterials for Biomedical Applications. *Nanomaterials*, 7, 10, 311, 2017.
63. Hing, K.A., Bioceramic bone graft substitutes: Influence of porosity and chemistry. *Int. J. Appl. Ceram. Technol.*, 2, 3, 184–199, 2005.
64. Kumar, R. and Prakash, K.H., Hydroxyapatite and composites: Versatile material for bone tissue engineering. *Front. Bioeng. Biotechnol.*, 7, 267, 2019.
65. Berzina-Cimdina, L. and Borodajenko, N., Research of calcium phosphates using Fourier transform infrared spectroscopy, in: *Infrared Spectroscopy – Materials Science, Engineering and Technology*, IntechOpen, London, UK, 2012.
66. Tamimi, F., *et al.*, Injectable bone substitutes: Current status and future perspectives. *Biomaterials*, 30, 27, 5279–5292, 2009.
67. Gopi, D., *et al.*, Development of hydroxyapatite coatings on biomedical implants: A review. *J. R. Soc. Interface*, 11, 100, 20131067, 2014.
68. Bohner, M., *et al.*, Effects of calcium phosphate nanoparticles on calcium phosphate cements. *Biomaterials*, 26, 17, 3542–3549, 2005.
69. Ma, P.X., Biomimetic materials for tissue engineering. *Adv. Drug Delivery Rev.*, 60, 2, 184–198, 2008.
70. Naik, A., Shepherd, D.V., Shepherd, J.H., Best, S.M., Cameron, R.E., The effect of the type of HA on the degradation of PLGA/HA composites. *Mater. Sci. Eng., C*, 70, 824–831, 2017.
71. Ang, K.C., Leong, K.F., Chua, C.K., Chandrasekaran, M., Compressive properties and degradability of poly (ε-caprolatone)/hydroxyapatite composites under accelerated hydrolytic degradation. *J. Biomed. Mater. Res. A*, 80, 3, 655–660, 2007.

72. Qian, G., Xiong, L., Ye, Q., Advancements in Hydroxyapatite Composites for Biomedical Applications. *RSC Adv.*, *13*, 16512, 2023.
73. Family, R., Solati-Hashjin, M., Nik, S.N., Nemati, A., Hydroxyapatite for Bone Regeneration: A Review. *Casp. J. Intern. Med.*, 3, 460, 2012.
74. Chen, Z.F., Darvell, B.W., Leung, V.W.H., Dental Applications of Hydroxyapatite. *Arch. Oral Biol.*, *49*, 359, 2004.
75. Codrea, C.I., Croitoru, A.-M., Baciu, C.C., Functionalized Hydroxyapatite in Biomedical Engineering. *J. Clin. Med.*, *10*, 253, 2021.
76. El-Kader, M.F.H.A., Ahmed, M.K., Elabbasy, M.T., Bioactive Surface Coatings of Hydroxyapatite. *Surf. Coat. Technol.*, *409*, 126861, 2021.
77. Sobczak-Kupiec, A., Drabczyk, A., Florkiewicz, W., Future Directions in Hydroxyapatite Compos. *Mater.*, *14*, 2096, 2021.
78. Albrektsson, T. and Johansson, C., Osseointegration and Orthopedic Applications of HA Composites. *Eur. Spine J.*, *10*, S96, 2001.
79. Ittycheria, P.G., George, T., John, M., Hydroxyapatite Biomaterials in Tissue Engineering, in: *Novel Biomaterials for Tissue Engineering*, IntechOpen, London, UK, 2024.
80. Bal, Z., Kaito, T., Korkusuz, F., Load-Bearing Applications of Hydroxyapatite Composites. *Emergent Mater.*, *3*, 521, 2020.
81. Paszynska, E., Pawinska, M., Enax, J., Innovations in HA for Dental Applications. *Front. Public Health*, *11*, 2023.
82. Niu, Y., Liu, J., Aymonier, C., Fermani, S., Kralj, D., Falini, G., Zhou, C., Calcium carbonate: controlled synthesis, surface functionalization, and nanostructured materials. *Chem. Soc. Rev.*, *51*, 7883–7943, 2022.
83. Yang, H., Qin, J., Liu, Y., Tang, R., Calcium carbonate nanostructures and biomimetic synthesis. *J. Mater. Chem.*, *21*, 29, 11135–11152, 2011, DOI: 10.1039/C1JM10363A.
84. Kim, J.E., Choi, J.H., Kweon, H.Y., Fabrication of biomimetic cellulose/calcium-deficient HA composite scaffolds for bone tissue engineering. *J. Biomed. Mater. Res. B Appl. Biomater.*, *106*, 4, 1626–1635, 2018.
85. Li, Q. and Wu, Z., Monodisperse rosette-like calcite mesocrystals in CMC by biomimetic gas-diffusion method. *CrystEngComm*, *11*, 7, 1356–1362, 2009, DOI: 10.1039/B812420A.
86. Lukasheva, N.V. and Tolmachev, D.A., Molecular dynamics simulation of HA/BC nanocomposites. *Mater. Sci. Forum*, *868*, 105–112, 2016, DOI: 10.4028/MSF201688.
87. Rauch, J., Wang, C., Zhao, S., Calcite synthesis with minor aragonite inclusion in composites. *Compos. Sci. Technol.*, *72*, 8, 983–988, 2012, DOI: 10.1016/j.compscitech.2012.03.015.
88. Kumar, M. and Biswas, D., Thermomechanical analyses of alkali-treated coconut husk-bagasse fiber-calcium carbonate hybrid composites. *Sustainability*, *15*, 2, 1686, 2023.
89. Xu, H., Zhang, X., Zhan, Y., Mechanical properties of calcium carbonate-based biodegradable polymers. *Front. Bioeng. Biotechnol.*, *9*, 937266, 2021, DOI: 10.3389/fbioe.2022.937266.

90. Okuda, T. and Kondo, T., Biomimetic approaches to preparing CMC/HA composites. *Mater. Lett.*, *290*, 129–135, 2022, DOI: 10.1016/j.matlet.2021.129032.
91. Berg, J.M., Tymoczko, J.L., Stryer, L., Biomaterials science: The role of calcium carbonate in advanced composites. *Nat. Rev. Mater.*, *10*, 1, 1–15, 2015, DOI: 10.1038/s41578-015-0001-y.
92. Kim, D.H., Kim, K.H., Seo, J.-H., Pack, S.P., Biomimetic Scaffolds of Calcium-Based Materials for Bone Regeneration. *Biomimetics*, 9, 9, 511, MDPI, 2024.
93. Schlegel, K.A. and Lutz, R., A Novel Resorbable Composite Material Containing Poly(ester-co-urethane) and Precipitated Calcium Carbonate Spherulites for Bone Augmentation—Development and Preclinical Pilot Trials. *Molecules*, 26, 1, 102, MDPI, 2021.
94. Umemoto, S., Furusawa, T., Unuma, H., Tajika, M., Sekino, T., *In vivo* bioresorbability and bone formation ability of sintered highly pure calcium carbonate granules. *Dent. Mater. J.*, 40, 5, 1202–1207, 2021.
95. Brunello, G., Panda, S., Schiavon, L., Sivolella, S., Biasetto, L., Del Fabbro, M., The impact of bioceramic scaffolds on bone regeneration in preclinical *in vivo* studies: A systematic review. *Materials*, 13, 7, 1500, 2020.
96. He, F., Yang, F., Zhu, J., Peng, Y., Tian, X., Chen, X., Fabrication of a novel calcium carbonate composite ceramic as bone substitute. *J. Am. Ceram. Soc.*, 98, 1, 223–228, 2015.
97. Lewandowska-Łańcucka, J., Fiejdasz, S., Rodzik, L., Łatkiewicz, A., Nowakowska, M., Novel hybrid materials for preparation of bone tissue engineering scaffolds. *J. Mater. Sci.: Mater. Med.*, 26, 1–15, 2015.
98. Olah, L., Filipczak, K., Jaegermann, Z., Czigany, T., Borbas, L., Sosnowski, S., Rosiak, J.M., Synthesis, structural and mechanical properties of porous polymeric scaffolds for bone tissue regeneration based on neat poly (ε-caprolactone) and its composites with calcium carbonate. *Polym. Adv. Technol.*, 17, 11-12, 889–897, 2006.
99. Rode, C., Wyrwa, R., Weisser, J., Schnabelrauch, M., Vučak, M., Grom, S., Lutz, R., A Novel Resorbable Composite Material Containing Poly (ester-co-urethane) and Precipitated Calcium Carbonate Spherulites for Bone Augmentation—Development and Preclinical Pilot Trials. *Molecules*, 26, 1, 102, 2020.
100. Campodoni, E., Montanari, M., Artusi, C., Bassi, G., Furlani, F., Montesi, M., Tampieri, A., Calcium-Based biomineralization: a smart approach for the design of novel multifunctional hybrid materials. *J. Compos. Sci.*, 5, 10, 278, 2021.
101. Wang, J., Fan, Z., Li, S., Liu, X., Shen, X., Su, F., Fabrication and characterization of composites composed of a bioresorbable polyester matrix and surface modified calcium carbonate whisker for bone regeneration. *Polym. Adv. Technol.*, 28, 12, 1892–190, 2017.
102. Yang, N., Zhong, Q., Zhou, Y., Kundu, S.C., Yao, J., Cai, Y., Controlled degradation pattern of hydroxyapatite/calcium carbonate composite microspheres. *Microsc. Res. Tech.*, 79, 6, 518–524, 2016.

12

Biological Characterization and Testing of Resorbable Composites

Udaya Vaka[1], Subhasree Panda[2] and M.C. Ramkumar[1]*

[1]*Department of Physics, School of Advanced Sciences, Vellore Institute of Technology, Chennai Campus, Chennai, Tamil Nadu, India*
[2]*Functional Nanomaterials and Polymer Nanocomposite Laboratory, Department of Physics, VIT-AP University, Amaravati, Guntur, Andhra Pradesh, India*

Abstract

Resorbable composites have emerged as a transformative class of biomaterials, extensively used in medical applications such as bone repair, tissue engineering, and drug delivery. Key considerations include the material's biocompatibility, degradation behavior, mechanical properties, and interactions with biological systems. The chapter explores into *in vitro* and *in vivo* testing protocols, emphasizing cell viability assays, and mechanical integrity assessments. The role of material composition, surface properties, and bioactive additives in influencing cellular responses and degradation kinetics is discussed in detail. Special attention is given to regulatory requirements and standardization of testing methodologies to ensure clinical translation. The biological characterization and testing methodologies essential for evaluating the safety, efficacy, and performance of resorbable composites are discussed in this chapter.

***Keywords*:** Resorbable composites, characterization, mechanical testing, *in vitro, in vivo* studies

**Corresponding author*: ramkumar.mc@vit.ac.in

Arbind Prasad (ed.) Resorbable Composites for Bioimplants and Fixation Devices, (355–380) © 2026 Scrivener Publishing LLC

12.1 Introduction

Resorbable composites have gained significant attention in biomedical applications due to their unique ability to degrade within the body and eliminate the need for surgical removal. These materials, typically composed of polymeric matrices, metallic alloys, carbon-fiber, and ceramics reinforced with bioactive fillers, are extensively used in tissue engineering, orthopedic implants, dental, drug delivery systems, etc. (Figure 12.1) [1, 2]. In recent studies, Magnesium-Zinc-Calcium (Mg-Zn-Ca) alloys exhibited significant attention due to their outstanding biocompatibility. These alloys are particularly promising as biodegradable materials for orthopedic internal fixation, because of their low specific density and elastic modulus. Additionally, their mechanical and corrosion properties can be finely tuned by adjusting the Zn/Ca ratio or applying heat treatments, further enhancing their versatility [3–5]. Cervical spine and lumbar interbody fusion cages are intended to employ bioresorbable devices that are made of hydroxyapatite (HA) and polymers because of their bioresorption and remarkable mechanical strength [6, 7].

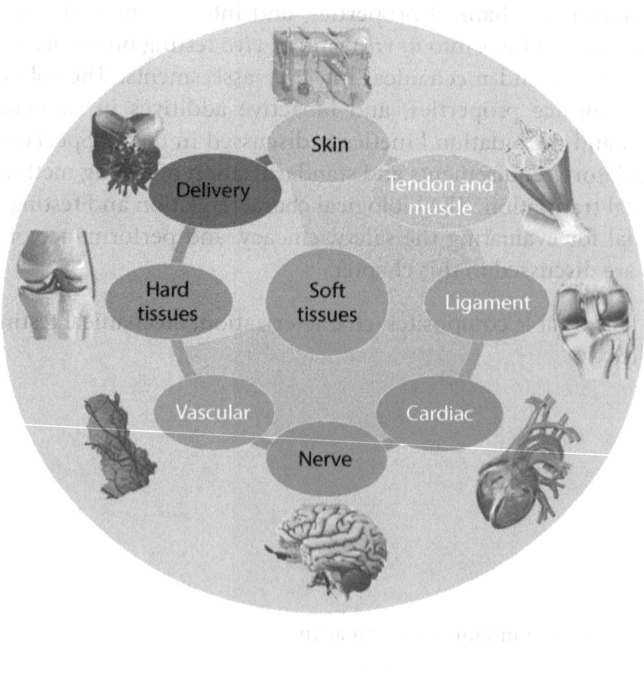

In a recent study, a three-phase composite material was developed using silk fibroin fibers and HA, designed specifically for load-bearing bone fracture fixation devices [8]. The characterization of resorbable composites involves understanding their interactions with biological systems, which is crucial for ensuring their safety and efficacy. This includes evaluating their structure, pore size, mechanical strength, degradation time, etc. as per their biomedical application [9]. In a study, the degradation rate of prepared composite was analyzed by testing their mechanical strength, weight loss and change in the pH [10]. Understanding the precise distribution of each component in the composite material is essential, it helps in determining the performance of the composite when it interacts with the body fluids, for which the FTIR spectroscopic imaging, SEM-EDX analysis was done [11]. Biological tests of resorbable composites such as *in vitro/in vivo* cytotoxicity, hemocompatibility, immune response, etc. are essential for evaluating their safety, biocompatibility, and functionality in biomedical applications. In a study the *in vivo* and protein release test demonstrated that the composites were biocompatible and promoted the development of natural bone [12]. In another study *in vitro* tests revealed that the prepared polymeric films were biocompatible and had no substantial toxicity to human cells [13]. The characterization and testing of the resorbable composite are discussed in detail later in this chapter.

12.2 Resorbable Composites

Resorbable composites, also known as bioresorbable or biodegradable composites, are a class of advanced materials (Figure 12.2) designed to degrade and be absorbed by the body over time. These composites combine a resorbable matrix material with reinforcing elements to provide mechanical strength, stability, and functionality. After fulfilling their purpose, they degrade into non-toxic byproducts, eliminating the need for surgical removal. Resorbable composites provide several advantages, including the elimination of secondary procedures for removal, which reduces patient risk and healthcare expenses and supports tissue regeneration by progressively shifting the load to healing tissues [14, 15].

12.2.1 Polymer Composites

Resorbable polymers are widely used in the production of many biomedical materials and devices, including bone implants, medication carriers, and sutures. Over time, the resorbable polymer in the human body breaks

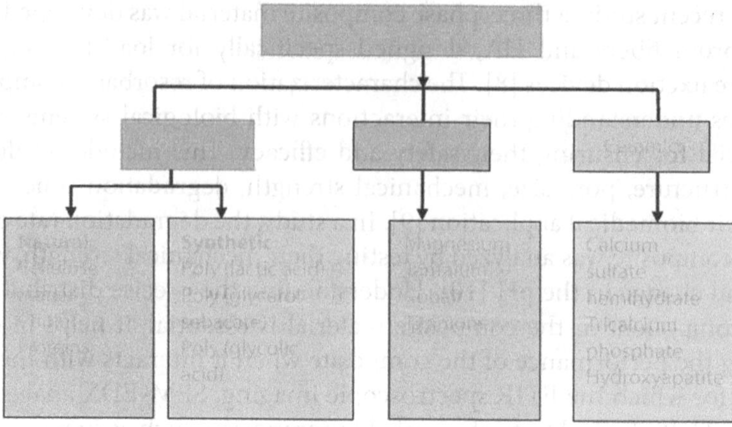

down into tiny pieces. Because there is no need for subsequent surgery, these resorbable materials and technologies are particularly well suited for *in vivo* implantation [16, 17]. There are two types of resorbable polymers; natural and synthetic. Resorbable synthetic polymers can be entirely produced from a synthetic monomer or modified natural polymers. The advantages of resorbable synthetic polymers, including their long shelf life, controllable formation and mechanical properties have drawn a lot of attention in the field of tissue engineering [18, 19]. A common option for any biomaterials is the polyester family known as poly (lactic acid)/poly (glycolic acid) (PLA/PGA) [20, 21]. It is possible to prepare PLA and PGA separately or copolymerize them to create poly (lactic acid-co-glycolic acid) (PLGA). They are suitable for severe degradation rate methods since PLA degrades slowly and PGA/PLGA degrades quickly. Clinically, the materials are recognized as orthopedic implants and resorbable sutures [22].

Resorbable polymer composites are useful for creating materials with a broad range of hydrolytic breakdown characteristics. The way the filler or reinforcing substance degrades is another consideration. Hydrolytic degradation in poly (a-hydroxy acids) is known to be impacted by a uniform dispersion of filler particles, such as magnesium oxide, tricalcium phosphate and Ca compounds [23, 24]. Composite made of a polymer matrix polylactide reinforced with phosphate glass fiber has been a considerable bioimplant. Because of their mechanical qualities and complete bio resorbability, they are perfect for bone repair applications, which release Ca and phosphate ions [25, 26]. Resorbable natural polymers are biodegradable materials derived from natural sources, such as collagen, starch, chitosan, and alginate etc.

Thermoplastic starch was synthesized by plasticizing starch with different concentrations of glycerol. To enhance the mechanical qualities, microcrystalline cellulose was utilized as a reinforcement filler. The produced composite exhibited potential use as a bioimplant for replacing trabecular bone [27].

12.2.2 Ceramic Composites

Resorbable ceramic composites are advanced materials combining bioactive ceramics with biodegradable polymers to create scaffolds or implants that gradually degrade in the body. The ceramic component, often calcium phosphate-based materials like HA or tricalcium phosphate, provides strength, bioactivity, and osteoconductivity, promoting bone regeneration. The polymer matrix, typically natural or synthetic resorbable polymers, offers flexibility and controls the composite's degradation rate. These composites are widely used in orthopedics and dental applications, such as bone grafts, due to their ability to integrate with natural tissues while eventually being replaced by new bone. Tricalcium phosphate and calcium sulfate hemihydrate are two examples of resorbable bioceramics that are commonly used as the components of implants. Tricalcium phosphate and calcium sulfate hemihydrate have the longest history of being used to treat bone abnormalities, and they quickly became extremely popular [28, 29]. Strong chemical bonds can be formed between bioactive ceramic materials and host bone tissue without leading to inflammation. Tricalcium phosphate, HA, bioglass and Ca-phosphorous-based ceramics are widely used as bone substitutes. These bioactive ceramics exhibit superior biological response, protein synthesis, and cell adhesion, but they have a much lower mechanical strength than ZrO_2 and Al_2O_3 [30–32].

The considerable brittleness of bioceramics significantly restricts their application in load-bearing implants. HA is frequently added to polymers as part of composite materials or coated on a metal core to alleviate this problem. Enhancing the bonding strength at the HA/metal contact has been the subject of extensive research. Improvements in coating techniques are thought to be essential to address this problem since the coating technique is a key factor in determining the HA layer's adhesion strength and dependability [33, 34]. The ease of processing of the PLA matrix and the improved resorbability of HA have drawn considerable interest in the PLA/HA composite. Nonetheless, interfacial failures in these composites are caused due to the lack of adhesion between the polymer matrix and the ceramic phase. In order to overcome this issue, silane derivatives are used as modifying agents [35].

12.2.3 Metal Composites

Resorbable metal composites are designed for temporary medical implants that degrade and are absorbed by the body over time. These materials, often based on metal alloys are combined with biocompatible reinforcements such as polymers or bio-ceramics to enhance mechanical properties and control degradation rates. They offer a significant advantage in applications like bone fixation and cardiovascular stents, eliminating the need for secondary surgeries to remove implants. Mg alloy's excellent mechanical qualities and perfect biocompatibility are the main characteristics that make them a promising material for orthopedic biodegradable implants. No other element has properties that are more similar to those of bone than Mg. However, one of the main drawbacks of Mg as a biocompatible and biodegradable material is thought to be its poor resistance to corrosion [36–38]. By alloying Mg and applying protective coatings, the rate of corrosion can be decreased. It is widely known that alloying Mg with Al, Zn, Mn, Zr, Ca, and rare earth metals improves both its mechanical qualities and corrosion resistance [39, 40].

One of the most often used cobalt alloys is Co-Cr-Mo because of its remarkable strength and ductility combination. Because of its higher density, rigidity, and elastic modulus value compared to other metallic materials, this specific alloy is perfect for use in implantation techniques [41, 42]. Hot compression molding is used to create polyethylene and diamond like carbon coated metal beads that can be used as hip joint prosthetic alternatives. The produced composite may be a promising combination for orthopedic applications, offering a remedy for wear debris and metal corrosion problems in orthopedic implants [43].

12.3 Characterization (Structural and Physicochemical)

The characterization and testing of resorbable materials are essential steps in ensuring their safety, efficacy, and functionality. These processes involve evaluating their physical, chemical, and mechanical properties to meet stringent regulatory and clinical requirements. Key parameters include the material's degradation kinetics, mechanical strength, thermal stability etc. Based on their applications, various research groups have conducted characterization studies in different ways (Table 12.1), as described below.

Table 12.1 Characterization/tests and applications of bioresorbable composites.

Composite	Characterization	Test	Application	Reference
Polycaprolactone/ Collagen	Three-point bending, ball burst, Contact angle	*In vitro* cytotoxicity, fibroblast migration	Guided bone regeneration	[69]
Carbonated hydroxyapatite	XRD, FTIR, TEM, porosity, thermal analysis, tensile strength	*In vitro* degradation	Bone regeneration	[70]
Tantalum-niobium oxide/Magnesium alloy	XRD, HR-TEM, XPS, SEM, wettability, corrosion	*In vitro* hemocompatibility and cytocompatibility	Biodegradable implant	[71]
Titanium/Eggshell	HR-SEM -EDS, XRD, TEM, tensile strength	*In vitro* cytotoxicity and bacterial study	Biomedical implant	[72]
Polyethylene/ Hydroxyapatite	SEM, porosity, FTIR, micro-CT analysis	*In vitro* hemocompatibility, mice craniotomy model	Reconstructive surgery	[73]
Hydroxyapatite/ Serum albumin/ Lysozyme	Pore size, porosity, tensile strength	Protein release, *in vivo* biocompatibility and resorption	Drug delivery for bone healing	[12]
PLA /PEG/ Triiodobenzoate-poly(ε-caprolactone)	Tensile strength, thermal analysis, degradation analysis	*In vivo* computed tomography imaging, human cadaveric implantability tests	Stent for liver transplantation	[75]
Molybdenum/Poly (D, L-lactic-co-glycolic acid)/ Polycaprolactone	Static immersion, SEM, degradation	*In vitro* cytotoxicity, *ex vivo*	Epicardial pacing wires	[76]

12.3.1 Tensile Strength

A Lloyd tensiometer is used to evaluate the sample's tensile strength and elongation at breakage. The sample was held between two clamps and pulled over a predetermined distance by the top clamp before being put back in place. The following formula was used to get the tensile strength and elongation at breakage [44].

$$\text{Tensile strength (MPa)} = \text{Breaking force (N)} / \text{Cross sectional area of sample (mm}^2\text{)} \quad (12.1)$$

$$\text{Elongation at breakage (\%)} = \text{Increase in length at breaking point (mm)} / \text{Original length (mm)} \times 100\% \quad (12.2)$$

Tensometer can also be used to conduct the adhesive and tensile tests. After applying pressure to the samples with a designated bonding region for few seconds at room temperature, the external pressure was promptly withdrawn. When attached to the tensometer, the adhering substrates were subjected to a shear stress. The adhesive strength, which is determined by dividing the maximum force by the initial bonding area, was determined by recording the maximum stress during shear adhesive testing [45].

12.3.2 pH

By submerging the sample in distilled water (DW), the pH was ascertained. After certain time, the sample was removed from the DW and its pH was measured. After then, the sample was kept back in the same DW and the pH was measured again after 1 day, to evaluate the changes in the value [46].

12.3.3 Thermal Analysis

A differential scanning calorimeter, operated in N_2 environment is used to examine the thermal behaviour of the samples. During the samples heating and cooling process, the melting, crystallization, and glass transition temperatures were noted [47]. A Thermogravimetry Differential Thermal analysis instrument can also be used to study the thermal behavior of the samples. An empty alumina crucible served as a reference, and the samples were heated in crucibles of alumina. Sample temperature, mass, and heat flow were continuously recorded and analyzed [48].

12.3.4 Degradation

After being initially weighed, the samples were put in the phosphate buffer solution (PBS) and sealed in the petri plates. Following incubation in these petri plates, the samples final weight was taken at various times and the degradation is calculated as follows [49].

$$\text{Weight loss} = ((W_i - W_f)/W_i) \times 100 \tag{12.3}$$

where, W_i- initial weight and W_f - final weight

Sample degradation is conducted in accordance with ISO 10993 Medical Biological Evaluation. After recording the sample's dry weights, they were soaked in PBS solution. The samples were degraded for a few days, and a UV-vis spectrometer was used to assess the possible degradation [50].

12.3.5 Swelling

The gravimetric method can be used to assess the swelling. After weighing the sample (W_i), it was submerged in PBS and incubated for different intervals of time. After taking the sample out of the medium for each interval, it was gently dried on filter paper, weighed (W_f), and the swelling rate was computed as follows [51].

$$\text{Swelling rate (\%)} = ((W_i - W_f)/ W_f) \text{ x } 100 \tag{12.4}$$

The samples dry weights (W_d) and diameters (D_d) were measured prior to testing. After that, the samples were submerged in PBS for varying lengths of time. The samples were cleaned to eliminate any surface moisture and wet weight (W_w) and diameter (D_w) were then determined. The following formulas were used to determine the water uptake (%) and swelling (%) [52].

$$\text{Water uptake (\%)} = ((W_w - W_d)/W_d) \times 100\% \tag{12.5}$$

$$\text{Swelling (\%)} = ((D_w - D_d)/D_d) \times 100\% \tag{12.6}$$

12.3.6 Porosity

According to mercury intrusion porosimetry, the porosity of every film was assessed. Every test was conducted using an Autopore device. To make sure the right mercury filled the spaces, all samples were dried and put into a penetrometer before examination. The Washburn equation, which uses the physical characteristics of a non-wetting material was used to define the relationship between applied pressure and pore size [53]. A scanning electron microscope was used to examine the samples surface structure and porosity. Platinum was used to coat all the samples for conductivity. The Digital Image Analysis Software was used to estimate the surface porosity of the images [54]. The Archimedes drainage technique can be used to determine the porosity (P) of the samples as follows.

$$P = 1 - (\rho/\rho_0) \tag{12.7}$$

where ρ and ρ_0 represent the samples bulk and theoretical densities. In the meantime, an indentation load was applied for a predetermined amount of time to perform the micro-hardness test [55].

12.3.7 Corrosion Testing

An electrochemical workstation was used to assess the samples corrosion behavior. A platinum sheet, a saturated calomel electrode, and disc samples with an exposed portion were employed as the, counter electrode, reference electrode, and working electrode in a three-electrode cell used for the electrochemical testing. The ASTM G102-89 standard was followed in calculating the corrosion rate [56]. Before the corrosion test, the samples were weighed. After that, the samples corroded for a predetermined amount of time were washed. To get rid of the corrosion products, they were cleansed with chromic acid after being rinsed. Following another weigh-in, the samples corrosion rates were computed as follows [57].

$$\text{Corrosion Rate} = (K \times W) / (A \times T \times D) \tag{12.8}$$

where, $K = 8.76\times 10^4$, W - mass loss (grams), A - area of sample (cm^2), T - exposure time (hours), and D - material density.

12.3.8 X-Ray Microtomography

X-ray microtomography is used to analyze the healing performance, surrounding tissue response to the implants/fixation devices. A desktop micro-CT Sky Scan was used to scan the samples, and various scanning settings, including rotation step, source voltage/current, and pixel size, were set. CTAn software was used for the analysis and measurements [58].

12.4 Bio-Studies

In vivo and *in vitro* biotesting play crucial roles in evaluating the suitability of resorbable composites for bioimplant applications. *In vitro* testing involves assessing the material's biocompatibility, degradation behaviour, and cellular responses, such as cell adhesion, proliferation, and cytotoxicity, under controlled laboratory conditions. These tests provide initial insights into the material's interactions with biological systems. *In vivo* testing, on the other hand, evaluates the performance of the composite within a living organism, focusing on its mechanical stability, degradation rate, tissue response, and overall integration with surrounding tissues. Together, these biotesting methods ensure that resorbable composites are safe, effective, and capable of meeting the functional requirements for bioimplants.

Hemolysis tests assess whether materials used in implants cause the destruction of red blood cells, which could lead to adverse reactions such as anemia or inflammatory responses. By identifying hemolytic properties, manufacturers can refine materials and coatings to minimize risks, ensuring the devices support healing without compromising blood integrity. After obtaining a red blood cells dispersion by centrifuging whole blood (animal/human) diluted with saline, the dispersion was combined with sample solution and allowed to incubate. Following incubation, the mixture was centrifuged, and a microplate reader was used to detect the absorbance of the top layers. Normal saline served as the negative control, and the Triton X-100 as the positive control [59].

In vitro degradation studies are critical for assessing the performance and safety of bioimplants and fixation devices. These studies simulate physiological conditions to evaluate material breakdown, mechanical integrity, and release of degradation by-products over time. Understanding degradation behavior ensures that the device maintains functionality during the healing process and biodegrades without adverse effects. The samples were submerged in blood plasma that had been diluted with PBS solution. The solution was then incubated and kept frozen until they were examined using inductively coupled plasma. SEM was also used to examine how the samples surface morphologies changed after being submerged in diluted plasma [60].

Drug release studies are crucial for bioimplants and fixation devices as they ensure the controlled and sustained delivery of therapeutic agents at the implant site. This helps in reducing infection risks, promoting tissue healing, and enhancing implant integration with surrounding biological structures. A UV-Vis Spectrophotometer was used to detect the samples absorption spectra. Samples were immersed in the PBS solution and at certain intervals, measurements were taken. The measured spectra were compared to the drug's reference spectra in PBS in order to determine the drug's loaded concentration [61].

Cytotoxicity test assesses whether materials release toxic substances that could harm cells, which is essential to prevent adverse biological responses when implanted in the body. After being cleaned with Milli-Q water, the samples were immersed in ethanol and exposed to UV light to sterilize them. Following that, samples were incubated using Dulbecco Eagle medium. The samples were then collected and combined with MC3T3-E1 preosteoblastic cells for the cytotoxicity assay. An Alamar Blue cell viability reagent was used to measure the rate of proliferation and a plate reader was used to measure the fluorescence at both the excitation and emission wavelengths [62]. NIH 3T3 fibroblast cells were used for the live/dead cells experiment and the cell counting kit to assess the samples cytotoxicity. The cell culture media was used to seed the cells, and they were then incubated. The cells were then incubated once more after the culture medium was swapped with CCK-8 solution. The sample solution was then moved to a cell plate, and a microplate reader was used to measure its absorbance. Using a fluorescence microscope, the living and dead cells were assessed [63]. Human endothelial cells and mouse fibroblast cells can also be used to assess the cytotoxicity of the sample using the MTT test for cell growth and survival. The absorbance was measured using a microplate reader to look into the cell viability [64]. After seeding and incubating MC3T3 mouse pre-osteoblast cells, the growth media was introduced and the sample was added to it. To assess cell attachment and proliferation, a Cell Counting Kit assay was conducted. A standard curve was used to translate the absorbance into live cell counts. Following an incubation period in phosphate buffered saline, the samples were imaged using a fluorescence microscope, revealing that dead cells were red and living cells were green [65]. Cell viability test determine whether the materials used, support the survival and function of living cells, which is a key factor in ensuring the implant to integrate well with the surrounding tissues. To test the cell vitality, an MTT experiment was conducted. MC3T3 cells were first plated and allowed to incubate. The cells were then exposed to the samples for a few days. To find the absorbance, a Multi-Scan Go spectrophotometer was used. The cell viability % was calculated as follows [66].

$$\text{Cell viability } (\%) = (T - B) / (C - B) \times 100 \tag{12.9}$$

where, T is the test data, B is blank data, and C is control data.

In general, *in vivo* testing for bioimplants and fixation devices involves implanting the device into a living organism, typically an animal model, under controlled conditions to assess its biological compatibility, functionality, and safety. The procedure starts with preclinical planning, including selecting a suitable animal model that closely mimics human physiology. The bioimplant is surgically implanted into the target site, such as bone or soft tissue, following ethical and regulatory guidelines. Post-implantation, parameters like biocompatibility, mechanical stability, tissue response, and integration are monitored over time using imaging techniques, histological analysis, and biomechanical testing. These results help evaluate the performance and potential clinical viability of the device. For a few months, the sample was inserted into the rat's femoral condyle. After soaking the femur in the decalcification solution for a month, the sample was then taken out of the femoral condyle. SEM was used to view the osteoblast like cells that were adhered to the sample. The tissue slides were stained using Masson's trichrome dye in order to identify bone development. The area of the new bone formed can be calculated as follows [67, 68].

Area% of the new bone formed = (new bone formed area in the bone defect/bone defect area) × 100,

Area% of host bone = (host bone area / bone defect area) × 100.

12.5 Cutting Edge Studies

Meshes of polycaprolactone (PCL) in a variety of patterns were made. Because of its ideal pore design and structural frame, the small dots pattern performed better. However, by considerably increasing porosity while preserving sufficient mechanical characteristics (Figure 12.3), the big dots pattern was able to strike a balance. In addition to offering mechanical stability against vertical stresses, PCL-collagen membranes serve as a barrier that stops cell migration through the membrane and promotes fibroblast attachment. These characteristics are advantageous for applications

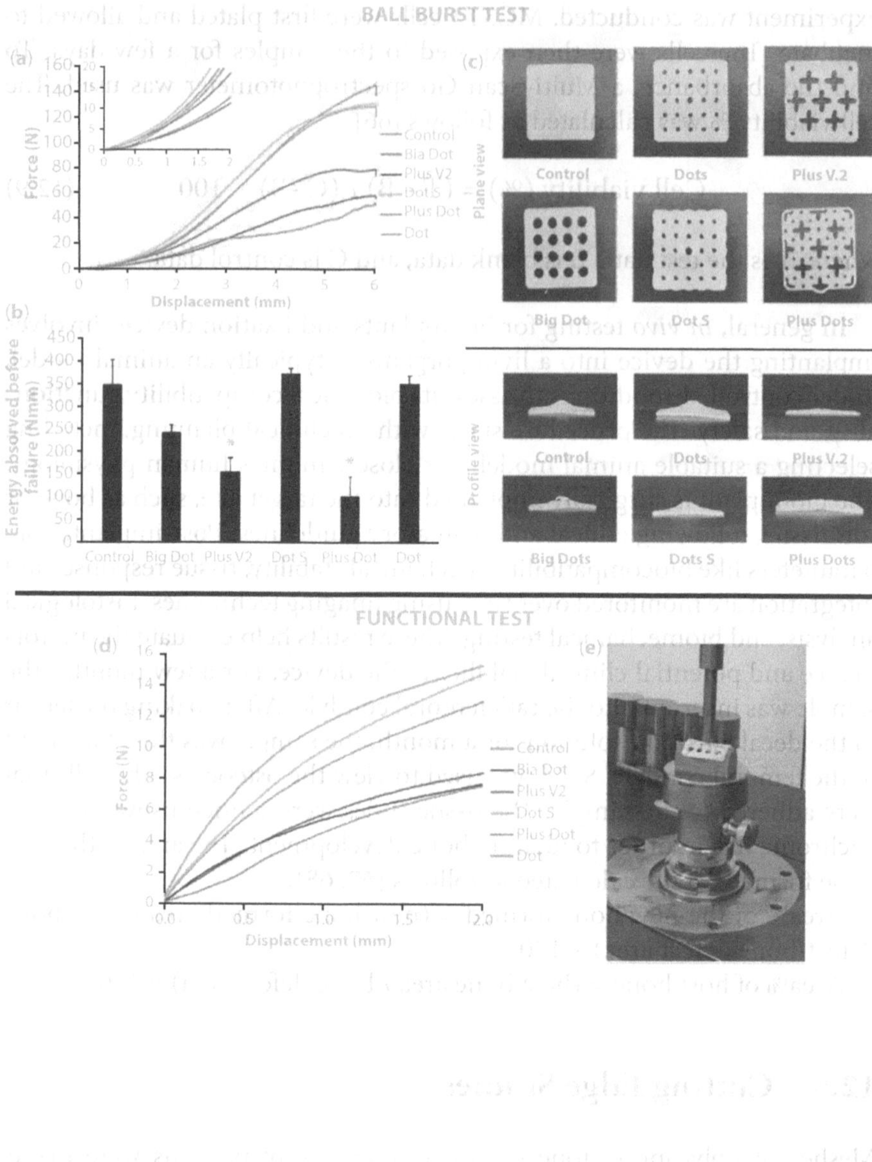

involving directed bone regeneration, in which the membrane is meant to serve as a barrier to stop non-osteogenic tissues from invading the defect location [69]. To achieve comparable porosity levels, carbonated hydroxyapatite (CHA) powders were created for the various compositions and thermally treated at the same temperature. Mechanical strength is enhanced by the addition of carbonate to HA. When compared to HA alone, the *in vitro* degradation rates in physiological and acidic conditions are likewise enhanced, rising with carbonate substitution, suggesting that CHA might be a viable composite for bone regeneration [70]. Biodegradable screws and pins made of regenerated silk fibroin/gelatin (RSF/G) were prepared with enough strength and little swelling. The screws and pins exhibited promising shear and pull out strength (Figure 12.4). Even after a few weeks of enzymatic breakdown, the devices maintained three fourth of their mass. Water absorption and diameter analyses showed that the RSF/G screws provided support throughout the implantation process. The RSF/G screws considerably increased osteogenic differentiation, adhesion, and proliferation while demonstrating high biocompatibility. X-ray, micro-CT, and MRI examinations performed after RSF/G screws were implanted in rat femoral condyles revealed that the fixed screws persisted with minimal inflammatory reaction and that new bone development was seen [52].

Thin films of tantalum niobium oxide were prepared and coated on magnesium alloy. The XPS examination validated the compositions of the materials. High corrosion resistance and protective effectiveness were

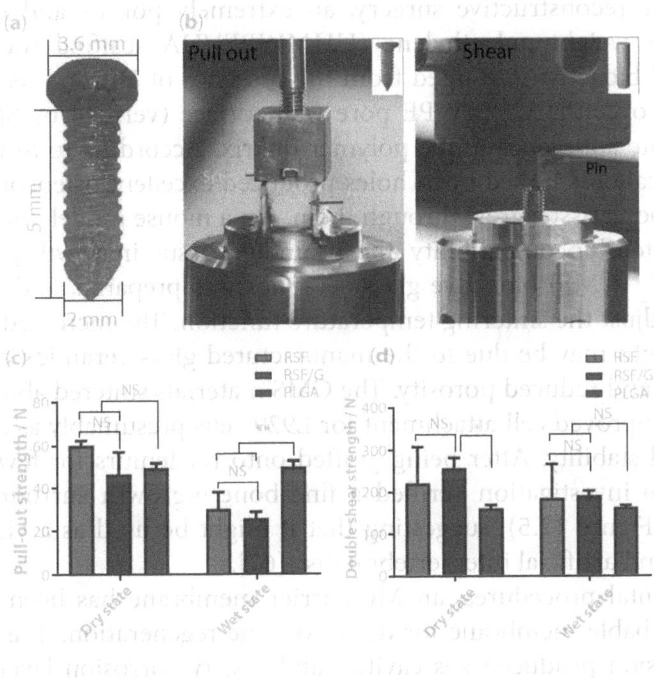

demonstrated by the composite which had greater niobium content. The MC3T3 cells morphology and MTT test verified the cells proliferation and growth. The outcome validated the viability of composite thin films as a potential bioactive coating on biodegradable implants [71]. A single-step hot direct extrusion procedure was used to create thin biodegradable wires from pure Mg and Mg-Zn alloys, which were then coated with polymer. Compared to pure Mg, composite having limited Zn concentration has better corrosion resistance. Cell adhesion and proliferation *in vitro* analysis on MG-63 cells was done. The results revealed that the cell behaviour of Mg-0.15Zn exhibited superior *in vitro* results. Mg-0.4Zn wires, both uncoated and polymer-coated, stimulated osteogenesis *in vivo* analysis. The breakdown process of Mg-0.4Zn wire is thought to be significantly influenced by preferential bone tissue development which represents its potential usage as biodegradable wires [58]. A composite made of metal (Ti) and ceramic (eggshell) showed compressive strength and Young's modulus that matched the human cortical bone. Additionally, the composite showed outstanding antibacterial activity and non-cytotoxicity. The study reports the effective monetization of eggshell waste in bio-implants, making it a potential material for bioimplant applications [72].

For the reconstructive surgery, an extremely porous and ultra-high molecular weight polyethylene (UHMWPE)/HA scaffold was created. Increased bioactivity resulted from the presence of HA and its even distribution over the UHMWPE pore wall surface (verified by SEM) with appropriate adherence to the polymer matrix. According to *in vivo* studies, the scaffolds linked open holes produced excellent osteoconductivity because bone tissue grew through them. On a mouse model, the composite exhibited biocompatibility and promoted tissue ingrowth [73]. CaO-MgO-SiO_2 (CMS) bioactive glass-ceramics was prepared, and ZrO_2 was used to adjust the sintering temperature function. The increased mechanical strength may be due to the manufactured glass-ceramic's crystalline structure and reduced porosity. The CMS materials sintered above 800 °C showed improved cell attachment for L929 cells presumably as a result of structural stability. After being grafted onto rat femurs for few months, an *in-vivo* investigation verified a fine bone-ingrowth surrounding the implant (Figure 12.5), suggesting that it might be used as a viable bone implant and artificial intervertebral disc [67].

For dental procedures, an Mg barrier membrane has been prepared as a resorbable membrane for directed bone regeneration. The Mg metal's corrosion produced gas cavities and a salty corrosion layer, both of which increased the membrane barrier life. Following implantation, the Mg membrane splits the rapidly growing soft tissue from the bone that

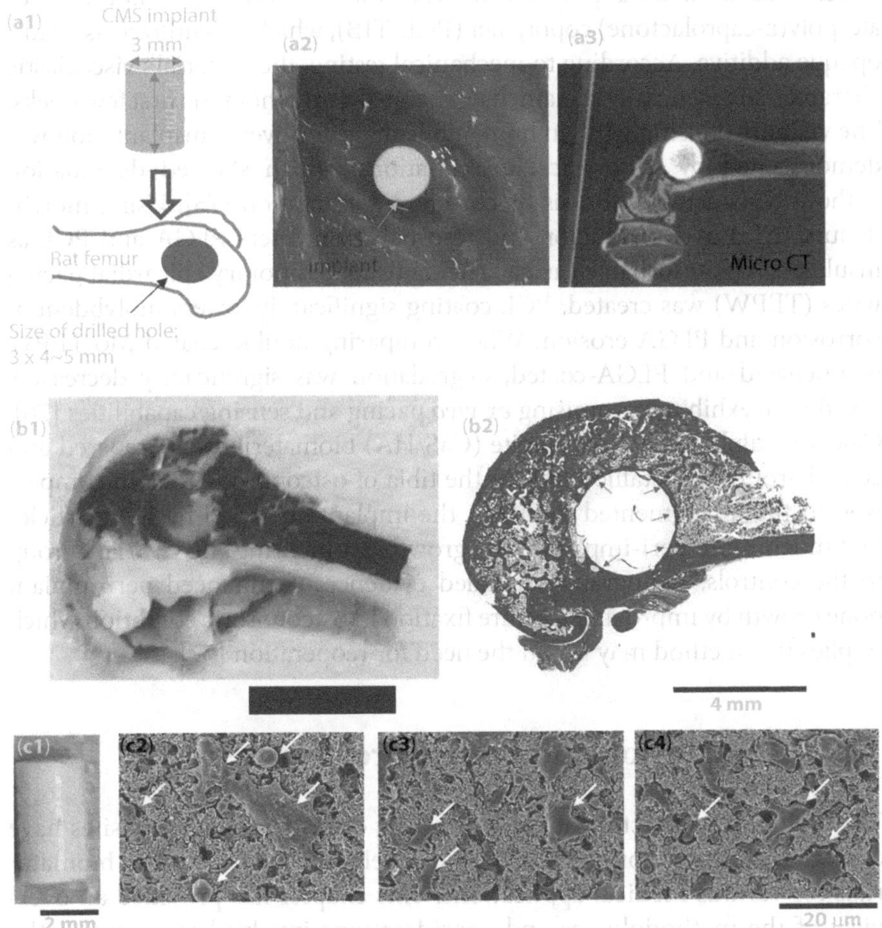

is healing. An *in vivo* investigation showed that the membrane preserved its barrier function during the first healing phase while keeping the bone graft material positioned inside the defect region. Only healthy tissue was left once the membrane had been reabsorbed, indicating that it may be employed as a barrier membrane [74].

X-ray-visible, resorbable, and radiopaque internal biliary stent (RIBS) was designed using a combination of PLA 50-PEG-PLA 50 triblock copolymer, which was used as a polymer matrix, and X-ray-visible triiodobenzoate-poly(ε-caprolactone) copolymer (PCL-TIB), which was utilized as a radiopaque additive. According to mechanical testing, the material's viscoelastic characteristics generally remained unchanged throughout the first few weeks. The viability of using RIBS in human bile ducts for liver transplantation was demonstrated by histological examinations, which showed degradation without considerable fibrosis or chronic inflammation [75]. Using molybdenum (Mo) as a conductor, and resorbable polymers PLGA and PCL as insulators a bioresorbable cardiac pacing using temporary epicardial pacing wires (TEPW) was created. PCL coating significantly lowers molybdenum corrosion and PLGA erosion. When comparing double-coated Mo TEPW to uncoated and PLGA-coated, degradation was significantly decreased. The device exhibited promising *ex vivo* pacing and sensing capabilities [76]. Calcium sulphate/Hydroxyapatite (CaS/HA) biomaterial was prepared and coated around a metallic screw in the tibia of osteoporotic rats. In comparison to the unaugmented implants, the implant encircled by HA particles had much more peri-implant bone growth. Comparing the CaS/HA group to the controls, radiography revealed evidence of enhanced peri-implant bone growth by improving fracture fixation in osteoporotic condition which implies this method may lessen the need for reoperation [77].

12.6 Conclusion and Future Prospective

The biological characterization and testing of resorbable composites have emerged as critical components in the development of advanced biomaterials for various medical applications. This chapter has provided an overview of the methodologies and considerations involved in assessing the biocompatibility, degradation behavior, and functional performance of resorbable composites. The integration of advanced analytical techniques, *in vitro*, and *in vivo* studies has proven essential in ensuring the safety and efficacy of these materials for clinical use. Looking ahead, the future of resorbable composites will be determined by the development of materials with customized degradation profiles, improved mechanical characteristics, and bioactive capabilities that can stimulate tissue regeneration and integration. Emerging technologies such as 3D bioprinting, nanotechnology, and bioinformatics are poised to revolutionize the design and testing of these materials, enabling unprecedented customization to meet patient-specific needs. Furthermore, the combination of sophisticated biomimetic methods with smart materials with stimuli-responsive features opens up new possibilities for developing next-generation resorbable composites. These improvements, along with a multidisciplinary strategy that draws on alliances from materials science, biology, and clinical disciplines, will definitely pave the way for novel solutions in regenerative medicine and implantable devices. However, addressing the challenges of regulatory compliance, scalability, and long-term clinical validation will remain paramount to translating these promising technologies into widespread medical applications. Thus, continued research and development in this field hold immense potential for transforming healthcare by offering safer, more effective, and personalized therapeutic options.

AI disclosure statement: The author confirms that no artificial intelligence (AI)-generated content or AI-assisted writing tools were used in the preparation of this chapter (including summaries and the abstract). All text, analysis, and conclusions are the result of the author's original work.

References

1. Said, S.S., Eltaher, H.M., El-Khordagui, L.K., Biomedical applications of composite resorbable fibers, in: *Materials for Biomedical Engineering*, pp. 127–155, Elsevier, Amsterdam, Netherlands, 2019.
2. Doppalapudi, S., Jain, A., Khan, W., Domb, A.J., Biodegradable polymers—an overview. *Polym. Adv. Technol.*, 25, 5, 427–435, 2014.
3. Ibrahim, H., Esfahani, S.N., Poorganji, B., Dean, D., Elahinia, M., Resorbable bone fixation alloys, forming, and post-fabrication treatments. *Mater. Sci. Eng.: C*, 70, 870–888, 2017.
4. Wu, G., Ibrahim, J.M., Chu, P.K., Surface design of biodegradable magnesium alloys — A review. *Surf. Coat. Technol.*, 233, 2–12, 2013.
5. Dorozhkin, S.V., Calcium orthophosphate coatings on magnesium and its biodegradable alloys. *Acta Biomater.*, 10, 7, 2919–2934, 2014.
6. Shikinami, Y. and Okuno, M., Mechanical evaluation of novel spinal interbody fusion cages made of bioactive, resorbable composites. *Biomaterials*, 24, 18, 3161–3170, 2003.
7. Shikinami, Y. and Okuno, M., Bioresorbable devices made of forged composites of hydroxyapatite (HA) particles and poly-L-lactide (PLLA): Part I. Basic characteristics. *Biomaterials*
8. Heimbach, B., Yu, L., Wei, M., *In vitro* behavior of silk fibroin-based composite resorbable bone fixation devices. *Materialia*, 9, 100587, 2020.
9. McMahon, S., Bertollo, N., Cearbhaill, E.D.O., Salber, J., Pierucci, L., Duffy, P., Dürig, T., Bi, V., Wang, W., Bio-resorbable polymer stents: a review of material progress and prospects. *Prog. Polym. Sci.*, 83, 79–96, 2018.

10. Zargar Kharazi, A., Fathi, M.H., Bahmani, F., Fanian, H., Partially resorbable composite bone plate with controlled degradation rate, desired mechanical properties and bioactivity. *Polym. Degrad. Stab.*, 96, 12, 2055–2063, 2011.
11. Kazarian, S.G., Andrew Chan, K.L., Maquet, V., Boccaccini, A.R., Characterisation of bioactive and resorbable polylactide/Bioglass® composites by FTIR spectroscopic imaging. *Biomaterials*, 25, 18, 3931–3938, 2004.
12. Mueller, B., Koch, D., Lutz, R., Schlegel, K.A., Treccani, L., Rezwan, K., A novel one-pot process for near-net-shape fabrication of open-porous resorbable hydroxyapatite/protein composites and *in vivo* assessment. *Mater. Sci. Eng.: C*, 42, 137–145, 2014.
13. Qin, M., Luo, J., Patel, B., Thong, K.X., Latefa, S., Shao, D., Tanner, A., Yu-Wai-Man, C., Developing a synergistic rate-retarding polymeric implant for controlling monoclonal antibody delivery in minimally invasive glaucoma surgery. *Int. J. Biol. Macromol.*, 272, 132655, 2024.
14. Shackelford, J.F., Bioceramics - Current Status and Future Trends. *MSF*, 293, 99–0, 1998.
15. Okazaki, Y. and Gotoh, E., Comparison of metal release from various metallic biomaterials *in vitro*. *Biomaterials*, 26, 1, 11–21, 2005.
16. Wu, T., Ding, M., Shi, C., Qiao, Y., Wang, P., Qiao, R., Wang, X., Zhong, J., Resorbable polymer electrospun nanofibers: History, shapes and application for tissue engineering. *Chin. Chem. Lett.*, 31, 3, 617–625, 2020.
17. Ullah, M.W., Fu, L., Lamboni, L., Shi, Z., Yang, G., Chapter 3: Current trends and biomedical applications of resorbable polymers, Materials for Biomedical Engineering, Elsevier, Amsterdam, 2019, 10.1016/B978-0-12-818415-8.00003-6.
18. Avérous, L. and Pollet, E., Biodegradable Polymers, in: *Environmental Silicate Nano-Biocomposites*, L. Avérous and E. Pollet (Eds.), pp. 13–39, Springer London, London, 2012.
19. Kuang, T., Chen, F., Chang, L., Zhao, Y., Fu, D., Gong, X., Peng, X., Facile preparation of open-cellular porous poly (l-lactic acid) scaffold by supercritical carbon dioxide foaming for potential tissue engineering applications. *Chem. Eng. J.*, 307, 1017–1025, 2017.
20. Stowell, C.E.T. and Wang, Y., Quickening: Translational design of resorbable synthetic vascular grafts. *Biomaterials*, 173, 71–86, 2018.
21. Treiser, M., Abramson, S., Langer, R., Kohn, J., Degradable and Resorbable Biomaterials, in: *Biomaterials Science*, pp. 179–195, Elsevier, Amsterdam; Boston; London; New York; San Diego; Sydney; Tokyo: Academic Press (an imprint of Elsevier), 2013.
22. Ayasaka, K., Ramanathan, M., Huy, N.X., Shijirbold, A., Okui, T., Tatsumi, H., Kotani, T., Shimamura, Y., Morioka, R., Kanno, T., Evaluation of Hard and Soft Tissue Responses to Four Different Generation Bioresorbable Materials-Poly-l-Lactic Acid (PLLA), Poly-l-Lactic Acid/Polyglycolic Acid (PLLA/PGA), Uncalcined/Unsintered Hydroxyapatite/Poly-l-Lactic Acid (u-HA/PLLA) and Uncalcined/Unsintered Hydroxyapatite/Poly-l-Lactic

Acid/Polyglycolic Acid (u-HA/PLLA/PGA) in Maxillofacial Surgery: An In-Vivo Animal Study. *Materials*, 16, 23, 7379, 2023.
23. Lehtonen, T.J., Tuominen, J.U., Hiekkanen, E., Resorbable composites with bioresorbable glass fibers for load-bearing applications. *In vitro* degradation and degradation mechanism. *Acta Biomater.*, 9, 1, 4868–4877, 2013.
24. Gloeckle, C., Konkol, T., Jacobs, O., Limberg, W., Ebel, T., Handge, U.A., Processing of Highly Filled Polymer–Metal Feedstocks for Fused Filament Fabrication and the Production of Metallic Implants. *Materials*, 13, 19, 4413, 2020.
25. Gao, X., Chen, M., Yang, X., Harper, L., Ahmed, I., Lu, J., Simulating damage onset and evolution in fully bio-resorbable composite under three-point bending. *J. Mech. Behav. Biomed. Mater.*, 81, 72–82, 2018.
26. Bitar, M., Salih, V., Mudera, V., Knowles, J.C., Lewis, M.P., Soluble phosphate glasses: *in vitro* studies using human cells of hard and soft tissue origin. *Biomaterials*, 25, 12, 2283–2292, 2004.
27. Rineksa, G., Whulanza, Y., Gozan, M., Preliminary Study of Potential Bioimplant from Glycerol Plasticized Starch-Microcrystalline Cellulose Composite. *J. Integr. Adv. Eng. (JIAE)*, 1, 1, 29–36, 2021.
28. Hsu, P.-Y., Kuo, H.-C., Syu, M.-L., Tuan, W.-H., Lai, P.-L., A head-to-head comparison of the degradation rate of resorbable bioceramics. *Mater. Sci. Eng.: C*, 106, 110175, 2020.
29. Chow, L.C. and Takagi, S., A natural bone cement - A laboratory novelty led to the development of revolutionary new biomaterials. *J. Res. Natl. Inst. Stand. Technol.*, 106, 6, 1029, 2001.
30. Liu, G., He, Y., Liu, P., Chen, Z., Chen, X., Wan, L., Li, Y., Lu, J., Development of Bioimplants with 2D, 3D, and 4D Additive Manufacturing Materials. *Engineering*, 6, 11, 1232–1243, 2020.
31. Chen, S., Shi, Y., Luo, Y., Ma, J., Layer-by-layer coated porous 3D printed hydroxyapatite composite scaffolds for controlled drug delivery. *Colloids Surf., B*, 179, 121–127, 2019.
32. Chen, Z., Li, Z., Li, J., Liu, C., Lao, C., Fu, Y., Liu, C., Li, Y., Wang, P., He, Y., 3D printing of ceramics: A review. *J. Eur. Ceram. Soc.*, 39, 4, 661–687, 2019.
33. Kang, C.-W. and Fang, F.-Z., State of the art of bioimplants manufacturing: part I. *Adv. Manuf.*, 6, 1, 20–40, 2018.
34. Bartolo, P., Kruth, J.-P., Silva, J., Levy, G., Malshe, A., Rajurkar, K., Mitsuishi, M., Ciurana, J., Leu, M., Biomedical production of implants by additive electro-chemical and physical processes. *CIRP Ann.*, 61, 2, 635–655, 2012.
35. Zhang, S.M., Liu, J., Zhou, W., Cheng, L., Guo, X.D., Interfacial fabrication and property of hydroxyapatite/polylactide resorbable bone fixation composites. *Curr. Appl Phys.*, 5, 5, 516–518, 2005.
36. Sumayli, A., Recent trends on bioimplant materials: A review. *Mater. Today: Proc.*, 46, 2726–2731, 2021.

37. Kirkland, N.T., Birbilis, N., Staiger, M.P., Assessing the corrosion of biodegradable magnesium implants: A critical review of current methodologies and their limitations. *Acta Biomater.*, 8, 3, 925–936, 2012.
38. Li, G.Y., Lian, J.S., Niu, L.Y., Jiang, Z.H., Jiang, Q., Growth of zinc phosphate coatings on AZ91D magnesium alloy. *Surf. Coat. Technol.*, 201, 3–4, 1814–1820, 2006.
39. Gnedenkov, S.V., Sinebryukhov, S.L., Zavidnaya, A.G., Egorkin, V.S., Puz', A.V., Mashtalyar, D.V., Sergienko, V.I., Yerokhin, A.L., Matthews, A., Composite hydroxyapatite–PTFE coatings on Mg–Mn–Ce alloy for resorbable implant applications via a plasma electrolytic oxidation-based route. Enhancing tribological performance of *J. Taiwan Inst. Chem. Eng.*, 45, 6, 3104–3109, 2014.
40. Purnama, A., Hermawan, H., Couet, J., Mantovani, D., Assessing the biocompatibility of degradable metallic materials: State-of-the-art and focus on the potential of genetic regulation☆. *Acta Biomater.*, 6, 5, 1800–1807, 2010.
41. Ralls, A., Kumar, P., Misra, M., Menezes, P.L., Material Design and Surface Engineering for Bio-implants. *JOM*, 72, 2, 684–696, 2020.
42. Prasad, K., Bazaka, O., Chua, M., Rochford, M., Fedrick, L., Spoor, J., Symes, R., Tieppo, M., Collins, C., Cao, A., Markwell, D., Ostrikov, K., Bazaka, K., Metallic Biomaterials: Current Challenges and Opportunities. *Materials*, 10, 8, 884, 2017.
43. Sidia, B. and Bensalah, W., Enhancing tribological performance of high-density polyethylene polymer with diamond-like carbon coating and mollusk shells filler. *Polym. Eng. Sci.*, 64, 10, 5218–5228, 2024.
44. Shen, S.-J., Feng, P.-C., Wu, R.-C., Kuo, Y.-H., Liu, S.-J., Ito, H., Resorbable nanofibrous membranes for local and sustained co-delivery of acyclovir and ketorolac in herpes therapy. *Int. J. Pharm.*, 654, 123988, 2024.
45. Wang, Y., Jing, W., Huang, Y., Wei, P., Ma, S., Yuan, Z., Zhao, B., Biodegradable all-natural hydrogel–mesh composite for wet tissues adhesion and infected wound healing. *Mater. Des.*, 241, 112902, 2024.
46. P, A., M, R.J., Joy, J.M., Visnuvinayagam, S., Remya, S., Mathew, S., Development of κ-carrageenan-based transparent and absorbent biodegradable films for wound dressing applications. *Int. J. Biol. Macromol.*, 282, 137084, 2024.
47. Bose, D., Prasannavenkadesan, V., Katiyar, V., Tailored material for bioresorbable internal fixation devices: a novel approach using nanohydroxyapatite and chitosan in polylactic acid. *J. Mater. Sci.*, 59, 1, 215–227, 2024.
48. Chrysafi, I., Ainali, N.M., Bikiaris, D.N., Thermal Degradation Mechanism and Decomposition Kinetic Studies of Poly(Lactic Acid) and Its Copolymers with Poly(Hexylene Succinate). *Polymers*, 13, 9, 1365, 2021.
49. Choudhary, N., Sharma, V., Kumar, P., Reinforcement of polylactic acid with bioceramics (alumina and YSZ composites) and their thermomechanical and physical properties for biomedical application. *Vinyl Addit. Technol.*, 27, 3, 612–625, 2021.

50. Dogan, D., Erdem, U., Bozer, B.M., Turkoz, M.B., Yıldırım, G., Metin, A.U., Resorbable membrane design: *In vitro* characterization of silver doped-hydroxyapatite-reinforced XG/PEI semi-IPN composite. *J. Mech. Behav. Biomed. Mater.*, 142, 105887, 2023.
51. Streza, A., Antoniac, A., Manescu (Paltanea), V., Paltanea, G., Robu, A., Dura, H., Verestiuc, L., Stanica, E., Voicu, S.I., Antoniac, I., Cristea, M.B., Dragomir, B.R., Rau, J.V., Manolea, M.-M., Effect of Filler Types on Cellulose-Acetate-Based Composite Used as Coatings for Biodegradable Magnesium Implants for Trauma. *Materials*, 16, 2, 554, 2023.
52. Zhang, Q.-C., Ding, W., Ding, S.-L., Meng, Q.-B., Su, D.-H., Zhang, T.-W., Chen, Q., Lian, R.-X., Zhao, M.-D., Yu, B.-Q., Li, X.-L., Li, Y.-L., Jiang, L.-B., Robust bioactive protein-based screws with dual crosslinked network for internal bone fixation. *Composites, Part B*, 238, 109884, 2022.
53. Stewart, S.A., Domínguez-Robles, J., McIlorum, V.J., Gonzalez, Z., Utomo, E., Mancuso, E., Lamprou, D.A., Donnelly, R.F., Larrañeta, E., Poly(caprolactone)-Based Coatings on 3D-Printed Biodegradable Implants: A Novel Strategy to Prolong Delivery of Hydrophilic Drugs. *Mol. Pharmaceutics*, 17, 9, 3487–3500, 2020.
54. Murueva, A.V., Dudaev, A.E., Shishatskaya, E.I., Ghorabe, F.D.E., Nemtsev, I.V., Lukyanenko, A.V., Volova, T.G., Biodegradable polymer casting films for drug delivery and cell culture. *Giant*, 19, 100314, 2024.
55. Alsakkaf, A., Md Yusop, A.H., Idris, H., Hassan, A.G., Iqbal, N., Unal, R., Sudin, I., Development of Fe-Mg alloys with intermediate degradation kinetics as potential biodegradable bone implants. *Mater. Today Commun.*, 41, 110609, 2024.
56. Tong, X., Shen, X., Lin, Z., Zhou, R., Han, Y., Zhu, L., Huang, S., Ma, J., Li, Y., Wen, C., Lin, J., *In situ* phosphorus-modified Mg2Ge/Zn-Cu composite with improved mechanical, degradation, biotribological properties, and *in vitro* and *in vivo* osteogenesis and osteointegration performance for biodegradable bone-implant applications. *Bioact. Mater.*, 43, 491–509, 2025.
57. Kačarević, ZP., Rider, P., Elad, A., Tadic, D., Rothamel, D., Sauer, G., Bornert, F., Windisch, P., Hangyási, D.B., Molnar, B., Kämmerer, T., Hesse, B., Bortel, E., Bartosch, M., Witte, F., Biodegradable magnesium fixation screw for barrier membranes used in guided bone regeneration. *Bioact. Mater.*, 14, 15–30, 2022.
58. Tesař, K., Luňáčková, J., Jex, M., Žaloudková, M., Vrbová, R., Bartoš, M., Singh, A., *In vivo* and *in vitro* study of resorbable magnesium wires for medical implants: Mg purity, surface quality, Zn alloying and polymer coating. *J. Magnesium Alloys*, 12, 6, 2472–2488, 2024.
59. Chen, Y., Chang, L., Zhang, Z., Zhou, M., Gao, Y., Wang, Y., Liu, Y., Qin, J., Biodegradable pectin-based thermo-responsive composite GO/hydrogel with mussel inspired tissue adhesion for NIR enhanced burn wound healing. *Chem. Eng. J.*, 480, 148067, 2024.

60. Elhadad, A., Mezour, M.A., Abu Nada, L., Shurbaji, S., Mansour, A., Smith, S., Moussa, H., Lee, L., Pérez-Soriano, E.M., Murshed, M., Chromik, R., Tamimi, F., 2D magnesium phosphate resorbable coating to enhance cell adhesion on titanium surfaces. *Mater. Chem. Phys.*, 316, 129114, 2024.
61. Kolour, A.K., Shahrousvand, M., Mohammadi-Rovshandeh, J., Puppi, D., Farzaneh, D., Absorbable and biodegradable enzyme-crosslinked gelatin/alginate semi-IPN hydrogel wound dressings containing curcumin. *Int. J. Biol. Macromol.*, 279, 134938, 2024.
62. Zaffora, A., Di Franco, F., Virtù, D., Carfì Pavia, F., Ghersi, G., Virtanen, S., Santamaria, M., Tuning of the Mg Alloy AZ31 Anodizing Process for Biodegradable Implants. *ACS Appl. Mater. Interfaces*, 13, 11, 12866–12876, 2021.
63. Zhou, X., Liu, C., Han, Y., Li, C., Liu, S., Li, X., Zhao, G., Jiang, Y., An anti-bacterial chitosan-based hydrogel as a potential degradable bio-scaffold for alveolar ridge preservation. *RSC Adv.*, 12, 50, 32219–32229, 2022.
64. Lei, X.-L., Cheng, K., Li, Y., Zhong, Z.-T., Hou, X.-L., Song, L.-B., Zhang, F., Wang, J.-H., Zhao, Y.-D., Xu, Q.-R., The eradication of biofilm for therapy of bacterial infected chronic wound based on pH-responsive micelle of antimicrobial peptide derived biodegradable microneedle patch. *Chem. Eng. J.*, 462, 142222, 2023.
65. Jones, S.E., Nichols, L., Elder, S.H., Priddy, L.B., Laser microgrooving and resorbable blast texturing for enhanced surface function of titanium alloy for dental implant applications. *Biomed. Eng. Adv.*, 5, 100090, 2023.
66. Barai, B., Boorgula, K., Begam, H., Sarkar, S., Barui, A., Kundu, S., Oraon, B., Mandal, T., Multiphysics simulation and experimental investigation of HA and Zn-doped HA coatings on magnesium alloys for resorbable implant applications. *Mater. Today Commun.*, 42, 111241, 2025.
67. Feng, K.-C., Wu, Y.-J., Wang, C.-Y., Tu, C.-S., Lin, Y.-L., Chen, C.-S., Lai, P.-L., Huang, Y.-T., Chen, P.-Y., Enhanced mechanical and biological performances of CaO-MgO-SiO2 glass-ceramics via the modulation of glass and ceramic phases. *Mater. Sci. Eng.: C*, 124, 112060, 2021.
68. Yanagisawa, T., Hayashi, K., Tsuchiya, A., Kishida, R., Ishikawa, K., *In vivo* trial of bioresorbable mesh cages contained bone graft granules in rabbit femoral bone defects. *Sci. Rep.*, 14, 1, 12449, 2024.
69. Vilanova-Corrales, P., Demiquels-Punzano, E., Caballé-Serrano, J., Hernández-Alfaro, F., Delgado, JA., Pérez, R.A., Gil, J., Delgado, L.M., Biodegradable and reinforced membranes based on polycaprolactone and collagen for guided bone regeneration. *Mater. Today Commun.*, 41, 111039, 2024.
70. Copete, H., López, E., Baudin, C., Synthesis and characterization of B-type carbonated hydroxyapatite materials: Effect of carbonate content on mechanical strength and *in vitro* degradation. *Bol. Soc. Esp. Ceram. Vidr.*, 63, 4, 255–267, 2024.

71. Rajan, S.T., Das, M., Arockiarajan, A., *In vitro* assessment of corrosion resistance and biocompatibility of tantalum-niobium oxide surface-functionalized Mg alloy. *Mater. Chem. Phys.*, 301, 127560, 2023.
72. Shukla, R., Yadav, M.K., Madruga, L.Y.C., Jayaraj, J., Popat, K., Wang, Z., Xi, L., Prashanth, K.G., A novel Ti-eggshell-based composite fabricated by combined additive manufacturing-powder metallurgical routes as bioimplants. *Ceram. Int.*, 2024, S0272884224057080.
73. Senatov, F., Amanbek, G., Orlova, P., Bartov, M., Grunina, T., Kolesnikov, E., Maksimkin, A., Kaloshkin, S., Poponova, M., Nikitin, K., Krivozubov, M., Strukova, N., Manskikh, V., Anisimova, N., Kiselevskiy, M., Scholz, R., Knyazeva, M., Walther, F., Lunin, V., Gromov, A., Karyagina, A., Biomimetic UHMWPE/HA scaffolds with rhBMP-2 and erythropoietin for reconstructive surgery. *Mater. Sci. Eng.: C*, 111, 110750, 2020.
74. Rider, P., Kačarević, ZP., Elad, A., Tadic, D., Rothamel, D., Sauer, G., Bornert, F., Windisch, P., Hangyási, D.B., Molnar, B., Bortel, E., Hesse, B., Witte, F., Biodegradable magnesium barrier membrane used for guided bone regeneration in dental surgery. *Bioact. Mater.*, 14, 152–168, 2022.
75. Girard, E., Chagnon, G., Broisat, A., Dejean, S., Soubies, A., Gil, H., Sharkawi, T., Boucher, F., Roth, G.S., Trilling, B., Nottelet, B., From *in vitro* evaluation to human postmortem pre-validation of a radiopaque and resorbable internal biliary stent for liver transplantation applications. *Acta Biomater.*, 106, 70–81, 2020.
76. Prieto Jarabo, M.-E., Redlich, C., Schauer, A., Alves, P.K.N., Guder, C., Poehle, G., Weissgaerber, T., Adams, V., Kappert, U., El-Armouche, A., Linke, A., Wagner, M., Bioresorbable molybdenum temporary epicardial pacing wires. *Acta Biomater.*, 178, 330–339, 2024.
77. Markeviciute, V., Puthia, M., Arvidsson, L., Liu, Y., Törnquist, E., Tengattini, A., Huang, J., Bai, Y., Vater, C., Petrolis, R., Zwingenberger, S., Krisciukaitis, A., Smailys, A., Lukosevicius, S., Stravinskas, M., Isaksson, H., Tarasevicius, S., Lidgren, L., Tägil, M., Raina, D.B., Systemically administered zoledronic acid activates locally implanted synthetic hydroxyapatite particles enhancing peri-implant bone formation: A regenerative medicine approach to improve fracture fixation. *Acta Biomater.*, 179, 354–370, 2024.

13

Tribological Studies of the Bioimplants and Fixation Devices

M. Vishnuvarthanan[1]*, Raja Venkatesan[2], Arbind Prasad[3] and B. Mouli Prasanth[4]

[1]*Department of Mechanical Engineering, Kalasalingam Academy of Research and Education, Srivilliputhur, Tamil Nadu, India*
[2]*School of Chemical Engineering, Yeungnam University, Daehak-Ro, Gyeongsan, Republic of Korea*
[3]*Mechanical Engineering Department, Katihar Engineering College (Under Department of Science, Technology and Technical Education, Government of Bihar), Katihar, Bihar, India*
[4]*School of Mechanical Engineering, Shandong University, Jinan, China*

Abstract

Tribological performance plays a crucial role in determining the durability and functionality of bioimplants and fixation devices, particularly in load-bearing applications. The interface between implant materials and surrounding biological tissues is subjected to continuous wear and friction, influencing the longevity and biocompatibility of these medical devices. This book chapter provides a comprehensive overview of tribological studies conducted on bioimplants and fixation devices, with a focus on resorbable composites. It explores the interplay between mechanical properties, surface modifications, and lubrication mechanisms in optimizing implant performance. The book chapter begins by reviewing key materials used in bioimplants such as polylactic acid (PLA), polyglycolic acid (PGA) and their composites with bioceramics like hydroxyapatite. The influence of material composition and structural design on wear resistance, friction coefficients, and material degradation rates is examined. Furthermore, the chapter highlights the role of advanced coatings such as bioactive ceramics and polymeric films in enhancing tribological behaviour while maintaining biocompatibility and resorb ability. Special attention is given to *in vitro* and *in vivo* studies evaluating the wear mechanisms of bioimplants in simulated body environments. The chapter also addresses the impact of biological factors such as synovial fluid composition and

**Corresponding author*: vishnuvarthanan.india@gmail.com; ORCID: 0000-0002-7821-2471

Arbind Prasad (ed.) Resorbable Composites for Bioimplants and Fixation Devices, (381–414) © 2026 Scrivener Publishing LLC

load dynamics on tribological performance. The integration of nanomaterials and 3D printing technologies are discussed as innovative approaches to improve the tribological properties of bioimplants and fixation devices. This chapter aims to provide valuable insights into the design and development of next-generation resorbable bioimplants. These advancements hold promise for reducing revision surgeries, improving patient outcomes, and fostering sustainable solutions in medical device manufacturing.

***Keywords*:** Tribological performance, bioimplants, resorbable composites, wear mechanisms, biocompatibility

13.1 Introduction to Tribology in Bioimplants and Fixation Devices

Tribology, the study of friction, wear, and lubrication, plays a critical role in the design and performance of bioimplants and fixation devices [1]. These medical devices used to restore or support biological functions, often face complex mechanical interactions at the interface of implant materials and biological tissues. Tribological performance directly influences the durability, functionality, and biocompatibility of bioimplants, particularly in load-bearing applications like joint replacements, spinal implants, and dental prosthetics. Bioimplants experience repetitive motion and stress within the human body, leading to wear and degradation over time. This can result in inflammation, implant failure and the need for revision surgeries. The understanding of underlying wear mechanisms, such as adhesive and abrasive wear, is crucial for developing materials and designs that can withstand these challenges. The surface modifications such as coatings and texturing have emerged as effective strategies to reduce friction and wear while maintaining compatibility with biological environments. The resorbable materials such as polylactic acid (PLA) and hydroxyapatite composites, have gained attention for their ability to degrade naturally, eliminating the need for surgical removal [2]. The advances in tribological studies aim to optimize these materials for improved performance, ensuring longer implant lifespans and better patient outcomes. It explores the foundational aspects of tribology in bioimplants, emphasizing its significance in medical applications. The overview of biomaterials in bio implants are shown in Figure 13.1.

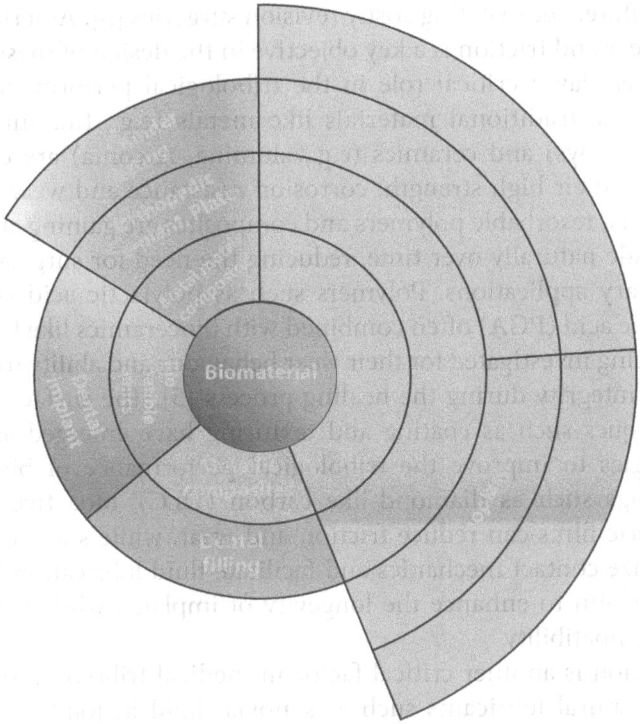

13.1.1 Overview of Tribology in Medical Application

Tribology, the science of friction, wear, and lubrication, is an essential field of study in medical applications, particularly in the development and performance evaluation of bioimplants and fixation devices. These medical devices, which include joint replacements, dental implants, spinal devices, and fracture fixation systems, are subjected to mechanical interactions and stresses within the body. The study of tribology is important to understanding how these devices perform under physiological conditions and to designing solutions that ensure their longevity and compatibility with the human body. Bioimplants and fixation devices often operate in challenging environments characterized by continuous motion, variable loading conditions, and exposure to biological fluids. In artificial joints such as hip or knee replacements, tribological interactions between the implant components dictate the wear behaviour, frictional forces, and energy dissipation during movement. The excessive wear of these components can lead to the release of wear debris, which may trigger inflammatory responses and implant failure, necessitating costly revision surgeries [3]. As a result, minimizing wear and friction is a key objective in the design of these devices.

Materials play a critical role in the tribological performance of bioimplants. The traditional materials like metals (e.g., titanium, cobalt-chromium alloys) and ceramics (e.g., alumina, zirconia) are commonly used due to their high strength, corrosion resistance and wear resistance [4]. However, resorbable polymers and composites are gaining attention as they degrade naturally over time, reducing the need for surgical removal in temporary applications. Polymers such as polylactic acid (PLA) and polyglycolic acid (PGA) often combined with bioceramics like hydroxyapatite are being investigated for their wear behaviour and ability to maintain structural integrity during the healing process [5]. The surface engineering techniques such as coating and texturing have emerged as promising strategies to improve the tribological performance of bioimplants. The coatings such as diamond-like carbon (DLC), bioactive ceramics, or polymeric films can reduce friction and wear, while surface texturing can optimize contact mechanics and facilitate fluid lubrication [6]. These approaches aim to enhance the longevity of implants while maintaining their biocompatibility.

Lubrication is another critical factor in medical tribology. The human body has natural lubricants such as synovial fluid in joints, which play a vital role in reducing wear and friction [7]. Tribological studies often involve simulating these biological environments to assess how implants interact with native lubrication systems. The innovations in bioinspired lubrication strategies, such as hydrogels and biomimetic lubricants, are being explored to further optimize implant performance. The integration of nanotechnology has introduced new possibilities for improving the tribological properties of medical devices. Nanoparticles such as graphene, carbon nanotubes, and nanodiamonds are being incorporated into coatings and composite materials to enhance wear resistance, reduce friction, and promote surface self-healing [8]. Additionally, 3D printing technologies are being leveraged to fabricate bioimplants with customized tribological properties and complex geometries tailored to individual patients. The tribology in medical applications is a multidisciplinary field that bridges material science, biomechanics and biomedical engineering. The advances in tribological studies are giving the way for the development of bioimplants and fixation devices that are not only more durable and efficient but also safer and more compatible with the human body.

13.1.2 Importance of Tribological Studies for Bioimplants

Tribological studies, focusing on the interplay of friction, wear, and lubrication are fundamental to the development and optimization of bioimplants. These medical devices, such as artificial joints, dental implants, and fracture fixation systems function in dynamic environments within the human body, where mechanical interactions and biological factors influence their performance and longevity. The understanding and improving the tribological properties of bioimplants is crucial for ensuring their effectiveness, durability, and biocompatibility thereby improving patient outcomes. One of the primary reasons of tribological studies is important for bioimplants is their role in minimizing wear. Bioimplants, especially in load-bearing applications like hip and knee replacements are subjected to repetitive motion and significant mechanical stress. This can lead to the generation of wear debris, which may accumulate in the surrounding tissues, causing inflammation and even implant failure.

Friction management is another critical aspect of tribological studies. The excessive friction between the surfaces of bioimplants can lead to increased energy loss, discomfort and accelerated wear. Tribological research helps identify materials and surface treatments that can lower friction coefficients, ensuring smoother motion and better functionality of the implant [9]. Material selection for bioimplants also relies heavily on tribological insights. The traditional materials like titanium alloys and ceramics offer high wear resistance and durability but may lack certain attributes like resorbability or flexibility. Tribological studies guide the development of resorbable materials such as polylactic acid (PLA) composites or hydroxyapatite, which provide the desired mechanical and tribological properties while degrading naturally in temporary applications. These materials are particularly significant in pediatric or fracture fixation implants, where removing the device would require additional surgeries.

13.2 Materials Used in Bioimplants and Fixation Devices

The selection of materials for bioimplants and fixation devices is a critical aspect that determines their functionality, biocompatibility, and longevity. These materials must withstand the mechanical loads and tribological interactions they encounter within the human body while maintaining compatibility with surrounding tissues. The ideal material for bioimplants should exhibit properties such as high strength, corrosion resistance, and wear resistance and in some cases, bioresorbability to avoid long-term complications. Traditional materials like metals, ceramics, and polymers are commonly used in bioimplants due to their superior mechanical and chemical properties. Metals such as titanium alloys and cobalt-chromium alloys are favored for their high strength and corrosion resistance making them suitable for load-bearing applications like joint replacements [10]. Ceramics, such as alumina and zirconia offer excellent wear resistance and biocompatibility, often used in articulating surfaces of implants [11]. Polymers, including polyethylene and polymethyl methacrylate are employed for their flexibility and low friction properties. Resorbable polymers like polylactic acid (PLA) and polyglycolic acid (PGA) are increasingly used in temporary fixation devices due to their ability to degrade naturally over time. The advances in biomaterials, including nanocomposites and bioactive coatings, are revolutionizing the field, enabling the development of implants with enhanced tribological and biological performance.

13.2.1 Polymers, Ceramics, and Composites

The choice of materials for bioimplants and fixation devices is critical to ensuring their success in biomedical applications. These materials must demonstrate excellent mechanical, tribological, and biological properties to meet the functional requirements of supporting and integrating with human tissues. Polymers, ceramics, and composites have emerged as the most widely used materials for these applications, each offering unique advantages tailored to specific medical needs.

13.2.1.1 Polymers

Polymers are extensively used in bioimplants due to their versatility, lightweight nature, and biocompatibility. Among them, ultra-high molecular weight polyethylene (UHMWPE) is a prominent material in orthopedic implants such as joint replacements, owing to its excellent wear resistance and low friction coefficient [12]. Polymethyl methacrylate (PMMA) is widely employed as bone cement, providing mechanical stability and enabling the fixation of implants. Biodegradable polymers like polylactic acid (PLA), polyglycolic acid (PGA) and their copolymers (PLGA) are particularly significant in resorbable fixation devices. These materials degrade into non-toxic byproducts, eliminating the need for removal surgeries and promoting natural healing. Polycaprolactone (PCL) is another biodegradable polymer used in scaffolds and drug delivery systems due to its tunable degradation rate and mechanical properties. However, polymers generally have lower mechanical strength and wear resistance compared to metals and ceramics, necessitating reinforcement for load-bearing applications.

13.2.1.2 Ceramics

Ceramic materials are known for their exceptional hardness, wear resistance, and biocompatibility. Alumina and zirconia are commonly used in articulating surfaces of joint implants such as hip and knee replacements, where their low friction and high wear resistance are critical. These ceramics also exhibit excellent chemical stability, reducing the risk of corrosion in the body. Hydroxyapatite (HA), a bioactive ceramic, plays a vital role in promoting osseointegration [13]. It is often used as a coating material on metallic implants to enhance bone bonding and accelerate tissue regeneration. The brittleness of ceramics limits their use in dynamic or highly stressed environments, but they remain indispensable for specific applications where wear resistance and bioactivity are paramount.

13.2.1.3 Composites

Composite materials combine the advantages of different components to achieve enhanced performance tailored to biomedical applications. Polymer-ceramic composites such as PLA-hydroxyapatite blends are extensively used in orthopedic implants for their ability to provide strength, bioactivity and controlled degradability [14]. These composites promote bone regeneration while offering mechanical properties suitable for load-bearing applications. Fiber-reinforced composites such as carbon fiber-reinforced polymers offer superior strength, stiffness and wear resistance, making them ideal for high-stress applications like spinal fixation devices. Glass fiber-reinforced composites are also used for their lightweight properties and biocompatibility. Nanocomposites have recently gained attention due to their improved tribological and mechanical properties. The incorporation of nanoparticles such as graphene, carbon nanotubes, or nano-hydroxyapatite, into polymer or ceramic matrices enhances wear resistance, mechanical strength, and self-lubrication. These advancements are particularly valuable in dynamic environments where friction and wear significantly impact implant performance.

13.2.2 Role of Resorbable Materials in Biomedical Applications

Resorbable materials have revolutionized the field of biomedical engineering by providing temporary support to the body while gradually degrading into biocompatible byproducts. Their ability to eliminate the need for surgical removal after fulfilling their purpose makes them invaluable in various medical applications, including bioimplants, fixation devices, drug delivery systems, and tissue engineering scaffolds. In bioimplants and fixation devices, resorbable materials play a critical role in reducing complications associated with permanent implants. For instance, resorbable screws, plates, and pins are extensively used in orthopedic and craniofacial surgeries. These devices provide mechanical stability to fractured bones or soft tissues during the healing process and degrade as the tissues regain strength, avoiding long-term issues such as stress shielding, implant migration or infection. Materials like polylactic acid (PLA), polyglycolic acid (PGA) and their copolymers (PLGA) are commonly used in such applications due to their predictable degradation rates and biocompatibility [15].

Resorbable materials also address challenges in pediatric applications where the growing anatomy of children necessitates temporary solutions. Additionally, resorbable stents in cardiovascular applications provide temporary support to blood vessels while preventing restenosis, eventually degrading as the vessel heals. Another significant application of resorbable materials is in drug delivery systems. These materials are used to fabricate carriers such as microspheres, hydrogels, and nanoparticles that encapsulate therapeutic agents and release them in a controlled manner. Resorbable polymers like polycaprolactone (PCL) and polydioxanone (PDO) enable localized drug delivery, minimizing systemic side effects while enhancing therapeutic efficacy. This approach is particularly beneficial in treating localized infections, cancers, and chronic wounds.

In tissue engineering, resorbable materials serve as scaffolds that mimic the extracellular matrix, providing structural support for cell attachment and proliferation. As the cells grow and form new tissue, the scaffold gradually degrades, leaving behind regenerated tissue. Natural polymers like collagen, chitosan, and alginate, as well as synthetic polymers like PLA and PGA are widely used in scaffold fabrication [16]. The resorbability of these materials ensures seamless integration with the host tissue and reduces long-term foreign body responses. From a tribological perspective, the role of resorbable materials is crucial in ensuring mechanical performance during their functional period. Resorbable fixation devices and implants often operate in dynamic environments, where friction and wear can impact their performance.

Nanostructured materials such as graphene-reinforced polymers or nano-hydroxyapatite composites provide enhanced functionality, paving the way for next-generation resorbable biomedical devices.

13.3 Mechanical and Tribological Properties of Bioimplant Materials

The mechanical and tribological properties of bioimplant materials are crucial factors in determining the performance, longevity, and biocompatibility of implants and fixation devices. These materials are subjected to complex mechanical loads and dynamic forces within the human body, requiring them to possess adequate strength, stiffness, and durability to ensure optimal function during the healing process and beyond. The mechanical properties such as tensile strength, fatigue resistance, and fracture toughness are essential to prevent implant failure due to mechanical stresses and strain. The friction between implant surfaces, as well as the wear of materials over time, can significantly affect the implant's lifespan and the surrounding tissue health. The reduced wear and low friction are critical to minimizing the risk of inflammation, osteolysis, and the need for premature revision surgeries.

13.3.1 Wear Resistance and Frictional Behaviour

Wear resistance and frictional behaviour are critical factors that influence the performance and longevity of bioimplants and fixation devices. These materials are often subjected to complex mechanical stresses and dynamic interactions with surrounding tissues, which can lead to wear, degradation and potential failure if not properly designed.

13.3.1.1 Wear Resistance in Bioimplants

Wear resistance refers to the ability of a material to withstand surface degradation or loss due to mechanical contact with other surfaces or tissues. In bioimplants, this property is especially important in joint replacement implants, where articulating surfaces are subjected to constant sliding contact. Materials such as ultra-high molecular weight polyethylene (UHMWPE), ceramics (like alumina and zirconia) and metal alloys are commonly used in joint replacements due to their excellent wear resistance [17]. However, over time, wear debris from these materials can cause inflammation and osteolysis, leading to implant failure.

13.3.1.2 Frictional Behaviour in Bioimplants

Friction is another significant factor affecting the performance of bioimplants. It refers to the resistance encountered when two surfaces move against each other. In joint replacements and fixation devices, high friction can lead to discomfort, reduced functionality and accelerated wear of both the implant and surrounding tissues. Ideally, bioimplant materials should exhibit low friction to minimize wear and provide smooth, pain-free motion for the patient. The frictional behaviour of bioimplants depends on various factors, including the material composition, surface finish and lubrication provided by synovial fluid or other natural body fluids [18]. Materials with smooth surfaces tend to have lower friction coefficients compared to rougher surfaces. For example, ceramic materials like zirconia exhibit lower frictional resistance than metal-based materials, making them more suitable for load-bearing applications in joint replacements. Furthermore, the lubrication from synovial fluid in the body plays a significant role in reducing friction at the articulating surfaces, helping maintain the function of the implant over time. Polymeric materials such as UHMWPE, are frequently used in combination with metals or ceramics to achieve lower friction in bearing surfaces. UHMWPE, when cross-linked or modified, can offer reduced friction and lower wear rates, making it a popular choice in hip and knee replacements. However, the frictional behaviour of polymers can be influenced by factors such as surface roughness, temperature and contact pressure which need to be carefully controlled during the design and manufacturing process to optimize implant performance.

13.3.1.3 Improving Wear and Frictional Properties

Hybrid composites, such as metal-ceramic or polymer-ceramic systems combine the strengths of different materials to achieve improved tribological performance. The coatings such as diamond-like carbon (DLC) have also been explored for their ability to reduce friction and wear in bioimplant materials. Furthermore, the surface roughness and geometry of the implant components play a critical role in minimizing wear and friction. The frictional behaviour can be reduced by optimizing the design of the implant and improving the surface finish leading to longer-lasting implants. The advances in 3D printing and additive manufacturing technologies are also contributing to the design of implants with tailored surface textures that optimize tribological properties.

13.3.2 Degradation Mechanisms in Resorbable Materials

Resorbable materials are increasingly used in bioimplants and fixation devices due to their unique ability to gradually degrade and be absorbed by the body, reducing the need for additional surgeries for implant removal. These materials offer distinct advantages such as the potential to support tissue regeneration and promote healing while avoiding long-term complications associated with permanent implants. However, the degradation mechanisms of resorbable materials in bioimplants are complex and influenced by various factors including the material composition, environmental conditions and mechanical stresses. The degradation mechanisms in resorbable materials are shown in Figure 13.2.

13.3.2.1 Hydrolytic Degradation

Hydrolysis is the primary degradation mechanism for many resorbable polymers and composites used in bioimplants. This process occurs when water molecules penetrate the polymer matrix and break the chemical bonds, leading to the material's gradual breakdown. In biodegradable polymers such as polylactic acid (PLA), polyglycolic acid (PGA) and their copolymers (PLGA), hydrolysis leads to the formation of oligomers and monomers, which are then absorbed or eliminated by the body through natural metabolic pathways. The rate of hydrolytic degradation can vary depending on several factors such as the polymer's chemical structure, crystallinity, molecular weight and the presence of additives [19]. For example, copolymer formulations that blend PLA and PGA offer a balance between mechanical strength and degradation rate, allowing for controlled resorption *in vivo*. The high crystallinity in polymers generally slows down the hydrolysis process, whereas more amorphous structures degrade faster. The hydrolytic degradation can also be influenced by the pH of the surrounding tissue, with more acidic environments typically accelerating the breakdown of polymer chains.

13.3.2.2 Enzymatic Degradation

Enzymatic degradation is another important mechanism especially for biopolymer-based resorbable materials such as chitosan, collagen and alginate. The enzymes present in the body such as proteases and lipases, break down these materials into smaller, biologically compatible fragments. This process is highly selective, as specific enzymes target certain chemical bonds within the polymer chains, leading to controlled degradation. In the

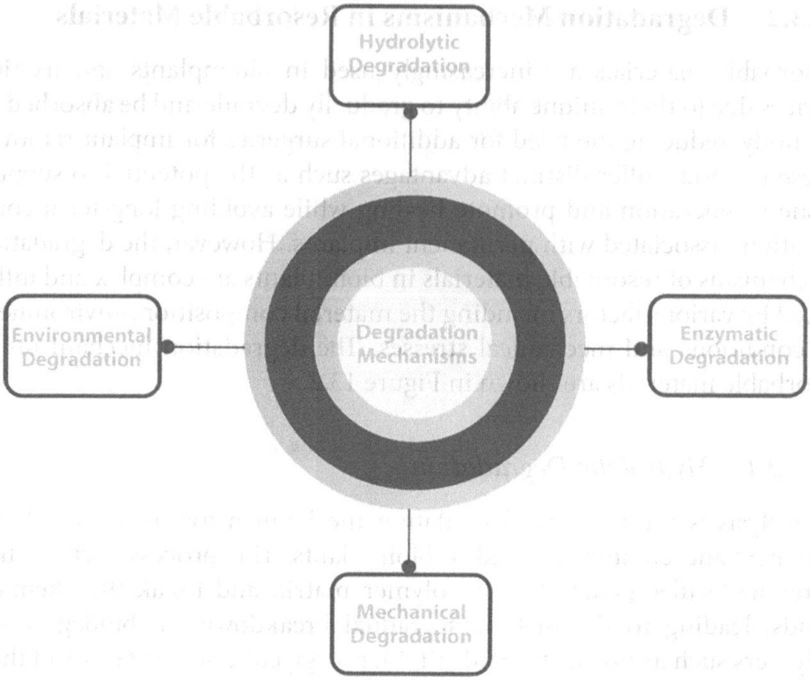

case of chitosan, the material undergoes enzymatic cleavage by enzymes like lysozyme, which break down the glycosidic linkages between the sugar units. This degradation process is typically slower than hydrolytic degradation but offers the advantage of more precise control over the resorption rate [20]. The enzymatic degradation plays a significant role in tissue-engineering scaffolds and drug delivery systems where the material's breakdown needs to be finely tuned to align with tissue regeneration processes.

13.3.2.3 *Mechanical Degradation*

Mechanical degradation is a critical consideration in bioimplants subjected to high loads or stress such as fixation devices used in bone healing. The resorbable materials may experience physical breakdown due to mechanical forces applied during the healing process, particularly when the mechanical properties of the material no longer match the requirements of the surrounding tissue [21]. The resorbable polymers may lose their structural integrity under cyclic loading or bending stresses, leading to fracture or failure. In some cases, the material may undergo fatigue degradation, where repeated mechanical stress results in the formation of microcracks that propagate over time, causing a gradual reduction in mechanical strength. This is particularly important in bioimplants designed for load-bearing applications such as bone fixation devices and joint replacements.

13.3.2.4 Environmental Degradation

Environmental factors, such as temperature, pH and the presence of biological fluids, significantly influence the degradation behaviour of resorbable materials. For example, the temperature of the body (around 37°C) can accelerate the degradation process by enhancing the rate of hydrolysis and enzymatic activity. The presence of inflammatory cells or oxidative stress can also contribute to accelerated degradation, particularly in the case of composite materials containing biodegradable and non-biodegradable phases [22]. In addition, the rate of degradation can vary across different anatomical sites. For example, bioimplants in areas with higher blood flow such as the femoral head or bone, may degrade at a different rate than implants in less vascularized tissues. This variation requires careful consideration in the design of resorbable materials to ensure that the degradation rate matches the healing time required for tissue regeneration.

13.4 Surface Modifications for Enhanced Tribological Performance

In the context of bioimplants and fixation devices, the tribological performance plays a critical role in ensuring their long-term functionality and the success of implantation. Tribology refers to the study of friction, wear and lubrication between interacting surfaces in relative motion. For bioimplants and fixation devices, the ability to reduce wear and minimize friction directly impacts the material's longevity, the body's response to the implant, and the overall success of the treatment. Surface modification techniques are essential for enhancing the tribological properties of these materials, ensuring optimal interaction with the surrounding tissues and reducing the risk of failure due to excessive wear or friction. Surface modification strategies aim to improve the mechanical, chemical and topographical properties of the implant surface without altering the bulk material properties [23]. Methods such as coating, surface texturing, laser treatment and plasma modification can significantly enhance the wear resistance, reduce friction and promote better bio-compatibility. The goal of these surface modifications is to optimize the interface between the implant and biological tissues, improving performance, reducing complications and promoting the overall success of the device.

13.4.1 Coatings and Treatments for Bioimplants

Coatings and surface treatments are crucial for enhancing the tribological performance of bioimplants and fixation devices. These modifications improve wear resistance, reduce friction, and increase the longevity of implants by minimizing damage to both the device and the surrounding tissue. Bioimplants are often subjected to harsh mechanical stresses and the tribological properties of these materials are vital for their performance and the prevention of complications such as inflammation, wear debris and implant loosening. Various coatings and surface treatments are used to modify the implant surface, offering improved wear resistance, biocompatibility and corrosion resistance, ensuring a favorable interface with biological tissues.

1. Ceramic Coatings

Ceramic coatings, such as hydroxyapatite (HA), zirconia and alumina are widely used in bioimplants for their excellent wear resistance and bio-compatibility [24]. Hydroxyapatite, in particular, is a calcium phosphate ceramic that mimics the natural mineral component of bone. When applied to implants, it promotes osseointegration and helps to reduce wear against bone or surrounding tissues. HA coatings are commonly used on titanium and its alloys, which are popular materials for orthopedic and dental implants. These coatings offer not only tribological benefits but also enhance the biological response of the implant, supporting better bonding with the surrounding bone tissue.

2. Metal-Based Coatings

Metal-based coatings such as titanium nitride (TiN) and titanium carbide (TiC) are employed for their hardness and wear resistance [25]. TiN coatings are especially beneficial in reducing friction between the implant and surrounding tissues, contributing to the longevity of the implant. These coatings also improve the corrosion resistance of the underlying material, preventing degradation over time. Similarly, TiC coatings provide excellent wear resistance, making them suitable for load-bearing implants such as joint replacements and bone fixation devices. The use of metal coatings also has the advantage of enhancing the implant's overall mechanical strength, which is particularly important in weight-bearing applications.

3. Polymer Coatings

Polymer coatings such as polyethylene (UHMWPE) and polytetrafluoroethylene (PTFE) are often applied to bioimplants for their lubricating properties and low friction coefficient [26]. UHMWPE is commonly used as a bearing material in joint replacements such as hip and knee prostheses. This polymer is chosen for its excellent wear resistance and low friction, which helps in reducing the wear rate of the implant and its counterfaces. PTFE coatings, known for their exceptional lubricating properties, can also be used to reduce friction between the implant and adjacent tissues, particularly in orthopedic applications where motion between the implant and bone or soft tissue is common.

4. Plasma Surface Treatments

Plasma treatments, including plasma spraying and plasma-enhanced chemical vapor deposition (PECVD) are utilized to improve the surface characteristics of bioimplants [27]. Plasma spraying involves the deposition of ceramic or metallic materials onto the implant surface, creating a coating that improves wear resistance and bio-compatibility. PECVD, on the other hand, is used to create thin films of organic or inorganic coatings that can modify the surface's chemical composition, improving both frictional behaviour and corrosion resistance.

5. Laser Surface Treatments

Laser surface treatment techniques such as laser remelting and laser texturing are used to modify the surface topography and hardness of bioimplant materials [28]. Laser remelting involves using a laser to melt and re-solidify the surface layer, which can improve surface hardness and wear resistance. This method can also help reduce surface defects such as porosity, which could lead to premature wear or failure. Laser texturing, on the other hand, involves creating micro- or nanoscale patterns on the implant surface to reduce friction and wear. These patterns can promote lubrication, either through the entrapment of synovial fluid or by reducing direct contact between the implant and surrounding tissues.

6. Anodizing

Anodizing is an electrochemical treatment often applied to titanium and its alloys to enhance surface properties. This process thickens the natural oxide layer on the material's surface, increasing corrosion resistance and improving the implant's wear resistance [29]. The anodized surface can also be further functionalized with bioactive molecules such as growth factors or drugs to promote tissue healing and reduce the risk of infection.

7. Hybrid Coatings

Hybrid coatings combine multiple materials to achieve optimal tribological and biological performance. For example, composite coatings that combine ceramics and polymers can be designed to offer both high wear resistance and low friction. These coatings are tailored to meet specific requirements, such as reducing wear and promoting tissue regeneration, which is particularly important for bioimplants used in orthopedic and dental applications.

13.5 Lubrication Mechanisms in Bioimplants

Lubrication plays a critical role in reducing friction and wear between bioimplants and surrounding biological tissues. In the human body, bioimplants are subjected to repetitive loading and movement, making effective lubrication mechanisms essential for ensuring long-term functionality and preventing damage to both the implant and the surrounding tissues. The interaction between bioimplants and biological fluids, such as synovial fluid in joints, is fundamental to minimizing friction and wear during the implant's service life. The lubrication mechanisms in bioimplants can be broadly categorized into two types: boundary lubrication and hydrodynamic lubrication. In boundary lubrication, a thin film of lubricant forms between the implant surface and the surrounding tissues, preventing direct contact and reducing friction [30]. In contrast, hydrodynamic lubrication occurs when the relative motion between surfaces generates a fluid film that separates the two surfaces, providing an effective lubrication layer and reducing wear [31].

13.5.1 Role of Synovial Fluid and Artificial Lubricants

Synovial fluid and artificial lubricants play a vital role in enhancing the tribological performance of bioimplants, particularly in joint replacements and fixation devices. The primary function of these lubricants is to reduce friction and wear between the implant surfaces and the surrounding biological tissues, such as bone and cartilage. In joint replacements, for example, synovial fluid naturally serves as a lubricant, facilitating smooth movement while minimizing damage to the implant and the surrounding tissues. The understanding of the behaviour of synovial fluid and developing effective artificial lubricants is crucial for the longevity and functionality of bioimplants.

13.5.1.1 Synovial Fluid and Its Role in Bioimplants

Synovial fluid is a viscous, non-Newtonian fluid found in synovial joints, where it acts as both a lubricant and a shock absorber [32]. It is primarily composed of water, electrolytes, proteins (such as hyaluronic acid) and cells, and its primary function is to reduce friction and wear between articulating surfaces in joints. In natural joints, the synovial fluid forms a thin film between the surfaces of the bones, preventing direct contact and minimizing wear. The high viscosity and shear-thinning properties of synovial fluid allow it to efficiently reduce friction during movement, even under high loads. In the case of bioimplants such as total joint replacements, synovial fluid plays a similar role in lubricating the implant surfaces. However, the performance of synovial fluid in bioimplants can be influenced by factors such as implant surface roughness, wear debris and the overall health of the surrounding tissues.

13.5.1.2 Artificial Lubricants for Bioimplants

The synovial fluid is effective in natural joints, its lubrication properties may not always be sufficient for bioimplants, particularly when they experience excessive loading or when the body's natural lubrication system is compromised. In such cases, artificial lubricants are developed to supplement or replace the body's natural lubrication. Artificial lubricants are specifically designed to mimic the properties of synovial fluid, improving the tribological performance of bioimplants by reducing friction and wear. Several types of artificial lubricants have been investigated for use in bioimplant applications, including polymer-based lubricants, protein-based lubricants and even lubricants derived from biocompatible oils [33]. Polymer-based lubricants are often used due to their ability to form protective films on the implant surface, reducing friction and wear. These lubricants can be applied to the surface of the implant or incorporated into the material itself. Protein-based lubricants such as lubricin or hyaluronic acid, are also studied because they are biocompatible and can effectively mimic the natural lubrication mechanisms of synovial fluid. These lubricants are typically applied to the implant surface, providing a layer of protection that reduces direct contact between the implant and surrounding tissues. In addition to providing friction reduction, artificial lubricants also help in reducing the formation of wear debris, which can lead to inflammation and osteolysis (bone resorption) around the implant. The ability of these lubricants to maintain a stable film under loading conditions is essential for ensuring the longevity of bioimplants, particularly in weight-bearing applications like hip and knee prostheses.

13.5.2 Bioinspired Lubrication Strategies

Bioinspired lubrication strategies in bioimplants draw inspiration from natural lubrication mechanisms found in the human body and other organisms to improve the tribological performance of implants. The primary objective of these strategies is to mimic the natural lubrication system of the body, especially the function of synovial fluid in joints, which reduces friction and wear between the surfaces of bones and cartilage. These bioinspired methods aim to extend the lifespan of bioimplants and enhance their performance by preventing wear, minimizing friction and reducing the risk of inflammatory responses or failure.

13.5.2.1 Mimicking Synovial Fluid Properties

Synovial fluid, which naturally lubricates joints, is a non-Newtonian fluid with unique properties such as high viscosity, shear-thinning behaviour and the ability to form a protective film on joint surfaces. Bioinspired lubrication strategies for bioimplants focus on replicating these properties to reduce friction between the implant and surrounding tissues. For instance, artificial lubricants have been developed that incorporate components like hyaluronic acid or lubricin, both of which are naturally found in synovial fluid and can mimic its lubricating effect. These lubricants help to reduce the friction between the bioimplant and the biological tissues, enhancing the performance and longevity of the implant. In addition, the incorporation of polysaccharides such as chondroitin sulfate, has been explored in bioimplants to replicate the gel-like consistency of synovial fluid [34]. These materials are hydrophilic, enabling them to absorb water and provide effective lubrication under loading conditions, thereby reducing wear on the implant surface. The ability to form a stable boundary layer between the implant surface and the surrounding tissues is a key characteristic of these bioinspired lubricants.

13.5.2.2 Surface Engineering Techniques

Bioinspired surface engineering techniques play a crucial role in improving the tribological performance of bioimplants. These strategies focus on modifying the implant surface to enhance its interaction with natural or artificial lubricants. One such approach is the development of superhydrophilic or superhydrophobic surfaces that can either attract or repel water and lubricants to improve lubrication. The surface texturing or micro/nano-patterning is used to create micro topographies on the implant surfaces, which can promote better retention of lubricants and reduce direct contact between the implant and tissues. This method can effectively reduce friction, minimize wear and promote a more durable lubrication film.

13.5.2.3 Nanotechnology and Tribology

Nanotechnology also facilitates the creation of nanostructured surfaces that mimic the natural microtextures found in cartilage and other biological tissues. These surfaces can significantly enhance lubricant retention, reduce wear and improve the overall tribological behaviour of bioimplants.

13.5.2.4 Self-Lubricating Materials

Another promising bioinspired lubrication strategy involves the development of self-lubricating materials. These materials are designed to release lubricants over time or as needed to maintain a stable lubrication film [35]. Bioinspired self-lubricating polymers, such as those containing embedded lubricating agents or those that undergo surface modification under mechanical loading, have been investigated. For example, incorporating materials like poly (lactic acid) or polycaprolactone, which can release lubricants as the implant is subjected to stress, allows for continuous lubrication without the need for external lubrication replenishment. These materials are also biodegradable and resorbable, making them ideal for use in bioimplants where long-term function is required but without causing adverse reactions in the body.

13.6 Wear Mechanisms in Bioimplants and Fixation Devices

Wear mechanisms in bioimplants and fixation devices are critical factors influencing the longevity, performance and biocompatibility of these medical devices. In the human body, bioimplants such as orthopedic devices, dental implants and fixation plates are subjected to complex mechanical loading conditions, which can lead to material degradation over time. These wear processes can result in the formation of debris, inflammation and potential failure of the implant, making it essential to understand the underlying wear mechanisms. The primary wear mechanisms in bioimplants include abrasive wear, adhesive wear, corrosive wear and fretting wear. Abrasive wear occurs when hard particles or surface roughness on the implant surface interact with opposing surfaces, leading to material removal.

Adhesive wear is characterized by the transfer of material between contacting surfaces, often resulting in surface deformation. Corrosive wear, which is particularly relevant for metal-based implants, involves the interaction of implant materials with the bodily fluids, leading to oxidation and material breakdown. Fretting wear arises due to small, oscillatory movements at the implant-tissue interface, contributing to surface damage and the generation of wear particles.

13.6.1 Adhesive, Abrasive, and Fatigue Wear

Bioimplants such as orthopedic devices, dental implants and fixation devices are subjected to various wear mechanisms when placed in the body. The understanding of the tribological behaviour of these implants is important for designing long-lasting and biocompatible materials. Among the different wear mechanisms, adhesive wear, abrasive wear, and fatigue wear play significant roles in the degradation of bioimplants and fixation devices. The different types of wear mechanism are shown in Figure 13.3.

13.6.1.1 Adhesive Wear

Adhesive wear occurs when two surfaces in contact transfer material from one surface to the other. This is particularly common in metallic implants, where friction between the implant and surrounding tissue or bone can lead to localized material bonding. When the stress exceeds the material's yield strength, micro-welding occurs at the contact points, causing material to detach from the surface. This process results in the generation of wear debris, which can lead to inflammation, adverse tissue responses, and implant failure. In bioimplants, adhesive wear is particularly relevant when

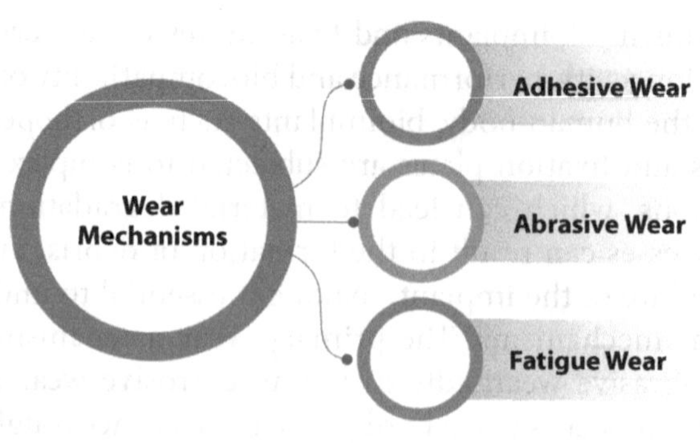

Figure 13.3 Types of wear mechanism.

implants experience high contact pressures during motion, such as in joint replacements [36]. The continuous movement, like flexion and extension, can cause wear on the interface between the implant and surrounding tissues. A common example is the frictional contact between the metal components in knee and hip replacements, where adhesive wear contributes to material loss and debris generation. To mitigate this, coatings like diamond-like carbon (DLC) and surface treatments such as anodizing or nitriding are employed to reduce friction and enhance the material's resistance to adhesive wear.

13.6.1.2 Abrasive Wear

Abrasive wear occurs when hard particles or surface asperities cause material to be mechanically removed from the implant surface. This type of wear is common in bioimplants exposed to rough biological environments such as bone or cartilage. Abrasive wear is particularly problematic when hard particles from the surrounding tissues or the implant itself interact with the implant surface [37]. For instance, in total joint replacements, the wear debris produced from the implant materials such as titanium or cobalt-chromium alloys can interact with the implant surface causing additional damage. Similarly, in dental implants, abrasion caused by food particles, plaque or even dental tools during installation can contribute to surface degradation over time. Abrasive wear can lead to surface roughening, which can further exacerbate adhesive wear as rough surfaces increase friction and material transfer between the implant and tissues. To minimize abrasive wear, bioimplants are often designed with smoother surface finishes or are coated with ceramic materials like alumina or zirconia, which provide enhanced wear resistance.

13.6.1.3 Fatigue Wear

Fatigue wear refers to the gradual degradation of materials due to repeated mechanical stress, often at lower magnitudes than those required to cause immediate failure. In bioimplants, fatigue wear is a critical concern as these implants experience cyclic loading throughout their functional life, especially in weight-bearing applications like joint replacements [38]. Over time, repeated loading can lead to the initiation of micro-cracks on the implant surface. These cracks propagate due to continuous mechanical stress, eventually leading to surface spalling or fracture. The repetitive loading in bioimplants such as hip or knee replacements induces fatigue wear, leading to failure of the material. For example, metallic implants made from titanium alloys or cobalt-chromium alloys are particularly vulnerable to fatigue wear under cyclic stress. The material's ability to withstand repeated loading without failure depends on its mechanical properties, including fatigue strength and resistance to crack propagation. Surface treatments such as shot peening or surface hardening, are commonly employed to improve the fatigue resistance of bioimplants by inducing compressive residual stresses on the surface, which help to prevent crack initiation and propagation.

13.6.2 Interplay of Wear and Material Degradation

In bioimplants and fixation devices, wear and material degradation are closely interlinked processes that significantly impact the longevity and performance of these devices in the human body. These mechanisms are critical in determining the functionality of medical implants such as orthopedic prosthetics, dental implants and fixation devices which are exposed to complex mechanical, chemical and biological environments. The interplay between wear and material degradation can result in failure modes that compromise the implant's structural integrity, biocompatibility and overall effectiveness.

13.6.2.1 Wear as a Driver of Material Degradation

Wear in bioimplants refers to the mechanical removal of material from the implant surface due to friction between the implant and surrounding tissues or other components. The wear process, whether abrasive, adhesive or fatigue-driven, generates debris that can contribute to further degradation of the material. This wear-induced degradation is particularly important in the context of metal, ceramic and polymer-based implants, which all face different wear challenges. In metallic implants like those made from titanium or cobalt-chromium alloys, wear debris generated by adhesive or abrasive wear can lead to oxidative degradation and corrosion at the wear site [39]. The debris can accumulate around the implant interface, creating a hostile environment that accelerates material degradation. Furthermore, wear debris can induce inflammatory responses in surrounding tissues, exacerbating the degradation process by triggering immune reactions and stimulating the release of free radicals, which promote corrosion. In ceramic implants, while wear resistance is generally higher, surface roughening due to abrasive wear can expose the material to aggressive biological fluids, accelerating degradation. In dental implants made of ceramics, wear particles generated during mastication can lead to micro-cracking and gradual degradation of the material. The interaction of wear particles with the biological environment can lead to hydrolysis or ion leaching, further compromising the implant's mechanical properties.

13.6.2.2 Material Degradation Contributing to Wear

Material degradation can also influence the wear behaviour of bioimplants [40]. Degradation mechanisms including corrosion, hydrolysis and surface oxidation, alter the mechanical properties of the implant material, making it more susceptible to wear. The corrosion in metallic bioimplants results in the formation of oxide layers that may alter surface hardness and reduce wear resistance. The corrosion process weakens the implant material, making it prone to fatigue failure and subsequent wear. This is particularly relevant for long-term implanted devices that are exposed to the body's fluids, which can cause localized corrosion or pitting, weakening the material over time. Polymeric bioimplants such as those used in soft tissue applications, can undergo hydrolytic degradation when exposed to bodily fluids, which weakens the polymer matrix and makes the implant more vulnerable to wear. As the polymer degrades, its ability to resist mechanical loading decreases, and it becomes more prone to the generation of wear debris. The combination of wear and material degradation in these polymers can shorten the lifespan of the implant and lead to the release of harmful degradation products into surrounding tissues.

13.6.2.3 Feedback Mechanisms between Wear and Degradation

The interaction between wear and material degradation in bioimplants is not linear but often exhibits a feedback loop. The wear can accelerate the degradation of materials by exposing fresh surfaces to aggressive biological environments, leading to faster material breakdown [41]. This, in turn, can increase the rate of wear as the material loses its original mechanical properties and hardness. Similarly, material degradation can increase the rate of wear by weakening the implant surface, making it more susceptible to abrasive and adhesive wear.

13.7 *In Vitro* and *In Vivo* Evaluation of Tribological Performance

The evaluation of the tribological performance of bioimplants and fixation devices is essential for ensuring their durability, functionality and biocompatibility. This assessment is conducted through both *in vitro* and *in vivo* testing, each offering unique insights into the behaviour of materials under different conditions. *In vitro* testing involves controlled laboratory experiments where bioimplant materials are subjected to simulated conditions that mimic the physiological environment. These tests typically assess wear, friction and lubrication properties of materials using mechanical simulators, friction testers and tribometers [42]. The replication of joint motion or other relevant mechanical loads, *in vitro* testing provides valuable data on how materials perform under controlled conditions, often focusing on factors like surface roughness, wear rate and frictional behaviour.

On the other hand, *in vivo* testing involves implanting the device into a living organism to study its performance in a natural biological environment. This type of testing provides real-world data on how the bioimplant behaves under dynamic loading, exposure to body fluids and tissue interactions [43]. It helps identify potential issues such as inflammatory responses, material degradation or failure modes that may not be evident in *in vitro* experiments. Both evaluation methods are crucial for understanding the tribological behaviour of bioimplants, ensuring they meet the required standards for clinical applications.

13.7.1 Testing Methods for Wear and Friction Analysis

The evaluation of wear and friction behaviour is crucial in assessing the performance and longevity of bioimplants and fixation devices. Various testing methods are employed to simulate the mechanical interactions that occur between the implant and surrounding tissues or joint surfaces in the human body. These tests help determine how materials will behave over time under different loading conditions and biological environments. The two primary categories of testing methods for wear and friction analysis are *in vitro* and *in vivo* evaluations.

13.7.1.1 In Vitro Testing Methods

In vitro testing involves replicating the mechanical and tribological conditions of the body using laboratory equipment. This allows for controlled experiments that provide quantitative data on the wear and friction properties of bioimplant materials. The common *in vitro* testing methods are shown in Figure 13.4.

1. **Pin-on-Disk Test**: One of the most widely used methods for wear and friction analysis, the pin-on-disk test involves pressing a pin (the sample material) against a rotating disk (often made of a material simulating bone

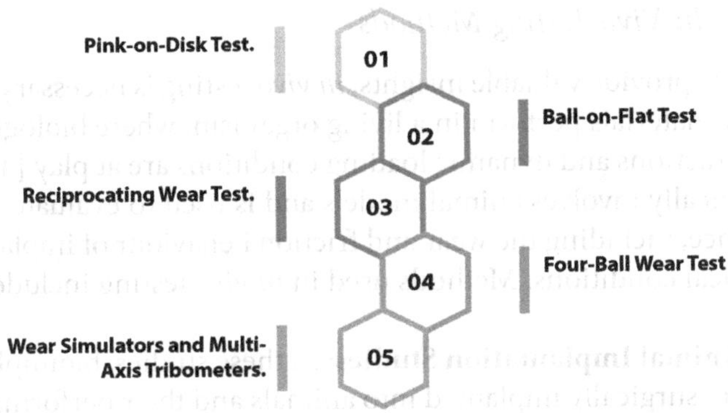

Figure 13.4 *In vitro* testing methods.

or cartilage). This test measures the friction coefficient and wear rate by analyzing the wear scars on the pin and disk surfaces. It is especially useful for simulating sliding contact in prosthetic joints.

2. **Ball-on-Flat Test**: This method is similar to the pin-on-disk test but uses a ball that rotates against a flat surface. It is commonly used to simulate the tribological interaction between joint components such as the femoral head and acetabular cup in hip implants. The wear and friction generated during the test can be measured through changes in surface topography.

3. **Reciprocating Wear Test**: This test simulates the repetitive back-and-forth motion of joints, such as the knee or hip. A sample material is subjected to a reciprocating motion against a counterface, mimicking the cyclic loading that bioimplants experience. It is particularly useful for studying the effects of wear and lubrication in joint replacements.

4. **Four-Ball Wear Test**: This method is used to evaluate the tribological properties of lubricants used in bioimplants. Four spherical balls are placed under a load and one ball is rotated against the others. The wear scars and frictional forces are analyzed to determine the lubricating properties and wear resistance of the material.

5. **Wear Simulators and Multi-Axis Tribometers**: These devices simulate more complex joint motions and loading conditions by applying variable loads and motions to the bioimplant material. They are particularly useful in simulating the actual conditions that implants would experience in the human body such as the multi-directional forces in knee or shoulder joints.

13.7.1.2 In Vivo Testing Methods

In vitro tests provide valuable insights, *in vivo* testing is necessary to understand how materials perform in a living organism, where biological fluids, tissue interactions and dynamic loading conditions are at play [44]. *In vivo* testing typically involves animal models and is used to evaluate long-term performance, including the wear and friction behaviour of implants in real physiological conditions. Methods used in *in vivo* testing include:

1. **Animal Implantation Studies**: In these studies, bioimplants are surgically implanted into animals and their performance is monitored over time. The wear and friction characteristics are analyzed by measuring the implant's mechanical integrity, surface degradation and potential tissue responses. This testing helps determine the long-term tribological behaviour of the implant, including how wear particles interact with surrounding tissues.
2. **Radiographic Analysis**: After implantation, radiographs (X-rays) are often used to observe the movement and wear of bioimplants *in vivo*. This can help identify implant loosening, wear patterns and any potential failure mechanisms in real biological environments.
3. **Histological Evaluation**: This method involves examining tissue samples surrounding the implant under a microscope after the implant has been removed. It provides valuable information about how wear debris and friction affect the surrounding tissues, such as inflammation, fibrosis or bone resorption.
4. **Wear Particle Monitoring**: *In vivo*, wear particles generated from the bioimplant can migrate to surrounding tissues. These particles can be analyzed to assess the type and number of debris generated which can provide insights into the wear rate and potential biological consequences of wear.

13.8 Integration of Nanomaterials in Tribological Applications

The integration of nanomaterials into tribological applications, particularly in bioimplants and fixation devices, has gained significant attention due to their ability to enhance wear resistance, reduce friction and improve the overall performance of implant materials. Nanomaterials, with their unique properties at the atomic or molecular level, offer substantial advantages over traditional materials in terms of mechanical strength, durability and surface interactions. The primary goal in incorporating nanomaterials into bioimplant design is to improve the longevity and functionality of implants, which are often subjected to complex loading conditions and harsh biological environments. Nanomaterials such as carbon nanotubes, nanoparticles and graphene can be integrated into polymeric, ceramic or metallic matrices to optimize their tribological performance [45]. These materials have demonstrated the potential to enhance wear resistance and friction properties by influencing surface morphology and mechanical properties at the nanoscale. Additionally, nanomaterials can be engineered to create self-lubricating systems or to improve biocompatibility by promoting better integration with surrounding tissues. The application of nanomaterials in tribological studies aims to overcome the challenges faced by conventional bioimplants such as wear-induced degradation, friction-induced inflammation and implant failure.

13.8.1 Impact of Nanoparticles on Wear and Friction Properties

Nanoparticles have emerged as crucial additives in enhancing the wear and friction properties of bioimplant materials, including those used in fixation devices. Their unique properties, derived from their nanoscale dimensions, enable significant improvements in the tribological performance of materials exposed to constant mechanical stress and wear in biological environments. These improvements are vital for extending the service life of bioimplants and reducing the risks of implant failure, such as loosening or wear debris formation. One of the primary ways nanoparticles affect wear resistance is by enhancing the hardness and surface durability of materials [46]. When incorporated into polymers, ceramics, or metals, nanoparticles like silica (SiO_2), alumina (Al_2O_3), titanium dioxide (TiO_2) and carbon-based nanoparticles (such as carbon nanotubes and graphene), they significantly reinforce the matrix. These nanoparticles create a protective barrier on the surface, reducing the contact area and minimizing the friction between the implant material and surrounding tissues or opposing surfaces. Nanoparticles also enhance the lubrication characteristics of bioimplants. Some nanoparticles like graphene oxide or nano-lubricants, exhibit self-lubricating properties that reduce friction and prevent the build-up of wear particles. These lubricating nanoparticles can form a thin, protective film on the surface of implants, which reduces the direct contact between surfaces, lowering friction and wear. This is especially beneficial in load-bearing applications such as orthopedic implants, where the constant movement of joints can lead to excessive wear over time.

13.9 Challenges in Tribological Studies

Tribological studies of bioimplants and fixation devices are essential for understanding the wear, friction and lubrication behaviors that directly affect the performance and longevity of medical implants [47]. These studies are crucial in ensuring that bioimplants such as joint replacements and bone fixation devices, function effectively within the human body, minimizing complications like implant loosening, wear debris formation and tissue inflammation. However, the integration of tribology in bioimplants presents several challenges, particularly due to the complex biological environment in which these devices operate. Factors such as the variability of human body fluids, the mechanical stresses experienced by implants, and the potential for biological degradation of materials complicate the accurate prediction of implant behaviour under real-life conditions. Moreover, the interplay between different materials used in bioimplants like metals, ceramics and polymers are further adds to the challenge, as each material exhibits distinct tribological properties. Despite these hurdles, there are significant opportunities to improve bioimplant design through the use of advanced materials like nanocomposites, coatings and surface modifications. These innovations offer the potential for enhanced wear resistance, reduced friction and better biocompatibility, providing promising solutions to the challenges faced in tribological studies of bioimplants and fixation devices.

13.9.1 Biological and Environmental Factors Affecting Performance

The performance of bioimplants and fixation devices is influenced by a variety of biological and environmental factors, which play a crucial role in their long-term functionality, wear resistance, and biocompatibility. The understanding of these factors is essential for optimizing the design and material selection of implants to ensure their successful integration with the human body. One of the key biological factors affecting bioimplant performance is the physiological environment within the human body. This environment includes factors such as body temperature, pH and the presence of synovial fluid or other bodily fluids all of which can impact the wear and friction characteristics of implant materials [48]. The mechanical stresses experienced by bioimplants also vary depending on the type of implant and its location in the body. These stresses include both static loads (such as body weight) and dynamic loads (from movement and physical activity). The implants subjected to high mechanical stress may experience accelerated wear, especially if the materials used do not possess adequate strength or fatigue resistance [49]. In addition to mechanical and biological factors, environmental conditions outside the body can also affect implant performance. For example, the sterilization process used during implant manufacturing can influence the material properties of implants. The techniques such as autoclaving or gamma irradiation may cause surface alterations, which can impact the tribological behaviour and long-term stability of the implant. Furthermore, the presence of corrosive elements in the environment such as chlorine in saline solutions, can lead to corrosion of metallic implants, compromising their mechanical properties and tribological performance [50].

13.10 Conclusion

Tribological studies of bioimplants and fixation devices are crucial for enhancing the longevity, performance and safety of medical implants. These studies explore the interaction between materials under friction and wear conditions, which directly impacts the functionality of implants such as orthopedic devices, dental implants and surgical fixation systems. Over the past decades, there has been significant progress in understanding the tribological behaviour of bioimplants, leading to the development of advanced materials and designs. However, despite the advances, challenges remain that must be addressed for improving the effectiveness and reliability of these devices. The tribological performance of bioimplants is highly influenced by factors such as material selection, surface properties, lubrication mechanisms and degradation behaviour. The materials like polymers, ceramics and metal alloys are commonly used in bioimplants, each offering unique advantages and limitations in terms of mechanical strength, wear resistance and biocompatibility. The surface modifications such as coatings and texturing, have shown promise in reducing wear rates and enhancing the tribological properties of implants. The role of synovial fluid and artificial lubricants is also significant, as they contribute to reducing friction and wear in joint replacements and other load-bearing implants.

AI disclosure statement: The author confirms that no artificial intelligence (AI)-generated content or AI-assisted writing tools were used in the preparation of this chapter (including summaries and the abstract). All text, analysis, and conclusions are the result of the author's original work.

References

1. Moghadasi, K., Isa, M.S.M., Ariffin, M.A., Raja, S., Wu, B., Yamani, M., bin Yusoff, N., A review on biomedical implant materials and the effect of friction stir based techniques on their mechanical and tribological properties. *J. Mater. Res. Technol.*, 17, 1054–1121, 2022.
2. Omigbodun, F.T., Oladapo, B.I., Osa-uwagboe, N., Exploring the frontier of Polylactic Acid/Hydroxyapatite composites in bone regeneration and their revolutionary biomedical applications–A review. *J. Reinf. Plast. Compos.*, 2024, 07316844241278045.
3. Goodman, S.B., Gallo, J., Gibon, E., Takagi, M., Diagnosis and management of implant debris-associated inflammation. *Expert Rev. Med. Devices*, 17, 1, 41–56, 2020.
4. Saha, S. and Roy, S., Metallic dental implants wear mechanisms, materials, and manufacturing processes: a literature review. *Materials*, 16, 1, 161, 2022.
5. Al Mahmud, M.Z., Mobarak, M.H., Hossain, N., Islam, M.A., Rayhan, M.T., Emerging breakthroughs in biomaterials for orthopedic applications: A comprehensive review. *Bioprinting*, e00323, 2023.
6. Kumar, A. and Singh, G., Surface modification of Ti6Al4V alloy *via* advanced coatings: Mechanical, Tribological, Corrosion, Wetting, and Biocompatibility studies. *J. Alloys Compd.*, 174418, 2024.
7. Ghosh, S., Choudhury, D., Das, N.S., Pingguan-Murphy, B., Tribological role of synovial fluid compositions on artificial joints—a systematic review of the last 10 years. *Lubr. Sci.*, 26, 6, 387–410, 2014.
8. Yin, X., Chen, H., Zhang, F., Yu, H., Li, S., Zhao, Z., Yang, K., Recent Progress on carbon nanomaterials for resisting the Wear damages. *J. Nanomater.*, 2022, 1, 1074133, 2022.
9. Wu, Z., Bao, H., Xing, Y., Liu, L., Tribological characteristics and advanced processing methods of textured surfaces: a review. *Int. J. Adv. Manuf. Technol.*, 114, 5, 1241–1277, 2021.
10. Kunčická, L., Kocich, R., Lowe, T.C., Advances in metals and alloys for joint replacement. *Prog. Mater Sci.*, 88, 232–280, 2017.
11. Skjöldebrand, C., Tipper, J.L., Hatto, P., Bryant, M., Hall, R.M., Persson, C., Current status and future potential of wear-resistant coatings and articulating surfaces for hip and knee implants. *Mater. Today Bio*, 15, 100270, 2022.

12. Baena, J.C., Wu, J., Peng, Z., Wear performance of UHMWPE and reinforced UHMWPE composites in arthroplasty applications: a review. *Lubricants*, 3, 2, 413–436, 2015.
13. Liu, X., Miao, Y., Liang, H., Diao, J., Hao, L., Shi, Z., Wang, Y., 3D-printed bioactive ceramic scaffolds with biomimetic micro/nano-HAp surfaces mediated cell fate and promoted bone augmentation of the bone–implant interface *in vivo*. *Bioact. Mater.*, 12, 120–132, 2022.
14. Monia, T. and Ridha, B.C., Polymer-ceramic composites for bone challenging applications: Materials and manufacturing processes. *J. Thermoplast. Compos. Mater.*, 37, 4, 1540–1557, 2024.
15. Jem, K.J., Tan, B., The development and challenges of poly (lactic acid) and poly (glycolic acid). *Adv. Ind. Eng. Polym. Res.*, 3, 2, 60–70, 2020.
16. Asti, A. and Gioglio, L., Natural and synthetic biodegradable polymers: different scaffolds for cell expansion and tissue formation. *Int. J. Artif. Organs*, 37, 3, 187–205, 2014.
17. Davidson, J.A., Characteristics of metal and ceramic total hip bearing surfaces and their effect on long-term ultra high molecular weight polyethylene wear. *Clin. Orthop. Relat. Res.*, 294, 361–378, 1993.
18. Ranuša, M., Odehnal, L., Kučera, O., Nečas, D., Hartl, M., Křupka, I., Vrbka, M., Effect of surface texturing on friction and lubrication of Ti6Al4V biomaterials for joint implants. *Tribol. Lett.*, 73, 1, 15, 2025.
19. Zhou, Q. and Xanthos, M., Nanoclay and crystallinity effects on the hydrolytic degradation of polylactides. *Polym. Degrad. Stab.*, 93, 8, 1450–1459, 2008.
20. Zaaba, N.F. and Jaafar, M., A review on degradation mechanisms of polylactic acid: Hydrolytic, photodegradative, microbial, and enzymatic degradation. *Polym. Eng. Sci.*, 60, 9, 2061–2075, 2020.
21. Mehboob, H. and Chang, S.H., Application of composites to orthopedic prostheses for effective bone healing: A review. *Compos. Struct.*, 118, 328–341, 2014.
22. Khodaei, T., Schmitzer, E., Suresh, A.P., Acharya, A.P., Immune response differences in degradable and non-degradable alloy implants. *Bioact. Mater.*, 24, 153–170, 2023.
23. Roach, P., Eglin, D., Rohde, K., Perry, C.C., Modern biomaterials: a review—bulk properties and implications of surface modifications. *J. Mater. Sci.: Mater. Med.*, 18, 1263–1277, 2007.
24. Kumar, R. and Agrawal, A., Tribological, mechanical, and cytocompatibility characteristics to study surface integrity of titanium–hydroxyapatite sintered bio-composites for implant applications. *J. Mater. Eng. Perform.*, 1–15, 2025.
25. Wang, H., Lin, N., Nouri, M., Liu, Z., Yu, Y., Zeng, Q., Wu, Y., Improvement in surface performance of stainless steel by nitride and carbon-based coatings prepared *via* physical vapor deposition for marine application. *J. Mater. Res. Technol.*, 27, 6021–6046, 2023.

26. Singh, D.K. and Verma, R.K., Contemporary development on the performance and functionalization of ultra high molecular weight polyethylene (UHMWPE) for biomedical implants. *Nano Life,* 11, 3, 2130009, 2021.
27. Hassan, S., Nadeem, A.Y., Qaiser, H., Kashif, A.S., Ahmed, A., Khan, K., Altaf, A., A review of carbon-based materials and their coating techniques for biomedical implants applications. *Carbon Lett.*, 33, 4, 1171–1188, 2023.
28. Tan, C.Y., Wen, C., Ang, H.Q., Influence of laser parameters on the microstructures and surface properties in laser surface modification of biomedical magnesium alloys. *J. Magnesium Alloys*, 2024.
29. Minagar, S., Berndt, C.C., Wang, J., Ivanova, E., Wen, C., A review of the application of anodization for the fabrication of nanotubes on metal implant surfaces. *Acta Biomater.*, 8, 8, 2875–2888, 2012.
30. Yuan, H. and Cui, W., The lubricated matter in body. *Prog. Mater Sci.*, 101334, 2024.
31. Han, J., Fang, L., Sun, J., Ge, S., Hydrodynamic lubrication of microdimple textured surface using three-dimensional CFD. *Tribol. Trans.,* 53, 6, 860–870, 2010.
32. More, S., Kotiya, A., Kotia, A., Ghosh, S.K., Spyrou, L.A., Sarris, I.E., Rheological properties of synovial fluid due to viscosupplements: a review for osteoarthritis remedy. *Comput. Methods Programs Biomed.,* 196, 105644, 2020.
33. Shit, S.C., Shah, P.M., Edible polymers: challenges and opportunities. *J. Polym.,* 2014, 427259, 2014.
34. Gericke, M., Amaral, A.J., Budtova, T., De Wever, P., Groth, T., Heinze, T., Fardim, P., The European Polysaccharide Network of Excellence (EPNOE) research roadmap 2040: Advanced strategies for exploiting the vast potential of polysaccharides as renewable bioresources. *Carbohydr. Polym.*, 326, 121633, 2024.
35. Gong, H., Yu, C., Zhang, L., Xie, G., Guo, D., Luo, J., Intelligent lubricating materials: A review. *Composites, Part B*, 202, 108450, 2020.
36. Shen, G., Fang, F., Kang, C., Tribological performance of bioimplants: a comprehensive review. *Nanotechnol. Precis. Eng. (NPE),* 1, 2, 107–122, 2018.
37. Bauer, S., Schmuki, P., Von Der Mark, K., Park, J., Engineering biocompatible implant surfaces: Part I: Materials and surfaces. *Prog. Mater Sci.*, 58, 3, 261–326, 2013.
38. Abraham, A.M., Venkatesan, S., A critical review on biomaterials using powder metallurgy method. *Eng. Res. Express,* 6, 1, 012508, 2024.
39. Saha, S., Kiran, K.U.V., Zhang, X., Hou, X., Roy, S., Investigating the tribological and corrosion behavior of Co–Cr alloy as an implant material for orthodontic applications. Wear, 523, 204755, 2023.
40. Ramacharyulu, D.A., Sonia, P., Kumar, A., Thakur, R., Khan, M.I., Abduvalieva, D., Recent advancement in surface modification techniques for bio-implants. *Results Eng.*, 105192, 2025.

41. Singh, R. and Dahotre, N.B., Corrosion degradation and prevention by surface modification of biometallic materials. *J. Mater. Sci.: Mater. Med.*, 18, 5, 725–751, 2007.
42. Muralidharan, K., Vignesh, V., Vignesh, R.V., Govindaraju, M., Baghad, A., Narassima, M.S., Siengchin, S., Comprehensive overview of nano, micro, and macro tribometers in practice. *J. Bio- Tribo-Corros.*, 10, 3, 44, 2024.
43. Singh, J.P. and Sharma, Y., Corrosion cracking in Mg alloys based bioimplants. *J. Electrochem. Sci. Eng.*, 13, 1, 193–214, 2023.
44. Yang, L., Pijuan-Galito, S., Rho, H.S., Vasilevich, A.S., Eren, A.D., Ge, L., Zhou, Q., High-throughput methods in the discovery and study of biomaterials and materiobiology. *Chem. Rev.*, 121, 8, 4561–4677, 2021.
45. Moghadam, A.D., Omrani, E., Menezes, P.L., Rohatgi, P.K., Mechanical and tribological properties of self-lubricating metal matrix nanocomposites reinforced by carbon nanotubes (CNTs) and graphene–a review. *Composites, Part B*, 77, 402–420, 2015.
46. Zhai, W., Bai, L., Zhou, R., Fan, X., Kang, G., Liu, Y., Zhou, K., Recent progress on wear-resistant materials: designs, properties, and applications. *Adv. Sci.*, 8, 11, 2003739, 2021.
47. Uwais, Z.A., Hussein, M.A., Samad, M.A., Al-Aqeeli, N., Surface modification of metallic biomaterials for better tribological properties: a review. *Arab. J. Sci. Eng.*, 42, 11, 4493–4512, 2017.
48. Cen, J., Chen, Q., Chen, K., Zhang, X., Feng, H., Feng, C., Ge, S., Coupling Influence between Environmental Factors and Synovial Fluid Components on Tribological Behavior of Hip Joint Materials. *Tribol. Int.*, 110464, 2024.
49. Chen, Q. and Thouas, G.A., Metallic implant biomaterials. *Mater. Sci. Eng.: R: Rep.*, 87, 1–57, 2015.
50. Unune, D.R., Brown, G.R., Reilly, G.C., Thermal based surface modification techniques for enhancing the corrosion and wear resistance of metallic implants: A review. *Vacuum*, 203, 111298, 2022.

14

Surface Coating and Clinical Translational Studies on Resorbable Composites for Implants

Abhinay Thakur

Division of Research and Development, Lovely Professional University, Phagwara, Punjab, India

Abstract

Resorbable composites have emerged as transformative materials in the field of biomedical implants due to their ability to degrade safely within the body, eliminating the need for secondary removal surgeries. This chapter explores the recent advancements in surface coating technologies and their clinical translational potential in enhancing the performance of resorbable composite implants. Focus is placed on bioactive coatings that improve osseointegration, corrosion resistance, and antimicrobial activity while maintaining the mechanical integrity and biocompatibility of the underlying material. Techniques such as plasma spraying, sol-gel deposition, and electrochemical coating are discussed for tailoring surface properties at the nano- and microscale. By bridging material science and clinical practice, this chapter provides a comprehensive understanding of how surface engineering strategies are driving the next generation of resorbable implants toward safer, more effective, and patient-specific orthopedic and dental solutions. The integration of smart coatings and biofunctional surfaces holds great promise for the future of personalized and regenerative implant therapies.

***Keywords*:** Resorbable composites, surface coatings, biomedical implants, osseointegration, biodegradability

Email: thakurabhinay96@gmail.com

Arbind Prasad (ed.) Resorbable Composites for Bioimplants and Fixation Devices, (415–448) © 2026 Scrivener Publishing LLC

14.1 Introduction

Implantable biomedical devices have revolutionized modern healthcare, offering critical support in orthopedic, dental, cardiovascular, and reconstructive surgeries. Traditional implant materials such as stainless steel, titanium, and cobalt-chromium alloys have demonstrated excellent mechanical strength and long-term biocompatibility [1, 2]. However, their permanent presence in the body often necessitates secondary surgeries for removal, can cause chronic inflammation, and may not integrate well with host tissues over extended periods. In response to these limitations, a new class of materials known as resorbable composites has emerged. These materials are designed to fulfill temporary mechanical and biological functions during the healing process and subsequently degrade harmlessly in the body, eliminating the need for surgical retrieval. Resorbable composites are engineered materials composed of biodegradable matrices reinforced with bioactive or inert fillers to improve their mechanical, biological, or chemical properties [3]. The base matrix can be composed of biodegradable polymers such as polylactic acid (PLA), polyglycolic acid (PGA), polycaprolactone (PCL), or natural polymers like collagen and chitosan [4–7]. Inorganic matrices may include metals such as magnesium and zinc, which gradually corrode in physiological environments. These matrices are often reinforced with fibers, particles, or other fillers to tailor their strength, degradation rate, and functionality. Magnesium-based composites, for instance, exhibit mechanical properties similar to natural bone and degrade into non-toxic ions, making them ideal candidates for load-bearing applications [8, 9]. However, their rapid corrosion in physiological environments poses challenges, which can be mitigated through surface modifications and alloying. Similarly, zinc-based composites provide moderate degradation rates and essential micronutrients but may suffer from insufficient mechanical strength in pure form. Polymer-based resorbable composites are widely used in sutures, orthopedic pins, and scaffolds due to their tunable degradation and processing ease but often require reinforcement to enhance their load-bearing capabilities. What makes resorbable composites especially significant is their synergistic design, which combines the mechanical performance of structural materials with bioactivity and controlled degradation. This allows for customized solutions that match the rate of tissue regeneration, supporting the body until natural healing is complete and the material safely dissolves [10].

The clinical motivation for using resorbable composites stems from the demand for materials that are both functionally supportive and biologically benign. Conventional permanent implants, while durable, pose various postoperative risks, including implant-related infections, stress shielding, allergic reactions, and long-term foreign body responses. Moreover, non-degradable implants are not suitable for pediatric patients whose bones are still growing or for applications requiring temporary support structures. Resorbable composites address these concerns effectively. Their ability to degrade *in situ* eliminates the need for secondary removal surgeries, significantly reducing patient morbidity, healthcare costs, and surgical risks [11, 12]. This is especially advantageous in pediatric orthopedic treatments, maxillofacial surgeries, and trauma cases, where repeat surgeries can be particularly burdensome. Furthermore, the degradation products of these materials, if appropriately designed, are non-toxic and may even promote local tissue regeneration through bioactive ion release or cellular signaling mechanisms. Another key advantage is the potential to incorporate therapeutic functionalities into these composites. For instance, drug-eluting coatings can be applied to release antibiotics or anti-inflammatory agents, and bioactive surface treatments can enhance cellular adhesion, proliferation, and differentiation. This multifunctionality enables the development of smart implants that not only provide mechanical support but also actively participate in the healing process. Resorbable composites also allow for customized degradation profiles, tailored to match the healing time of specific tissues. This adaptability ensures that the mechanical support gradually diminishes as the new tissue regenerates, maintaining a harmonious load distribution and minimizing complications associated with premature material failure or prolonged presence in the body [13–15]. This chapter provides a comprehensive exploration of the surface engineering strategies employed to optimize the performance of resorbable composites for implantable medical applications. It focuses on the role of surface coatings in addressing critical challenges such as rapid degradation, insufficient biocompatibility, and lack of antimicrobial protection. Figure 14.1 depicts the surface modification and functional coating of dental implants that enhance osseointegration by improving bioactivity, corrosion resistance, and cellular responses at the bone-implant interface. These advanced treatments enable better implant stability, longevity, and success rates, especially in compromised clinical conditions.

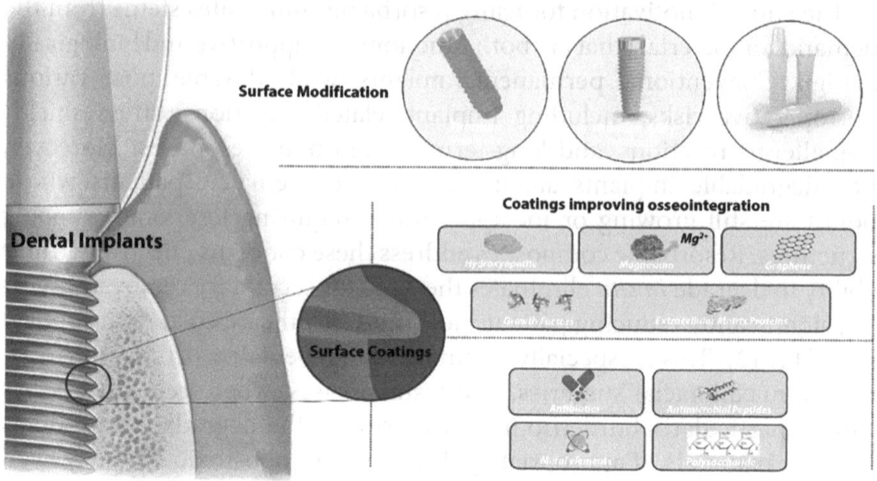

14.2 Fundamentals of Resorbable Composite Materials

14.2.1 Material Classes: Magnesium, Zinc, and Polymer-Based Composites

Resorbable composites are typically classified based on the nature of the matrix material—metals such as magnesium and zinc, or polymers, each with distinct advantages and limitations. The matrix is often combined with reinforcing agents such as fibers, particles, or secondary phases to enhance specific properties, including strength, degradation rate, and biological activity. Magnesium and its alloys are among the most widely studied metallic resorbable materials due to their mechanical similarity to natural bone and their ability to degrade in physiological environments [17, 18]. With a density and elastic modulus close to that of cortical bone, magnesium offers ideal load-sharing characteristics, minimizing stress shielding and promoting bone remodeling. Magnesium degrades through a corrosion process that results in the release of magnesium ions and hydrogen gas. While magnesium ions are biocompatible and beneficial for bone regeneration, the rapid release of hydrogen gas and alkalinization of the local environment can lead to tissue irritation and impaired healing. To address this, alloying with elements such as calcium, zinc, or rare earth metals, as well as the incorporation of ceramic reinforcements (e.g., hydroxyapatite or β-tricalcium phosphate), are used to modulate degradation kinetics and enhance osteoconductivity [19–21]. Zinc is emerging as a promising alternative to magnesium due to its slower and more uniform degradation profile. Zinc plays a vital role in enzyme function and wound healing, making it inherently biofunctional. However, pure zinc has relatively low mechanical strength and ductility, which limits its standalone application in load-bearing implants. To overcome this, zinc is alloyed with elements such as magnesium, calcium, or manganese, and often reinforced with bioceramics or biodegradable polymers to form composite structures that exhibit improved strength, corrosion resistance, and biological performance. Zinc-based composites demonstrate reduced hydrogen evolution compared to magnesium and offer a degradation timeline more compatible with long-term healing processes, making them suitable for vascular stents and low-load orthopedic applications. Biodegradable polymers, both synthetic and natural, are extensively used in resorbable medical devices. Synthetic polymers such as polylactic acid (PLA), polyglycolic acid (PGA), polycaprolactone (PCL), and their copolymers (e.g., PLGA) are widely used due to their tunable degradation rates, processability, and FDA approval history. Natural polymers like collagen, gelatin, chitosan, and alginate offer superior biocompatibility and bioactivity but may lack mechanical robustness and require crosslinking or reinforcement [16, 22]. Polymer matrices can be reinforced with ceramic particles (e.g., hydroxyapatite, silica) or biodegradable metal particles to enhance mechanical strength, osteoconductivity, and biofunctionality. Polymer-based composites are ideal for applications requiring flexibility and bioactivity over extended degradation periods, such as soft tissue scaffolds, resorbable sutures, and drug delivery systems.

14.2.2 Biodegradation Mechanisms

The degradation behavior of resorbable composites is central to their functionality. Understanding the mechanisms of biodegradation is crucial for tailoring implant materials to specific clinical applications. Biodegradation occurs through various pathways depending on the material class, environmental conditions, and surface modifications [23]. In physiological environments, magnesium and zinc degrade through electrochemical corrosion. For magnesium, the general corrosion reaction involves the formation of magnesium hydroxide and hydrogen gas:

$$Mg + 2H_2O \rightarrow Mg(OH)_2 + H_2\uparrow$$

This reaction raises the local pH and generates gas pockets that can interfere with tissue integration. Magnesium hydroxide can further react with chloride ions in body fluids, forming soluble magnesium chloride and accelerating degradation. The corrosion process can be influenced by several factors, including alloy composition, grain structure, and the presence of impurities.

Zinc follows a similar electrochemical degradation route but at a much slower rate:

$$Zn + 2H_2O \rightarrow Zn(OH)_2 + H_2\uparrow$$

Zinc hydroxide is less soluble and forms a more stable passivating layer than magnesium hydroxide, reducing the degradation rate. Additionally, the lack of significant hydrogen evolution makes zinc-based materials more tissue-friendly in some applications. Surface coatings, alloying, and microstructural refinement are commonly employed to control the corrosion behavior of both metals [24, 25]. Biodegradable polymers typically degrade *via* hydrolytic or enzymatic mechanisms. Synthetic polymers like PLA and PGA degrade through hydrolysis of ester bonds, forming oligomers and eventually monomers (lactic and glycolic acid), which are metabolized *via* natural pathways. Factors influencing polymer degradation include molecular weight, crystallinity, hydrophilicity, and environmental pH. Natural polymers, on the other hand, often degrade *via* enzymatic cleavage by naturally occurring enzymes such as collagenase (for collagen) or lysozyme (for chitosan). The degradation products of natural polymers are usually non-toxic and bioresorbable, making them ideal for applications involving sensitive tissues. The degradation behavior of composites depends not only on the matrix but also on the interaction between the matrix and filler materials. For example, incorporating a fast-degrading ceramic into a slower-degrading polymer matrix can lead to early exposure of the filler, altering mechanical properties and biological interactions over time. Therefore, a careful balance between components is essential for predictable and functional degradation profiles.

14.2.3 Mechanical and Biological Properties

The mechanical and biological properties of resorbable composites must be optimized to meet the specific demands of various clinical applications. Key mechanical parameters include tensile strength, compressive strength, elastic modulus, fracture toughness, and fatigue resistance [26–28]. These properties should be sufficient to provide initial mechanical support and gradually transfer load to the regenerating tissue as the material degrades. Magnesium-based composites exhibit mechanical properties close to those of cortical bone (Young's modulus ~40–45 GPa), which helps avoid stress shielding—a phenomenon where the implant bears too much load, leading to bone resorption. However, pure magnesium is relatively brittle and prone to rapid corrosion-induced embrittlement. Alloying and reinforcement with ceramics or polymers improve its mechanical resilience and corrosion tolerance. Zinc-based composites are generally softer than magnesium, with a lower elastic modulus (~100 GPa for pure zinc) and tensile strength. Their mechanical properties must be enhanced through alloying or reinforcement for use in load-bearing applications. Zinc's slow degradation makes it particularly suitable where prolonged mechanical support is necessary, such as in vascular scaffolds or intermediate load-bearing orthopedic devices. Polymer-based composites offer excellent flexibility, making them suitable for applications where compliance with surrounding tissues is essential [29, 30]. However, their mechanical strength is often insufficient for major load-bearing roles unless heavily reinforced with ceramics or biodegradable metals. The incorporation of fibers (e.g., bioactive glass, carbon, silk) or particles improves their tensile and compressive strength, allowing their use in bone fixation devices, pins, and screws. Beyond mechanical performance, biocompatibility and bioactivity are crucial for the success of resorbable implants. A key advantage of resorbable composites is their potential to be biofunctional, actively supporting cellular processes such as adhesion, proliferation, differentiation, and tissue regeneration. Magnesium ions are known to stimulate osteoblastic activity and angiogenesis, promoting bone formation and vascularization. However, excessive ion release and pH elevation due to rapid corrosion can induce cytotoxicity. Therefore, surface treatments and alloying are used to modulate magnesium ion release.

Zinc plays a pivotal role in enzymatic activity, cell signaling, and immune function. Controlled release of zinc ions supports wound healing and tissue regeneration, making zinc-based composites inherently bioactive. However, the biological response is highly dependent on the degradation rate and ion concentration. Polymer matrices, especially those derived from natural sources, support excellent biocompatibility and often exhibit intrinsic bioactivity. For example, collagen promotes cell attachment and differentiation, while chitosan exhibits antimicrobial properties. Synthetic polymers can be rendered bioactive by incorporating peptides, growth factors, or ceramic particles like hydroxyapatite, which enhances osseointegration. The incorporation of bioactive fillers, such as bioactive glass, calcium phosphates, or antimicrobial agents, further enhances the biological performance of resorbable composites. These fillers release therapeutic ions or modulate the local microenvironment to encourage favorable tissue responses and prevent infection [31, 32].

14.3 Surface Coating Technologies for Resorbable Implants

14.3.1 Objectives of Surface Modification

The primary objective of surface modification in resorbable implants is to tailor the implant's surface characteristics to achieve optimal interaction with the biological environment. This involves enhancing properties such as surface roughness, wettability, chemical composition, and topography to promote favorable cellular responses. One of the key goals is to improve osseointegration, the direct structural and functional connection between living bone and the surface of the implant. By modifying the surface to mimic the natural bone environment, implants can achieve faster and more robust integration. Additionally, surface modifications aim to reduce bacterial adhesion, thereby minimizing the risk of infections that can compromise implant success [33]. Another critical objective is to control the degradation rate of resorbable implants. By applying specific coatings, the corrosion behavior of materials like magnesium and zinc can be modulated, ensuring that the implant maintains its mechanical integrity throughout the healing process before safely degrading. Furthermore, surface modifications can enhance the bioactivity of implants by incorporating bioactive molecules or ions that stimulate specific cellular responses, such as osteogenesis or angiogenesis. This biofunctionalization transforms passive implants into active participants in the healing process.

14.3.2 Plasma Spraying

Plasma spraying is a thermal coating process that involves the projection of molten or semi-molten materials onto a substrate to form a coating. In the context of resorbable implants, plasma spraying is commonly used to deposit bioactive ceramics like hydroxyapatite (HA) onto metallic substrates. The high temperatures involved in plasma spraying allow for the formation of dense and adherent coatings that can significantly enhance the bioactivity of implants. Hydroxyapatite coatings, for instance, provide a favorable surface for bone cell attachment and proliferation, promoting osseointegration.

However, plasma spraying has its limitations. The high processing temperatures can lead to phase transformations in the coating material, potentially affecting its bioactivity. Additionally, achieving uniform coating thickness and composition can be challenging, which may influence the consistency of the implant's performance [34, 35]. Despite these challenges, advancements in plasma spraying techniques, such as the use of controlled atmospheres and optimized process parameters, have improved the quality and reliability of the coatings, making plasma spraying a valuable method for enhancing the surface properties of resorbable implants. Sol-gel deposition is a versatile and cost-effective method for creating thin, uniform coatings on implant surfaces. This process involves the transition of a system from a liquid "sol" into a solid "gel" phase, allowing for precise control over the coating's composition and microstructure. One of the significant advantages of sol-gel coatings is their ability to incorporate bioactive materials, such as calcium phosphates, into the coating matrix. These bioactive coatings can enhance the implant's integration with bone tissue by promoting the formation of a bone-like apatite layer on the implant surface. Moreover, sol-gel coatings can be engineered to release therapeutic agents, such as antibiotics or growth factors, providing additional functionalities to the implant. The low processing temperatures associated with sol-gel techniques also preserve the integrity of heat-sensitive bioactive molecules, making it an attractive method for multifunctional implant coatings. However, the sol-gel process requires careful control of parameters such as pH, temperature, and precursor concentrations to achieve the desired coating properties. Inconsistent processing conditions can lead to defects in the coating, affecting its performance. Nevertheless, with optimized protocols, sol-gel deposition remains a powerful tool for enhancing the surface characteristics of resorbable implants. In an experiment, Wu *et al.* [36] investigated the physicochemical behavior and osteogenic capabilities of a novel bone cement formulation—gelatin-incorporated calcium silicate (CSG), comprising 5 wt% gelatin. The performance of this experimental cement was systematically compared to that of unmodified calcium silicate (CS) and conventional calcium sulfate hemihydrate (CSH) cements. Multiple assessments were conducted, including characterization of phase composition and microstructure, determination of setting time, washout resistance, and diametral tensile strength to evaluate their mechanical integrity and clinical handling properties. Biological evaluations involved *in vitro* studies utilizing murine macrophages (RAW 264.7), MG63 osteoblast-like cells, and human mesenchymal stem cells (hMSCs) to assess cell proliferation, osteogenic differentiation, and mineral deposition. For *in vivo* analysis, a mini-pig model with mandibular alveolar bone defects was employed. The bone regeneration efficacy of the cements was examined histologically and through histomorphometric analysis at 6 and 12 weeks post-implantation. Results revealed that CS and CSG powders primarily consisted of poorly crystalline β-dicalcium silicate, while CSH exhibited a highly crystalline structure with irregular morphology. Upon setting, CS and CSG formed a calcium-silicate-hydrate (C-S-H) gel phase, whereas CSH transformed into a plate-like gypsum (calcium sulfate dihydrate) crystalline structure. The initial setting times recorded were 19 minutes for CS, 35 minutes for CSG, and 10 minutes for CSH. Notably, the incorporation of gelatin significantly enhanced the mechanical and handling properties of CS cement, improving its diametral tensile strength from 2.4 MPa to 3.4 MPa and conferring strong washout resistance—a property absent in CSH, which nonetheless exhibited the highest strength at 7.6 MPa. Cell culture studies demonstrated that the CSG cement supported superior osteogenic activity in MG63 and hMSC cultures compared to CSH. Additionally, both CS and CSG cements elicited a lower pro-inflammatory response in RAW 264.7 macrophages than CSH, suggesting better biocompatibility. After 12 weeks of implantation, histological analysis indicated slightly greater new bone formation in the CS group compared to both CSG and CSH groups. Similarly, Xu *et al.* [37] addressed that limitation by integrating strontium-enriched amorphous calcium phosphate (Sr-ACP) granules into the existing scaffold system to promote enhanced subchondral bone healing. The Sr-ACP granules, with particle sizes ranging from 100 to 150 μm, were synthesized and confirmed to contain 2.49 ± 0.04 wt% strontium through inductively coupled plasma mass spectrometry (ICP-MS). Experimental scaffold formulations were developed by embedding 30 wt% of either Sr-ACP or undoped ACP granules into the collagen-based matrices. Micro-computed tomography (micro-CT), scanning electron microscopy (SEM), and energy-dispersive X-ray spectroscopy (EDS) demonstrated uniform integration and dispersion of the granules within the collagenous framework. *In vitro* cytotoxicity assays confirmed the biocompatibility of both the granules and the composite scaffolds, indicating that the materials support cellular viability. Initial *in vivo* biocompatibility and tissue response assessments using a subcutaneous implantation model in mice revealed favorable interactions with host tissue for both ACP and Sr-ACP-modified scaffolds. Critically, in a preclinical large-animal model involving goats with osteochondral defects in the femoral condyle, scaffolds enriched with Sr-ACP demonstrated significantly greater bone regeneration at the six-month evaluation point compared to scaffolds without granules. The most pronounced improvements were observed in the subchondral bone within the load-bearing regions of the joint. In conclusion, this study establishes that the incorporation of strontium-containing amorphous calcium phosphate granules into collagen/collagen-Mg-hydroxyapatite scaffolds substantially enhances subchondral bone repair. The strategy presents a viable approach to improving the long-term success of osteochondral tissue engineering, particularly by reinforcing the structural integrity of the regenerated subchondral bone and protecting overlying cartilage regeneration.

Kraisuriyawong *et al.* [38] presented the design and characterization of a novel guided tissue regeneration (GTR) membrane derived from bacterial cellulose (BC), a biocompatible and biodegradable polysaccharide known for its superior mechanical properties and structural similarity to extracellular matrix. To enhance its functionality and suitability for bone tissue engineering applications, a bilayer composite membrane was fabricated by depositing a nanofibrous coating composed of carboxymethyl cellulose (CMC) and poly(ethylene oxide) (PEO) onto the BC substrate *via* electrospinning. Aqueous CMC/PEO solutions at 8% (wt/v) and a 1:1 mass ratio were optimized to yield uniform, bead-free nanofibers with an average diameter of 125 ± 10 nm. The resulting CMC/PEO-coated BC membranes exhibited desirable mechanical strength and swelling capacity—critical properties for GTR applications. Furthermore, *in vitro* assays confirmed the cytocompatibility of the composite membranes with human periodontal ligament (hPDL) cells, affirming their potential as regenerative scaffolds. The membrane's surface was chemically functionalized *via* the carboxyl groups present in CMC, enabling the covalent immobilization of a bioactive molecule—specifically, the calcium-binding domain of plant-derived recombinant human osteopontin (p-rhOPN-C122). High immobilization efficiency (98–99%) was achieved, as quantified using enzyme-linked immunosorbent assay (ELISA), with a concentration-dependent immobilization profile observed across the tested range of 7.5–30 ng/mL. To assess the bioactivity of the functionalized membranes, osteogenic differentiation of hPDL cells was evaluated. Quantitative real-time reverse transcription polymerase chain reaction (qRT-PCR) revealed upregulation of key osteogenic gene markers, while mineralization assays confirmed enhanced calcium deposition, indicating successful induction of osteogenic pathways. In summary, the engineered CMC/PEO-coated bacterial cellulose membrane demonstrated favorable physical properties, excellent cytocompatibility, and the ability to be biofunctionalized with osteoinductive agents. These characteristics underscore its strong potential as an advanced GTR membrane suitable for promoting bone regeneration in periodontal and craniofacial tissue engineering applications.

14.3.3 Electrochemical Coating

Electrochemical coating, including techniques like electrochemical deposition and anodization, involves the use of electrical currents to deposit materials onto a substrate. This method allows for precise control over the coating's thickness, composition, and morphology. In the realm of resorbable implants, electrochemical coatings are employed to deposit bioactive materials such as hydroxyapatite or to create oxide layers that enhance corrosion resistance. For instance, anodization of magnesium implants can form a protective oxide layer that slows down the degradation rate, ensuring the implant maintains its mechanical support during the critical healing period [39, 40]. Additionally, electrochemical techniques can be used to incorporate therapeutic agents into the coating, enabling controlled drug release from the implant surface. This multifunctionality can be particularly beneficial in preventing infections or promoting tissue regeneration. The scalability and cost-effectiveness of electrochemical coating methods make them suitable for industrial applications. However, the success of these coatings depends on factors such as electrolyte composition, current density, and deposition time, which must be meticulously optimized to achieve the desired outcomes. In an experiment, Torroni *et al.* [41] investigated the biocompatibility and degradation characteristics of WE43 alloys, comparing the untreated (as-cast) and thermally modified (T5-aged) forms in a subperiosteal maxillofacial sheep model. In those earlier evaluations, both variants demonstrated commendable biocompatibility; however, the as-cast alloy exhibited a more rapid degradation rate than its T5-treated counterpart. The current research extends this inquiry by evaluating the same WE43 alloy variants as endosteal implants in a more mechanically demanding anatomical location—a cranial bone sheep model. Cylindrical disc-shaped implants were fabricated from both the as-cast and T5 forms of WE43 alloy. A total of six adult sheep were used in the *in vivo* study, with each animal receiving two implants (one of each alloy type) placed in monocortical defects surgically created in the parietal bone using a high-speed trephine burr. After a six-week implantation period, the animals were euthanized, and the bone–implant complexes were harvested for detailed histomorphological assessment. The analysis revealed that the as-cast WE43 implants experienced accelerated degradation, accompanied by notable osteolytic activity, formation of gas pockets, and enhanced bone remodeling in the surrounding tissue. In contrast, the T5-treated WE43 implants exhibited slower and more controlled degradation, with minimal gas evolution and better structural integration at the bone–implant interface. These findings suggest that thermal treatment of WE43 alloys significantly improves implant stability by modulating the degradation rate and enhancing osseointegration. The T5-treated variant of the WE43 alloy demonstrated superior performance as an endosteal implant material, making it a more viable candidate for use in orthopedic or craniofacial applications where gradual resorption and long-term biomechanical support are essential.

Steyl *et al.* [42] aimed to investigate the potential of zinc-doped fluorapatite (ZnFA) as a synthetic bone graft substitute with both antimicrobial and regenerative capabilities for use in dental applications. The performance of ZnFA was compared against traditional bone fillers, including autografts and demineralized bone matrix (DBM), in promoting bone regeneration in mandibular defects. Fluorapatite (FA) and a 2% zinc-doped variant (2ZnFA) were synthesized in-house and subjected to comprehensive physicochemical characterization using Fourier Transform Infrared Spectroscopy (FTIR), Inductively Coupled Plasma Mass Spectrometry (ICP-MS), fluoride ion selective electrode analysis, and X-ray Diffraction (XRD). Both materials were then fabricated into compressed, sintered disc-shaped scaffolds. To assess antimicrobial activity, the scaffolds were exposed to bacterial cultures, and their resistance to surface colonization was compared with titanium control discs. Adipose-derived mesenchymal stem cells (ADSCs) were cultured on both FA and 2ZnFA surfaces to evaluate their cytocompatibility and capacity to support osteogenic differentiation. The expression levels of osteogenic markers, particularly osteopontin and osteocalcin, were used to determine scaffold bioactivity. The top-performing materials were selected for a preclinical *in vivo* study. In the animal model phase, 20 rats were allocated into five treatment groups, each receiving a 5 mm surgically created mandibular defect. The defects were treated with either 2ZnFA, FA, autograft, DBM, or left unfilled as a control. Twelve weeks post-implantation, the harvested specimens were subjected to micro-computed tomography (microCT) and histological analysis to evaluate bone formation and tissue response.

The structural and compositional assessments confirmed successful synthesis of both FA and ZnFA. Antibacterial evaluations showed that although FA and ZnFA did not significantly differ from each other in terms of bacterial adhesion, both exhibited superior resistance to bacterial colonization compared to titanium reference discs ($p < 0.05$). *In vitro* cell culture studies demonstrated that both FA and ZnFA scaffolds supported robust cell attachment, proliferation, and osteogenic differentiation, as evidenced by elevated levels of osteopontin and osteocalcin. MicroCT and histological analyses of the *in vivo* samples revealed that both FA and ZnFA achieved bone regeneration comparable to that of the autograft group, which remains the clinical gold standard. Furthermore, both materials showed superior regenerative outcomes compared to DBM, a widely used allogenic bone substitute in dental surgeries. The results indicate that sintered FA and ZnFA scaffolds possess promising dual functionalities—supporting osteogenesis while offering antimicrobial properties, making them viable alternatives to conventional autograft materials. These findings highlight the potential of FA-based ceramics, particularly ZnFA, as advanced synthetic bone fillers in dental regenerative procedures. Future translational studies are warranted to further validate their clinical applicability. In the treatment of mandibular bone defects, FA and ZnFA scaffolds demonstrated regenerative capabilities on par with autografts and surpassed DBM in promoting new bone formation. These outcomes position ZnFA as a compelling candidate for next-generation antimicrobial and osteoinductive synthetic graft materials in oral and maxillofacial surgery. Similarly, Fang *et al.* [40] fabricated a novel bioactive construct using a specially processed demineralized extracellular cancellous bone matrix (SDECM), which was chemically and electrostatically functionalized with vancomycin (Van)—a potent antibiotic commonly used in treating resistant bacterial infections. The resulting vancomycin-conjugated scaffold (Van-SDECM) was designed to offer both therapeutic and regenerative benefits. *In vitro* characterization confirmed the scaffold's ability to release vancomycin in a pH-responsive manner, enabling targeted drug delivery within the acidic microenvironments typical of infected bone tissue. This pH-triggered mechanism ensured sustained antibiotic release, enhancing antimicrobial activity over time. The material also demonstrated effective inhibition of biofilm formation and maintained bactericidal properties throughout the degradation process. Importantly, the incorporation of vancomycin did not interfere with the scaffold's inherent capacity to support osteogenic differentiation. An animal model of infectious bone defects was utilized to further validate the scaffold's dual functionality. The Van-SDECM scaffold significantly reduced bacterial load and inflammation at the defect site while simultaneously facilitating new bone formation. These therapeutic effects were attributed to multiple synergistic mechanisms, including sustained antibacterial activity at infection-relevant pH, direct scaffold-mediated microbial inhibition, suppression of osteoclast-mediated bone resorption, stimulation of osteoblast activity, and modulation of local immune responses. The findings underscore the clinical potential of Van-SDECM as a next-generation therapeutic scaffold that not only combats infection through controlled, localized antibiotic delivery and biofilm suppression but also actively promotes bone repair. This innovative approach leverages a biologically derived extracellular matrix scaffold as a drug carrier, offering a promising strategy for the integrated treatment of infected bone defects. The Van-SDECM scaffold represents a significant advancement in the field of orthopedic biomaterials, providing a dual-function platform that addresses both infection control and bone regeneration. Its pH-sensitive drug release, immunomodulatory action, and compatibility with bone healing make it a promising candidate for future clinical applications in the management of complex bone infections.

14.3.4 Emerging Techniques: Layer-by-Layer Assembly and Atomic Layer Deposition

Advancements in nanotechnology have introduced layer-by-layer (LbL) assembly and atomic layer deposition (ALD) as cutting-edge techniques for surface modification of resorbable implants. LbL assembly involves the sequential adsorption of oppositely charged materials, such as polyelectrolytes, onto a substrate, forming multilayered coatings with nanometer-scale precision [41, 42]. This technique allows for the incorporation of various bioactive molecules, including proteins, peptides, and drugs, into the coating layers. The modularity of LbL assembly enables the design of coatings with tailored properties, such as controlled degradation rates and targeted drug release profiles. Moreover, the gentle processing conditions preserve the activity of sensitive biomolecules, making LbL assembly ideal for creating biofunctional coatings on resorbable implants. ALD is a vapor-phase technique that deposits thin films onto substrates through sequential, self-limiting chemical reactions. This method offers unparalleled control over film thickness and composition at the atomic level, resulting in highly uniform and conformal coatings. In the context of resorbable implants, ALD can be used to deposit protective layers that modulate the degradation rate of the implant material. For example, applying an ultrathin oxide layer on a magnesium implant can significantly reduce its corrosion rate, ensuring mechanical stability during tissue healing. Furthermore, ALD coatings can serve as diffusion barriers or functional layers that enhance biocompatibility and reduce inflammatory responses. The precise control offered by ALD makes it a promising technique for developing next-generation resorbable implants with customized surface properties. In an experiment, Lyu *et al.* [39] developed an innovative tissue engineering strategy combining a biomimetic, multi-component scaffold with the sustained release of PFI-2, a small molecule known to modulate epigenetic regulators involved in tissue regeneration. The engineered construct consists of an electrospun scaffold made from a poly(lactic-co-glycolic acid)/polycaprolactone (PLGA-PCL) blend, which was further functionalized with a polydopamine coating and a collagen layer to enhance cell adhesion and mimic natural extracellular matrix cues. The scaffold was embedded with PFI-2 both within its core structure and on the surface layer, allowing for dual-stage, long-acting release (termed PP-PFI-pDA-COL-PFI). To evaluate the regenerative potential of this system, 3D *in vitro* assays were conducted using human periodontal ligament progenitor cells. The PFI-2-loaded scaffold demonstrated significant enhancement in mineralization, as evidenced by a fourfold increase in calcium-to-phosphorus ratios and pronounced Alizarin Red staining. Moreover, the scaffold promoted cell proliferation by over 12 times compared to control scaffolds lacking PFI-2, indicating robust biological activity suitable for mineralized tissue regeneration. When tested in a preclinical periodontitis model, the scaffold facilitated remarkable tissue repair. The treatment led to a 34% reduction in vertical bone loss and a 25% increase in the length of new periodontal attachment along the root surface. Importantly, regenerated acellular cementum was observed to be approximately twice as thick as that found in control groups, and was structurally aligned in an orderly manner—an achievement rarely observed in regenerative approaches for periodontal disease. Molecular analysis revealed that PFI-2 plays a pivotal role in modulating epigenetic regulation during regeneration. Specifically, PFI-2 was shown to inhibit the activity of SETD7, an enzyme that methylates β-Catenin, a key regulator of osteogenic signaling. This inhibition resulted in increased nuclear translocation of β-Catenin, thereby enhancing gene expression programs associated with mineralized tissue formation and regenerative signaling. The dual-functional scaffold, featuring sustained PFI-2 delivery and a biomimetic collagen/polydopamine-modified surface, demonstrated unprecedented precision in regenerating the full tripartite periodontal structure—including bone, ligament, and cementum—in an inflammatory disease environment. This study introduces a promising, multifaceted therapeutic platform for advanced periodontal tissue engineering, offering significant clinical potential for the treatment of severe periodontal defects. Similarly, Chen *et al.* [34] presented a novel strategy to functionalize titanium alloy (Ti6Al7Nb) implants with bioactive molecular reservoirs capable of delivering dual therapeutic agents in a controlled manner, thereby enhancing bone regeneration and implant integration in osteoporotic environments. The approach involves the use of β-cyclodextrin (β-CD), a cyclic oligosaccharide with a hydrophobic inner cavity, which is covalently linked to chitosan to serve as a host matrix for encapsulating calcitriol (VD3)—the hormonally active form of vitamin D. This chitosan-β-CD conjugate not only provides structural stability but also enables efficient loading and gradual release of VD3. Additionally, the implant surface is co-functionalized with the anti-osteoporotic peptide hormone calcitonin (CT), using a layer-by-layer (LbL) self-assembly technique to build a multilayered bioactive film. This engineered coating enables sequential and sustained release of both VD3 and CT directly at the bone-implant interface. Experimental data revealed that the release of VD3 and CT individually stimulates osteoblast differentiation and mineralization. Specifically, the expression levels of calcium-binding proteins—Calbindin D9k and D28k—as well as bone morphogenetic protein 2 (BMP2) were significantly upregulated in osteoblasts cultured around the implant, facilitating calcium deposition and osteogenic differentiation. Additionally, when murine macrophage RAW264.7 cells were exposed to the dual-drug-loaded surface, a notable polarization towards the anti-inflammatory M2 phenotype was observed, as indicated by increased expression of BMP2 and vascular endothelial growth factor (VEGF). Concurrently, pro-inflammatory M1 markers and genes associated with osteoclast activation (such as MCF and TRAP) were markedly downregulated. In an osteoporotic rat model, micro-computed tomography (micro-CT) imaging and histological assessments demonstrated that implants functionalized with both VD3 and CT significantly enhanced new bone formation and bone-implant interfacial strength. The co-delivery system accelerated bone remodeling and achieved superior osseointegration compared to control implants or those loaded with a single agent. This study introduces a synergistic therapeutic platform by integrating a dual-drug delivery system into titanium alloy implants *via* molecularly engineered multilayer coatings. The combination of VD3 and CT not only supports osteoblast activity but also modulates the immune response and osteoclast function, collectively contributing to improved bone healing in osteoporotic conditions. These findings pave the way for next-generation implant technologies capable of overcoming the limitations of skeletal repair in osteoporotic patients.

Sarkar *et al.* [28] developed a three-dimensional (3D) printed, biodegradable scaffold system incorporating calcium peroxide (CPO), an oxygen-generating compound, within a polycaprolactone (PCL) matrix. The hydrophobic nature of PCL serves to modulate the interaction between water and CPO, thereby regulating the rate of oxygen release and minimizing initial burst effects. Upon exposure to aqueous environments, the CPO undergoes hydrolysis to generate molecular oxygen in a sustained, concentration-dependent manner. Under experimentally induced anoxic conditions, scaffolds embedded with CPO supported significantly greater viability and proliferation of human adipose-derived stem/stromal cells (hASCs) compared to scaffolds composed solely of PCL. This suggests that the sustained release of oxygen from the scaffold can mitigate hypoxia-induced cell death and promote cell growth during the early stages of healing. To assess osteogenic potential, hASC-seeded CPO-PCL scaffolds were implanted into 4-mm non-healing calvarial defects in nude mice and monitored over 8 weeks. Despite the absence of additional osteoinductive agents, the oxygen-generating scaffolds significantly enhanced new bone formation relative to control scaffolds, as evidenced by micro-computed tomography (micro-CT) imaging and histological analyses. Advanced 3D lightsheet fluorescence microscopy was used to quantitatively analyze the vascular networks within the defect area. Although total vascular volume was comparable between groups, the CPO-PCL scaffolds exhibited a modest increase in pro-osteogenic type H blood vessels, characterized by high expression of CD31 and endomucin (Emcn), which are known to support bone-forming environments. Furthermore, a higher number of Osterix-positive skeletal progenitor cells were detected around the oxygen-releasing scaffolds, indicating enhanced recruitment of osteoprogenitor populations. This study demonstrates that 3D-printed, CPO-loaded scaffolds with controlled oxygen delivery capabilities provide a promising, customizable platform for improving bone healing in large craniofacial defects. By simultaneously supporting cellular survival, promoting angiogenesis, and enhancing progenitor cell recruitment, these bioengineered constructs offer a versatile and efficient approach to regenerating bone tissue without relying on exogenous growth factors.

14.4 Bioactive Coatings: Enhancing Osseointegration and Biocompatibility

14.4.1 Hydroxyapatite and Calcium Phosphate Coatings

Hydroxyapatite (HAp), a crystalline form of calcium phosphate, has garnered significant attention in biomedical applications, particularly as a surface coating for resorbable and permanent implants. Its chemical composition—$Ca_{10}(PO_4)_6(OH)_2$—closely mirrors that of the mineral phase of human bone, making it inherently bioactive and osteoconductive [43]. This close resemblance enables HAp-coated implants to establish a direct biochemical bond with native bone tissue, facilitating rapid osseointegration and minimizing fibrous tissue formation at the bone–implant interface. As a result, hydroxyapatite serves not merely as a physical barrier but as a biointeractive medium that supports cellular adhesion, proliferation, and differentiation of osteoblasts, thereby expediting the healing process post-implantation. The broader class of calcium phosphate-based coatings, including tricalcium phosphate (TCP) and biphasic calcium phosphates, shares similar bioactive properties and has been applied extensively to metallic substrates such as titanium, magnesium, and their alloys. These coatings mitigate the biological inertness and rapid degradation behavior of bare metal implants, especially in biodegradable systems like magnesium-based composites, which are prone to aggressive corrosion in physiological conditions. Among the various techniques employed for coating deposition, plasma spraying remains one of the most widely used due to its ability to produce thick, adherent coatings with controlled porosity and crystallinity [38]. Plasma-sprayed HAp coatings can create a roughened surface topography that not only enhances mechanical interlocking with bone but also increases the surface area for protein adsorption and cell attachment—key factors in early-stage bone regeneration. Another popular method, the sol-gel technique, offers advantages such as low processing temperatures, better compositional control, and the potential for incorporating therapeutic ions or bioactive molecules into the HAp matrix. The sol-gel route typically yields thin, nanostructured coatings with excellent homogeneity and the possibility of functionalizing the surface with organic or inorganic dopants. Electrochemical deposition, on the other hand, allows for precise control over coating thickness and morphology, often at room temperature, making it compatible with temperature-sensitive materials or drug incorporation strategies. These methods collectively provide flexibility in tailoring the surface characteristics of implants to specific clinical needs.

In addition to traditional HAp coatings, researchers have increasingly explored doped variants to enhance the biological efficacy of the material. Doping HAp with trace elements such as strontium (Sr), zinc (Zn), magnesium (Mg), silicon (Si), or copper (Cu) has shown promising results in modulating cellular responses and promoting bone remodeling. Strontium-doped HAp is particularly noteworthy for its dual functionality: it stimulates osteoblastic activity while simultaneously inhibiting osteoclastic resorption. This synergistic effect helps maintain bone mass around the implant, which is crucial in elderly patients or those suffering from osteoporosis [37]. Zinc, a known cofactor for many enzymes involved in cellular metabolism, enhances antibacterial activity and accelerates bone formation. Magnesium incorporation not only improves the mechanical properties of HAp coatings but also positively influences cellular adhesion and differentiation. These bioinspired doping strategies aim to replicate the complex biochemical environment of natural bone, thereby enhancing the implant's integration with host tissue. Despite these advancements, several challenges remain in the development and clinical application of HAp and calcium phosphate coatings. One of the persistent issues is coating delamination, particularly in mechanically demanding environments such as load-bearing orthopedic implants. Poor interfacial bonding between the coating and the substrate, differences in thermal expansion coefficients, and inadequate post-deposition treatment can contribute to this problem. Furthermore, the crystallinity of HAp coatings significantly influences their dissolution rate and biological behavior. Highly crystalline coatings degrade slowly, potentially impeding complete integration with the surrounding bone, while amorphous coatings dissolve rapidly and may not provide sufficient long-term support. Therefore, achieving an optimal balance between crystallinity, porosity, and coating thickness is essential to ensure sustained performance [36]. To address these limitations, current research is increasingly focusing on composite coatings that combine hydroxyapatite with polymers (such as chitosan, polylactic acid, or collagen) or with other ceramics and bioactive glasses. These hybrid coatings offer improved mechanical flexibility, enhanced bioactivity, and controlled degradation rates. For example, HAp-chitosan composites not only improve coating adhesion and toughness but also offer antibacterial properties, expanding their utility in infection-prone implant sites. Additionally, the integration of growth factors, peptides, or antibiotics into these composite matrices allows for multifunctional coatings that can simultaneously support bone healing and prevent post-surgical infections. Advancements in nanotechnology have also enabled the fabrication of nano-HAp coatings that more closely mimic the natural nanostructure of bone [44–50]. These coatings promote better cellular responses due to their higher surface area and more favorable topographical cues for cell attachment and proliferation. Layer-by-layer assembly and atomic layer deposition techniques are now being employed to construct such ultrathin, precisely engineered coatings, allowing for the inclusion of smart features such as pH-responsive or enzyme-degradable layers.

14.4.2 Antimicrobial and Anti-Inflammatory Coatings

Implant-associated infections and inflammatory responses continue to pose formidable challenges in both orthopedic and dental implantology. These complications not only compromise the success of the implant but also extend recovery times, increase patient discomfort, and may necessitate revision surgeries. The implant surface, being a foreign material, often becomes a nidus for bacterial colonization and biofilm formation, which protects pathogens from the host immune response and systemic antibiotics. Simultaneously, the body's inflammatory reaction to the implant can lead to adverse outcomes such as peri-implantitis, fibrosis, or osteolysis, which may jeopardize long-term implant stability. In this context, surface coatings engineered with antimicrobial and anti-inflammatory properties have emerged as critical tools to improve implant integration and clinical outcomes [51–53]. Antimicrobial coatings are specifically designed to prevent the initial adhesion and proliferation of bacteria at the implant site. One of the most effective strategies involves the incorporation of metallic nanoparticles—particularly silver (Ag), copper (Cu), and zinc (Zn)—into coating matrices. Silver nanoparticles are among the most widely studied due to their broad-spectrum antibacterial efficacy, low risk of developing resistance, and high compatibility with bone tissue. These nanoparticles disrupt microbial cell walls, interfere with metabolic enzymes, and generate reactive oxygen species (ROS), ultimately leading to bacterial death. Coatings embedded with silver nanoparticles have demonstrated excellent antibacterial performance against common pathogens such as Staphylococcus aureus and Escherichia coli, which are frequently implicated in implant infections. In addition to metal-based agents, antibiotic-loaded coatings offer targeted bacterial control [54–56]. Antibiotics such as vancomycin, gentamicin, and rifampicin can be incorporated into biopolymer or ceramic matrices, allowing for controlled, sustained release directly at the implant interface. This localized delivery system significantly reduces systemic toxicity and enhances local drug concentration, offering an effective means to combat early postoperative infections. However, a growing concern with antibiotic-based coatings is the potential for antibiotic resistance, especially with long-term exposure or sub-lethal drug concentrations. To counter this, researchers are exploring antimicrobial peptides (AMPs) as alternatives. These peptides, derived from natural immune defense systems, exhibit strong bactericidal action through mechanisms like membrane disruption and interference with bacterial intracellular targets. Moreover, AMPs often show low cytotoxicity and reduced resistance development, making them promising candidates for next-generation coatings. Alongside infection control, managing the host inflammatory response is equally critical. Upon implantation, the body's immune system identifies the implant as a foreign entity, leading to the activation of immune cells and the release of pro-inflammatory cytokines. If left unchecked, this can evolve into chronic inflammation, hampering tissue regeneration and leading to implant failure. To mitigate this, anti-inflammatory coatings have been developed using pharmacologically active agents such as dexamethasone, indomethacin, and other non-steroidal anti-inflammatory drugs (NSAIDs). These compounds can be embedded within biodegradable polymers like poly(lactic-co-glycolic acid) (PLGA) or chitosan to enable gradual and sustained release over time.

Moreover, emerging bioengineering approaches are focusing on modulating the immune response at the cellular level. For instance, surface coatings that release cytokine inhibitors (e.g., IL-1 or TNF-α blockers) can suppress pro-inflammatory pathways, while coatings that promote the polarization of macrophages towards the M2 (anti-inflammatory) phenotype can create a more regenerative microenvironment. This shift from a pro-inflammatory (M1) to anti-inflammatory (M2) macrophage response has been shown to facilitate angiogenesis, reduce fibrosis, and enhance tissue integration with the implant. The convergence of antimicrobial and anti-inflammatory functionalities into a single multifunctional coating is particularly advantageous in scenarios involving immunocompromised patients, individuals with diabetes, or revision surgeries where the risk of infection and inflammation is elevated. For example, composite coatings that incorporate both silver nanoparticles and dexamethasone can simultaneously suppress bacterial growth and dampen excessive immune responses. These dual-action systems not only reduce the incidence of post-surgical infections but also promote quicker tissue healing and osseointegration [57, 58]. Despite their promise, the development and clinical translation of multifunctional coatings require meticulous design to balance efficacy and safety. Several critical factors must be considered. First, the dosage and release profile of antimicrobial or anti-inflammatory agents must be optimized to ensure effective concentrations at the implant site without inducing cytotoxicity to surrounding tissues. Uncontrolled release or burst effects may lead to local tissue damage or systemic side effects. Second, the long-term stability and adhesion of the coating under physiological conditions and mechanical stress are crucial to prevent delamination or premature degradation. Third, the risk of inducing bacterial resistance remains a concern, especially when conventional antibiotics are employed. This underscores the importance of integrating non-traditional antimicrobials, such as AMPs or metal ions, with synergistic mechanisms of action. Advanced fabrication techniques such as electrospinning, layer-by-layer (LbL) assembly, and 3D printing are being harnessed to produce coatings with fine-tuned structural and functional properties. These techniques allow precise control over the spatial distribution of active agents, enabling gradient or multi-layered release profiles that better match the physiological healing timeline. Additionally, smart coatings that respond to environmental stimuli, such as pH, temperature, or the presence of specific enzymes, are under investigation. These systems can be programmed to release therapeutic agents only when pathological conditions arise, minimizing unnecessary exposure and enhancing therapeutic precision.

14.4.3 Smart and Stimuli-Responsive Coatings

Smart and stimuli-responsive coatings represent a transformative innovation in biomedical implant technology, pushing the boundaries of traditional passive surfaces toward dynamic, interactive systems that respond to the body's environment or externally applied signals. These coatings are engineered to sense and respond to specific biochemical or physical stimuli, such as changes in pH, temperature, enzymatic activity, redox potential, or externally applied magnetic, optical, or ultrasonic fields, enabling on-demand or site-specific therapeutic functions [59, 60]. This capability is particularly crucial in the context of resorbable implants, where the temporal sequence of biological events, including inflammation, infection, tissue regeneration, and eventual material resorption, can be synchronized with intelligent material responses for optimized clinical outcomes. One of the most widely explored mechanisms in smart coating systems is pH-responsiveness. The localized pH environment around an implant can vary significantly during infection or inflammation, often becoming more acidic. Smart coatings embedded with pH-sensitive linkers or matrices can leverage this shift to trigger the release of encapsulated antibiotics or anti-inflammatory drugs specifically at infected sites. Such a targeted response not only enhances the therapeutic efficacy by delivering high concentrations of drugs where needed but also reduces systemic toxicity and the risk of antibiotic resistance due to unnecessary exposure of healthy tissues. Polymers like poly(acrylic acid), chitosan derivatives, and polyaniline are commonly utilized in these systems for their pH-responsiveness and biocompatibility. Temperature-responsive coatings, on the other hand, employ materials that undergo physical or chemical changes at physiologically relevant temperatures. A classic example is poly(N-isopropylacrylamide) (PNIPAM), which exhibits a sharp phase transition around 32°C. When used in implant coatings, such polymers can be designed to release drugs or alter surface characteristics in response to slight temperature variations associated with inflammation or fever [61, 62]. These systems are particularly advantageous for modulating drug release in febrile conditions, allowing for a self-regulating therapeutic effect that mirrors the body's pathological signals. Enzyme-responsive coatings introduce an even greater level of specificity. These coatings contain moieties or polymer backbones that are cleaved or degraded by enzymes overexpressed in pathological environments. For example, coatings sensitive to matrix metalloproteinases (MMPs)—enzymes typically elevated in infected or inflamed tissues—can release embedded therapeutics only upon encountering these enzymes. This ensures that drug delivery is activated strictly in diseased tissues, preserving healthy areas and optimizing resource utilization. Other examples include coatings that respond to lysozyme, phosphatases, or proteases, depending on the target tissue and desired therapeutic response.

In addition to endogenous biological stimuli, externally triggered coatings offer the possibility of remote control over therapeutic actions, thereby affording clinicians greater precision in implant management. Light-responsive systems, for instance, utilize photo-cleavable linkers or photosensitive polymers that disintegrate or change conformation upon exposure to specific wavelengths of light, releasing their therapeutic load or altering surface characteristics such as hydrophilicity or cell adhesiveness. This spatial and temporal control is particularly useful in delicate surgical scenarios or for post-operative management through non-invasive light exposure. Magnetic field-responsive coatings integrate magnetic nanoparticles or materials that heat up or vibrate upon exposure to alternating magnetic fields. These effects can be used to trigger drug release or to induce mild hyperthermia for therapeutic purposes, such as disrupting bacterial biofilms. Similarly, ultrasound-responsive coatings can facilitate controlled delivery through acoustic cavitation or mechanical disruption of polymer matrices, offering a non-invasive, deeply penetrating external trigger for drug deployment. These technologies are still evolving but present significant potential for applications where precision and minimal invasiveness are critical. The incorporation of such smart coatings into resorbable implants adds a functional, responsive layer to materials that are already engineered to degrade and be absorbed by the body over time [44, 63–65]. These implants are increasingly being used in orthopedic, cardiovascular, and dental applications where temporary support is needed without long-term foreign body presence. Smart coatings can be designed to align with the different stages of healing, offering phase-specific therapeutic benefits. For instance, antimicrobial agents can be released immediately upon implantation to prevent early infections, followed by anti-inflammatory drugs to modulate immune response and finally growth factors like bone morphogenetic proteins (BMPs) or vascular endothelial growth factor (VEGF) to stimulate tissue regeneration. Moreover, by tuning the degradation kinetics of both the coating and the underlying implant material, researchers can create highly synchronized systems that adapt to the body's repair cycle. This spatiotemporal control allows for the sequential or staged release of therapeutics based on real-time biological cues, vastly improving the prospects for complete and complication-free healing.

Despite these compelling advantages, several technical and translational challenges must be addressed to enable the broader clinical application of smart and stimuli-responsive coatings. One major concern is the biocompatibility and long-term stability of the materials used. While many smart polymers and nanoparticles demonstrate excellent functionality *in vitro*, their interactions with the immune system, potential for off-target effects, or toxic degradation products must be thoroughly evaluated *in vivo*. Additionally, ensuring uniformity, reproducibility, and scalability in the fabrication of such coatings remains a hurdle, especially when transitioning from laboratory-scale synthesis to industrial manufacturing. Another critical factor is the control and predictability of the stimulus-response relationship. The coatings must not only respond to stimuli within a narrowly defined and physiologically relevant range but also maintain this sensitivity over the functional lifespan of the implant. External stimulus-triggered systems also require the development of safe and efficient devices for stimulus delivery (e.g., light sources or magnetic field generators) and protocols for clinical use that are feasible and patient-friendly. To address these limitations, ongoing research is focusing on multi-responsive systems that integrate several stimulus-sensing mechanisms into a single platform, thereby enhancing specificity and redundancy [66–69]. Researchers are also exploring biomimetic approaches, using natural materials or inspired designs to improve compatibility and performance. Additionally, the integration of nanotechnology, bioelectronics, and machine learning is opening new frontiers in smart coatings, such as self-reporting implants that can monitor biological conditions and adjust their therapeutic output in real time.

14.5 Conclusion and Future Perspectives

Resorbable composite materials have emerged as a transformative solution for temporary implantable devices in orthopedics, dentistry, and other medical fields. Unlike permanent implants, these materials are designed to degrade safely within the physiological environment, eliminating the need for secondary removal surgeries and reducing long-term complications. This chapter has explored the fundamental aspects of resorbable composites, including their material classes (magnesium, zinc, and polymer-based systems), biodegradation mechanisms, and mechanical and biological properties. A thorough understanding of these characteristics is essential to tailor implants that balance mechanical strength with biocompatibility and controlled degradation. Surface engineering, particularly *via* advanced coating technologies, plays a pivotal role in enhancing the clinical performance of resorbable implants. Techniques such as plasma spraying, sol-gel deposition, and electrochemical coating have demonstrated their ability to modify surface topography, enhance corrosion resistance, and enable the controlled delivery of therapeutic agents. Emerging technologies like layer-by-layer (LbL) assembly and atomic layer deposition (ALD) further extend the capacity to precisely engineer surface functionalities at the nanoscale. These techniques open new avenues for producing ultrathin, uniform, and multifunctional coatings that respond intelligently to the surrounding biological milieu.

Bioactive coatings, in particular, hold exceptional promise for improving implant integration and therapeutic efficacy. Coatings based on hydroxyapatite and other calcium phosphates foster strong osseointegration by mimicking the mineral component of bone. The development of antimicrobial and anti-inflammatory coatings addresses two major challenges—post-operative infections and inflammation—that often compromise implant success. Meanwhile, smart coatings that respond to stimuli such as pH, enzymes, or external triggers like magnetic fields offer the potential for real-time, patient-specific therapeutic responses. Despite significant progress, the clinical translation of these technologies still faces numerous challenges. One major hurdle is achieving the optimal balance between biodegradability and mechanical stability. Rapid degradation may compromise structural integrity before the healing process is complete, while slow degradation may provoke chronic inflammation or fibrotic encapsulation. Additionally, ensuring long-term safety and efficacy of surface coatings under *in vivo* conditions remains a critical requirement, especially for smart and drug-loaded systems that involve complex chemistries and release kinetics.

From a regulatory standpoint, the complexity of composite materials and multifunctional coatings necessitates robust preclinical evaluations, standardized protocols, and clear regulatory frameworks to guide their approval and use. The potential variability in material behavior due to patient-specific factors such as age, comorbidities, or local tissue conditions further complicates their clinical deployment. Therefore, interdisciplinary collaboration among materials scientists, biomedical engineers, clinicians, and regulatory experts is essential to ensure the safe and effective integration of these advanced materials into routine clinical practice. Looking forward, the future of resorbable implants lies in the convergence of nanotechnology, biotechnology, and advanced manufacturing. 3D printing and additive manufacturing are expected to revolutionize implant design by enabling the production of patient-specific geometries with integrated surface coatings tailored for individual healing needs. Moreover, the incorporation of bioactive molecules such as growth factors, peptides, or even gene therapy agents into surface layers may transform implants into intelligent therapeutic platforms that not only support tissue regeneration but also actively guide it.

AI disclosure statment: The author confirms that no artificial intelligence (AI)-generated content or AI-assisted writing tools were used in the preparation of this chapter (including summaries and the abstract). All text, analysis, and conclusions are the result of the author's original work.

References

1. Pugliese, R. and Graziosi, S., Biomimetic scaffolds using triply periodic minimal surface-based porous structures for biomedical applications. *SLAS Technol.*, 28, 165–182, 2023, https://doi.org/10.1016/j.slast.2023.04.004.
2. Torroni, A., Xiang, C., Witek, L., Rodriguez, E.D., Coelho, P.G., Gupta, N., Biocompatibility and degradation properties of WE43 Mg alloys with and without heat treatment: *In vivo* evaluation and comparison in a cranial bone sheep model. *J. Craniomaxillofac. Surg.*, 45, 2075–2083, 2017, https://doi.org/10.1016/j.jcms.2017.09.016.
3. Xu, B., Gao, R., Yang, Y., Cao, X., Qin, L., Li, Y., Li, Z., Li, X., Lin, H., Guo, Y., Ma, Y., Wang, J., Nie, S., Xu, L., Cao, E., Guan, C., Stone, G.W., Biodegradable Polymer-Based Sirolimus-Eluting Stents with Differing Elution and Absorption Kinetics the PANDA III Trial. *J. Am. Coll. Cardiol.*, 67, 2249–2258, 2016, https://doi.org/10.1016/j.jacc.2016.03.475.
4. Allen, M.J., Leone, K.A., Dunbar, M.J., Race, A., Rosenbaum, P.F., Sacks, J.M., Tibial Component Fixation with a Peri-Apatite Coating. Evaluation by Radiostereometric Analysis in a Canine Total Knee Arthroplasty Model. *J. Arthroplasty*, 27, 1138–1148, 2012, https://doi.org/10.1016/j.arth.2011.12.029.
5. Furko, M., Horváth, Z.E., Czömpöly, O., Balázsi, K., Balázsi, C., Biominerals Added Bioresorbable Calcium Phosphate Loaded Biopolymer Composites. *Int. J. Mol. Sci.*, 23, 2022, https://doi.org/10.3390/ijms232415737.
6. Mosas, K.K.A., Chandrasekar, A.R., Dasan, A., Pakseresht, A., Galusek, D., Recent Advancements in Materials and Coatings for Biomedical Implants. *Gels*, 8, 1–35, 2022, https://doi.org/10.3390/gels8050323.
7. Förster, Y., Schulze, S., Penk, A., Neuber, C., Möller, S., Hintze, V., Scharnweber, D., Schnabelrauch, M., Pietzsch, J., Huster, D., Rammelt, S., The influence of different artificial extracellular matrix implant coatings on the regeneration of a critical size femur defect in rats. *Mater. Sci. Eng., C*, 116, 2020, https://doi.org/10.1016/j.msec.2020.111157.

8. Kyllönen, L., D'Este, M., Alini, M., Eglin, D., Local drug delivery for enhancing fracture healing in osteoporotic bone. *Acta Biomater.*, 11, 412–434, 2015, https://doi.org/10.1016/j.actbio.2014.09.006.
9. Ramakrishna, S., Daroonparvar, M., Berto, F., A Comprehensive Review on Surface Modifications of Coatings Opportunities and Challenges. *Coatings*, 2021.
10. Zeng, N., van Leeuwen, A.C., Grijpma, D.W., Bos, R.R.M., Kuijer, R., Poly(trimethylene carbonate)-based composite materials for reconstruction of critical-sized cranial bone defects in sheep. *J. Craniomaxillofac. Surg.*, 45, 338–346, 2017, https://doi.org/10.1016/j.jcms.2016.12.008.
11. Dang, H.P., Vaquette, C., Shabab, T., Pérez, R.A., Yang, Y., Dargaville, T.R., Shafiee, A., Tran, P.A., Porous 3D Printed Scaffolds for Guided Bone Regeneration in a Rat Calvarial Defect Model. *Appl. Mater. Today*, 20, 100706, 2020, https://doi.org/10.1016/j.apmt.2020.100706.
12. Barachini, S., Danti, S., Pacini, S., D'Alessandro, D., Carnicelli, V., Trombi, L., Moscato, S., Mannari, C., Cei, S., Petrini, M., Plasticity of human dental pulp stromal cells with bioengineering platforms: A versatile tool for regenerative medicine. *Micron*, 67, 155–168, 2014, https://doi.org/10.1016/j.micron.2014.07.003.
13. Kaliannagounder, V.K., Raj, N.P.M.J., Unnithan, A.R., Park, J., Park, S.S., Kim, S.J., Park, C.H., Kim, C.S., Sasikala, A.R.K., Remotely controlled self-powering electrical stimulators for osteogenic differentiation using bone inspired bioactive piezoelectric whitlockite nanoparticles. *Nano Energy*, 85, 105901, 2021, https://doi.org/10.1016/j.nanoen.2021.105901.
14. Pountos, I., Corscadden, D., Emery, P., Giannoudis, P.V., Mesenchymal stem cell tissue engineering: Techniques for isolation, expansion and application. *Injury*, 38, 2007, https://doi.org/10.1016/S0020-1383(08)70006-8.
15. Emechebe, G.A., Obiweluozor, F.O., Jeong, I.S., June kyu, P., Park, C.H., Kim, C.S., Merging 3D printing with electrospun biodegradable small-caliber vascular grafts immobilized with VEGF. *Nanomed. Nanotechnol. Biol. Med.*, 30, 1–17, 2020, https://doi.org/10.1016/j.nano.2020.102306.
16. Dong, H., Liu, H., Zhou, N., Li, Q., Yang, G., Chen, L., Mou, Y., Surface modified techniques and emerging functional coating of dental implants. *Coatings*, 10, 1–25, 2020, https://doi.org/10.3390/coatings10111012.
17. Streza, A., Antoniac, A., Manescu, V., Paltanea, G., Robu, A., Dura, H., Verestiuc, L., Stanica, E., Voicu, S.I., Antoniac, I., Cristea, M.B., Dragomir, B.R., Rau, J.V., Manolea, M.M., Effect of Filler Types on Cellulose-Acetate-Based Composite Used as Coatings for Biodegradable Magnesium Implants for Trauma. *Materials (Basel)*, 16, 2023, https://doi.org/10.3390/ma16020554.
18. Roohani-Esfahani, S.I., Dunstan, C.R., Davies, B., Pearce, S., Williams, R., Zreiqat, H., Repairing a critical-sized bone defect with highly porous modified and unmodified baghdadite scaffolds. *Acta Biomater.*, 8, 4162–4172, 2012, https://doi.org/10.1016/j.actbio.2012.07.036.

19. Tomala, A.M., Słota, D., Florkiewicz, W., Piętak, K., Dylag, M., Sobczak-Kupiec, A., Tribological Properties and Physiochemical Analysis of Polymer-Ceramic Composite Coatings for Bone Regeneration. *Lubricants*, 10, 2022, https://doi.org/10.3390/lubricants10040058.
20. Oriňaková, R., Gorejová, R., Králová, Z.O., Oriňak, A., Surface Modifications of Biodegradable Metallic. *Coatings*, 10, 819, 2020.
21. Krishnamoorthy, E. and Subramanian, B., Synergistic effects of silica-enriched bioactive glass and tri-calcium phosphate nanocomposites on BMP2 gene expression for bone repair and regeneration applications. *Int. J. Pharm.*, 669, 125026, 2025, https://doi.org/10.1016/j.ijpharm.2024.125026.
22. Parfenova, L., Galimshina, Z., Parfenov, E., Organic-Inorganic Biocompatible Coatings for Temporary and Permanent Metal Implants. *SSRN Electron. J.*, 2024, https://doi.org/10.2139/ssrn.4825085.
23. Rehner, A.M.G., Moldoveanu, E.T., Niculescu, A.G., Bîclesanu, F.C., Pangică, A.M., Grumezescu, A.M., Croitoru, G.A., Advances in Dental Implants: A Review of *In Vitro* and *In Vivo* Testing with Nanoparticle Coatings. *J. Compos. Sci.*, 9, 2025, https://doi.org/10.3390/jcs9030140.
24. Lu, Y., Zhang, T., Chen, K., Canavese, F., Huang, C., Yang, H., Shi, J., He, W., Zheng, Y., Chen, S., Application of biodegradable implants in pediatric orthopedics: shifting from absorbable polymers to biodegradable metals. *Bioact. Mater.*, 50, 189–214, 2025, https://doi.org/10.1016/j.bioactmat.2025.04.001.
25. Zhou, Y., Reshmy, R., Philip, E., Thomas, D., Sindhu, R., Bhargava, P.C., Tiwari, A., Ruiz, H.A., Madhavan, A., Pandey, A., Binod, P., Awasthi, M.K., Bacterial nanocellulose: Optimized synthesis and biomedical applications. *Ind. Crops Prod.*, 205, 117589, 2023, https://doi.org/10.1016/j.indcrop.2023.117589.
26. Friess, W., Collagen - Biomaterial for drug delivery. *Eur. J. Pharm. Biopharm.*, 45, 113–136, 1998, https://doi.org/10.1016/S0939-6411(98)00017-4.
27. Kim, D.H., Huegel, J., Taylor, B.L., Nuss, C.A., Weiss, S.N., Soslowsky, L.J., Mauck, R.L., Kuntz, A.F., Biocompatibility and bioactivity of an FGF-loaded microsphere-based bilayer delivery system. *Acta Biomater.*, 111, 341–348, 2020, https://doi.org/10.1016/j.actbio.2020.04.048.
28. Sarkar, N., Zhao, J., Zhang, N.Y., Horenberg, A.L., Grayson, W.L., 3D printed O2-generating scaffolds enhance osteoprogenitor- and type H vessel recruitment during bone healing. *Acta Biomater.*, 185, 126–143, 2024, https://doi.org/10.1016/j.actbio.2024.07.011.
29. Entezari, A., Swain, M.V., Gooding, J.J., Roohani, I., Li, Q., A modular design strategy to integrate mechanotransduction concepts in scaffold-based bone tissue engineering. *Acta Biomater.*, 118, 100–112, 2020, https://doi.org/10.1016/j.actbio.2020.10.012.
30. Tian, X., Raina, D.B., Oberländer, J.T., Liu, Y., Goronzy, J., Apolle, R., Vater, C., Richter, R.F., Tägil, M., Lidgren, L., Thomas, A., Platz, U., Schaser, K.D., Disch, A.C., Zwingenberger, S., Comparison of immediate anchoring effectiveness of two different techniques of bioresorbable ceramic application for

pedicle screw augmentation. *Ceram. Int.*, 50, 12877–12889, 2024, https://doi.org/10.1016/j.ceramint.2024.01.195.
31. Norton, M.R., Kay, G.W., Brown, M.C., Cochran, D.L., Bone glue - The final frontier for fracture repair and implantable device stabilization. *Int. J. Adhes. Adhes.*, 102, 102647, 2020, https://doi.org/10.1016/j.ijadhadh.2020.102647.
32. Vach Agocsova, S., Culenova, M., Birova, I., Omanikova, L., Moncmanova, B., Danisovic, L., Ziaran, S., Bakos, D., Alexy, P., Resorbable Biomaterials Used for 3D Scaffolds in Tissue Engineering: A Review. *Materials (Basel)*, 16, 2023, https://doi.org/10.3390/ma16124267.
33. Nikolova, M.P. and Apostolova, M.D., Advances in Multifunctional Bioactive Coatings for Metallic Bone Implants. *Materials (Basel)*, 16, 2023, https://doi.org/10.3390/ma16010183.
34. Chen, M., Huang, L., Shen, X., Li, M., Luo, Z., Cai, K., Hu, Y., Construction of multilayered molecular reservoirs on a titanium alloy implant for combinational drug delivery to promote osseointegration in osteoporotic conditions. *Acta Biomater.*, 105, 304–318, 2020, https://doi.org/10.1016/j.actbio.2020.01.029.
35. Zhao, D., Dong, H., Niu, Y., Fan, W., Jiang, M., Li, K., Wei, Q., Palin, W.M., Zhang, Z., Electrophoretic deposition of novel semi-permeable coatings on 3D-printed Ti-Nb alloy meshes for guided alveolar bone regeneration. *Dent. Mater.*, 38, 431–443, 2022, https://doi.org/10.1016/j.dental.2021.12.026.
36. Wu, I.T., Kao, P.F., Huang, Y.R., Ding, S.J., *In vitro* and *in vivo* osteogenesis of gelatin-modified calcium silicate cement with washout resistance. *Mater. Sci. Eng., C*, 117, 111297, 2020, https://doi.org/10.1016/j.msec.2020.111297.
37. Xu, J., Vecstaudza, J., Wesdorp, M.A., Labberté, M., Kops, N., Salerno, M., Kok, J., Simon, M., Harmand, M.F., Vancíková, K., van Rietbergen, B., Misciagna, M.M., Dolcini, L., Filardo, G., Farrell, E., van Osch, G.J.V.M., Locs, J., Brama, P.A.J., Incorporating strontium enriched amorphous calcium phosphate granules in collagen/collagen-magnesium-hydroxyapatite osteochondral scaffolds improves subchondral bone repair. *Mater. Today Bio*, 25, 2024, https://doi.org/10.1016/j.mtbio.2024.100959.
38. Kraisuriyawong, P., Kornsuthisopon, C., Pavasant, P., Rattanapisit, K., Phoolcharoen, W., Hoven, V.P., Functionalizable bacterial cellulose composite membrane for guided tissue regeneration. *Int. J. Biol. Macromol.*, 268, 131655, 2024, https://doi.org/10.1016/j.ijbiomac.2024.131655.
39. Lyu, H., Zhou, X., Qian, Y., Liu, X., Gopinathan, G., Pandya, M., Qin, C., Luan, X., Diekwisch, T.G.H., Long-acting PFI-2 small molecule release and multilayer scaffold design achieve extensive new formation of complex periodontal tissues with unprecedented fidelity. *Biomaterials*, 290, 121819, 2022, https://doi.org/10.1016/j.biomaterials.2022.121819.
40. Fang, B., Qiu, P., Xia, C., Cai, D., Zhao, C., Chen, Y., Wang, H., Liu, S., Cheng, H., Tang, Z., Wang, B., Fan, S., Lin, X., Extracellular matrix scaffold crosslinked with vancomycin for multifunctional antibacterial bone

infection therapy. *Biomaterials*, 268, 120603, 2021, https://doi.org/10.1016/j.biomaterials.2020.120603.

41. Torroni, A., Xiang, C., Witek, L., Rodriguez, E.D., Flores, R.L., Gupta, N., Coelho, P.G., Histo-morphologic characteristics of intra-osseous implants of WE43 Mg alloys with and without heat treatment in an *in vivo* cranial bone sheep model. *J. Craniomaxillofac. Surg.*, 46, 473–478, 2018, https://doi.org/10.1016/j.jcms.2017.12.028.
42. Steyl, S.K., Jeyapalina, S., Griffin, A., Krishnamoorthi, V., Beck, J.P., Agarwal, J., Shea, J., Efficacy of sintered Zinc-doped fluorapatite scaffold as an antimicrobial regenerative bone filler for dental applications. *J. Dent.*, 146, 105070, 2024, https://doi.org/10.1016/j.jdent.2024.105070.
43. Liu, H., Xuan Gan, M., Zhai, W., Song, X., Design and additive manufacturing of root analogue dental implants: A comprehensive review. *Mater. Des.*, 236, 112462, 2023, https://doi.org/10.1016/j.matdes.2023.112462.
44. Sharma, D., Thakur, A., Sharma, M.K., Kumar, A., Jakhar, K., Kumar, S., Sihmar, A., Dahiya, H., Kumar, A., Sharma, A.K., Om, H., A convenient synthesis, electrochemical profiling, and morphological studies of a pyridine-based 1,3,4-oxadiazole hybrid: A promising study for corrosion mitigation of mild steel in strongly acidic environment. *Inorg. Chem. Commun.*, 158, 111554, 2023, https://doi.org/10.1016/j.inoche.2023.111554.
45. Amaka Chikaodili, N., Sunday Ume, C., Chukwudi Nnaji, P., Bartholomew Iroha, N., Dagdag, O., Okechukwu Ezeugo, J., Thakur, A., Chikaodili Anadebe, V., Dominic Onukwuli, O., Physisorption or chemisorption: Insight from AI computing model based on DFT, MC/MD-simulation for prediction of MOF-based inhibitor adsorption on Cu in brine solution. *Comput. Theor. Chem.*, 1238, 114730, 2024, https://doi.org/10.1016/j.comptc.2024.114730.
46. Chahboun, N., Timoudan, N., Belhoussaine, O., Barrahi, A., Thakur, A., Warad, I., Flouchi, R., Zaroual, A., Benhiba, F., El Faydy, M., Karrouchi, K., Harhar, H., Abuelizz, H.A., Zarrouk, A., Assessment of ethanolic extract of Dittrichia viscosa from Kardoussa Douar region of Taza in Morocco as antioxidant and green inhibitor for carbon steel corrosion in acidic medium. *Int. J. Electrochem. Sci.*, 19, 100812, 2024, https://doi.org/10.1016/j.ijoes.2024.100812.
47. Banu, A.M., Farzana, B.A., Mujafarkani, N., Thakur, A., Ahamed, K.R., Microwave-assisted synthesis of carbon nanodots from Bombax ceiba leaves for enhanced Tribo-corrosion resistance: An experimental and computational analysis. *Results Surf. Interfaces*, 17, 100329, 2024, https://doi.org/10.1016/j.rsurfi.2024.100329.
48. Gapsari, F., Hadisaputra, S., Mujib, A., Ebenso, E., Thakur, A., Kumar, A., Eco-friendly corrosion inhibition of copper in NaCl media using Perseana americana extract: Insights from quantum and electrochemical studies. *Results Surf. Interfaces*, 17, 2024, https://doi.org/https://doi.org/10.1016/j.rsurfi.2024.100327.

49. Abeng, F.E., Ita, B.I., Ikpi, M.E., Chukwuike, V.I., Ikeuba, A.I., Edim, M.M., Chidiebere, M.A., Thakur, A., Anadebe, V.C., Millettia aboensis leaves extract as eco-friendly corrosion inhibitor for mild steel in acidizing solution: From experimental to molecular level prediction. *Results Eng.*, 24, 102950, 2024, https://doi.org/10.1016/j.rineng.2024.102950.
50. Subramanyam, S., Trabelsi, Y., Mahajan, P., Khanna, V., Thakur, A., Chahal, S., Singh, S., Singh, A., Kumar, S., Advancing sustainable energy storage with MXenes: A comprehensive analysis of fabrication techniques and applications. *J. Energy Storage*, 101, 114067, 2024, https://doi.org/10.1016/j.est.2024.114067.
51. Chikaodili, V., Abeng, F.E., Thakur, A., Ebenso, E., Surface interaction and inhibition mechanism prediction of Aciclovir molecule on Fe (110) using computational model based on DFT, RDF and MD simulation. *Comput. Theor. Chem.*, 1238, 114702, 2024, https://doi.org/10.1016/j.comptc.2024.114702.
52. A.T.A.K.D.Ö.I.N.K., Kaya, S., Lgaz, H., Ben Ahmed, S., Molecular insights into the corrosion inhibition mechanism of omeprazole and tinidazole: a theoretical investigation. *Mol. Simul.*, 49, 1632–1646, 2023, https://doi.org/10.1080/08927022.2023.2256888.
53. Sharma, D., Thakur, A., Sharma, M.K., Jakhar, K., Kumar, A., Sharma, A.K., Om, H., Synthesis, Electrochemical, Morphological, Computational and Corrosion Inhibition Studies of 3-(5-Naphthalen-2-yl-[1,3,4]oxadiazol-2-yl)-pyridine against Mild Steel in 1 M HCl. *Asian J. Chem.*, 35, 1079–1088, 2023.
54. Samide, A., Stoean, R., Stoean, C., Tutunaru, B., Grecu, R., Cioatera, N., Investigation of polymer coatings formed by polyvinyl alcohol and silver nanoparticles on copper surface in acid medium by means of deep convolutional neural networks. *Coatings*, 9, 1–14, 2019, https://doi.org/10.3390/COATINGS9020105.
55. Birdeanu, M., Epuran, C., Fratilescu, I., Fagadar-Cosma, E., Structured thin films based on synergistic effects of mnta2 o6 oxide and bis-carboxy-phenyl-substituted porphyrins, capable to inhibit steel corrosion. *Processes*, 9, 2021, https://doi.org/10.3390/pr9111890.
56. Ress, J., Martin, U., Bosch, J., Bastidas, D.M., Protection of carbon steel rebars by epoxy coating with smart environmentally friendly microcapsules. *Coatings*, 11, 1–12, 2021, https://doi.org/10.3390/coatings11020113.
57. Afzali, M., Mostafavi, A., Nekooie, R., Jahromi, Z., A novel voltammetric sensor based on palladium nanoparticles/carbon nanofibers/ionic liquid modified carbon paste electrode for sensitive determination of anti-cancer drug pemetrexed. *J. Mol. Liq.*, 282, 456–465, 2019, https://doi.org/10.1016/j.molliq.2019.03.041.

58. Jawad, Q.A., Zinad, D.S., Salim, R.D., Al-Amiery, A.A., Gaaz, T.S., Takriff, M.S., Kadhum, A.A.H., Synthesis, characterization, and corrosion inhibition potential of novel thiosemicarbazone on mild steel in sulfuric acid environment. *Coatings*, 9, 2019, https://doi.org/10.3390/coatings9110729.
59. Singh, A.K., Understanding of corrosion inhibition behavior of expired isoniazid modified benzothiazole 2-carboxaldehyde at mild steel/0.5 M H2SO4 interface. *Mater. Chem. Phys.*, 319, 129323, 2024, https://doi.org/10.1016/j.matchemphys.2024.129323.
60. Murphy, M., Theyagarajan, K., Ganesan, P., Senthilkumar, S., Thenmozhi, K., Electrochemical biosensor for the detection of hydrogen peroxide using cytochrome c covalently immobilized on carboxyl functionalized ionic liquid/multiwalled carbon nanotube hybrid. *Appl. Surf. Sci.*, 492, 718–725, 2019, https://doi.org/10.1016/j.apsusc.2019.06.283.
61. Wang, Y., Qiang, Y., Zhi, H., Ran, B., Zhang, D., Evaluating the synergistic effect of maple leaves extract and iodide ions on corrosion inhibition of Q235 steel in H2SO4 solution. *J. Ind. Eng. Chem.*, 117, 422–433, 2023, https://doi.org/10.1016/j.jiec.2022.10.030.
62. Jalalvand, A.R., Fabrication of a novel and ultrasensitive label-free electrochemical aptasensor for detection of biomarker prostate specific antigen. *Int. J. Biol. Macromol.*, 126, 1065–1073, 2019, https://doi.org/10.1016/j.ijbiomac.2019.01.012.
63. Thakur, A., Kaya, S., Abousalem, A.S., Sharma, S., Ganjoo, R., Assad, H., Kumar, A., Computational and experimental studies on the corrosion inhibition performance of an aerial extract of Cnicus Benedictus weed on the acidic corrosion of mild steel. *Process Saf. Environ. Prot.*, 161, 801–818, 2022, https://doi.org/https://doi.org/10.1016/j.psep.2022.03.082.
64. Thakur, A., Kumar, A., Sharma, S., Ganjoo, R., Assad, H., Computational and experimental studies on the efficiency of Sonchus arvensis as green corrosion inhibitor for mild steel in 0.5 M HCl solution. *Mater. Today Proc.*, 66, 609–621, 2022, https://doi.org/https://doi.org/10.1016/j.matpr.2022.06.479.
65. Thakur, A., Dagdag, O., Berisha, A., Ebenso, E.E., Kumar, A., Sharma, S., Ganjoo, R., Assad, H., Mechanistic insights into the corrosion inhibition of mild steel by eco-benign Asphodelus Tenuifolius aerial extract in acidic environment: Electrochemical and computational analysis. *Surf. Coat. Technol.*, 480, 130568, 2024, https://doi.org/https://doi.org/10.1016/j.surfcoat.2024.130568.
66. Hu, S.W., Chen, A.C., Wu, S.F., Drug-induced esophageal ulcer in adolescent population: Experience at a single medical center in central Taiwan. *Med.*, 57, 1–7, 2021, https://doi.org/10.3390/medicina57121286.
67. Rožanc, J., Žižek, M., Milojević, M., Maver, U., Finšgar, M., Dexamethasone-loaded bioactive coatings on medical grade stainless steel promote osteointegration. *Pharmaceutics*, 13, 1–20, 2021, https://doi.org/10.3390/pharmaceutics13040568.

68. Al Jabri, H., Devi, M.G., Al-Shukaili, M.A., Development of polyaniline – TiO2 nano composite films and its application in corrosion inhibition of oil pipelines. *J. Indian Chem. Soc.*, 100, 100826, 2023, https://doi.org/10.1016/j.jics.2022.100826.
69. Ukaga, I.C., Okafor, P.C., Onyeachu, I.B., Ikeuba, A.I., Njoku, D.I., The inhibitive performance of 2,3-pyrazine dicarboxylic acid and synergistic impact of KI during acid corrosion of 70/30 and 90/10 copper–nickel alloys. *Mater. Chem. Phys.*, 296, 127313, 2023, https://doi.org/10.1016/j.matchemphys.2023.127313.

15

Challenges and Opportunities of Artificial Intelligence Technologies for Developing Implants and Fixation Devices

Atanu Kumar Paul[1]*, Gourhari Chakraborty[2] and Arbind Prasad[3]

[1]*Department of Chemical Engineering, National Institute of Technology Andhra Pradesh, Tadepalligudem, Andhra Pradesh, India*
[2]*School of Chemical and Biotechnology, SASTRA Deemed University, Tamil Nadu, India*
[3]*Mechanical Engineering Department, Katihar Engineering College (Under Department of Science, Technology and Technical Education, Government of Bihar), Katihar, Bihar, India*

Abstract

The application of Artificial Intelligence (AI) in the design of prosthetic limbs, implants, and fixation devices is drastically changing the field of medical engineering. The power of AI allows advanced imaging techniques to be used, so patient-specific anatomical data can be obtained. That way improved longevity, function, and fit are offered through the personalization of the implants. Biocompatibility and durability of materials used in implants are improved through Machine Learning, while predictive analytics anticipate the device's performance and associated issues lowering the chances of Revision Surgeries. The invention of smart implants enables continuous health monitoring and assessment which allows for timely clinical interventions elevating safety and patient outcomes.

Despite these revolutionary advancements, much still needs to be done to facilitate the safe and effective integration of AI with implant technology. There are significant barriers to overcome such as privacy issues related to data, security issues, the significant requirement for larger and more diverse proper zac linebreak data sets, and the potential for algorithmic bias that could affect the performance of the device in distinct populations. Regulatory environments are shifting, but they are complicated, creating obstacles for the regulation and oversight of AI-enabled implants. Finally, transparency and interpretability of AI algorithms are also paramount for clinician trust and

**Corresponding author*: atanukpaul@outlook.com

Arbind Prasad (ed.) Resorbable Composites for Bioimplants and Fixation Devices, (449–488) © 2026 Scrivener Publishing LLC

patient acceptance. Costs, a constantly evolving model, and the need for teamwork between medical professionals, authorities, and developers add complexity.

In summary, while AI technologies present substantial opportunities for advancing the design and functionality of implants and fixation devices, realizing their full potential requires overcoming ethical, technical, and regulatory challenges. Ongoing innovation, coupled with robust governance and stakeholder collaboration, will be essential to harness AI's benefits and deliver safer, more effective, and accessible implantable medical solutions for diverse patient populations.

Keywords: Artificial intelligence, medical implants, fixation devices, machine learning, patient-specific design, additive manufacturing, clinical decision support, digital twin technology

15.1 Introduction

15.1.1 Background on AI in Healthcare and Biomedical Engineering

Artificial Intelligence (AI) is rapidly transforming healthcare and biomedical engineering, offering innovative solutions for various medical challenges [1]. AI techniques, including machine learning, deep learning, and natural language processing, are being applied to analyze structured and unstructured medical data [2]. These technologies are particularly useful in areas such as cancer detection, cardiology, and neurology [2]. AI assists in medical imaging analysis, disease classification, and early diagnosis, enhancing the efficiency and accuracy of healthcare processes [3]. Furthermore, AI-powered virtual assistants and cognitive computing systems are being developed to improve clinical decision-making and patient care [2]. The integration of AI with cognitive science in biomedical engineering is also advancing, with the development of AI-Cognitive Computing Agent Systems (AI-CCAS) to detect cognitive human-like abilities and aid in neurological disease diagnosis [4].

15.1.2 Importance of Implants and Fixation Devices

Orthopedic fixation devices play a crucial role in treating fractures, soft-tissue injuries, and reconstructive surgery by providing stability and maintaining alignment during healing [5]. These devices include internal, external, and intramedullary fixation methods, with screws, plates, pins, and wires being common components [5, 6]. Bioabsorbable fixation devices, made from polylactic and polyglycolic acid polymers, have gained popularity in trauma, orthopedic, and craniomaxillofacial surgery due to their ability to be completely absorbed, eliminating the need for removal surgeries and reducing long-term complications [7]. The evolution of shoulder stabilization techniques has seen a shift from open procedures to arthroscopic approaches, with implant designs continuously improving to address complications associated with earlier versions [8]. Understanding the history and development of these implants provides valuable context for guiding future designs and clinical applications in orthopedic surgery [8].

15.1.3 Scope and Objectives of the Chapter

This chapter focuses on examining the current and emerging roles of artificial intelligence (AI) technologies in the development of medical implants and fixation devices. It aims to provide a clear and structured understanding of how AI is transforming traditional design, material selection, manufacturing, and clinical decision-making processes within this specialized area of biomedical engineering.

The scope of this chapter includes:

- Exploring foundational AI methods and their specific applications to implant and fixation device development.
- Highlighting practical uses of AI in the customization, optimization, and validation of these medical devices.
- Addressing the technical, regulatory, and ethical challenges that arise when integrating AI into healthcare product development.
- Identifying future trends, research directions, and opportunities where AI can add significant value to clinical and engineering outcomes.

The objectives of this chapter are to:

- Introduce the fundamental AI technologies relevant to the design and production of implants and fixation systems.
- Demonstrate how AI contributes to innovation in material science, imaging analysis, and surgical planning.
- Analyze current limitations and barriers to adoption, including data constraints, regulatory issues, and algorithm transparency.
- Present real-world examples and case studies that illustrate successful AI applications in the field.
- Encourage interdisciplinary collaboration by outlining strategic pathways for integrating AI into both research and clinical practice.

Through this discussion, the chapter seeks to guide researchers, engineers, clinicians, and policymakers in understanding the dual nature of AI—as both a disruptive tool and a promising opportunity—for advancing the next generation of medical implants and fixation devices.

15.2 Overview of Artificial Intelligence Technologies

15.2.1 Machine Learning and Deep Learning

Machine learning (ML) and deep learning (DL) have shown significant potential in orthopedic and dental implant applications. In orthopedics, ML has demonstrated effectiveness in risk assessment, outcomes prediction, and imaging diagnostics, often outperforming traditional methods. For cochlear implants, ML has been applied to optimize signal processing, automate evoked potential measurements, and predict postoperative performance [9]. In dental implantology, DL algorithms have shown promise in identifying implant systems and abutments [10]. However, the integration of wireless connectivity in implantable medical devices has raised security concerns, which can be addressed using DL-based security schemes [11]. As these technologies continue to evolve, they offer opportunities to improve diagnostic accuracy, reduce human error, and enhance patient care across various implant and fixation device applications. Figure 15.1 presents an academic classification of artificial intelligence techniques, highlighting their diverse applications across various sectors of healthcare [12].

15.2.2 Computer Vision and Image Analysis

Computer vision and image analysis have evolved significantly since the 1950s, progressing from basic digital image processing to advanced scene analysis and object recognition [13]. The field encompasses various techniques, including texture analysis, segmentation, motion tracking, and pattern recognition [14]. Applications have expanded across multiple domains, particularly in animal and veterinary sciences, where computer vision systems (CVS) are used for animal identification, behavior monitoring, and phenotype measurement [15]. CVS shows potential for precision livestock farming and high-throughput phenotyping, although challenges remain in developing autonomous solutions [15]. Recent advancements integrate digital image processing, pattern recognition, machine learning, and computer

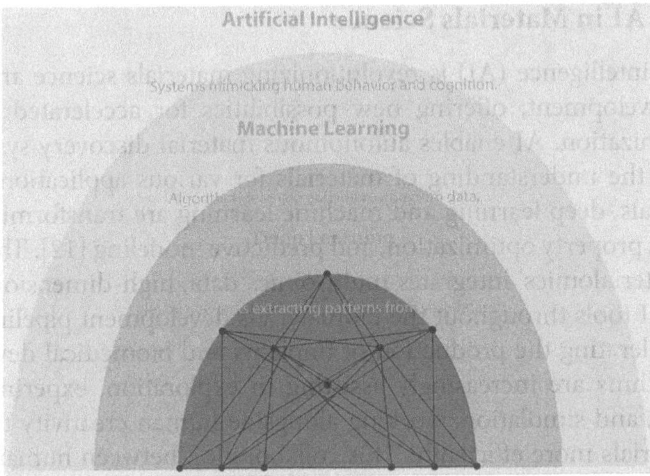

graphics [16]. The field continues to grow, with applications in diverse areas such as medical imaging, document analysis, and object recognition, driven by improvements in sensor technology and data analysis methods [14, 16].

15.2.3 Generative AI and Design Automation

Recent advancements in generative AI and design automation are revolutionizing the development of patient-specific implants and fixation devices. Automated design processes can significantly reduce the time and cost associated with creating custom orthopedic implants [17]. Artificial intelligence techniques, including deep learning and machine learning, are being applied to design patient-specific implants, improving efficiency and accessibility [18]. Generative design, combined with additive manufacturing, enables the creation of complex, optimized implant designs that can be fabricated using techniques like electron beam powder bed fusion [19]. Recent innovations in point cloud diffusion models allow for the automatic generation of high-quality implants, offering physicians multiple options to choose from for each patient [20, 21]. These advancements are making patient-specific implants more accessible, safer, and better performing compared to traditional approaches.

15.2.4 AI in Materials Science

Artificial intelligence (AI) is revolutionizing materials science and biomaterials development, offering new possibilities for accelerated discovery and optimization. AI enables autonomous material discovery systems and enhances the understanding of materials for various applications [22]. In biomaterials, deep learning and machine learning are transforming design processes, property optimization, and predictive modeling [12]. The concept of 'biomaterialomics' integrates multi-omics data, high-dimensional analysis, and AI tools throughout the biomaterials development pipeline, potentially accelerating the production of implants and biomedical devices [23]. AI algorithms are increasingly assisting in exploration, experimentation, modeling, and simulation, working alongside human creativity to develop new materials more efficiently. This collaboration between humans and AI could lead to creative leaps beyond human intuition, potentially redefining the relationship as a partnership rather than a user-tool dynamic [24].

15.3 Applications of AI in Implant and Fixation Device Development

15.3.1 Design Optimization and Topology Generation

Artificial intelligence (AI) applications in implant dentistry and design optimization are showing promising results. AI models have demonstrated high accuracy (93.8–98%) in recognizing implant types from radiographic images and moderate success (62.4–80.5%) in predicting implant outcomes [25]. In implant design optimization, AI integration with finite element analysis has led to significant improvements, including a 36.6% reduction in implant-bone interface stress. Topology optimization techniques, combined with additive manufacturing, are being used to address challenges such as stress shielding, osseointegration, and personalization in implant design [26]. Specific applications include optimizing interbody fusion implants for lumbar spine fixation, where topology and shape optimization methods are employed to generate designs that can withstand various loading conditions while maintaining proper intervertebral spacing [27]. These advancements are driving the development of more effective and patient-oriented implantable devices.

The reviewed literature presents a comprehensive overview of the applications of artificial intelligence (AI) in the design optimization and topology generation of orthopedic implants and fixation devices. Across ten studies, various AI techniques were utilized to enhance the performance, accuracy, and efficiency of implant design. Neural networks were the most frequently employed method, featured in four studies, while genetic algorithms and machine learning techniques each appeared in three. Other AI approaches such as support vector regression, clustering, and random forests were also explored, albeit less commonly. This diversity in methodology reflects the expanding role of AI in the biomedical engineering domain, particularly in orthopedic applications.

A central goal across most studies was to improve implant performance, as reported elsewhere [28–32]. These performance gains included improvements in biomechanical integration, reduced strain at the bone-implant interface, and increased energy efficiency in specialized implants. Accuracy was another important outcome, emphasized in two studies focusing on preoperative planning and manufacturing precision. Efficiency, although less frequently discussed, was noted in one study which reported a significant reduction in design time. While not all literature explicitly mentioned efficiency or material usage, many included constraints and objectives indicative of a drive toward more resource-conscious design practices.

Design optimization was one of the most prominent themes. Genetic algorithms were often integrated with other AI tools to enhance the robustness and adaptability of the optimization process. For instance, Chatterjee *et al.* [30] and Biswas *et al.* [29] used genetic algorithms alongside neural networks to tailor implant geometry based on varying bone conditions. Roy *et al.* [32] applied a hybrid model combining both techniques for dental implant optimization, addressing critical parameters such as micro-strain at the bone-implant interface. Neural networks also played a significant role in predicting implant behavior and enabling inverse design. Jiao *et al.* [33] employed a neural network with an evolutionary algorithm to facilitate the design of deployable implants, exemplifying the potential of AI in creating innovative, functional solutions.

Topology generation and material efficiency were also explored, particularly through lattice structure optimization. Guariento *et al.* [17] automated the design of lattice-based orthopedic implants to customize internal architecture while minimizing material usage. Although explicit discussions of material reduction were limited, several studies incorporated constraints on stress and strain, implying an effort to optimize material usage without compromising mechanical performance. Roy *et al.* [32], for example, set a stress threshold of 350 MPa and optimized the strain range for biomechanical compatibility.

Another major trend identified was the movement toward personalized, patient-specific implant solutions. Multiple studies utilized anatomical data from CT and MRI scans to tailor implants to individual patients. Akhtar *et al.* [28] used imaging data to customize implant dimensions, while Lambrechts *et al.* [34] demonstrated improved accuracy in preoperative planning for knee arthroplasty. Moscol-Albañil *et al.* [35] developed patient-specific short femoral stems, further reflecting the industry's shift toward customized treatment approaches. Table 15.1 presents a comprehensive summary of artificial intelligence techniques, their respective application areas, and key outcomes related to the design of orthopedic implants and fixation devices.

In summary, AI is playing an increasingly vital role in orthopedic implant design by enhancing performance, enabling customization, and contributing to more efficient and precise design processes. The integration of various AI techniques—especially neural networks, genetic algorithms, and hybrid models—has opened new possibilities for creating implants that are both functionally superior and tailored to individual patient needs.

Table 15.1 Summary of AI techniques, application areas, and key outcomes in orthopedic implant and fixation device design.

Study	AI techniques used	Application area	Key outcomes	References
Lambrechts *et al.*, 2022	Machine Learning Model (non-linear regression models), Support Vector Regression (SVR), Least Absolute Deviation Support Vector Machines (LAD-SVR)	Total knee arthroplasty preoperative planning	39.7% reduction in corrections, 82.2% and 85.0% accuracy for femoral and tibial implant sizes	[34]
Oladapo *et al.*, 2024	Machine Learning	Piezoelectric implants for orthopedic applications	Maximum power output of 4.10 Watts, surpassing traditional solutions by over 100%	[36]

(*Continued*)

Table 15.1 Summary of AI techniques, application areas, and key outcomes in orthopedic implant and fixation device design. (*Continued*)

Study	AI techniques used	Application area	Key outcomes	References
Jiao *et al.*, 2025	Evolutionary Algorithm, Neural Network	Deployable thermo-metamaterial implants	Over 200% increase in volume or cross-sectional area in fully deployed states	[33]
Moscol-Albañil *et al.*, 2024	Clustering algorithms, Random forests, Stacking	Patient-specific short femoral stems	No mention found	[35]
Kang *et al.*, 2012	Topology Optimization Algorithm	Bone fixation implant design	Error in fabricated volume to CAD design volume < 8%, differences in cross-sectional areas < 9%	[31]
Chatterjee *et al.*, 2019	Genetic Algorithm, Artificial Neural Network	Implant design for varying bone conditions	Minimization of strain deviation to zero	[30]
Biswas *et al.*, 2020	Genetic Algorithm, Artificial Neural Network	Patient-specific spinal implants (pedicle screws)	Minimization of strain differences between natural and implanted bone	[29]
Akhtar *et al.*, 2024	AI-driven algorithm, machine learning model	Humerus bone implants	Enhanced integration with host bone, improved patient outcomes	[28]
Roy *et al.*, 2018	Genetic Algorithm, Neural Network, Hybrid Model	Patient-specific dental implants	Micro strain at bone-implant interface optimized to 1500–3000, implant stress limited to 350 MPa	[32]
Guariento *et al.*, 2022	Topology Optimization Algorithm, Parametric Modeling, Computational Intelligence Technique	Lattice-based customized orthopedic implants	Reduction in design time from 40 minutes to 3 minutes	[17]

15.3.2 Personalized Implants Using AI and Imaging Data

Artificial intelligence (AI) is revolutionizing implant dentistry by enhancing various aspects of the field. AI applications include automated implant planning, image analysis, predictive modeling, and virtual surgical simulation [37]. These advancements optimize treatment processes and improve accuracy. AI facilitates efficient systematic reviews, predicts implant failure rates, and aids in detecting complications using radiographs and CT scans [37]. Personalized implants can be manufactured using CT image data and computer-aided design tools [38, 39]. A novel protocol combining AI and augmented reality (AR) for authentic 3D planning of dental implants has been proposed, offering an effective and time-efficient approach for simple cases of static guided implant surgery [40]. This protocol may replace conventional radiological guided surgery software, potentially changing the perspectives of modern guided implant surgery [40].

15.3.3 Additive Manufacturing and AI-Based Quality Control

Recent research highlights the integration of artificial intelligence (AI) for quality control in additive manufacturing (AM). Deep learning models, such as convolutional neural networks, have been employed for real-time quality assessment using point cloud data [41] and image-based defect detection [42]. Machine learning techniques, including support vector machines and reinforcement learning, have shown promise in classifying material properties and optimizing process parameters [43]. Innovative approaches utilize acoustic emissions and AI to monitor in-depth process dynamics and classify defects in real time [44]. These AI-driven methodologies have demonstrated significant improvements in defect detection accuracy, material property classification reliability, and process optimization efficiency [43]. The integration of AI in AM quality control enables automated, consistent, and non-contact monitoring, potentially eliminating the need for sophisticated post-production inspections and improving overall production efficiency [42].

15.4 Clinical Decision Support and AI Integration

15.4.1 AI for Surgical Planning and Outcome Prediction

Artificial Intelligence (AI) is revolutionizing surgical planning and outcome prediction across various specialties, including neurosurgery, emergency procedures, and orthognathic treatment. AI enhances precision, reduces errors, and offers personalized patient care [45, 46]. In neurosurgery,

AI applications span preoperative imaging, intraoperative navigation, and postoperative outcome forecasting [47]. For emergency surgeries, AI models outperform traditional methods in predicting complications, mortality, and hospital stay length. In orthognathic treatment, AI demonstrates high agreement with human decision-making for treatment planning and shows promise in soft tissue outcome prediction, although challenges remain in accurately predicting lip and chin areas [48]. While AI tools are proving essential in improving surgical care quality, further research is needed for comprehensive integration into clinical practice, addressing challenges such as interpretability and ethical considerations.

15.4.2 AI Integration with Robotic Surgery, Navigation Systems, and Patient-Specific Risk Modeling

The integration of artificial intelligence (AI) and robotic technologies in orthopedic surgery is gaining substantial attention, with applications spanning surgical planning, outcome prediction, intraoperative guidance, and patient-specific monitoring. A comprehensive review of recent literature revealed varied study designs and technological implementations, highlighting both the potential and challenges of AI in this field [34, 49–51]. Out of the studies examined, systematic reviews were the most common design type, followed by retrospective reviews, machine learning model development studies, prospective cohort studies, and systematic review protocols. This range indicates a growing but methodologically diverse body of research. In terms of AI approaches, machine learning was the dominant technique, followed by general AI frameworks, robotics, deep learning, artificial neural networks, and navigation technologies. These technologies were applied across several clinical areas, with knee arthroplasty being the most frequently studied application, followed by hip arthroplasty, spine surgery, anterior cruciate ligament (ACL) reconstruction, and broader orthopedic procedures [51, 52].

Thematically, the studies demonstrated a strong emphasis on AI integration with surgical technologies. This includes machine learning algorithms used for surgical planning, AI-powered decision support systems, and robotic-assisted procedures [53]. Many studies explored the use of AI in preoperative planning and intraoperative decision-making, showcasing benefits like improved surgical accuracy, better planning outcomes, and reduced complications. Table 15.2 provides a detailed overview of artificial intelligence and robotic technologies utilized in orthopedic surgical applications, highlighting study designs, employed technologies, clinical focus areas, and observed outcomes. Other clinical advantages included personalized treatment plans, enhanced risk assessment, early complication detection, and tailored rehabilitation protocols. Notably, each study appeared to explore distinct AI use cases, underscoring a trend of exploratory innovation rather than standardized application. This also meant that no specific AI technology or benefit was consistently reported across multiple studies, reflecting either the current early-stage diversity of approaches or the limited sample size of the review.

Table 15.2 Summary of AI and robotic technologies in orthopedic surgical applications: study designs, technologies, clinical areas, and outcomes.

Study	Study design	Technology focus	Clinical application	Primary outcomes	References
Batailler *et al.*, 2022	Systematic review	Machine learning, deep learning, robotic-assisted systems	Knee arthroplasty	Improved surgical precision, reduced rehospitalization rates	[54]
Lambrechts *et al.*, 2022	Retrospective review, Machine learning model development	Non-linear regression models, Support Vector Regression	Total Knee Arthroplasty (TKA)	Reduction in corrections needed by surgeons, improved implant size prediction accuracy	[34]
Ramkumar *et al.*, 2019	Machine learning model development, Retrospective review, Prospective cohort	Machine learning algorithms, artificial neural networks	Total knee arthroplasty (TKA), total hip arthroplasty (THA)	Prediction of length of stay (LOS), charges, discharge disposition, therapy compliance	[49]
Ames *et al.*, 2019	Retrospective review, Machine learning model development	Hierarchical clustering	Adult spinal deformity (ASD) surgery	Normalized 2-year improvement, major complication rates	[55]
Andriollo *et al.*, 2025	Systematic review	Deep learning models, AI-powered systems, robotic-assisted THA	Total Hip Arthroplasty (THA)	Precise implant positioning, personalized risk stratification, reduced complications	[53]

(*Continued*)

Table 15.2 Summary of AI and robotic technologies in orthopedic surgical applications: study designs, technologies, clinical areas, and outcomes. (*Continued*)

Study	Study design	Technology focus	Clinical application	Primary outcomes	References
Lopez *et al.*, 2021	Systematic review	AI and machine learning applications	Total hip and knee arthroplasty	Accuracy of AI/ML applications for decision-making	[50]
Wang, 2024	Systematic review	AI and semi-autonomous robotics	Spine surgery	Surgical precision, accuracy, safety	[56]
Purnomo *et al.*, 2021	Systematic review	Machine learning, deep learning, robotic-assisted surgery	Arthroplasty (total knee and hip)	Length of stay prediction, cost prediction, patient satisfaction	[52]
Maffulli *et al.*, 2020	Systematic review protocol	AI and machine learning applications	Orthopedic surgery	Accuracy of clinical diagnosis, prediction of post-operative outcomes	[51]

Patient-specific adaptations were a consistent theme throughout the reviewed literature. Nearly all studies emphasized personalized aspects such as individualized implant sizing and positioning, customized risk prediction models, and rehabilitation strategies tailored to specific patient needs [50]. These findings support the broader movement toward precision medicine in orthopedics, enabled by AI's ability to process complex datasets and generate patient-centered insights [51]. Table 15.3 presents an overview of the implementation of artificial intelligence technologies in surgical practice, emphasizing their potential benefits and the challenges associated with their adoption.

Despite these promising outcomes, the studies also highlighted several integration challenges. Workflow disruption was a commonly cited issue, while other studies mentioned concerns related to data quality, ethical implications, implementation costs, and legal liability [55]. These barriers indicate that while AI integration in orthopedic surgery holds great promise, it requires careful planning and systemic support for successful adoption. Figure 15.2 illustrates the integration of AI and emerging technologies across six stages of personalized orthopedic treatment and patient care [57].

Table 15.3 Overview of AI technologies in surgical practice: implementation, benefits, and challenges [34,49–51, 58].

Technology type	Implementation method	Clinical benefits	Integration challenges
Machine learning algorithms	Predictive modeling for patient-specific outcomes	Improved preoperative planning, reduced complications	Data quality and standardization
Deep learning models	Automated image analysis for diagnosis and planning	Enhanced diagnostic accuracy, personalized treatment plans	Integration with existing workflows
AI-powered decision support systems	Risk stratification and outcome prediction	Personalized risk assessment, improved patient counseling	Ethical considerations and transparency
Robotic-assisted surgery systems	Intraoperative guidance and precision enhancement	Increased surgical accuracy, reduced complications	Learning curve for surgeons, cost considerations
Navigation technologies	Real-time feedback during surgery	Improved identification of anatomical landmarks	Integration with existing surgical techniques
AI-driven simulation platforms	Preoperative planning and surgical training	Enhanced surgeon preparedness, optimized procedural approach	Development of realistic simulations
Machine learning-based clustering	Patient stratification for tailored treatments	Optimized surgical strategies, improved outcome prediction	Complexity in interpreting clustered data
AI-enhanced remote monitoring	Postoperative patient tracking and intervention	Early detection of complications, improved rehabilitation	Patient compliance and data privacy concerns
AI-based preoperative planning tools	Automated surgical plan generation	Reduced planning time, increased plan accuracy	Acceptance by surgeons, validation of AI-generated plans
AI-driven clinical decision support	Integration of multiple data sources for decision-making	Enhanced clinical judgment, standardized care protocols	Validation of AI recommendations, liability concerns

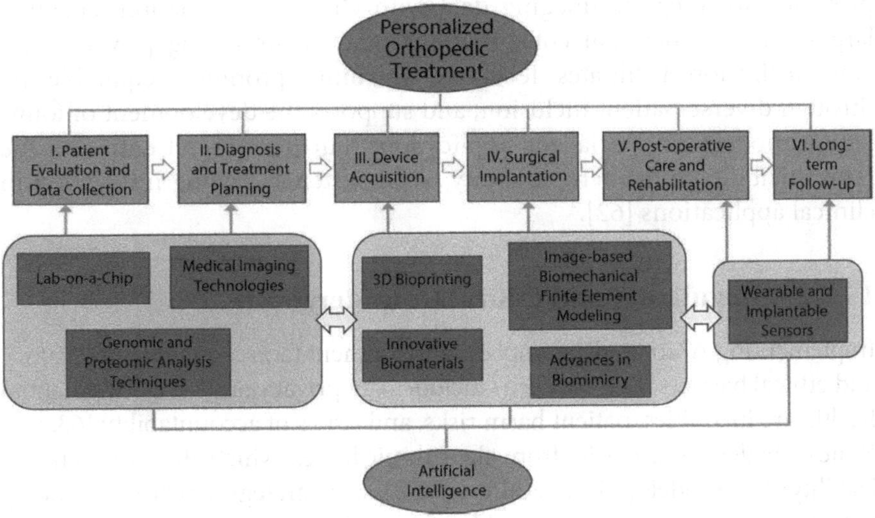

In summary, the literature reveals a rapidly evolving but heterogeneous landscape in the use of AI and robotics in orthopedic surgery. The diversity of AI applications—from surgical assistance to post-operative monitoring—demonstrates its versatility, while the emphasis on patient-specific interventions points toward a future of more personalized and efficient orthopedic care. However, methodological inconsistencies and implementation challenges must be addressed to move from exploratory studies to standardized clinical practice.

15.5 Challenges in Adopting AI for Implant Development

15.5.1 Data Availability, Quality, and Standardization

The effective adoption of AI in medical implant design and development is significantly impacted by data availability, quality, and standardization. Large, diverse datasets are crucial for AI implementation in healthcare, particularly in fields like otolaryngology which rely heavily on medical imagery [59]. Standardization plays a vital role in ensuring consistent approaches, reducing regulatory burdens, and promoting trust in AI applications [60, 61]. The Medical Imaging Common Data Model has been proposed to integrate imaging data with clinical information, enabling large-scale international collaborations while maintaining privacy [62]. Standardization facilitates federated learning, promotes equitable AI through diverse patient inclusion, and supports the development of foundation models for specialized applications. Furthermore, it enhances the reproducibility and interoperability of AI systems, driving innovation in clinical applications [62].

15.5.2 Regulatory and Ethical Considerations

Implementing AI in medical implant development faces significant regulatory and ethical barriers. Key concerns include data privacy and security, potential healthcare inequities, patient harm risks, and issues of accountability [63, 64]. Ethical challenges also arise from algorithmic biases, which affect the generalizability of AI models [65]. To address these issues, strategies such as enhancing regulatory oversight, promoting AI safety and transparency, and improving patient and healthcare provider engagement are recommended [63]. The EU and the US have different approaches to regulating AI medical devices, with the EU implementing new legislation like GDPR and the Cybersecurity Directive [64]. Developing unbiased AI models and ethical data-sharing frameworks is crucial for safe implementation [65]. Additionally, evolving regulatory pathways must balance patient safety with innovation, particularly considering the adaptive nature of AI technologies [66].

15.5.3 Model Interpretability and Clinical Trust

Interpretability and explainability of AI models are crucial factors influencing clinical trust and decision-making in healthcare applications, including medical implants. Key elements promoting trust in medical AI include explainability, transparency, interpretability, usability, and education [67]. Explainable AI (XAI) techniques are being developed to enhance transparency and accountability in various medical specialties, such as radiology, pathology, cardiology, and oncology [68]. These interpretable models aim to maintain high diagnostic performance while providing insights into decision-making processes, potentially revolutionizing the synergy between AI and human expertise [69]. A systematic review proposes a framework for interpretability in clinical decision support systems, emphasizing the importance of understandable and explainable AI results to end-users, particularly clinicians [70]. Implementing XAI in clinical settings can improve communication between AI systems and healthcare professionals, ultimately enhancing trust and adoption.

15.5.4 Cost, Infrastructure, and Workforce Readiness

The implementation of AI in healthcare faces numerous barriers across economic, infrastructural, and workforce-related domains. Key economic challenges include financial constraints and resource allocation issues [71, 72]. Infrastructural barriers encompass data management complexities, technological limitations, and integration difficulties with existing workflows [72, 73]. Workforce-related obstacles involve the need for specialized training, resistance to change, and concerns about job displacement [72, 74]. To address these barriers, healthcare systems can adopt strategies such as securing leadership buy-in, implementing change management processes, and providing comprehensive training programs [72]. Additionally, developing clear ethical guidelines, establishing robust legal frameworks, and fostering collaboration among stakeholders is crucial for overcoming implementation challenges [71, 72, 74]. Addressing these barriers is essential for enabling the widespread adoption of AI in healthcare.

15.6 Opportunities and Future Directions

15.6.1 AI for Real-Time Intraoperative Support

Artificial intelligence (AI) is revolutionizing surgical decision-making and precision, particularly in implant placement procedures. AI-driven tools enhance preoperative planning, intraoperative navigation, and postoperative care, leading to improved surgical accuracy and patient outcomes [75]. Real-time AI systems can analyze surgical videos and imaging, helping surgeons identify critical structures and guide precise incisions during operations [76]. These systems can also monitor physiological indicators to detect early signs of complications and predict outcomes, improving intraoperative decision-making [76]. AI-powered surgical robots further enhance precision and efficiency, enabling surgeons to perform intricate procedures with unparalleled accuracy [75]. Integration of AI with surgical decision-making has the potential to transform care by augmenting the decision to operate, informed consent process, and postoperative management [77]. However, challenges such as data quality, algorithm interpretability, and ethical considerations must be addressed for successful implementation.

15.6.2 Digital Twins and Simulation Environments

Digital twin technology is revolutionizing healthcare by creating virtual replicas of patients and systems, enabling personalized medicine and optimized operations [78]. This technology integrates real-time data from IoT devices and sensors with artificial intelligence to simulate and predict health outcomes [79, 80]. In medical implants, finite element analysis is used to create digital instances of organs, allowing for personalized diagnosis and therapy selection [81]. Digital twins can also model an individual's genomic makeup and physiological characteristics to develop tailored medications and treatment plans [80]. Furthermore, digital twins facilitate the optimization of clinical workflows, resource allocation, and training simulations for healthcare professionals [78]. As technology advances, it promises to enhance patient care, enable predictive analytics, and drive innovation in precision medicine [82]. However, challenges remain in implementing digital twin technology in healthcare systems [83].

15.6.3 AI-Enhanced Biocompatibility Testing and Longevity Prediction

Recent research highlights the potential of AI in advancing biomaterials and implant development. Machine learning models can predict the mechanical properties of biocompatible alloys, optimizing composition for improved strength [84]. The concept of 'biomaterialomics' integrates multi-omics data, AI tools, and advanced manufacturing to accelerate personalized implant development and predict clinical performance using digital twins [85]. AI-driven predictive modeling enables rapid screening of materials, property enhancement, and *in vivo* response forecasting, potentially expediting clinical translation [86]. However, challenges remain in establishing an AI-compatible definition of biocompatibility for automated data extraction and safety profiling [87]. These advancements suggest a paradigm shift in biomaterials research, leveraging AI to improve accuracy, efficiency, and personalization in implant design and performance prediction based on preclinical data and patient-specific variables.

15.6.4 Collaborative AI-Human Design Systems

Emerging AI technologies enhance implant development by refining surgical execution and implant design. In dental and joint replacement procedures, machine-learning systems paired with mixed or augmented reality and robotics have reported improved surgical precision. For instance, one study achieved implant-type recognition accuracies between 93.8% and 98%, while another integrating AI with robotic systems noted enhanced placement accuracy and improved treatment planning [33, 88, 89]. Digital twin approaches and AI-guided inverse design methods have yielded substantial design improvements. One investigation demonstrated a more than 200% increase in implant volume or cross-sectional area when applying an AI-guided inverse design paradigm, and data-driven strategies accelerated the production of implantable biomaterials—suggesting potential gains in biocompatibility [40]. Collaborative design systems, such as those combining AI with computer-aided design for patient-specific facial and cranial implants, have improved efficiency in digital data acquisition and planning. Collectively, the studies indicate that emerging AI techniques can advance implant development by improving surgical precision and design optimization, while also laying the groundwork for enhanced biocompatibility and long-term monitoring [25, 33, 89].

This study synthesizes findings from different literature exploring the integration of artificial intelligence (AI) in implant development and fixation devices. The studies encompassed a range of methodological approaches, including computational modeling, simulation studies, and case studies, though some lacked clear methodological descriptions [40]. AI applications were most frequently seen in dental implants, followed by joint replacements, facial/cranial implants, and other specific contexts like spinal fusion and tracheal stents [33]. Table 15.4 provides a summary of selected studies investigating the application of artificial intelligence in implant design and development, with a focus on key methodologies and research outcomes.

Machine learning was the most commonly used AI technology, while deep learning, CAD systems, digital twins, and natural language processing appeared less frequently. Key areas of application included AI-enhanced surgical planning and implementation, with technologies like augmented reality and robotic systems showing improved surgical precision and treatment planning. However, most studies lacked long-term outcome data and had limited clinical validation [25].

Digital twin technologies were discussed as promising tools for implant optimization through virtual testing and simulation, while collaborative AI-human systems demonstrated efficiency in design and planning processes [40, 92]. Despite these innovations, recurring challenges were noted in several studies. These included the need for large-scale clinical validation, high implementation costs, limited accessibility, training and adaptability issues, ethical concerns, and data interoperability problems [53].

Table 15.4 Summary of selected studies on artificial intelligence applications in implant design and development.

Study	Study focus	AI technology type	Application area	Key outcomes	References
Revilla-León *et al.*, 2021	AI applications in implant dentistry	Machine learning algorithms	Dental implants	Implant type recognition (93.8–98% accuracy), implant success prediction (62.4–80.5% accuracy), implant design optimization	[25]
Jiao *et al.*, 2025	AI-guided inverse design of implants	Machine learning (evolutionary algorithm and neural network)	Spinal fusion implants and tracheal stents	Over 200% increase in volume or cross-sectional area in fully deployed states	[33]
Andriollo *et al.*, 2025	AI and emerging technologies in total hip arthroplasty	Deep learning techniques	Joint replacement implants (hip)	Enhanced diagnosis, preoperative planning, surgical precision, and postoperative rehabilitation	[53]
Mangano *et al.*, 2024	AI and mixed reality for dental implant planning	Machine learning algorithms for segmentation and alignment	Dental implants	Clinically acceptable precision with minor deviations	[89]
Mangano *et al.*, 2023	AI and augmented reality for guided implant surgery planning	Machine learning algorithms for segmentation and alignment; Computer-Aided Design (CAD) systems	Dental implants	Clinically acceptable precision with minor deviations	[40]

(*Continued*)

Table 15.4 Summary of selected studies on artificial intelligence applications in implant design and development. (*Continued*)

Study	Study focus	AI technology type	Application area	Key outcomes	References
Revilla-León *et al.*, 2024	AI-based applications for digital data acquisition and implant planning	Machine learning algorithms; CAD systems	Dental implants	Improved efficiency and accuracy in digital data acquisition and implant planning	[90]
Basu *et al.*, 2022	Data science-driven pathways for biomaterials development	Machine learning algorithms; Digital twin technologies	No mention found	Accelerated production of implantable biomaterials, implants, and biomedical devices	[88]
Saeed *et al.*, 2023	Robotic and AI applications in implant dentistry	Machine learning algorithms	Dental implants	Improved implant placement accuracy, treatment planning, and patient satisfaction	[91]
Dhalluin *et al.*, 2021	Real-world digital data for orthopedic implant surveillance	Natural language processing	Joint replacement implants	Improved automation of monitoring in orthopedics	[92]
Memon *et al.*, 2021	AI techniques for patient-specific facial and cranial implant design	Machine learning algorithms; Deep learning techniques	Facial and cranial implants	Improved design of patient-specific implants	[18]

Performance metrics varied widely. Some studies reported clinically acceptable precision and improved implant placement accuracy, while others highlighted gains in implant volume, production efficiency, and design accuracy. However, many did not specify clear metrics or relied on qualitative descriptions [18, 90]. Table 15.5 presents an overview of artificial intelligence integration strategies in implant development, highlighting associated performance improvements, clinical advantages, and implementation challenges.

Overall, the findings reflect an early yet rapidly evolving landscape of AI in implant technology. While the potential benefits are significant, particularly in enhancing precision, personalization, and efficiency, substantial barriers—especially related to clinical integration, data quality, and regulatory acceptance—must be addressed for widespread adoption.

15.7 Case Studies and Real-World Implementations

Integrating artificial intelligence (AI) into implant design has led to notable improvements in both precision and manufacturing efficiency compared to traditional engineering methods [35]. For example, in total knee arthroplasty, one study showed femoral implant accuracy increasing from 68.4% to 82.2%—a 13.8% improvement—while tibial implant accuracy rose from 73.1% to 85.0%, an 11.9% gain [93]. Additional studies highlighted enhancements such as a twofold increase in the accuracy of bionic design and production, implant recognition rates reaching the high 90% range, a mean CAD deviation of just 44 μm, and volume or cross-sectional growth exceeding 200%, although these lacked direct comparisons to conventional techniques [28, 33]. Figure 15.3 presents a structured roadmap for the development of databases essential for implementing AI and machine learning solutions in bone tissue engineering [94].

Efficiency in manufacturing also saw significant gains. One study reported a drop in required preoperative corrections from 7.13 to 3.76—a 47.3% reduction. Other research noted over 70% cuts in simulation time, more than double the power output, and a full 100% improvement in overall design and production efficiency [28]. In terms of clinical outcomes, one study documented a 99.0% implant survival rate and a 91.3% success rate over three years for AI-designed implants, though no direct comparisons with traditional implants were made [25, 33, 93]. Table 15.6 provides a comprehensive summary of studies exploring the application of artificial intelligence technologies in medical implant procedures, detailing study designs, AI methodologies, clinical application areas, and sample characteristics.

Table 15.5 Summary of AI integration approaches, performance improvements, clinical benefits, and implementation challenges in implant development.

Study	Integration approach	Performance improvements	Clinical benefits	Implementation challenges	References
Revilla-León *et al.*, 2023	AI models for implant type recognition, success prediction, and design optimization	Implant type recognition: 93.8–98% accuracy; Success prediction: 62.4–80.5% accuracy	Potential for improved implant selection and outcomes	Models still in development, need for clinical validation	[25]
Jiao *et al.*, 2025	AI-guided inverse design integrating evolutionary algorithm with neural network	Over 200% increase in implant volume/cross-sectional area	Potential for minimally invasive and personalized surgery	Need for clinical validation of the approach	[33]
Andriollo *et al.*, 2025	Integration of AI in diagnosis, planning, and surgical execution	Enhanced surgical precision and personalized risk stratification	Improved patient outcomes, reduced complications	Cost, accessibility, steep learning curve for surgeons	[53]
Mangano *et al.*, 2024	Combination of AI and mixed reality for implant planning	Clinically acceptable precision with minor deviations	Authentic 3D planning without traditional software	Need for further clinical studies to validate the protocol	[89]
Mangano *et al.*, 2023	Integration of AI and augmented reality for 3D implant planning	Clinically acceptable precision with minor deviations	Authentic 3D planning without radiological guided surgery software	Need for further clinical studies to validate the protocol	[40]

(*Continued*)

Table 15.5 Summary of AI integration approaches, performance improvements, clinical benefits, and implementation challenges in implant development. (*Continued*)

Study	Integration approach	Performance improvements	Clinical benefits	Implementation challenges	References
Revilla-León *et al.*, 2024	AI integration in various digital data acquisition and planning technologies	Improved efficiency and accuracy in data acquisition and planning	Enhanced implant planning and design process	No mention found	[90]
Basu *et al.*, 2022	Integration of multi-omics data and AI tools throughout biomaterials development	Accelerated production of implantable biomaterials and devices	Potential for improved biocompatibility and performance	Need for increased adaptability of AI/ML tools and data science training	[88]
Saeed *et al.*, 2023	Integration of AI algorithms with robotic systems	Improved implant placement accuracy and treatment planning	Enhanced patient satisfaction and aesthetic outcomes	Balancing human expertise with technology reliance, ethical concerns	[91]
Dhalluin *et al.*, 2021	Use of natural language processing for automated implant surveillance	Improved automation of monitoring in orthopedics	Potential for better long-term implant performance tracking	Challenges in data quality and interoperability	[92]
Memon *et al.*, 2021	Application of AI techniques in the implant design process	Improved design of patient-specific implants	Potential for better-fitting and more effective implants	No mention found	[18]

Overall, the findings suggest that AI-enhanced implant design leads to measurable improvements in accuracy and efficiency across various medical applications.

15.7.1 Successful AI-Driven Implant Design Projects

Artificial intelligence (AI) is increasingly being applied in implant dentistry, offering improved precision and efficiency. AI-assisted tools have demonstrated comparable or better accuracy than conventional methods in implant planning, while significantly reducing operation time for both experienced and novice users [96]. AI algorithms have shown promise in automating tasks such as bone detection, maxillary sinus identification, and mandibular canal segmentation [97]. Studies have reported high accuracy (93.8–98%) in implant-type recognition using AI models on radiographic images [25]. AI has also been utilized to predict implant success rates (62.4–80.5%) and optimize implant designs, potentially reducing stress at the implant-bone interface by 36.6% compared to traditional finite element models [25]. Despite these advancements, challenges remain, including the need for high-quality training data and standardized protocols [91, 97].

Table 15.6 Summary of studies on AI technologies applied in medical implant procedures.

Study	Study type	AI technology used	Application area	Sample size	References
Revilla-León *et al.*, 2023	Systematic review	No mention found	Dental implants	17 studies included	[25]
Lerner *et al.*, 2020	Retrospective study	Machine learning for margin line detection	Dental implants	90 patients, 106 implants	[95]
Jiao *et al.*, 2025	Theoretical study	Evolutionary algorithm with neural network	Spinal fusion implants, tracheal stents	No mention found	[33]
Lambrechts *et al.*, 2022	Retrospective study	Non-linear regression models (Support Vector Regression (SVR), Least Absolute Deviation Support Vector Regression (LAD-SVR))	Total knee arthroplasty	5409 preoperative plans	[34]
Valente *et al.*, 2023	Systematic review	Artificial Neural Network (ANN) with Genetic Algorithm (GA), Deep Learning Network	Dental implants	6 studies included	[93]
Oladapo *et al.*, 2024	Theoretical study	Machine learning and AI-driven techniques	Orthopedic piezoelectric implants	No mention found	[36]
Moscol *et al.*, 2022	A systematic review and computational study	Clustering algorithms	Hip implants	No mention found	[35]
Akhtar *et al.*, 2024	Theoretical study	Python-based DICOM library with 3D reconstruction algorithm	Humerus bone implants	No mention found	[28]

15.7.2 Industrial and Academic Collaborations

Successful development and implementation of AI-driven implant design rely heavily on collaborations between industry, academia, and government. These partnerships are crucial for innovation, scalability, and clinical adoption [98]. Standardization, interdisciplinary teams, and aligned incentives are key to overcoming barriers in AI integration, particularly in fields like otolaryngology [59]. In implant dentistry, robotic technology and AI algorithms enhance precision, efficiency, and treatment outcomes, necessitating collaboration among practitioners, researchers, and industry experts [91]. For patient-specific implants, AI techniques such as deep learning and machine learning are employed in design processes, potentially reducing costs and improving patient health [18]. However, ethical considerations and balancing human expertise with technological reliance remain important challenges. Overall, these partnerships drive progress in AI-driven implant design while ensuring responsible implementation and maintaining high standards of patient care.

15.7.3 Lessons Learned from AI Implant Design Cases

Artificial Intelligence (AI) is revolutionizing implant dentistry, offering improved precision, efficiency, and personalized care. AI-driven image analysis and deep learning algorithms enhance implant placement accuracy and optimize aesthetics [99]. Integration of AI with Finite Element Analysis (FEA) improves parameter optimization and computational efficiency in implant design [93]. AI algorithms analyze patient data to assist in diagnosis, treatment planning, and implant design, leading to improved implant survival rates and patient satisfaction [91]. Machine learning techniques are being applied to design patient-specific implants, potentially reducing costs and improving patient outcomes [18]. However, challenges remain, including ethical considerations and the need to balance human expertise with technological reliance [91]. Future research should focus on incorporating factors like bone homeostasis and conducting long-term clinical studies to validate AI model predictions in real-world scenarios [93].

15.7.3.1 Enhanced Precision and Accuracy

AI-driven systems have significantly improved the precision of implant placement by automating the detection of anatomical landmarks (such as the mandibular canal and maxillary sinus) and accurately measuring bone dimensions. This reduces the risk of complications like nerve damage and sinus perforation, and studies show AI-based analysis can achieve agreement rates comparable to expert clinicians [100, 101].

AI-generated surgical guides and real-time data analysis during procedures have made surgeries more predictable, reducing procedural errors and improving the alignment and fit of implants and prosthetics [18].

15.7.3.2 Personalized and Optimized Treatment Planning

AI models excel at integrating diverse patient-specific data (demographics, medical history, bone quality, radiographic findings) to tailor implant design and placement. This enables clinicians to predict implant stability, survival rates, and risk of complications such as peri-implant bone loss with greater accuracy than traditional methods [100, 102].

Machine learning models can identify high-risk patients early, allowing for preventive strategies that improve long-term outcomes [18].

15.7.3.3 Improved Efficiency and Workflow

AI and mixed reality (MR) protocols streamline the implant planning process, allowing clinicians to work in authentic 3D environments without relying solely on traditional radiology-guided software. This can save time and resources, especially in straightforward cases [18, 100].

Automated segmentation and model alignment reduce manual workload and the potential for human error in data processing and surgical planning [103].

15.7.3.4 Cost Reduction and Accessibility

The integration of AI in implant design and manufacturing can decrease overall costs, making advanced implant solutions more accessible and efficient for both clinicians and patients [18].

15.7.3.5 Data-Driven Decision Making

AI's ability to analyze large, complex datasets uncovers subtle patterns and risk factors that may not be readily apparent to human clinicians. This supports evidence-based decision-making and helps optimize patient selection and treatment protocols [100].

15.7.4 Best Practices for Future Projects

15.7.4.1 Clinical Context

Integrate Multimodal Data: Use AI models that combine clinical, radiographic, and patient-specific data for comprehensive risk assessment and personalized treatment planning [100, 102].

Validate AI Tools Clinically: Ensure AI-driven protocols and predictive models are validated in real-world clinical settings for accuracy, sensitivity, and specificity before widespread adoption [18].

Early Risk Identification: Implement AI-driven risk profiling to identify high-risk patients and tailor preventive interventions accordingly [102].

Continuous Monitoring: Use AI for ongoing monitoring of implant stability and early detection of complications, enabling timely interventions [104].

15.7.4.2 Engineering Context

Automate Segmentation and Alignment: Employ AI for automatic segmentation and alignment of 3D models to improve workflow efficiency and reduce manual errors [105].

Leverage Open-Source Tools: Combine AI with open-source software for designing and 3D printing surgical guides, promoting flexibility and scalability [106, 107].

Iterative Model Training: Continuously update and refine AI models with new patient data and clinical outcomes to enhance predictive performance over time [25].

User-Centric Design: Develop AI interfaces and protocols that are intuitive for clinicians, minimizing the learning curve and facilitating adoption in diverse clinical environments [108].

15.8 Conclusion

Artificial intelligence (AI) is poised to revolutionize the development of implants and fixation devices, offering unprecedented opportunities across the healthcare ecosystem. As our global demographics shift and demand for healthcare surges, AI provides powerful tools to address pressing challenges—ranging from rising costs to the need for personalized, timely, and effective treatments. The increasing digitization of healthcare and the exponential growth of biomedical data offer fertile ground for AI technologies such as machine learning and deep learning to deliver clinically actionable insights. These advancements span the entire spectrum of care, from prevention and diagnosis to personalized treatment and rehabilitation.

The field of orthopedics, in particular, is undergoing a transformation driven by the convergence of AI, materials science, genomics, and advanced imaging. Intelligent systems, such as CRIGNet, demonstrate how AI can significantly reduce implant design time, improve mechanical and anatomical accuracy, and increase accessibility to care—especially in underserved regions. Nonetheless, realizing the full potential of AI in implant development demands overcoming substantial challenges. These include data-related issues like bias, availability, and standardization, as well as concerns surrounding regulatory compliance, clinical validation, and ethical use of patient data.

Model interpretability and clinical trust remain central hurdles, as the successful integration of AI hinges on the confidence of its end-users: the clinicians. The complexity of AI systems necessitates collaboration across diverse disciplines, including biologists, clinicians, materials scientists, and engineers, to ensure that these technologies are not only technically sound but also clinically relevant and ethically responsible.

In conclusion, while the path to fully integrated, AI-driven implant development is complex, it is also rich with opportunity. With strategic collaboration, rigorous validation, and continued technological refinement, AI holds the promise to redefine how implants and fixation devices are conceived, designed, and delivered—ultimately transforming patient care on a global scale.

AI disclosure statement: The author confirms that no artificial intelligence (AI)-generated content or AI-assisted writingtools were used in the preparation of this chapter (including summaries and the abstract). All text,analysis, and conclusions are the result of the author's original work.

References

1. Sadiku, M.N.O., Fagbohungbe, O.I., Musa, S.M., Artificial Intelligence in Healthcare: An Overview. *Int. J. Eng. Res. Adv. Technol.*, 06, 38–45, 2020, https://doi.org/10.31695/ijerat.2020.3670.
2. Prasad, P.S., Gunjan, V.K., Pathak, R., Mukherjee, S., Applications of Artificial Intelligence in Biomedical Engineering, in: *Handb. Artif. Intell. Biomed. Eng.*, pp. 125–145, Apple Academic Press, Taylor & Francis Group, New York, 2020, https://doi.org/10.1201/9781003045564-6.
3. Padmavathi, K. and Saranya, A.S., Biomedical Engineering and Informatics Using Artificial Intelligence, in: *Handb. Artif. Intell. Biomed. Eng.*, pp. 51–72, Apple Academic Press, Taylor & Francis Group, New York, 2020, https://doi.org/10.1201/9781003045564-3.
4. Garza-Ulloa, J., Biomedical engineering and the evolution of artificial intelligence, in: *Appl. Biomed. Eng. Using Artif. Intell. Cogn. Model.*, pp. 1–37, Academic Press, Elsevier, United Kingdom, 2022, https://doi.org/10.1016/b978-0-12-820718-5.00009-x.
5. Slone, R.M., Heare, M.M., Vander Griend, R.A., Montgomery, W.J., Orthopedic fixation devices. *Radiographics*, 11, 823–847, 1991, https://doi.org/10.1148/radiographics.11.5.1947319.
6. Paul, A.K., Migration Studies of Biodegradable Composites, in: *Biodegrad. Compos. Packag. Appl.*, pp. 193–212, CRC Press, Boca Raton, 2022, https://doi.org/10.1201/9781003227908-11.
7. Waris, E., Konttinen, Y.T., Ashammakhi, N., Suuronen, R., Santavirta, S., Bioabsorbable fixation devices in trauma and bone surgery: Current clinical standing. *Expert Rev. Med. Devices*, 1, 229–240, 2004, https://doi.org/10.1586/17434440.1.2.229.
8. Kramer, J.D., Robinson, S., Hohn, E., Purviance, C., Wolf, E.M., Fixation methods and implants in shoulder stabilization: A historical perspective. *J. Orthop.*, 15, 630–635, 2018, https://doi.org/10.1016/j.jor.2018.05.029.
9. Crowson, M.G., Lin, V., Chen, J.M., Chan, T.C.Y., Machine Learning and Cochlear Implantation - A Structured Review of Opportunities and Challenges. *Otol. Neurotol.*, 41, E36–E45, 2020, https://doi.org/10.1097/MAO.0000000000002440.
10. Anbazhagan, I., Ravichandran, R., H.K. K, Nair, V.V., Identifying implant abutments and fixtures by application of deep learning concept through artificial intelligence – A review. *IP Ann. Prosthodont. Restor. Dent.*, 10, 1–6, 2024, https://doi.org/10.18231/j.aprd.2024.001.
11. Rathore, H., Mohamed, A., Guizani, M., Deep learning-based security schemes for implantable medical devices, in: *Energy Effic. Med. Devices Healthc. Appl.*, pp. 109–130, Academic Press, Elsevier, United Kingdom, 2020, https://doi.org/10.1016/B978-0-12-819045-6.00006-6.

12. Gokcekuyu, Y., Ekinci, F., Guzel, M.S., Acici, K., Aydin, S., Asuroglu, T., Artificial Intelligence in Biomaterials: A Comprehensive Review. *Appl. Sci.*, 14, 2024, https://doi.org/10.3390/app14156590.
13. Rosenfeld, A., From image analysis to computer vision: An annotated bibliography, 1955-1979. *Comput. Vis. Image Underst.*, 84, 298–324, 2001, https://doi.org/10.1006/cviu.2001.0953.
14. Bunke, H., Villanueva, J.J., Sanchez, G., Otazu, X., *Progress in Computer Vision and Image Analysis*, 1st ed., World Scientific, Singapore, 2008, https://doi.org/https://doi.org/10.1142/7003.
15. Fernandes, A.F.A., Dórea, J.R.R., de M. Rosa, G.J., Image Analysis and Computer Vision Applications in Animal Sciences: An Overview. *Front. Vet. Sci.*, 7, 2020, https://doi.org/10.3389/fvets.2020.551269.
16. Wiley, V. and Lucas, T., Computer Vision and Image Processing: A Paper Review. *Int. J. Artif. Intell. Res.*, 2, 22, 2018, https://doi.org/10.29099/ijair.v2i1.42.
17. Guariento, L., Buonamici, F., Marzola, A., Scorianz, M., Volpe, Y., Design Automation of Lattice-based Customized Orthopedic for Load-bearing Implants, pp. 308–312, 2022, https://doi.org/10.14733/cadconfp.2022.308-312.
18. Memon, A.R., Li, J., Egger, J., Chen, X., A review on patient-specific facial and cranial implant design using Artificial Intelligence (AI) techniques. *Expert Rev. Med. Devices*, 18, 985–994, 2021, https://doi.org/10.1080/17434440.2021.1969914.
19. Kanagalingam, S., Dalton, C., Champneys, P., Boutefnouchet, T., Fernandez-Vicente, M., Shepherd, D.E.T., Wimpenny, D., Thomas-Seale, L.E.J., Detailed design for additive manufacturing and post processing of generatively designed high tibial osteotomy fixation plates. *Prog. Addit. Manuf.*, 8, 409–426, 2023, https://doi.org/10.1007/s40964-022-00342-2.
20. Friedrich, P., Wolleb, J., Bieder, F., Thieringer, F.M., Cattin, P.C., Point Cloud Diffusion Models for Automatic Implant Generation. *Lect. Notes Comput. Sci. (Including Subser. Lect. Notes Artif. Intell. Lect. Notes Bioinformatics)*, 14228 LNCS, 112–122, 2023, https://doi.org/10.1007/978-3-031-43996-4_11.
21. Paul, A.K., Borkotoky, S.S., Prasad, A., Nanohydroxyapatite-Based Composite Materials and Processing, in: *Adv. Mater. Manuf. Tech. Biomed. Appl.*, pp. 157–183, Wiley, New Jersey, United States, 2023, https://doi.org/10.1002/9781394166985.ch7.
22. Zafar, S. and Rana, N., Artificial intelligence in material science, in: *Sustain. Mater. Role Artif. Intell. Mach. Learn.*, pp. 1–18, CRC Press (Taylor & Francis Group), Boca Raton, 2024, https://doi.org/10.1201/9781003437369-1
23. Kilian, K.A. and Moghe, P.V., High throughput strategies for the design, discovery, and analysis of biomaterials. *Acta Biomater.*, 34, v–vi, 2016, https://doi.org/10.1016/j.actbio.2016.03.019.
24. Ball, P., Using artificial intelligence to accelerate materials development. *MRS Bull.*, 44, 335–343, 2019, https://doi.org/10.1557/mrs.2019.113.

25. Revilla-León, M., Gómez-Polo, M., Vyas, S., Barmak, B.A., Galluci, G.O., Att, W., Krishnamurthy, V.R., Artificial intelligence applications in implant dentistry: A systematic review. *J. Prosthet. Dent.*, 129, 293–300, 2023, https://doi.org/10.1016/j.prosdent.2021.05.008.
26. Peto, M., García-Ávila, J., Rodriguez, C.A., Siller, H.R., da Silva, J.V.L., Ramírez-Cedillo, E., Review on structural optimization techniques for additively manufactured implantable medical devices. *Front. Mech. Eng.*, 10, 2024, https://doi.org/10.3389/fmech.2024.1353108.
27. Tovar, A., Renaud, J.E., Mason, J.J., Optimum topology design of an interbody fusion implant for lumbar spine fixation. *Struct. Mater.*, 13, 329–339, 2003.
28. Akhtar, M.N., Haleem, A., Javaid, M., Mathur, S., Vaish, A., Vaishya, R., Artificial intelligence-based orthopaedic perpetual design. *J. Clin. Orthop. Trauma*, 49, 2024, https://doi.org/10.1016/j.jcot.2024.102356.
29. Biswas, J.K., Dey, S., Karmakar, S.K., Roychowdhury, A., Datta, S., Design of Patient Specific Spinal Implant (Pedicle Screw Fixation) using FE Analysis and Soft Computing Techniques. *Curr. Med. Imaging Former. Curr. Med. Imaging Rev.*, 16, 371–382, 2018, https://doi.org/10.2174/157340561466618 1018122538.
30. Chatterjee, S., Dey, S., Majumder, S., RoyChowdhury, A., Datta, S., Computational intelligence based design of implant for varying bone conditions. *Int. J. Numer. Method. Biomed. Eng.*, 35, 2019, https://doi.org/10.1002/cnm.3191.
31. Kang, H., Long, J.P., Urbiel Goldner, G.D., Goldstein, S.A., Hollister, S.J., A paradigm for the development and evaluation of novel implant topologies for bone fixation: Implant design and fabrication. *J. Biomech.*, 45, 2241–2247, 2012, https://doi.org/10.1016/j.jbiomech.2012.06.011.
32. Roy, S., Dey, S., Khutia, N., Roy Chowdhury, A., Datta, S., Design of patient specific dental implant using FE analysis and computational intelligence techniques. *Appl. Soft Comput. J.*, 65, 272–279, 2018, https://doi.org/10.1016/j.asoc.2018.01.025.
33. Jiao, P., Zhang, C., Meng, W., Wang, J., Jang, D., Wu, Z., Agarwal, N., Alavi, A.H., Artificial Intelligence-Guided Inverse Design of Deployable Thermo-Metamaterial Implants. *ACS Appl. Mater. Interfaces*, 17, 2991–3001, 2025, https://doi.org/10.1021/acsami.4c17625.
34. Lambrechts, A., Wirix-Speetjens, R., Maes, F., Van Huffel, S., Artificial Intelligence Based Patient-Specific Preoperative Planning Algorithm for Total Knee Arthroplasty. *Front. Robot. AI*, 9, 2022, https://doi.org/10.3389/frobt.2022.840282.
35. Moscol-Albañil, I., Solórzano-Requejo, W., Rodriguez, C., Ojeda, C., Díaz Lantada, A., Innovative AI-driven design of patient-specific short femoral stems in primary hip arthroplasty. *Mater. Des.*, 240, 2024, https://doi.org/10.1016/j.matdes.2024.112868.

36. Oladapo, B.I., Olawumi, M.A., Olugbade, T.O., Innovative Orthopedic Solutions for AI-Optimized Piezoelectric Implants for Superior Patient Care. *Appl. Sci.*, 14, 2024, https://doi.org/10.3390/app14177457.
37. Amiri, H., Shariati, A., Ghorbani, S., The application of artificial intelligence in implant dentistry. *World J. Biol. Pharm. Heal. Sci.*, 15, 082–084, 2023, https://doi.org/10.30574/wjbphs.2023.15.2.0330.
38. Rhodes, M.L., Glenn, Jr., W.V., Quinn, J.F., Rothman, S.L.G., Pratt, C., Brewster, R.B., Computer Aided Prosthetic Implant Manufacturing Using CT Image Data. Med. Imaging Image Interpret., vol. 0375, pp. 585–588, 1982, https://doi.org/10.1117/12.934712.
39. Khairkar, A., Kadam, S., Tandel, S., Ola, S., Future Trend: Review of Artificial Intelligence Applications in Prosthodontics. *Proc. 18th INDIAcom; 2024 11th Int. Conf. Comput. Sustain. Glob. Dev. INDIACom 2024*, pp. 1617–1621, 2024, https://doi.org/10.23919/INDIACom61295.2024.10498696.
40. Mangano, F.G., Admakin, O., Lerner, H., Mangano, C., Artificial intelligence and augmented reality for guided implant surgery planning: A proof of concept. *J. Dent.*, 133, 2023, https://doi.org/10.1016/j.jdent.2023.104485.
41. Akhavan, J., Lyu, J., Manoochehri, S., A deep learning solution for real-time quality assessment and control in additive manufacturing using point cloud data. *J. Intell. Manuf.*, 35, 1389–1406, 2024, https://doi.org/10.1007/s10845-023-02121-4.
42. Banadaki, Y., Razaviarab, N., Fekrmandi, H., Li, G., Mensah, P., Bai, S., Sharifi, S., Automated Quality and Process Control for Additive Manufacturing using Deep Convolutional Neural Networks. *Recent Prog. Mater.*, 4, 1–1, 2021, https://doi.org/10.21926/rpm.2201005.
43. Mondal, S. and Shubhra Goswami, S., Machine learning techniques for quality assurance in additive manufacturing processes. *Int. J. AI Mater. Des.*, 3455, 2024, https://doi.org/10.36922/ijamd.3455.
44. Masinelli, G., Shevchik, S.A., Pandiyan, V., Quang-Le, T., Wasmer, K., Artificial Intelligence for Monitoring and Control of Metal Additive Manufacturing. Ind. Addit. Manuf., pp. 205–220, 2021, https://doi.org/10.1007/978-3-030-54334-1_15.
45. Awuah, W.A., Adebusoye, F.T., Wellington, J., David, L., Salam, A., Weng Yee, A.L., Lansiaux, E., Yarlagadda, R., Garg, T., Abdul-Rahman, T., Kalmanovich, J., Miteu, G.D., Kundu, M., Mykolaivna, N.I., Recent Outcomes and Challenges of Artificial Intelligence, Machine Learning, and Deep Learning in Neurosurgery. *World Neurosurg.*, X 23, 2024, https://doi.org/10.1016/j.wnsx.2024.100301.
46. Alieksieieva, H.M., Khomenko, V.N., Khomenko, V.V., The Role of Artificial Intelligence in Enhancing Online Learning. *Innov. Pedagog.*, 2, 172–175, 2023, https://doi.org/10.32782/2663-6085/2023/63.2.36.
47. Hashimoto, D.A., Rosman, G., Rus, D., Meireles, O.R., Artificial Intelligence in Surgery: Promises and Perils. *Ann. Surg.*, 268, 70–76, 2018, https://doi.org/10.1097/SLA.0000000000002693.

48. Salazar, D., Rossouw, P.E., Javed, F., Michelogiannakis, D., Artificial intelligence for treatment planning and soft tissue outcome prediction of orthognathic treatment: A systematic review. *J. Orthod.*, 2023, https://doi.org/10.1177/14653125231203743.
49. Ramkumar, P.N., Haeberle, H.S., Bloomfield, M.R., Schaffer, J.L., Kamath, A.F., Patterson, B.M., Krebs, V.E., Artificial Intelligence and Arthroplasty at a Single Institution: Real-World Applications of Machine Learning to Big Data, Value-Based Care, Mobile Health, and Remote Patient Monitoring. *J. Arthroplasty*, 34, 2204–2209, 2019, https://doi.org/10.1016/j.arth.2019.06.018.
50. Lopez, C.D., Gazgalis, A., Boddapati, V., Shah, R.P., Cooper, H.J., Geller, J.A., Artificial Learning and Machine Learning Decision Guidance Applications in Total Hip and Knee Arthroplasty: A Systematic Review. *Arthroplast. Today*, 11, 103–112, 2021, https://doi.org/10.1016/j.artd.2021.07.012.
51. Maffulli, N., Rodriguez, H.C., Stone, I.W., Nam, A., Song, A., Gupta, M., Alvarado, R., Ramon, D., Gupta, A., Artificial intelligence and machine learning in orthopedic surgery: A systematic review protocol. *J. Orthop. Surg. Res.*, 15, 2020, https://doi.org/10.1186/s13018-020-02002-z.
52. Purnomo, G., Yeo, S.J., Liow, M.H.L., Artificial intelligence in arthroplasty. *Arthroplasty*, 3, 2021, https://doi.org/10.1186/s42836-021-00095-3.
53. Andriollo, L., Picchi, A., Iademarco, G., Fidanza, A., Perticarini, L., Rossi, S.M.P., Logroscino, G., Benazzo, F., The Role of Artificial Intelligence and Emerging Technologies in Advancing Total Hip Arthroplasty. *J. Pers. Med.*, 15, 2025, https://doi.org/10.3390/jpm15010021.
54. Batailler, C., Shatrov, J., Sappey-Marinier, E., Servien, E., Parratte, S., Lustig, S., Artificial intelligence in knee arthroplasty: Current concept of the available clinical applications. *Arthroplasty*, 4, 2022, https://doi.org/10.1186/s42836-022-00119-6.
55. Ames, C.P., Smith, J.S., Pellisé, F., Kelly, M., Alanay, A., Acaroğlu, E., Pérez-Grueso, F.J.S., Kleinstück, F., Obeid, I., Vila-Casademunt, A., Burton, D., Lafage, V., Schwab, F., Shaffrey, C.I., Bess, S., Serra-Burriel, M., Artificial Intelligence Based Hierarchical Clustering of Patient Types and Intervention Categories in Adult Spinal Deformity Surgery: Towards a New Classification Scheme that Predicts Quality and Value. *Spine (Phila. Pa. 1976)*, 44, 915–926, 2019, https://doi.org/10.1097/BRS.0000000000002974.
56. Wang, S., Application of AI integrated semi-autonomous robotics in spine surgery: A review. *Appl. Comput. Eng.*, 71, 231–235, 2024, https://doi.org/10.54254/2755-2721/71/20241670.
57. Luo, Y., Toward Fully Automated Personalized Orthopedic Treatments: Innovations and Interdisciplinary Gaps. *Bioengineering*, 11, 2024, https://doi.org/10.3390/bioengineering11080817.
58. Andriollo, L., Picchi, A., Sangaletti, R., Perticarini, L., Rossi, S.M.P., Logroscino, G., Benazzo, F., The Role of Artificial Intelligence in Anterior

Cruciate Ligament Injuries: Current Concepts and Future Perspectives. *Healthc.*, 12, 2024, https://doi.org/10.3390/healthcare12030300.

59. Evangelista, E. and Bensoussan, Y., Standardization, Collaboration, and Education in the Implementation of Artificial Intelligence in Otolaryngology: The Key to Scalable Impact. *Otolaryngol. Clin. North Am.*, 57, 897–908, 2024, https://doi.org/10.1016/j.otc.2024.04.005.
60. Lebedev, G., Artificial Intelligence in Healthcare: Directions of Standardization. *Intell. Syst. Ref. Libr.*, 212, 231–257, 2022, https://doi.org/10.1007/978-3-030-83620-7_10.
61. Ma, J., Schneider, L., Lapuschkin, S., Achtibat, R., Duchrau, M., Krois, J., Schwendicke, F., Samek, W., Towards Trustworthy AI in Dentistry. *J. Dent. Res.*, 101, 1263–1268, 2022, https://doi.org/10.1177/00220345221106086.
62. Jeon, K., Park, W.Y., Kahn, C.E., Nagy, P., You, S.C., Yoon, S.H., Advancing medical imaging research through standardization the path to rapid development, rigorous validation, and robust reproducibility. *Invest. Radiol.*, 2024, https://doi.org/10.1097/RLI.0000000000001106.
63. Mooghali, M., Stroud, A.M., Yoo, D.W., Barry, B.A., Grimshaw, A.A., Ross, J.S., Zhu, X., Miller, J.E., Barriers and Facilitators to Trustworthy and Ethical AI-enabled Medical Care from Patient's and Healthcare Provider's Perspectives: A Literature Review, MedRxiv. 2023, https://doi.org/10.1101/2023.10.02.23296447.
64. Pesapane, F., Volonté, C., Codari, M., Sardanelli, F., Artificial intelligence as a medical device in radiology: Ethical and regulatory issues in Europe and the United States. *Insights Imaging*, 9, 745–753, 2018, https://doi.org/10.1007/s13244-018-0645-y.
65. Tripathi, S. and Musiolik, T.H., Fairness and Ethics in Artificial Intelligence-Based Medical Imaging, pp. 71–85, 2021, https://doi.org/10.4018/978-1-7998-7888-9.ch004.
66. Ahmad, O.F., Stoyanov, D., Lovat, L.B., Barriers and pitfalls for artificial intelligence in gastroenterology: Ethical and regulatory issues. *Tech. Innov. Gastrointest. Endosc.*, 22, 80–84, 2020, https://doi.org/10.1016/j.tgie.2019.150636.
67. Tucci, V., Saary, J., Doyle, T.E., Factors influencing trust in medical artificial intelligence for healthcare professionals: A narrative review. *J. Med. Artif. Intell.*, 5, 2022, https://doi.org/10.21037/jmai-21-25.
68. Rane, N., Choudhary, S., Rane, J., Explainable Artificial Intelligence (XAI) in healthcare: Interpretable Models for Clinical Decision Support. *SSRN Electron. J.*, 2023, https://doi.org/10.2139/ssrn.4637897.
69. Thakur, R., Explainable AI: Developing Interpretable Deep Learning Models for Medical Diagnosis. *Int. J. Multidiscip. Res.*, 6, 2024, https://doi.org/10.36948/ijfmr.2024.v06i04.25281.
70. Nasarian, E., Alizadehsani, R., Acharya, U.R., Tsui, K.L., Designing interpretable ML system to enhance trust in healthcare: A systematic review to

proposed responsible clinician-AI-collaboration framework. *Inf. Fusion*, 108, 2024, https://doi.org/10.1016/j.inffus.2024.102412.
71. Joshi, S., Sharma, M., Das, R.P., Rosak-Szyrocka, J., Żywiołek, J., Muduli, K., Prasad, M., Modeling Conceptual Framework for Implementing Barriers of AI in Public Healthcare for Improving Operational Excellence: Experiences from Developing Countries. *Sustain.*, 14, 2022, https://doi.org/10.3390/su141811698.
72. Nair, M., Svedberg, P., Larsson, I., Nygren, J.M., A comprehensive overview of barriers and strategies for AI implementation in healthcare: Mixed-method design. *PLoS One*, 19, 2024, https://doi.org/10.1371/journal.pone.0305949.
73. Lanka, S. and Madala, P., AI in Healthcare: Bridging the Gap between Research and Clinical Implementation. *Int. J. Innov. Sci. Res. Technol.*, 500–507, 2024, https://doi.org/10.38124/ijisrt/IJISRT24OCT089.
74. Ahmed, M.I., Spooner, B., Isherwood, J., Lane, M., Orrock, E., Dennison, A., Systematic Review of the Barriers to the Implementation of Artificial Intelligence in Healthcare. *Cureus*, 2023, https://doi.org/10.7759/cureus.46454.
75. Abbasi, N. and Hussain, H.K., Integration of Artificial Intelligence and Smart Technology: AI-Driven Robotics in Surgery: Precision and Efficiency. *J. Artif. Intell. Gen. Sci.*, ISSN3006-4023 5, 381–390, 2024, https://doi.org/10.60087/jaigs.v5i1.207.
76. Zarghami, A., Role of Artificial Intelligence in Surgical Decision-Making: A Comprehensive Review: Role of AI in SDM. *Galen Med. J.*, 13, e3332, 2024, https://doi.org/10.31661/gmj.v13i.3332.
77. Loftus, T.J., Tighe, P.J., Filiberto, A.C., Efron, P.A., Brakenridge, S.C., Mohr, A.M., Rashidi, P., Upchurch, G.R., Bihorac, A., Artificial Intelligence and Surgical Decision-making. *JAMA Surg.*, 155, 148–158, 2020, https://doi.org/10.1001/jamasurg.2019.4917.
78. Vallée, A., Digital twin for healthcare systems. *Front. Digit. Health*, 5, 2023, https://doi.org/10.3389/fdgth.2023.1253050.
79. Paul, A.K., Prasad, A., Kumar, A., Review on Artificial Neural Network and its Application in the Field of Engineering. *J. Mech. Eng. Prakash*, 01, 53–61, 2022, https://doi.org/10.56697/jmep.2022.1107.
80. Raghavan, S., Arunachalam, O., Babu, A., Govindaraju, S., Leveraging the benefits of digital twin in delivering personalized medicine using IOT. *AIP Conf. Proc.*, 3035, 2024, https://doi.org/10.1063/5.0194752.
81. Bende, A. and Gochhait, S., Leveraging digital twin technology in the healthcare industry—a machine learning based approach. *Eur. J. Mol. Clin. Med.*, 7, 2547–2557, 2020.
82. Sun, T., He, X., Li, Z., Digital twin in healthcare: Recent updates and challenges. *Digit. Health*, 9, 2023, https://doi.org/10.1177/20552076221149651.
83. Sahal, R., Alsamhi, S.H., Brown, K.N., Personal Digital Twin: A Close Look into the Present and a Step towards the Future of Personalised Healthcare Industry. *Sensors*, 22, 2022, https://doi.org/10.3390/s22155918.

84. Valipoorsalimi, P., Sari, Y.A., Pekguleryuz, M., Mechanical Property Design of Bio-compatible Mg alloys using Machine-Learning Algorithms, *ArXiv Prepr. ArXiv2305.12060*, 2023.
85. Basu, B., Gowtham, N.H., Xiao, Y., Kalidindi, S.R., Leong, K.W., Biomaterialomics: Data-Driven Pathways to Next-Generation Biomaterials. *SSRN Electron. J.*, 2021, https://doi.org/10.2139/ssrn.3988072.
86. Rajitha, A., Kansal, L., Raj, G., Kalra, R., Dhamija, K., Abdul-Zahra, D.S., Biomaterials and Artificial Intelligence: Predictive Modeling and Design. *E3S Web Conf.*, 505, 2024, https://doi.org/10.1051/e3sconf/202450501003.
87. Mateu-Sanz, M., Fuenteslópez, C.V., Uribe-Gomez, J., Haugen, H.J., Pandit, A., Ginebra, M.P., Hakimi, O., Krallinger, M., Samara, A., Redefining biomaterial biocompatibility: Challenges for artificial intelligence and text mining. *Trends Biotechnol.*, 42, 402–417, 2024, https://doi.org/10.1016/j.tibtech.2023.09.015.
88. Basu, B., Gowtham, N.H., Xiao, Y., Kalidindi, S.R., Leong, K.W., Biomaterialomics: Data science-driven pathways to develop fourth-generation biomaterials. *Acta Biomater.*, 143, 1–25, 2022, https://doi.org/10.1016/j.actbio.2022.02.027.
89. Mangano, F.G., Yang, K.R., Lerner, H., Admakin, O., Mangano, C., Artificial intelligence and mixed reality for dental implant planning: A technical note. *Clin. Implant Dent. Relat. Res.*, 2024, https://doi.org/10.1111/cid.13357.
90. Revilla-León, M., Gómez-Polo, M., Sailer, I., Kois, J.C., Rokhshad, R., An overview of artificial intelligence based applications for assisting digital data acquisition and implant planning procedures. *J. Esthet. Restor. Dent.*, 2024, https://doi.org/10.1111/jerd.13249.
91. Saeed, A., Alkhurays, M., AlMutlaqah, M., AlAzbah, M., Alajlan, S.A., Future of Using Robotic and Artificial Intelligence in Implant Dentistry. *Cureus*, 2023, https://doi.org/10.7759/cureus.43209.
92. Dhalluin, T., Fakhiri, S., Bouzillé, G., Herbert, J., Rosset, P., Cuggia, M., Grammatico-Guillon, L., Role of real-world digital data for orthopedic implant automated surveillance: a systematic review. *Expert Rev. Med. Devices*, 18, 799–810, 2021, https://doi.org/10.1080/17434440.2021.1943361.
93. Valente, F., Falconio, L., Falcinelli, C., Roy, S., Trubiani, O., Traini, T., Artificial intelligence and finite element analysis: Applications in implant dentistry. *Ital. J. Anat. Embryol.*, 127, 99–103, 2023, https://doi.org/10.36253/ijae-14708.
94. Kolomenskaya, E., Butova, V., Poltavskiy, A., Soldatov, A., Butakova, M., Application of Artificial Intelligence at All Stages of Bone Tissue Engineering. *Biomedicines*, 12, 2024, https://doi.org/10.3390/biomedicines12010076.
95. Lerner, H., Mouhyi, J., Admakin, O., Mangano, F., Artificial intelligence in fixed implant prosthodontics: A retrospective study of 106 implant-supported monolithic zirconia crowns inserted in the posterior jaws of 90 patients. *BMC Oral Health*, 20, 2020, https://doi.org/10.1186/s12903-020-1062-4.

96. Ntovas, P., Laurent, M., Schnappauf, A., Matthew, F., Revilla-Leon, M., Att, W., Comparison Between Conventional and Artificial Intelligence-Assisted Setup for Digital Implant Planning: Accuracy, Time-Efficiency, and User Experience. *Clin. Oral Implants Res.*, 2024, https://doi.org/10.1111/clr.14382.
97. Macrì, M., D'Albis, V., D'Albis, G., Forte, M., Capodiferro, S., Favia, G., Alrashadah, A.O., García, V.D.F., Festa, F., The Role and Applications of Artificial Intelligence in Dental Implant Planning: A Systematic Review. *Bioengineering*, 11, 2024, https://doi.org/10.3390/bioengineering11080778.
98. Tanaka, M.L. and Lopez, O., Outlook on Industry-Academia-Government Collaborations Impacting Medical Device Innovation. *J. Eng. Sci. Med. Diagn. Ther.*, 7, 2024, https://doi.org/10.1115/1.4063464.
99. Altalhi, A.M., Alharbi, F.S., Alhodaithy, M.A., Almarshedy, B.S., Al-saaib, M.Y., Al jfshar, R.M., Aljohani, A.S., Alshareef, A.H., Muhayya, M., AL-harbi, N.H., The Impact of Artificial Intelligence on Dental Implantology: A Narrative Review. *Cureus*, 2023, https://doi.org/10.7759/cureus.47941.
100. Rokaya, D., Al Jaghsi, A., Jagtap, R., Srimaneepong, V., Artificial intelligence in dentistry and dental biomaterials. *Front. Dent. Med.*, 5, 2024, https://doi.org/10.3389/fdmed.2024.1525505.
101. Panahi, O., Artificial intelligence in oral implantology, its applications, impact and challenges. *Adv. Dent. Oral Health*, 17, 555966, 2024.
102. Nazari, Y., Lngeroodi, P.F., Maddahi, M., Kobravi, S., Amin, M.R., Bargrizaneh, A.A., Fouladi, S., Artificial intelligence models and predicting implant success. *Biomed. Res. Ther.*, 12, 7029–7038, 2025, https://doi.org/10.15419/bmrat.v12i1.949.
103. Ghaffari, M., Zhu, Y., Shrestha, A., A review of advancements of artificial intelligence in dentistry. *Dent. Rev.*, 4, 100081, 2024, https://doi.org/10.1016/j.dentre.2024.100081.
104. Karnik, A.P., Chhajer, H., Venkatesh, S.B., Transforming Prosthodontics and oral implantology using robotics and artificial intelligence. *Front. Oral Health*, 5, 2024, https://doi.org/10.3389/froh.2024.1442100.
105. Memon, A.R., Wang, E., Hu, J., Egger, J., Chen, X., A review on computer-aided design and manufacturing of patient-specific maxillofacial implants. *Expert Rev. Med. Devices*, 17, 345–356, 2020, https://doi.org/10.1080/17434440.2020.1736040.
106. Chakraborty, G. and Paul, A.K., 3D Printing in Biomedical Applications, in: *Adv. Mater. Manuf. Tech. Biomed. Appl*, pp. 275–301, Wiley, New Jersey, United States, 2023, https://doi.org/10.1002/9781394166985.ch11.
107. Harrison, L.M., Edison, R.L., Hallac, R.R., Artificial Intelligence Applications in Pediatric Craniofacial Surgery. *Diagnostics*, 15, 2025, https://doi.org/10.3390/diagnostics15070829.
108. Banerjee, T.N., Paul, P., Debnath, A., Banerjee, S., Unveiling the prospects and challenges of artificial intelligence in implant dentistry. A systematic review. *J. Osseointegration*, 16, 53–60, 2024, https://doi.org/10.23805/JO.2024.605.

Index

Printed in the USA/Agawam, MA
September 3, 2026

910067.013